# MODERN PIG PRODUCTION TECHNOLOGY

## A practical guide to profit

*John Gadd*

**Nottingham**
University Press

First published by Nottingham University Press

This reissued original edition published 2023 by 5m Books Ltd www.5mbooks.com

**British Library Cataloguing in Publication Data**
Modern Pig Production Technology
A practical guide to profit

ISBN 9781789182927

*'isclaimer*

'y reasonable effort has been made to ensure that the material in this book is true, correct, complete
'ppropriate at the time of writing. Nevertheless the publishers and the author do not accept
'ibility for any omission or error, or for any injury, damage, loss or financial consequences arising
use of the book. Views expressed in the articles are those of the author and not of the Editor or

'ty Press, Nottingham

'e
ssin, 17000, LA ROCHELLE, France

# CONTENTS

# FOREWORD

A good foreword should tell you a lot about the book
and a little about the author.

## ABOUT THE BOOK

When Nottingham University Press told me that stocks of my first pig textbook 'Pig Production Problems…' were running low and suggested a second edition, I was happy to oblige, thinking that a revision would not involve too much work.

Six months later and 800 hours of work on a complete overhaul has certainly proved me wrong!  It also shows us all how far pig technology has progressed in only 7 short years since 2003. So a completely new book has emerged.

Existing pig raising strategies and advice given required quite extensive updating to bring them into line with modern economics. New subjects which had developed significantly over the period needed whole chapters to describe them properly – such as baby piglet feeding and management; the importance of the hyperprolific gilt; modern man management; mycotoxins as a growing threat; CWF feeding and parity segregation.  In addition, all the costings and paybacks required updating and new cost-effective advice supplied. Acknowleged experts had to be located and consulted to ensure some of my conclusions were likely to be correct, and I thank them for their patient advice so freely given.

What surprised me before I had gone 10% into the project was how much new information I had in my databank, which had grown to comprise 15 steel filing cabinets, each with 4 drawers containing 30 files in hard copy let alone a growing number of computer disks with technical information obtained through the internet.

Retrieving and selecting the information took up 80% of the time spent on the ensuing text – writing it up was no problem for someone like myself who has scribbled away in various journals on pig subjects for 46 years.

## A new book

By the time I had got a fifth of the way through the project I realized that rather than being just a revised second edition, this was a whole new book – so much new information on modern **profitable** pig production needed to be discussed. I stress the word 'profitable' because as a self-employed pig adviser working for myself for 26 years (and for another 26 years before that as an employee for others) then just advising producers and their staff on how to increase their physical performance was not enough - in two main ways....

First, Superb performance can cost too much, eroding the promise of more profit which such praiseworthy effort, skill and courageous capitalisation might secure.

Second, Low-cost pig production (outdoors, temporary cheap but well-designed housing, franchising needing much less capital) where performance can even be below that achieved by the much-lauded top producers, can nevertheless make more profit for the investment in fixed and working capital. I have several such astute operators on my books who, though bruised, are still there after two pig price crises while some of the top performers have gone.

Thus, because potential profit is to my mind even more important than potential performance (I go so far as to suggest where the critical thresholds may occur later in the book) I have tried wherever possible to frame each subject along the following lines.

For the subject discussed....

1. What has been effective in the past?
2. What might be done to make it even more effective, in both performance and profit terms, under modern pig production conditions?
3. What can be, among the confusing number of fiscal choices available to the pig farmer today, those worth looking at? Why, and what are the likely cost benefits? This can be boiled down to one phrase "What is potentially the best way of using your money?
4. What might we be doing in future in this area, and why?

## Different times need different terms

As I see it, the last two are the most important. But it needs - as I discovered from sharp-end advisory work 30 years ago - a whole new range of terminology, not only to measure physical performance as of old (and still, sadly, as of now) but to include as far as possible the profit aspect as well in the self-same term. This must be done without altering the value of the old performance-dominated terms of food conversion, cost/kg gain, % mortality and pigs/sow/year etc., which. although desperately familiar, must be well past their sell-by date. To replace them there are profit-related performance terms and I describe them in the Business Section along with a couple of recent additions.

I've said many times that we must stop saying we are 'pig producers' and call ourselves '**meat** producers' In future we should think of ourselves not only as pig meat producers but also as pig **profit** producers.

Our job is to **produce the maximum amount of good quality lean meat at the lowest possible cost**. Using these new terms will help towards this goal, therefore please be sure to read this section in the book.

So…. many changes are in progress needing a new book to describe them.

Considerable changes; exhilarating changes; challenging changes. I hope you find my suggestions worthwhile on how to deal with,and then profit from, the changes.

# THE ORIGINS OF THIS BOOK

Have I always been successful in my advice? No, not always. Of course not. I know this because if I had the time I've always followed the visit up. "Did it work? What do you think of things now?" Again I was trained to do this by my business colleagues. In my early days as an on-farm adviser supporting a range of products, complaints had to be investigated promptly. They said "A well-handled complaint is a potential customer for life". How true this is! I also did it because I'm a writer-holic – I write everything down. Always have done. It helps keep track of things when you are busy, and I've a rotten memory anyway! I have kept a farm diary since 1950, which subsequently became a general (social) diary in the 1960's and has been what I call an "Omnium Gatherum" – in which anything which interests me is recorded and, if needs be, photographed or copied – for the past 32 years. There are over 30,000 photographs and 4 million words in it now, many to do with pig production – a feat which astonishes me as much as anyone else, but it is surprising what you accumulate over time if you keep at it. Eccentric maybe – but useful!

It is useful in the present context of writing this book because the diary, annually indexed and cross-referenced as it is (I take 10 days off every Christmas to do this) it is a databank of experience to draw upon – to refresh memory and correct the distortions of time.

## Checklists

And it is because I never have been able to remember things too well that I soon found that consulting a written-down check-list of my own was a useful thing to carry with me when a pig problem had to be solved. After all, even the pilot of an aircraft has his checklists to ensure he doesn't miss something before he leaves terra firma, and we advisers have just as many things to take into account, even if our feet should be firmly on the ground, or in the muck in my case!

Some of the experiences I've gathered in this way are in the following pages, presented as far as I am able in a checklist format, to which I've added the probable reasons why, and the evidence for them if I can locate it, as well as the economics – costs and benefits. Quite rightly, the modern pig producer wants evidence; a lifetime's experience is all very well, but it needs to be supported by fact, and what has gone before - experience, yours and my own experiences.

# MOTIVATING PEOPLE

An important part of this book is about people.

" People make or break businesses". I don't know who first said this, but how true it is!

I have had a long life in and around pig production (over 50 years) and have been both motivated – first by a schoolmaster and then two successive managers – and been demotivated by two people I had to report to during my career.

I've also had to try to enthuse people myself, from my gardener to several teams of pig stockpeople, to fellow committee members of societies to whom I belong and to engage in periodical fund-raising for our community. This has provided a wealth of experience on the subject, all learned the hard way.

In the chapter 'All about people' I have set most of it down in detail, the result of my own experience of how I was 'managed' by others and from the privilege of visiting so many pig farms in those 50 years, it must be well over 3000 now in 30 countries.

**It is clear that the majority of the most profitable units have had the most contented and settled staff.**

I was heartened to learn from my recent questioning of several CEOs of very successful retail, i.e. high street, organizations that all of them put people motivation at the top of their list to ensure profitability – not necessarily accountants, or buyers, or warehousemen, or order/dispatch personnel or technologists etc., although these are 'people' too. They did put 'front of house' people at the top of their people list and spent much of their time motivating and training them, something many of our agricultural supply companies can improve upon as in my experience they can be very casual correspondents when being asked about their products, often never replying to a phone call or email, probably because so many of their in-contact people are demotivated through bad managers.

Briefly then, this is what I have picked up….

1.  **Teamwork** is vital to any business
2.  **Keep the team informed.** My father's mantra, (a senior army officer) was "keep the troops informed" He did ; they loved him and would follow him anywhere - my family still has the glowing letters they wrote to him when he was promoted and moved on.
3.  **If staff don't have the full picture** they may get demotivated.
4.  Give your team the **chance to speak up and influence your decisions**. If things cannot be done, come clean and say why, not just 'I/we can't afford it' but explain why.
5.  Ask them **what they expect from you** as well as telling what you expect from them. The latter is always done – the former rarely, a bad error.
6.  If the business for the time being cannot raise wages, for example, there are **other things you can do** such as….

Start by praising them. It is amazing what a little praise can do – even an off-hand remark ( "Oh – by the way, you injected that struggling piglet very neatly!") lasts in the memory for days. Identify their efforts/achieving targets. Keep them motivated by incentives – recognizing extra hours put in, giving a day's holiday or paid excursion. Recognizing one person's or group's achievement motivates not only them but their colleagues, too. Identify and arrange training opportunities. Create a job progression dossier based on training and achievement. Provide a simple gift for the family – especially the children. Things like this have a motivating value far beyond financial reward until the business can afford a wages rise – and anyway are much cheaper!

**Above all – communicate!**

# ABOUT THE AUTHOR

I confess that as an on-farm adviser – I prefer this old term to today's more fashionable 'consultant' – I am very lowly-qualified, just a couple of Scottish agricultural diplomas obtained sixty years ago. At the time I started my interest in farm livestock these were enough to get me an entrance ticket into the advisory arena, a junior in a commercial company who had to do all the mundane tasks the senior advisers didn't want to tackle. So I dipped 20,000 sheep, chemically caponized as many chickens and injected about 1,000 litters of piglets in my role as a (very) junior adviser in a drug company – all in one year. Quite a start! I got my hands dirty, which set me on course to marry technology with mud-on-your-boots practice, something I've held to be paramount all my life. Even now - 60 years later - I still ask clients if I can help demonstrate some tasks I've learned over a lifetime's practice. I love it!

But it was the pigs I fell for and identified with, and it was pigs I decided, happily, with whom to throw in my future lot. I loved the practical work involved in testing new pig products and gradually became confident enough to make suggestions of my own. When back in the company office I was given a deskload of enquiries and grumbles from real pig producers to answer, not being too proud to telephone my team of expert pig farmer friends for the answer if my colleagues in the advisory office were stumped or just disagreed with each other, as all experts do, and gave me contradictory answers. Whenever I could, I went out of the office and did the jobs stockpeople have to do. I've done them all - some not very well, I admit!

My seniors, who were patient with me and very helpful, had enormous knowledge as specialists in nutrition, animal disease, chemotherapy and all the other scientific '-isms', and so were a tower of strength to a callow youngster. But I see now with hindsight, and many years of pig advisory work behind me, that they may have lacked two things when giving advice. First, a practical knowledge of how solutions based on science **might or might not** work in a real farm situation from the practical point of view. Secondly, being experts and specialists to a man, they might not have been good enough at seeing how the problem might impinge on, or be linked together by several scientific areas. What the scientist today calls 'multi-factorial'. The broad canvas. As specialist experts, they weren't so good at the broad canvas which was something else I thought I'd better remember wherever I went on farms.

In my days as a salaried commercial adviser making sure that my employers' products, first in the drug trade and then in the animal feed trade, worked as they were designed to do – or finding out why they didn't if complaints arose – I learned several vital pre-requisites necessary to help come to the right answer in either situation, and then the final one, which was how to make the advice 'stick'!

# APOLOGIA

There are bound to be – only a few, I hope – errors and omissions. These will be entirely my fault and I apologise in advance. No man knows it all, especially me! I learn something new about pigs every day, something important once a month, and something of revolutionary importance once a year. And you learn more as you get older; but then become increasingly disturbed and ashamed at what you don't know about the subject of your life's work. One good thing with age is that you are then quite happy to confess you don't know, and are content to pass the enquirer on to a specialist who probably does. As I said, I'm a broad-brush man, which is why I've found my 50 or more checklists so useful to keep me straight on the detail.

I hope you do, too.

# ACKNOWLEDGEMENTS

Although one man has written this book, it wouldn't have been possible without a huge number of people who have been so patient and helpful to him across 50 years.

Chronologically from the present to my student days, I owe much to ....

Barbara, my wife of 55 golden years, for the use of our dining room table for 18 months, and for this re-write a further 8 months! As well as innumerable cups of coffee or something stronger when the writing goes badly.

To my daughter Alison who has prevented my computer from driving me mad.

Next, the wise counsel and hard work of the Nottingham University Press team, especially Sarah Keeling for her skills in layout and presentation, who have produced such an attractive and readable volume.

Then, in the years I've been self-employed as a pig consultant since 1984, 38 loyal and persevering British and overseas pig producers, most of whom prefer to remain anonymous, who have generously allowed me to use their pigs and facilities to do farm trials on products and problems on which there seemed to be no published information. We all owe them a debt of thanks for their trouble and patience. This includes a farmer in Eastern Europe who agreed to a suggestion of mine that a management error could be continued for a while so we could measure the difference in another nursery which was doing this much better. What forbearance!

Next, it must be around 3000 pig producers and their stockpeople who, across the years, have opened their premises to a visitor who criticised too much and praised too little. How much all of you have taught me over four decades about pigs – and people!

Then there are my erstwhile colleagues in the feed and allied industries. While we had shared interests on our employer's behalf, we didn't always agree on how best to attain them, but we have, I hope remained friends as our careers went their separate ways. And I include my peers in 'the competition' too. One of the pleasant memories of my days in commerce was how you could ring up a competitor and just ask a technical question of their adviser or nutritionist.

The answer was nearly always friendlily and quietly given – I'm sure it is no longer like that in these more flurried, intensively competitive and – yes – less gentlemanly days!

Then there are the many academics and scientists who know so much more about matters technical than I do, some of whom I have had the temerity to disagree with or even upset from time to time, and I trust will continue to do – but still in a constructive way! Where would we be – and where would this book be – without them. I certainly owe them a debt, especially to those whose advice or work are gratefully acknowledged in the references, and in the 'further reading' appendix. I hope I have traced you all - if not, forgive me.

Then there are about 40 commercial firms who have given me work to do. As an ex-commercial man myself I have been able to appreciate their problems, and this econometric background has influenced and I think has also improved much of the advice given in these pages, centred on the bottom line as it should always be.

We now arrive at my early, formative years, thus names can be mentioned either because they are sadly no longer with us, or must have retired, so this will not cause them any embarrassment.

The late Stephen Williams, that original thinker on farm management. Very wise, a bit cynical but always right, none the less. My early mentors, veterinarians Norman Black and Ollie Murch, who convinced me of the continuing value of the pig veterinarian to us all. Another original thinker in pigs and pig farmer extraordinary, David Taylor, MBE (Taymix); maybe difficult to work for, but a marvellous practical tutor and so generous with his knowledge to a young agriculturist finding his way among real pig producers.

We are all salesmen these days. Producers sell pigmeat while I sell information about producing pigmeat effectively. A couple of individuals may not have noticed it, but I have learned much from travelling to pig farms with two superb professional pig-feed and pig-product salesmen still actively working I believe – Reg Hardy (Wiltshire, UK) and Mike King (Iowa, USA). If I get some messages across in this book it is partly due to them and their exemplary persuasive techniques.

And three college tutors – my 'prof' at Aberdeen, New Zealander Professor Neil Cooper. As he was a sheep man we actually argued about pigs – and also we shared an interest in mountain climbing, my other lifelong passion. Then Dr. R.V. Jones, fresh from his secret work on Radar Beams in WW2, who got me interested in physics – how air moved and about thermodynamics, a vital groundwork subject for anyone finding himself in pig advisory work. R.V. Jones shot real bullets into a sandbox to wake up us students after lunch – he was that sort of teacher. What would the health and safety 'police' think of that today! And in contrast Dr. Tom Dodsworth at the College's Craibstone Farm, a quiet, friendly cattle researcher who taught me about the lessons and the pitfalls of achieving accurate trial work. And through all my working life, the great Dr Gordon Rosen, a stickler for accuracy, has kept me straight many times - thank you Gordon.

Finally farmer Bob Milne, up in Laurencekirk, Scotland, my first employer. Another original thinker. It was he who ordered me (aged 19) to keep a farm diary and demanded it be initialled every month during my 18 month farm apprenticeship and thus started me writing professionally. I still have his pencilled 'No!' on many pages! Thankfully the ten editors I have written for to date didn't continue the practice.

I wrote a monthly column for four successive editors of the UK 'Pig Farming' magazine over a period of exactly 35 years. That's 413 articles in all, just for one journal (among the half dozen or so others I write for) which is, I suppose, quite an achievement and a tribute to the editors' patience in allowing me the space to write about pig production 'As I See It' for three and a half decades. The Japanese pig journal Yoton Kai have now presented nearly 250 of my articles on pigs since 1990 and Pig Progress over 150 pieces - good going. Long associations like this encourage much feedback from readers, and I learn a lot from it, thank you all for writing to me across the years.

I thank you all, producers and experts alike, because every one of you has contributed to this book. Without you I couldn't have written it.

I have consulted several people when writing this new book - all experts in their own fields.

Dr Grant Walling* (JSR Genetics) for advice on the Gilt Chapter.

Dr Jules Taylor-Pickard* (Alltech UK) for information used in the Mycotoxin Chapter.

Andy Deeks* and Simon Walburn (Du Pont International) for reading the Biosecurity Chapter.

Dr William Close (Close Consultancy) for advice on mycotoxins and organic trace elements.

Stephen Hall (Microsoft Ltd) for discussions on computer recording.

Luc Ledoux and James Auchincloss (Cidlines bv) for information and costings on water use.

Steve Stokes (Hampshire Pipelines) for CWF costings.

Helen Thoday (BPEX UK) on expected growth rates.

Nick McIvor (AM Warkup Ltd) on current housing costs.

Thank you all very much for your specialist advice.

*Have also taken time to read the relevant chapters.

**This book is dedicated to:**

The 5,000 or so hardworking, dedicated, patient, enthusiastic, generally under-rewarded people in pig production worldwide whom I've met, sometimes argued with, and always listened to – all of whom helped make this book possible.

"As a general rule the most successful man in life
is the man who has the best information"

*Benjamin Disraeli*
*British Prime Minister (1874-1880 and 1868)*

# 1

# CREEP FEEDING SCIENCE – THE LATEST

A 'creep'. The provision of a dedicated area for suckling piglets where by means of a railed-off area the sow cannot gain access to the baby pig starter food, a 'creep feed', which is placed therein.

Why a chapter on creep feeding in a book describing some pig problems?

For two reasons.

**Problem 1.** Creep feeding is by no means universal. It should be, as the benefits are now fully proven even for the smallest pig breeding farmer. I provide cost-effective evidence for this on page 7.

From now on, due to the major progress achieved by the geneticists in providing hyperprolificacy, we now have sows, and especially gilts, which give birth to such large litters that they are unable to rear them without risking considerable damage to their subsequent reproductive capability.

**Problem 2.** Creep feeding is not carried out anything like well enough on many farms, particularly with regard to cleanliness, keeping stored creep feeds fresh and frequent enough replenishment.

There are also supplementary problems in that too many producers are not prepared to pay the high price of a modern gut-friendly creep feed. Others try to avoid this by making their own, which all too often causes digestive upset because insufficiently digestible ingredients are used, the correct ones being largely unavailable to the farmer.

# WHY CREEP FEED?

Since Dr English's work of 2002 – the first results from the 'sophisticated creep feed' era – I have some 30 published trials in my files.

All but two say the same thing, revealing improvements varying from 210g to 1.15kg heavier weaners between 18 to 28 day weaning ranges.

# CREEP FEEDING IS NOW ESSENTIAL

Look at it this way; in the 1980s we had litters of 10 putting on around 43 kg total litter bodyweight by 28 days.

Now we have litters of 13 putting on 86 kg – double that of 20 years ago.

Previously the sow needed to eat 3.5 kg feed/day (14.6 MJ/DE and 10.2 g lysine/kg) at 4 days from farrowing to commence sustaining that 43 kg of litter growth of two decades ago.

Now she needs to eat 5.6 to 5.8 kg of food per day (14.8 MJ/DE and 10.5 g/kg lysine) by day 4 from farrowing. The modern sow still provides about 4.5 g of milk to provide 1g of piglet growth - that has barely changed.

**Most sows cannot consume this amount of food by day 4.**

So she needs help to avoid her' milking off her back' and putting her into a **body condition nosedive** down through lactation of the past. This was/is relatively easy to see by touch and/or by sight (Condition Scoring) and to rectify as it emerged, but a **reproductive breeding hormone** nose-dive today, is much more difficult to recognize - except from the records when the damage has already been done, and is now the main problem with modern high-performing sows and gilts. Creep feeding so as to take some of the load of her back is one of the strategies employed to prevent this, thus a very important one.

Some farmers suggest that just increasing the nutrient density of the lactation ration could be the answer – so that she acquires more nutrients with every mouthful she consumes.

The problem here is that we have got pretty well as far as we can get along this line of action (although wetting the food does increase consumption and is a part-solution although it does not preclude the need for a creep feed as well). Trouble is that where energy is concerned, increasing the energy density of the lactation feed beyond a certain level merely results in the sow eating less and we are no better off.

**Thus creep feeding today is essential for three main reasons:**

The figures I quote come from my clients' records of the past decade

1.  Creep feeding helps the litter achieve the genetic growth potential now built into it to weaning, at whatever age weaning is desired. Not to creep feed reduces the 28-day weaning weight of a genetically-sophisticated piglet by a likely 17%. This fallaway can easily be doubled by slaughter – and has been trebled where the post-weaning check is more than 6 days.

2.  **To defend the sow**. A sow struggling to feed a big litter will provide a litter with each member, on average, being 1kg lighter at 28 day weaning than if she had fewer to feed.

    **To prolong her productive life.** In addition a sow helped by her progeny eating a well-designed creep feed on average extends her productive life by 1.7 parities.

3.  **To boost her immune competence.** An otherwise healthy sow which is nevertheless productively-stressed during lactation reduces her immune competence. Her IgA and IgM levels in particular can fall by 18% and she has to direct about 16% of her energy and about 10% of her protein intakes into reinforcing her challenged immune shield. Not into her milk supply.

    **She** gets the nutrient benefit – not her litter. As a result 28 day weaning weight can reduce by 0.6 kg and days to slaughter increase by 6 to 8 days.

    In less than good hygiene conditions in the farrowing pen the demand on the sow's immunity is much more severe and disease can overwhelm both the sow and her litter. Good hygiene has a major effect on birthweights.

# MODERN PIGLET FEED DESIGN

Why do creep feeds have to cost so much? A complaint I hear every week.

**Digestibility** is the key factor. Sows milk does not aggravate the very delicate and sensitive villous structure lining the absorbtive portion of the gut – see Fig 4, page 67. Milk contains lactose so initially similarly high lactose levels are required in creep feeds and lower levels in the immediate post-weaning feed.

This is one factor needed both for palatability and also to get the piglets growing away with the minimum check.

However, there are two snags to lactose.

**First,** it is expensive due to competition from the human baby food industry, also the further you are away from a dairy manufacturing plant the more expensive it is.

**Second,** you can feed too much of it. Too much and the pigs get loose which can let in pathogenic scour.

 Levels of 20%-30% lactose are now advised for the first few days in a Prestarter feed, then down to about 10-12% for about a week in a Starter, and maybe if needed down to 5% for a further day or two if looseness is noticed. This – and for other nutrient reasons - is why you see several grades of early creep feeds on the market, which tends to irritate farmers who think is just a sales ploy to sell more expensive feeds over a period of 3 weeks or so - when it is not! I appreciate that this sequence of ration-swapping is onerous, but the top breeders train their staff to carry it out – and they give them enough time to do so!

The Early Grower feed containing no lactose then takes over from about 14 – 18 days.

## Two things to remember:

Farms differ and the timing of the lactose reduction in itself will vary from farm to farm, so you are wise to experiment as we did on our Taymix farm. Being regional distributors of skim and whey, both full of lactose, we knew all about that particular nutrient! Table 1 gives the latest advice on this subject

**Table 1. Recommended lactose specifications in diets for piglets**

|  | *Minimum (%)[1]* | *Optimum (%)[2]* | *Maximum (%)[3]* |
|---|---|---|---|
| Sow's milk content | - | 25 | - |
| Piglet weight |  |  |  |
| 4-6kg | 20 | 30 | 40 |
| 6-8kg | 10 | 20 | 30 |
| 8-12kg | 5 | 10 | 15 |
| 12-20kg | 0 | 5 | 7.5 |
| 20-30kg | 0 | 0 | 0 |

[1]For acceptable growth performance in low-cost production systems
[2]For balanced diet cost and growth performance under average conditions
[3]For accelerated growth performance under high-health conditions
Source: Mavromichalis (2008)

Secondly, start by following the specialist creep feed manufacturers' advice. They know what they are doing and it is a very sophisticated field of nutrition these days, so listen and learn from them.

# THE AGE-OLD PROBLEM OF COST

I now return to the old problem of the cost of creep feeds. It really is holding things up and for this reason alone consumer resistance is still rife.

## Do they really need to be so expensive?

Yes they do. I must hasten to say that no manufacturer pays me to take this line; I work for none of them. The science involved, together with the results from my most successful clients who have bitten the cost bullet, have convinced me that investment in  the latest nutritional technology, restricted choice of ingredients and safe, i.e.. very careful and skilful manufacture of baby pig diets, **pays back handsomely at slaughter.** I emphasise – at <u>slaughter</u>

That so many producers have not yet taken on board this truism must stem largely **from researchers not taking the improved performance results at the end of the nursery stage <u>on to slaughter weights</u>** where they should pay back handsomely in food and overheads saved.  We do not sell end –of –nursery pigs! An example of how important this can be, is given in Table 2.

 The extra cost of really good creep feeds and the investment in equipment, time and labour necessary to feed them properly (mainly hygienically) tends to put  producers off. Of course it does.

It all looks so expensive, but the proportion of an admittedly very costly feed at the baby pig stage is tiny in relation to the total costs of producing a slaughter pig, which as a result of this early investment in the growth stage should pay back threefold in having to feed much less of a cheaper finishing food.  And then there are 7 to 12 days - even 20 days - fewer overheads the costs of which are climbing rapidly these days.

## So what are the reasons for this high cost?

1.    Some common ingredients are banned

*Soya for one*. In its raw state, which is perfectly suitable for older pigs, this ingredient aggravates the baby pigs very delicate gut lining. Use soy protein and extruded soya bean if you have to, under a nutritionists supervision.

*'Cheap' fish and meat meals*. Yes, use 10% fish meal, but only the best grades which will be 'low temperature' processed raw fish which has been cooked <u>slowly</u>.

*Meat meals* . Steer well clear of any grade of meat meal for small pigs. In some countries they will be banned anyway for all pigs due to past human food scares, right or wrong.

*Groundnut* A nice cheap protein which in my experience can be dangerous to young pigs, and because so many parcels are so full of mycotoxins - to any pig for that matter.

2.    Too many non-starch polysaccharides (NSPs).

These contain anti-nutrient factors which can interfere with digestibility by as much as 50%. They cause looseness as the piglet tries to flush them down the gut out of harm's way. This can progress into full scouring as pathogens take advantage of insufficiently-processed nutrients in the gut causing a digestive traffic-jam in which the pathogens thrive. The nutritionist can damp down the presence of anti-nutrients in essential cereals like wheat and especially barley, while parcels of rye and oats are best avoided for baby pigs.

Corn (maize) is relatively innocuous in this area of nutrition but may be a source of dangerous mycotoxins. NSPs can be neutralized by skilled analysis and a variety of carefully-chosen enzymes. Counteracting NSPs especially in baby pigs is not cheap.

3.    As well as unwelcome ingredients, there are many which are essential but these, while costly, are cost effective nonetheless.

- Specialised egg protein powder
- Nucleotides (forerunners of amino-acids)
- Refined milk by-products
- Specialised fats and oils
- Specially protected vitamins
- Bioplexed trace minerals
- Selected enzymes

4.    Using a specialist manufacturer.

This must be left to piglet and calf food firms, as supplies of about 8 critical raw materials can usually only be bought in bulk, and some are scarce enough to be limited to those that are able to do so.

Also in order not to damage the ingredients during manufacture (denature the proteins for example - see Glossary), specialized plant is needed. One Dutch plant I visited last year was akin to an operating theatre in cleanliness – we had to view it through a window. Because of the high level of milk-based ingredients, making very small pellets - rather like 3 to 5 mm pencil leads - of the correct hardness (meal tends to cake the piglets mouth) is very difficult to do and needs specialized plant.

**Because of all these manufacturing constraints,not just any producer can make modern creep feeds.** Sorry about this – you will be most unlikely indeed to get comparative results if you try to make your own creep feeds.

Since 2005, I have recorded the performance from those breeders – all large ones – who did and did not make their own creep feeds. Table 2 shows the difference.

Unfortunately due to lack of time on the farms I could only record one of each in terms of the difference in MTF at slaughter - which was nevertheless impressive and allows a true cost effectiveness comparison to be made.

**Table 2. Home-made v. bought-in creep feeds**

All farms were large and competent breeders and the nutrient specs of the home made feeds excellent. Pigs involved, several thousand in all cases.

| *1. Home-mixed creep feed all through* | | *2. Purchased creep feeds used sequentially throughout* | |
|---|---|---|---|
| No of farms | Av.weaning weight (kg) | No of farms | Av. weaning weight (kg) |
| 4 | 6.05 | 3 | 7.10 |

And two farms where the pigs were followed through to slaughter....

| Av weaning weight per pig (kg) | Av.weaning weight per pig (kg) |
|---|---|
| 6.21 | 6.85 |

| | | | |
|---|---|---|---|
| Cost of creep feed/weaner | £1.36 | £4.22 | ( £ 2.86/pig more) |
| MTF at slaughter (kg) | 304 | 333 | ( 29 kg/ tonne more) |
| Value of this extra income per tonne fed | | £44.37. | |
| Value per pig | | £10.30 | |
| **REO** (less the extra cost of the expensive creep feeds) | | **3.58: 1** | |

**Comment:**
1. On feed costs alone, excluding saved overheads, on this result the bought-in creep feeds can be at least 3.5 times more expensive than home-made formulae - fed in the same way by equally-competent staff.
2. Overheads are very variable from farm to farm, but assuming an average of 37% of feed costs, the break-even figure rises to 4.7 times due to the overheads per day saved.
3. On all farms, while the home made creep feeds were much cheaper, the pigs **ate far less of them** despite equal competence by the staff in charge. While this put the comparative costs of the manufactured products up substantially (which is what puts farmers off) the savings at slaughter from them were much higher.
4. Of course a few swallows don't make a summer, but the results noted in Table 2 are pretty typical. I mention them because the farmers, their buildings and their staff were of a high order - but still the difference was there.

## Pellets or meal?

Weaners don't take to large pellets and especially to pellets with too many fines. They won't eat hard pellets any sense, either. Ideally they should be <u>just</u> be able to be crushed between fingers and thumb. Can't do it? We sent them back!

The trouble with meals is wastage and dust (fines) and congealed mouths. Yes, wetting meal sorts out these snags but the real problem with meal is that few stockpeople can keep pace with the very high and <u>consistent</u> level of cleanliness needed to avoid scouring. Bacterial degredation takes minutes rather than hours or days.

## Pellet quality and the danger of dust

Unable to find any published data on the acceptance/ refusal rates between dusty and well-made pellets (understandably the feed manufacturers are not keen on making these known, if they exist!) one of my on-farm mixer 'cooperative clients' agreed to do a trial. We sieved out fines from a batch of creep pellets and had them added by a local compounder into same-formula pellets in increasing amounts of 5%, 10% and 15% (the original pellets contained 2% fines anyway).

Feed eaten to weaning at 21 days dropped by 16%-20% in proportion. The results of the groups carried on to slaughter at 104 kg on normal non-dusty feed were exactly as my previous experience suggested, - the 15% 'dusty creep'-fed pigs took 10 days longer to slaughter and the 5% pigs took 3 days longer. My client was a videorecording enthusiast and he set up equipment which showed that 'nosing' i.e. just messing about at the trough, compared to eating behaviour at the 15% hoppers was + 20% and at the 5% hoppers + 6% compared to the controls on the original feed with only 2% fines.

There is much work published on pellet size and hardness for all ages of pig and to cut the papers in my files down to size, I now follow my own two simple criteria- creep pellets should resemble pencil leads in diameter (1.5mm) and be about 5 mm long. As to hardness, they should just be crushable with some effort between thumb and forefinger. A subjective measurement but one which corresponded well with the Holmen pellet hardness tester I had in my feed compounding days, so I have never bothered with one since I left mine behind somewhere.

I am not going to go into the complexities of creep feed particle-size as this is very much the province of the feed manufacturer –and from the digestibility standpoint it is quite important. You shouldn't be making your own creep pellets or crumbs or granules anyway – leave it to them!

## Storing creep pellets

Having ordered not too much (3 weeks supply is quite enough, less in a hot summer) and made certain that the supplier hasn't had them in his warehouse for longer than 7 to 10 days (oh, for the unlikelihood of legislation of a sell-by date with creep and weaner feeds!) remember that creep feeds contain a lot of amino-acids, lactose and fats. These are prone to deterioration if kept at farrowing house (or nursery) conditions of 28-32 °C and 70-95% humidity.

In these conditions proteins are affected by the Maillard (Browning) reaction which binds amino-acids, particularly the critical one of lysine. 10% is locked up at first and then this bio-unavailability rapidly increases with time.
Fats are rapidly oxidized (rancidity) and while the manufacturer does guard against this by adding anti-oxidants like ethoxyquin, these only delay the reaction which is why freshness and not overordering is so important.

Rancidity of lipids also hastens the deterioration of methionine and tryptophane and produces foul-tasting and objectionable odours from butyric and malic acids - the same additives which are used in a fox, mole and cats-in-the-garden deterrents!

For a piglet which nature endows with the ability to recognise its dam and even a preferred teat by smell, then food which is even the slightest bit stale must be a deterrent too – much greater then we humans with our comparatively insensitive taste and smell faculties realize.

**A creep feed doesn't have to be 'off" to reduce uptake – just stale**

On the whole farmers tend to store feedstuffs carelessly, I find, which is why I am grumbling away like this - especially relevant as creep feeds are so expensive due to the great knowledge and care the manufacturer puts into them these days.

# CREEP FEED INTAKES

Assuming an average birthweight of 1.35 kg, BPEX in the UK – a reliable and knowledgeable source – suggest a cumulative creep feed intake as shown in Table 3. However this does not meet the solid food consumption of 400g by the still-common weaning age at 21 days. 400g cumulative consumption is thought to be a reasonable safety threshold so as to put the weaner well on the way to a quick getaway after the sow has gone, by having its gut surface fully primed before it has to be solely reliant on solid food. This is borne out by some of my best clients and therefore, with due deference to BPEX, I include a rather more desirable upper target in a second column which is still attainable at the time of writing.

Table 3. Typical and target creep feed intakes

| Weaning age (days) | Cumulative creep feed intake (g/pig) | | |
|---|---|---|---|
| | BPEX target | Top producer target | Typical |
| 21 | 275 | 400 | 125 |
| 24 | 350 | 550 | 200 |
| 28 | 700 | 850 | 400 |
| 32 | 1200 | 1500 | 800 |
| 35 | 2500 | 3500 | 1300 |

**Comment:**
In the 'typical column the 400g is only reached  at 28 days, which could be a main reason why at the time of writing so many producers are moving back to 4 week weaning or even later. They find that their piglets scour far less at this later weaning age probably because they have eaten enough solid food to prime their gut surface. It also suggests that the more skilled creep feeders, converted to the cost-effectiveness of modern creep feed design, can revert back to earlier weaning once again as they achieve the 400g required considerably sooner and so take advantage of the saved production costs from the resultant earlier weaning.

# THE FUTURE - IMMUNOGLOBULINS

A really major advance of the past 10 years which, with improvements in housing and stockmanship, has revolutionized the lot of the baby pig.

You are all aware of the classic graph which shows the fall-off in antibody protection between the highs of maternal protection when the suckler is on the sow and the delay in the piglet establishing sufficient of its own  immune defences once it has been weaned.

Adding 3 to 5% selected **animal plasma** to creep and post-weaning foods, which is especially rich in IgG antibody so as to cover this gap, can be of great benefit at this time and helps counteract that deficiency. But it is expensive – even 3% can raise the cost of a link feed by 35%. Also supplies can get short and the public scare about adding animal by-products to farm feeds hasn't helped this useful commodity one bit.  Lastly, plasma is not very protective against those bacteria which cause scouring.

But there is an alternative – **pasteurised egg powder**. This is obtained from hens specially immunised against those pathogens which affect newly-born pigs.
 Such eggs (and they are considered 'politically correct', too) are rich in those useful immunoglobulins to such an extent that about 1kg. of this special egg powder can replace 50 kg of animal plasma – a lot cheaper.

It is also more effective against plasma in the more common diarrhetic scours.

## Other options and additives

There are dozens of products which the makers claim have a positive effect on immunity as firms compete to jump on this particular topical bandwagon. There are ingredients like herbs, prebiotics and even DDGS additives (which if used with care) are said to enhance the pigs immune status. Extra tryptophane is another one and it will take time to sort out the cost-effective 'men from the boys' in this area of nutrition. I am not going to attempt to go further until more independently proven results and usage experience accumulate.

This said, there is one group of nutrients for which there is a growing amount of evidence - minerals that influence electrolyte balance (dEB) which improves amino-acid availability to strengthen immunity and also mollify some of the effect of NSPs in the weaner diet. Yet again, the chapter on Mycotoxins show how these fungal residues interfere with immune status, so products which help control them have another rather roundabout influence on establishing a stronger immune shield.

## Heat–treated cereals

Cooking cereals makes them more digestible and palatable. It also discourages looseness especially if the cereals have been finely ground. Cooking is essential in a Link Feed, see my next chapter.

# MANAGING THE CREEP FEEDING PROCESS

A recap. As I've already mentioned, the scientists tell us that the suckling piglet needs to have eaten – in total – around 400g of a well designed creep feed before it is weaned.

We've seen that this is necessary to prime and condition the absorbtive surface of the digestive tract  so as to give  the weaner the best chance of moving smoothly on to just solid food alone, without any help from its dam's milk. The following remarks will help towards this goal

# START CREEP FEEDING EARLY

Using  the better creep feeds of course. The reason why this advice to start as soon as possible has been controversial is that with the old-fashioned, poorer-designed creep feeds the earlier you start/ the more they eat of them then the more disappointing tend to be the results – a mainly looseness or worse. The worry over these kick-backs is as old as the hills! But we are in a totally new era now.

It is because of these historical worries that most people don't start creep feeding soon enough.

Some of the reasons given to me are…

"They only waste this expensive gold dust"
"It makes them loose"
" They consume so little – it doesn't seem worth the hassle"
"The feed is just so expensive – I'll feed it once I'm sure they'll eat it"

The first thing to appreciate is that the initial uptake of creep feed is enormously variable between litters. (Fig 1) This happens less if the creep feeds are of modern design and offered properly, but variation will still occur.

So don't be discouraged by this as some of the 'slow starters' could have very milky dams, be in a very warm creep or not able to drink enough water. Apart from the latter circumstance they will still do alright. Even if initially they do waste the food, under reasonably clean conditions it can always be given to the sow.

If the creep feed is causing scour then it is not well enough designed, or the scouring is due to something else.

Well-designed, well-fed modern creep feeds do not cause scour in themselves.

Figure 2, from work we did on the Deans Grove farm, shows clearly how feeding a well designed creep feed really early moves piglets towards the critical 400g total consumption advised by the experts from weaning, in our case at 21 days. Both groups reached the target but the early started pigs were 1kg heavier at 46 days and got to slaughter weight of 93 kg 6 days sooner. Same creep feed, same conditions.

Figure 3 shows the very small proportion of the creep feed contribution to the overall amount of feed consumed to slaughter.

In proportionate cost terms - even at the price asked these days for the sophisticated formulae described earlier – it is still well under 10% of the food cost to slaughter, paid back by about 3 days food and overhead costs saved to shipping. A good creep feed, fed well, even on the most modest of farms, should result in 6 to 7 days quicker to slaughter over the old formulae used by many farmers, which is a 2: to 2.3:1 return. The better clients of mine achieve double that as can be seen from Table 1.

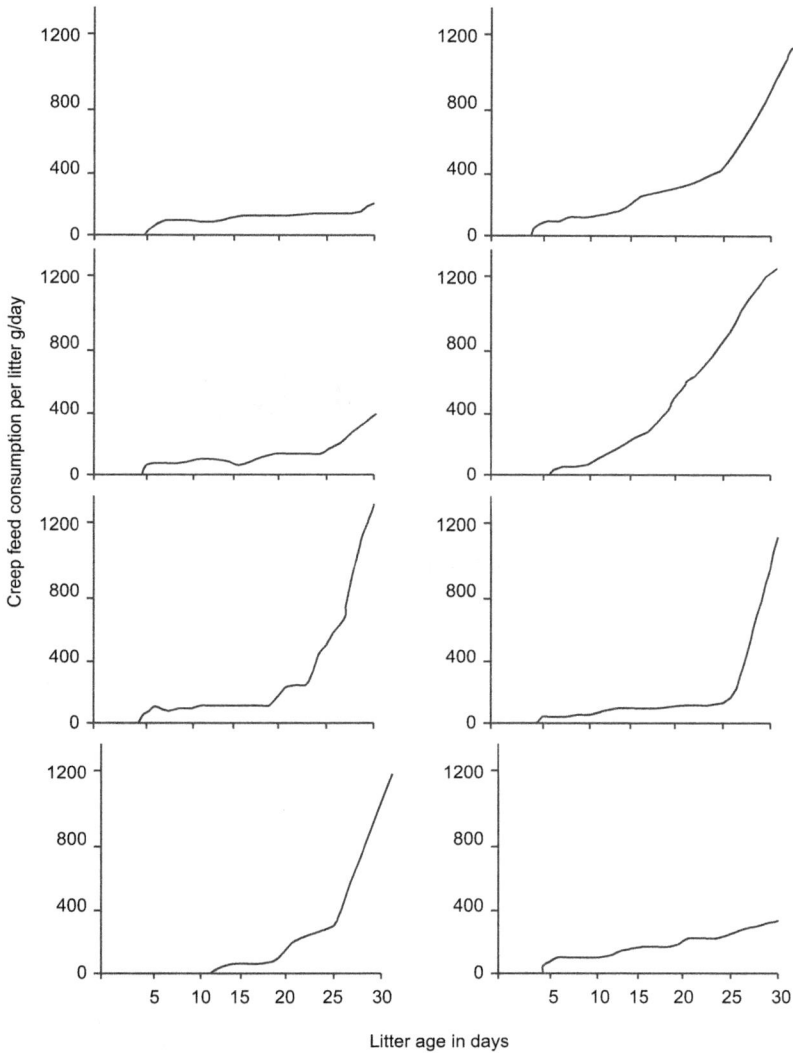

**Figure 1.** There can be great variation in creep consumption between litters.
Same creep feed, same environment.

## Presentation

Provide a light scattering of creep on a flat plastic tray with a 1 cm. edge-flange for a couple of days, but you will need two of these per crate space, one to be removed daily to be cleaned **and dried.** Complete freshness of the surface area is vital as well as the creep feed itself to get early uptake. So order the food frequently and store well – we used the insulated section of a redundant ice-cream van to keep the bags cool. And keep the feed troughs utensils <u>clean</u>. Piglets have a keen sense of smell.

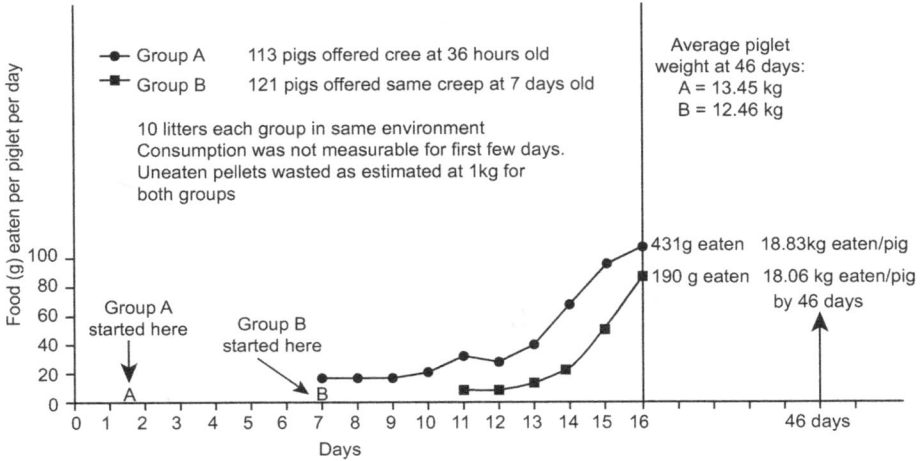

**Figure 2.** Consumption of pre-starter creep feed offered at 36 hours old and the same food offered at 7 days old (Source: RHM Agriculture).

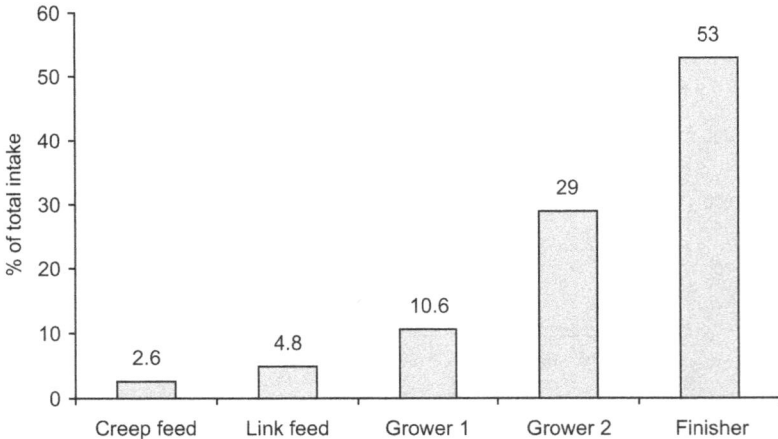

**Figure 3.** The proportion of feed inputs at different stages to produce a bacon pig (Source: Varley, Aust. Pork Jnl (2003).

# THE 'THREE-THREES' APPROACH

## So how much to offer?

Freshness and quick consumption of a modern creep feed is helped by my 'Three-threes' idea.

For the first *three days*, i.e. from (i.e. after) day 3 to 4 from birth, the creep feed must be offered *three times a day* and only enough is given to last *three hours.*

Any creep feed not eaten must be given to sows in least good condition, or being heavily milked, or one suspected of a low milk yield

This way its cost will not be wasted.

Now I know this is a chore – a darned nuisance – but a survey I published some while ago ( see below) showed that skilled pig technicians must spend more 'quality time' with the young pigs and less on the heavy duty tasks which can be done adequately enough by less-skilled or contract labour. During these 'intensive care days' the small, first-stage creep receptacles must be taken up once a day and cleaned.

As staff are busy enough at that time, spares are a boon so that a daily bulk cleaning and drying period can be accommodated with the minimum of work and disturbance to routine work with the pigs. This is one reason why I am strongly in favour of batch-farrowing, which allows this important routine to be completed with enough time to do it well, the attention to the very smallest pigs being largely concentrated into one week in the month.

### Time spent with the baby pigs is 'Golden Time'

Time is at a premium with baby pigs. The average farrow-to-finish breeder is only according 6% (about 132 man-hours/sow/year) of his total labour force availability. It should be twice to three times as much. This is borne out by six of my clients who consistently wean 28 pigs per sow per year and who on average spend 375 manhours per year (17%) in just looking after baby pigs.

## CREEP FEEDERS

Fortunately creep feeders can be small and inexpensive. I illustrate three of them. Another, not shown, is a cast-iron bowl with metal rod dividers falling from a central carrying handle, But they are heavy to carry around and keep clean. A concrete/resin heavy bowl (illustrated) is more convenient but also really needs to be cleaned in situ due to its weight.

Plastic or steel designs are cheaper and lighter (illustrated below) but need to be anchored in the perforated floor of the farrowing pen, in which case a central, spring-loaded handle is depressed and twisted to lock a small 'T-piece' under the slat and keep it from being overturned. Those with solid dividers should preferably be avoided as piglets like to see others eating and the more timid will start eating that much sooner.

Two similar creep starter bowls.
Left: fixed to the floor; Right: heavy freestanding plasticised concrete.

Another more costly but intelligent design which does not need such frequent replenishment compared to those where the creep feed loses its attractiveness when exposed to the smelly air in the farrowing house, is one I've seen used in the USA and made by Osborne, Kansas.(illustrated). This has a mini-tray under a small adjustable dispensing hopper well-sealed by a snap-on lid, which container keeps a modest supply of feed away from flies and odours and helps preserve its 'nose' which is so important for quick uptake when solid food is first introduced in competition to the sows milk.

The larger removable tray under the smaller fixed  tray can be taken off and washed, but ask for an extra  number of these so that the device can be kept as clean as possible by cleaning and replacing with the spares.

I hope they are still available, as we always got good results with them.

Many creep feed hoppers are what I call 'permanent' - heavy, well-made from steel, galvanized metal or plastic, the trough partitioned off by dividers(see photo above). Fine – but they tend to be overfilled and thus the feeding surfaces are not cleaned frequently enough during those vital 4 to 7 days of use, and there is the risk of food becoming stale. Labour-saving certainly, but labour should not be 'saved ' where the very small pigs are concerned. On the contrary, it should be intensified as my survey indicated.

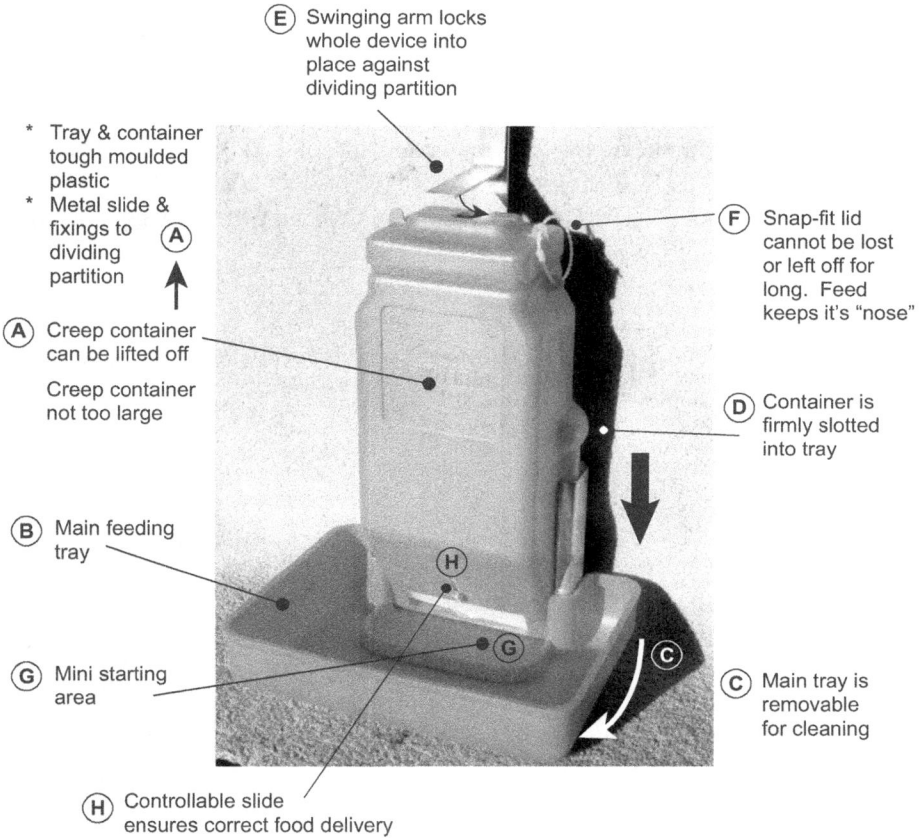

(E) Swinging arm locks whole device into place against dividing partition

\* Tray & container tough moulded plastic
\* Metal slide & fixings to dividing partition (A)

(A) Creep container can be lifted off

Creep container not too large

(B) Main feeding tray

(G) Mini starting area

(H) Controllable slide ensures correct food delivery

(F) Snap-fit lid cannot be lost or left off for long. Feed keeps it's "nose"

(D) Container is firmly slotted into tray

(C) Main tray is removable for cleaning

An excellent creep feeder. Notice the detachable base tray for cleaning.

*Spotless cleanliness is the keystone of a successful early start to creep feeding –* once you have summoned the courage to buy really advanced creep feeds, of course.

*Plenty of spare small creep feed receptacles, frequently sanitised and refilled,* enables this to be done.

*Owners and managers must allow sufficient time for their stockpeople to accomplish all that is needed.*

Because the larger, conventional creep hoppers are so permanent they are not placed into the farrowing pen until too late, in other words when the technician thinks the piglets will eat the creep feed willingly, which is about 10-14 days in his /her opinion. Yes, use them later by all means, but commence with the open flat tray idea or with special mini-feeders and you will help to secure the benefits demonstrated in Figure 2, at least from sufficient litters to justify the extra cost by the time the pigs so treated get to finish weight. The 'Three Threes' concept depends on their use, of course.

## Women or men?

Iincidentally I find women are much better at this than men, with a few notable exceptions - Gordon our stockman at our Deans Grove farm was a superb baby pig manager and I could name others. In my travels on some 2000 or more breeding farms I find men – for some reason I cannot fathom - are better at breeding than women but women are the best at farrowing and establishing the baby pig– but here the mothering genes must be responsible.

But I digress.

## PLACING THE CREEP FEED DISPENSER

Thee are 5 basic farrowing crate floor designs…

1.    Central crate/side creep
2.    Central crate/forward creep
3.    Central crate/ triangular (corner) creep
4.    Diagonal crate/corner creep
5.    'Freedom' crate/pen – corner creep.

Design No.1  is the most common across the world and Figure 4 shows the layouts I prefer. The idea of a sliding cover is a good one and preferable to the fixed cover usually seen. Figure 5 illustrates the concept.

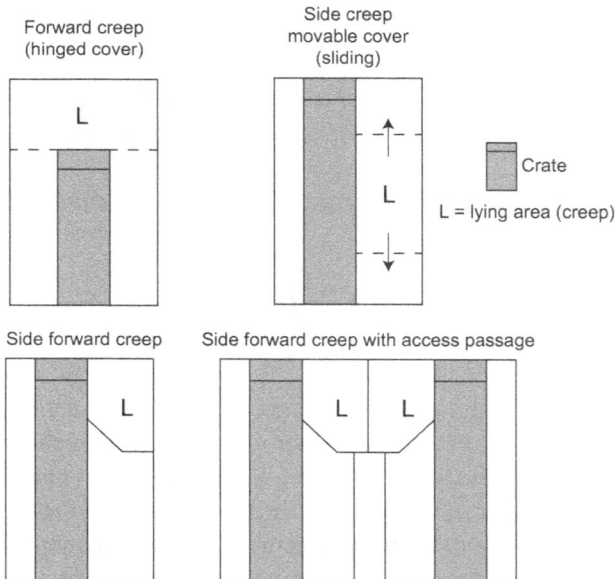

**Figure 4.** These four creep layouts work best.

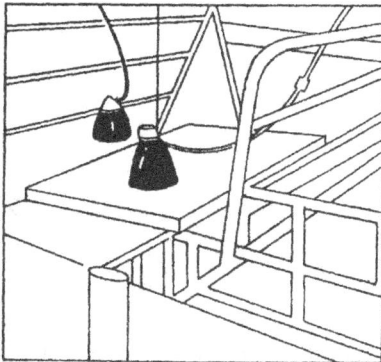

This variable-geometry insulated creep cover solves, at a stroke, the problem of retaining the heat down where it is needed. It coaxes the piglets towards the udder after farrowing, and keeps them out of harm's way subsequently.

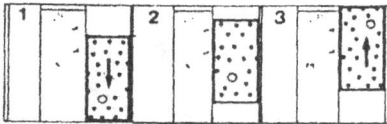

Position 1 – At farrowing: Lamp draws newborn piglets towards udder when cover is pulled to rear.

Position 2 – After farrowing: Cover pulled forward opposite udder leaves "dirty" end of creep cooler.

Position 3 – 7-10 days later: Board is now reversed and pulled forwards to coax piglets out of resting sow's way.

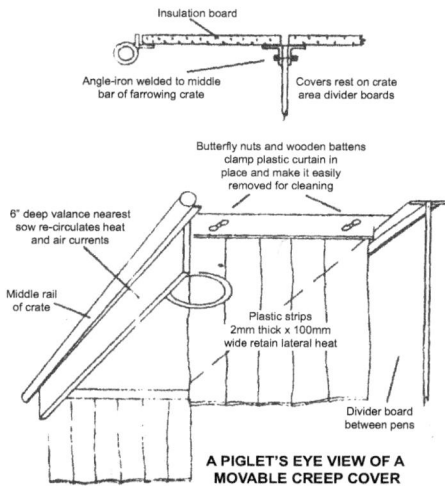

**A PIGLET'S EYE VIEW OF A MOVABLE CREEP COVER**

Another tip: If the rear area is slatted, use a square of hardboard to cover the slats temporarily so as to coax the piglets to the udder

**Figure 5.** Practical low-cost ideas

# CHECKLIST - SOME GROUND RULES ON PLACING CREEP DISPENSERS

✓     Try to make them easily accessible by the stockperson, avoiding unnecessary entry into the pen.

✓     Never under a heat lamp, on a heated floor pad or inside a warm covered creep area.

✓     Away from the sow's urine splashings.

✓     Out of reach of the sow lying supine when she can get her head under the crate bottom rail – attempts to get at the creep feed increase the incidence of shoulder sores.

✓     Not too far away from the piglets separate water source.

✓     Make them difficult to shift by the piglets. Either site them securely; make them heavy, or fix them to a base plate so that the piglet's own weight stops them from moving

✓    Ensure they are well lit (100 lux) For how to measure light intensity simply and inexpensively, see page 40.

The reason why Design 1 is favoured is because the piglets like to lie close to the udder, and if this area is gently heated the sow tends to lie with the udder facing the heat source anyway.

# CREEP AREAS – HOW MUCH SPACE?

Under optimal heat conditions, provide a piglet lying area of $0.4m^2$/per litter rising to $1m^2$ at 21 days and up to $1.7 m^2$ at 35 days, a litter being 10 to 12. Larger litters need a proportionate increase.

To conserve heat, provide an easily-lifted, or slid-forwards, cover (lid) with a 6 cm valance to conserve heat, but out of the sows chewing range! And put either a starter tray and subsequent follow-on hopper towards the sow's head, but place a small circular type of dispenser for the "3x3's" period on the same side and rather further back towards the posterior teats.

Keep all three types of dispensers away form a heat source which is where the piglets should lie and suckle.  Piglets prefer side creeps (Table 4).

Table 4. Piglets prefer side to forward creeps

|  | Proportion of total resting time spent in creep areas (%) | |
| --- | --- | --- |
| *Hours after birth* | *Front creep* | *Side creep* |
| No.of litters | 26 | 19 |
| 0-12 | 13.8 | 41.8 |
| 12-24 | 22.6 | 75.6 |
| 24-48 | 46.5 | 92.0 |

Source: Pope (1992)

I cite this interesting piece of work, despite it involving relatively few litters, because NOSCA reported very similar results to these several years before and I myself have noticed the differences between farms which had more active piglets (when they should be resting and putting on weight) where the creeps were of the forward type.

Lidded forward creeps are much easier to access, which is fine, but pig technicians tend to allow the neonates too much living room which encourages them to void in an unused corner when very young. This then causes a tendency for them to wrong-muck in their resting areas when they are older.

This bad habit originating in the creep area is easily avoided by nailing vertical battens on the creep sides so that a shut-off board can be lowered when the litter is young or small in numbers and removed when they get bigger and need more space.

This ensures all the area is used for sleeping only and 'house-trains' the piglets to go outside to do their business.

## A great improvement

Table 5 was one of my great successes, which surprised everyone including myself, in how dramatic was the improvement. The animal behaviourists say it is dfficult to obtain significant results in their field due to the variables encountered in behavioural responses.

**Table 5. Wrong mucking in growers 35 - 80 kg influenced by creep area allowance before weaning**

| Period | No. of very dirty grower pens before attention to creep areas | Period. | No of pens affected after attention to creep areas |
|---|---|---|---|
| 12-18 months before | 47% | 3 months after | 3% |

Source: Gadd (1984)

**Comment:** The wrong-mucking in the growers pens did not seem to be due to under- or overcrowding, or to temperature, or gases, or to incorrect air placement especially at night, which had all been checked out by others and then myself. We resorted to the creep shut-off board idea to clean up the messing the creep areas in the forward covered creeps. Virtually no wrong-mucking occurred from those pigs as growers.

## Feeding creep feed as gruel

Small piglets prefer gruel to meal or pellets. Gruel is a mixture of whole milk/skimmed milk, lightly molassed water or just plain water made into a 'porridge' consistency of about 2 by weight of creep feed solids to 1 by weight of liquid.
 This does increase dry matter intake and performance when properly constituted i.e.. *freshly made* and the receptacles *well designed* (to minimize spillage) and kept *spotlessly clean.*

Fail to achieve these 3 essentials and it will probably not work.

It is particularly valuable for the "Three-Threes" stage and for 10 days thereafter when the sucklers should be eating normal creep feed well (100-150g/day) see Table 6.

**Caution:** But not if the dispenser and the surrounding area are not kept clean. Why? because baby pigs are messy eaters just like human babies. This takes a lot of care and effort to do properly and does not take the place of adequate water provision either. Gruel needs plain water to be alongside just as CWF wet feeding does later in life.

**Table 6. Benefit from offering a good dry feed pellet made into a gruel with skimmed milk, fed x 4/day with meticulously clean conditions, compared to the same pellets offered dry.**

|  | *Gruel creep* | *Non-gruel creep* |
|---|---|---|
| No. of litters | 36 | 36 |
| First offered (days) | 5 | 3 |
| Av birthweight (kg) | 1.34 | 1.35 |
| Weight at 24 day weaning (kg) | 6.8 | 6.2 |
| Days from weaning to 105 kg | 121 | 131 |
| *Benefits* |  |  |
| Extra saleable meat/tonne of grower/finisher feed (kg) | 9 | - |
| Extra value from (353) pigs finished at slaughter weight | €95 | - |
| Less cost of gruel and labour | €186 |  |
| **Return ( REO)  5.1:1** |  |  |

Source: Clients records.

**Comment:** Gruel creep feeding gives the piglets a better start which is carried on significantly to final slaughter weight, providing the management is exemplary, as in this case.

Did the skimmed milk make a difference? Very probably, but added nutrients in the liquid fraction is a bonus from gruel or 'porridge' feeding anyway.

## The snags

Gruel feeding is nevertheless quite a difficult process with small sucking piglets. On our Taymix farm we were big distributors of skim milk and sold millions of litres annually, so we knew all about the product. Even so on our huge pig unit (partly created to soak up any skim milk not sold as well as to benefit from the very cheap price we could buy it in huge quantities) I found it quite a trial to do the job sufficiently well as a  method of creep feeding creep and transferred my efforts to feeding it post-weaning, which was easier and more successful in my case.

Some baby pigs can get 'sticky mouths' if the gruel is too thick or a separate source of water is not available Try it on a couple of litters first.

# EQUIPMENT -THE FUTURE

There are now machines (illustrated) which will do the job very successfully if you want to invest the capital in this way. The fortified milks to go with the machines are very good.

At the time of writing the equipment is used to take underprivileged neonates (after a period of colostral intake of course) and the stronger pigs from hyperprolific litters so as to relieve the sow or gilt of a punishingly large family.

A 'rescue deck'. At the time of writing, the latest in a succession of artificial foster mothers.

Thus the development of this seemingly 'unnecessarily expensive and futuristic' idea as some initial critics are remarking, is quite likely. Indeed I wish these forward-thinking manufacturers well - they have realized the importance of the baby pig as well as the threat to SPL, (Sow Productive Life), from the dangers of placing too high a demand on the young sow in particular, which the worthy skills of the geneticist has recently provided for our industry through hyperprolificacy.

# IMPRINTING

Here is a good idea which I've not seen in any textbook to date – despite it having been promoted 25 years ago. Some really good old ideas never seem to catch on, and this one is ideal for the 21st century when creep feeding is re-emerging as an essential strategy.

Imprinting involves the inclusion of a specific flavour to the sow`s diet just prior to farrowing and throughout lactation. The fat-soluble flavour is taken up by the sow`s mammary glands and secreted in her milk. When the same familiar flavour the piglets have experienced from birth is added to the creep feed, the piglets are encouraged to eat it sooner, and more of it too.

## But does it work?

Seems so. Quite a bit of reliable work was done in the 1970s and 80s. Campbell showed that imprinted piglets ate 63g/day more creep feed to weaning and grew 65g/day faster. Hunt in 1979 showed 66g/day more food eaten from 27- 60days old and they grew 38 g/day faster. The Firmenich company showed creep consumption doubled and the imprinted pigs were recorded on video to be twice as long at the feed trough, etc.

## So why look again at this intriguing idea?

1.  Compared to 25 years ago the modern sow produces, on average, at least one more piglet per litter – say 11 rather than 10. Or if you prefer, more than this, but between 1 to 2 weaners more is expected now as routine.
2.  Weaning date in certain industries is moving towards 26 to 30 days rather than the 21 days of tradition.  A week later and the weaner`s digestive capability is just that bit more developed to withstand the transition of milk to solid food .
3.  This means the modern sow is having to feed one to two more piglets, and all in the litter are at least 1 kg heavier towards the end of lactation. A total additional litterweight to sustain of around 17 to 18 kg - much more in the case of hyperprolific gilts and sows.

Working on the premise that the sow needs to provide 4.25g of milk for each g of piglet growth, the piglets of 25 years ago growing at 120 g/day in the four days from birth required 6.38 kg milk per litter per day, while today`s little monsters growing at 138g/day over the same four days need 7.65 kg of milk/day per litter.
The secret is to get the piglets eating creep feed as soon as possible which is as difficult a problem as getting the sow to eat more in lactation. Thus imprinting justifies a further look, I guess, because it does just that. In fact it could do both of them.

## But why didn't Imprinting catch on?

Producers felt in those days that creep feeds if eaten too soon caused gut upset. Nutritional science has now largely overcome that threat - if the breeder is prepared to pay the justifiable extra cost of such specialised modern diets -as he should. For the reasons I listed earlier.

In those days cost/tonne was much to the fore in peoples` minds, and if the breeder had to provide a separate gestation diet for the last 3 weeks of pregnancy the two together tended to get the thumbs down.

## So what about cost?

In 1990 we worked on imprinting adding on a further 3% to18% of the cost of the dry sow feed. At today's prices this comes to €0.33. Add in the cost of the lactation feed (€1.00). And the extra cost of adding the flavour to the preweaning creep feed (€1.45), a total of €2.78 per litter.

Taking the evidence of improved weight gain from the 1980 era research trials, one can expect imprinted pigs to be 0.5 kg heavier at weaning. Extrapolating this to 2 days saved in food and overheads at slaughter, which is a modest enough estimate, then for a litter of 11 piglets, this is a saving of €1.38/pig or €15.18/litter. This provides a REO ( Return on Extra Outlay) of 15.18 divided by the cost/litter of 2.78, or 5.5:1 – very encouraging.

Even more encouraging is the AIV (Annual Investment Value, see the Business Section, page 249 for the description) which is what the bank manager looks at to see how hard your monetary decisions are working for you –and him! The AIV in this case takes the turnover of a typical Farrowing Index today of, say 2.3:1 litters/ year multiplied by the REO of 5.5, which is12.65. Anything over AIV 6 gladdens his hard old heart.

No, the textbooks mention nothing about imprinting, and I guess it is time they did!

## *SUMMARY*

This has been a long and involved chapter. Sorry! But don't you agree - a fascinating one? We are on the cusp of really getting creep feeding right at last, and it is all so interesting, isn't it?
In summary, the key factors to success are fairly straightforward….

✓    You must creep feed now that litters are so large.
✓    You must be prepared to pay for the sophistication now included in these redesigned feeds.
✓    Keep things much, much cleaner.
✓    Stockpeople – spend more <u>time</u> in this area.
✓    Owners and managers – ensure they have enough time to do what is necessary.

# 2

# GETTING BIGGER LITTERS

## LITTER SIZE

The number of piglets born. Normally expressed as 'total-borns' i.e. the number of fully-formed individuals expelled at farrowing, alive and dead. Also sometimes classified as 'born-alives' i.e. those piglets known to have drawn breath immediately after expulsion.

The former figure is preferred, as in an examination of poor litter size, the number of foetuses carried to full term could be important in establishing causes.

## TARGETS

Are of course variable across the world, hence a range of targets are given:–

Table 1. Litter size targets

| | Total piglets born and piglets born alive per litter (It is advisable to sample a minimum of 50 to 100 litters) | | | |
|---|---|---|---|---|
| | *Poor* | *Typical worldwide* | *Good/Target* | *Target (Hyperprolific genes)* * |
| Total-born | 9.5 | 9.9 | 11.8 | 13.1 |
| Born-alive | 9.0 | 9.4 | 11.25 | 12.5 |

Generally, target for no more than 5% piglets born dead.

*Some paragons exceed this but find it difficult to maintain this level of productivity, and birthweights (page 47) and weaning capacity (page 252) suffer.

## THE PROBLEM AND ITS COST

Low litter size is a major problem world-wide. The breeder is usually alerted when periods occur where only 8.5 pigs per litter or less are weaned and/or where the numbers weaned per sow per year drop below 20 at any weaning age.

It is impracticable to estimate what the financial penalty is for a drop in litter size, as costs, incomes and margins vary across the world, and within one pig industry across, say a five year period, or even among one farm's financial picture over as little as one year's output.

However, as it costs an appreciable amount just to get one piglet born, whether dead or alive, my experience of the financial cost between the 'Typical world wide' and 'Good/Target' figures cited in Table 1 of 1.6 piglets total born per litter reduces sow income by 20% and has affected my clients' gross margin by 18% to as much as 45%

In many pig industries a 20% gross margin on sow output is considered a minimal baseline, then litter size must become a major influence on breeding efficiency.

The problem of reduced litter size can be due to many factors, complicated  by possible interactions between them.

A list of primary factors encountered by the author is:–

# CHECKLIST - PRIMARY FACTORS ALLIED TO LOW LITTER SIZE

*Gilts*
- ✓ Serving too light, grown too fast
- ✓ Stress rather than stimulation
- ✓ Lack of flushing
- ✓ Outdated diets pre-service and in first lactation (see Gilt Section)

*Sows*
- ✓ Lactation nose-dive
- ✓ Incorrect lighting pre-service
- ✓ Adequate feed weaning to service
- ✓ Boar presence
- ✓ Poor AI technique
- ✓ Not practising parity segregation
- ✓ Poor and late heat detection
- ✓ Clean floors and rear-ends
- ✓ Rest and quiet after service
- ✓ Bullying
- ✓ Litter scatter not analysed & acted on
- ✓ Disease

| | | |
|---|---|---|
| ***Sows*** | ✓ | Poor checking for returns |
| (contd) | ✓ | Herd age profile |
| | ✓ | Stress, discomfort, anxiety |
| | ✓ | Genetics |
| | ✓ | Short lactation |
| ***Boars*** | ✓ | Poor AI technique |
| | ✓ | Not monitoring boar records |
| | ✓ | Overuse of favourites |
| | ✓ | Lethal genes in some boars |
| | ✓ | Not sanitizing boar's sheath |
| | ✓ | No separate insemination/breeding area |
| | ✓ | Nutrition |
| | ✓ | Lack of exercise |
| ***General*** | ✓ | Lack of the 'Feel Good Factor' |
| | ✓ | Poor use of pig specialist veterinarian |
| | ✓ | Too few man-hours/sow/year |

# TACKLING A LOW LITTER SIZE PROBLEM

Good records are vital, particularly on sow and boar use. While mixed semen can be an aid to improved AI results, the inability to identify problems associated with individual males is a distinct drawback.

# GENERAL FACTORS

## The 'feel good' factor

A difficult term to define, but any pigs which seem harried or stressed rather than comfortable and contented are prone to poorer performance.

In litter size problems the Feel Good Factor is particularly important in the lead-up to breeding gilts, and in the post service period in sows. Producers should do a stress, anxiety and comfort audit of their breeding stock especially at these times.

Stockmen should distinguish between stimulation and stress before and during breeding, and do everything they can to keep the sows calm and contented in a restful atmosphere once mating is over.

Stimulation encourages the reproductive hormones. Stress inhibits/neutralises them.

## Too few man-hours per sow per year

Experience suggests that stockmen (especially experienced stockmen as distinct from trainees) who spend more time with their breeding stock enjoy better litter sizes. It is difficult to suggest a minimal number of man-hours devoted to the breeding sow per year, but a threshold of around 20 is suggested. Some massive units, due to economy of scale, are run as low as 10 or less, but the performance of such units is not high even if the profit is considered 'adequate'.

The major labour problem among medium to small farms with reduced litter size is always the time spent on urgent repairs & precautionary maintenance at the expense of breeding & farrowing time or adequate hygiene (see Table 2). Some specialist help/contract labour can help.

I have done some survey work on the apportionment of labour on some 56 farms in 7 countries over the past four years. (Table 2)

**Table 2. Workload – from stockpersons estimates – affecting litter size expressed as manhours /sow/year**

The total time spent on breeding by the farm's workforce was 14 hrs/sow/year for the smaller units (50.7%) and 11 hrs for the larger farms (57.5%).

|  | *40 farms 120-550 sows* | *16 farms 875-2500 sows* |
|---|---|---|
| Feeding | 4.2 | 2.1 |
| * Serving | 3.5 | 3.1 |
| * Farrowing and post-natal | 2.5 | 1.8 |
| Moving and records | 2.0 | 1.9 |
| Cleaning and disinfection | 1.8 | 1.9 |

* Areas impacting on litter size. These figures indicate to me that just under half of the breeding farm's time-load spent on the two vital tasks of breeding and farrowing is not enough, especially as repairs and maintenance took up around 2 manhours sow/ year which approached the time spent on farrowing and raising the litters.

## Lack of pig specialist veterinary advice and presence

Disease, particularly viruses at a low level like PRRS, can be a cause of poor litter size. All farms, whatever their size, should have a regular input from a pig specialist veterinarian. In the author's records there are several cases where the disease level fell, adding generally more than 20% to the net income. However the veterinarian's time only increased vet/med costs by 4% and increased preventive protocols (drugs, vaccines, hygiene products) by another 5%, an REO of 20÷9 or 2.2:1.

In these cases litter size based on born-alives rose by an average of 0.9 pigs/litter.

# GILTS

Gilts at a 39% - 48% replacement rate (the latter is too high but commonplace) can constitute a large part of the modern breeding herd.

Modern gilts have the capability to give large litters (over 12 total-borns) but too often give 9 or less. Even so a gilt (*i.e.* 1st parity sow) should not be culled on poor litter size. [*See Culling Checklist.*]

# FACTORS AFFECTING GILT LITTER SIZE

## Serving too light

135 kg is the current advised threshold, though some producers prefer 130 kg. Consult your supplier, and then a nutritionist, see below. In any event, 240 days of age is the critical factor, not so much weight or fat cover [*See Gilt Section*].

## Grown too fast

The modern gilt can grow over 1000 g/day, and can be in danger of her hormone system falling behind her precocious growth. Thus she may be heavy enough and even display enthusiasm alongside or in with companions showing vigorous signs of oestrus, but still give you a poor conception rate or low first litter size due to immature sexual hormones.

Table 3 gives the latest weight-for-age table to suit most genotypes. Check with a nutritionist that his gilt developer feed will achieve this level of weight for age.

Table 4 shows the effect of this steadier approach to puberty in the modern gilt.

# WEANER OR 'JUNIOR' GILTS

Buying the gilt at 35–40 kg and raising it yourself avoids any possibility of the multiplier 'forcing' the gilt before delivery in order to save costs. It has other significant beneficial effects apart from that of a possible rise in gilt litter size, but the higher cost/lower selection/increased failure rate should be taken into account. [*See 'Choosing a Gilt' Section*].

**Table 3. Gilts : suggested typical weights for age for modern high lean gain European breeds***

| *Gilt growth rate* | *Aim to achieve 100 kg in 170-180 days, gilt growth-rate at 550 g/day, rising to 750 g/day towards puberty* | | |
|---|---|---|---|
| 100 kg | 180 days | 25th or 26th week | 6½ months+ old |
| 104 kg | 187 days | week 27 | |
| 108 kg | 194 days | week 28 | 7 months old |
| 112 kg | 201 days | week 29 | |
| 116 kg | 209 days | week 30 | |
| 121 kg | 216 days | week 31 | |
| 126 kg | 223 days | week 32 | 8 months old |
| 131 kg | 230 days | week 33 | |
| 136 kg | 240 days | week 34 | 8½ months old |

\* *Consult your seedstock supplier for actual targets.*

**Table 4. Gilt litter size by age at first serivce**

| *Age at 1st Service (days)* | *200-210* | *215-225* | *230-240* | *245-255* | *260-270* |
|---|---|---|---|---|---|
| Percentage of gilts | 28 | 27 | 21 | 16 | 8 |
| Total borns | 10.58 | 12.27 | 12.92 | 12.87 | 10.44 |

Source : Easton Lodge Pigs (UK) March 2000
Note that this early work forecasts the advantage of serving at 240 days [*See Gilt Section*]

# FLUSHING

Flushing is particularly valuable if gilts are delivered rather lean and light, as is common today, because multiplying farms want to get their finished gilts off their hands as quickly as possible. Table 5 gives a programme I've found valuable for many years. Table 6 gives the sort of results it can give when allied to later mating.

## What does flushing do?

In gilts which have a medium to low ovulation rate, it should improve things.

In gilts which already have a high ovulation rate, flushing will probably make little difference. As there is no real way for the farmer to determine this is it best to flush all gilts as routine. The response seems to be due to increased energy availability stimulating follicle growth just before oestrus (7 to 10 days) resulting hopefully in one to one and a half more ovae being shed.

**Table 5. Typical gilt flushing regime**

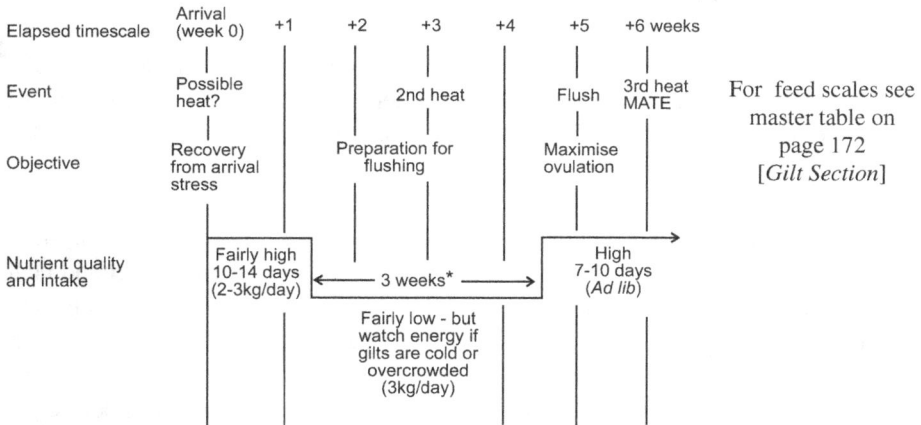

| Elapsed timescale | Arrival (week 0) | +1 | +2 | +3 | +4 | +5 | +6 weeks |
|---|---|---|---|---|---|---|---|
| Event | Possible heat? | | | 2nd heat | | Flush | 3rd heat MATE |
| Objective | Recovery from arrival stress | | | Preparation for flushing | | Maximise ovulation | |

For feed scales see
master table on
page 172
[*Gilt Section*]

| Nutrient quality and intake | Fairly high 10-14 days (2-3kg/day) | ← 3 weeks* → | | | | High 7-10 days (*Ad lib*) | |

Fairly low - but
watch energy if
gilts are cold or
overcrowded
(3kg/day)

Note: If (a) virus disease is present or (b) low arrival weights (<100kg) then for (a) buy weaner gilts and for (b) delay final service by one heat.

I have found the above programme to be particularly valuable for young or lean gilts. Even so this period of relatively modest nutrient intake should be shortened to 10-14 days if the gilts are still lightish or thinish at 2-3 weeks after arrival. Be observant; be flexible. Nothing in gilt management is carved in stone!

**Table 6.  Results of the above technique (same genotype, same farm)**

| | | *Flushing / later mating* Cost (from previous table) £30/gilt*. Served at 3$^{rd}$ heat, 240 days and flushed. | | *Old system* Gilts rested for 2 weeks, fed well (to appetite). Served at 2$^{nd}$ heat. | |
|---|---|---|---|---|---|
| | *Results* | *Pigs born alive* | *Weaned* | *Pigs born alive* | *Weaned* |
| Parity | 1 | 10.8 | 9.4 | 10.3 | 9.4 |
| | 2 | 11.1 | 10.0 | 9.2 | 8.2 |
| | 3 | 11.7 | 10.4 | 10.7 | 9.4 |
| | 4 | 11.6 | 10.8 | 10.9 | 9.4 |
| | 5 | 12.1 | 10.9 | 10.8 | 9.6 |
| **TOTALS** | | | **51.5** | | **46.0** |

Extra Cost – £ 30      Extra Productivity   5.5 weaners  worth £70 margin in 2008
REO (2008) 70 ÷ 30 = 2.33:1. *Cost at the time.

Source : Author's Records (2009)

Flushing the sow is not usually practised as there is not enough time to do this, unless the animal is run down in condition and a heat is skipped to allow her time to recover her rebreeding capability. However, maintain the lactation diet after weaning.

Flushing is also thought to encourage the 'feel-good' factor in the gilt. So what is the 'feel-good' factor?

Gilts can arrive light, lean, stressed (nervous) with some of them quarrelsome. Along with capturing their interest (e.g. straw), giving them a good high-nutrient, specialised diet settles them down quickly into the 'feel good' factor which is so necessary for successful first litter formation.

We now have to keep them quiet for quite a time before eventual service at their 3rd oestrus at least, and just grow them steadily and quietly across the second heat period on a diet of modest nutrient density. How long this modest plane of nutrition should last depends on the gilt's weight increase and condition. If they are thought to be still thin, light, cold or stressed, it is better not to continue for the full 3 weeks in Table 4. Shorten it to a few days before the sudden increase 7-10 days before the oestrus chosen for mating.

**Economics**

In the UK the extra cost of food is about £11 per gilt (6-8% of purchase cost) but the total cost/gilt, including labour and housing costs, plus other overheads, can reach £23/gilt (+14%). But this is a lifetime cost, and Table 6 suggests that the resultant 5-6 more weaners per lifetime can raise margin by £50 - an REO of 2.5:1. Money spent at the gilt rearing stage is always a good investment. The younger and lighter the gilt on arrival, the more flushing will pay.

# STIMULATION, NOT STRESS

Hormones encouraged by fear and anxiety tend to neutralize the pro-oestrus hormones and affect litter size. Stockmen also tend to approve noisy mounting and chattering behaviour in the gilt pens during the run-up period to service, especially if boars are nearby to encourage this.

However some of this noise can be due to fear and protest. So check for:–

- Adequate space, especially fleeing space (3.5 m²/gilt, approx 38ft²/gilt)
- Pen shape – as square as possible
- Numbers together. This depends on space adequacy but normally 6 is enough. Very big gilt pools (e.g. yards on bedding can run up to 30 gilts per section)
- Evenness. Excited heavier gilts can be dominant and aggressive. Also it is a cardinal error to run gilts with multiparous sows; even to pen them adjacently at oestrus is unwise.
- Adequate food and water access.

# SOWS

***The Nose Dive*** – a term given to muscle and body-fat fallaway through lactation (Figure 1).

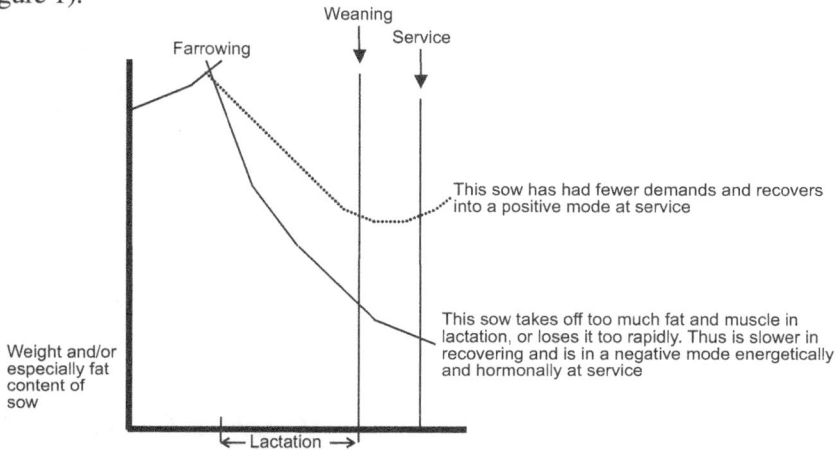

Weaning

Service

Farrowing

This sow has had fewer demands and recovers into a positive mode at service

This sow takes off too much fat and muscle in lactation, or loses it too rapidly. Thus is slower in recovering and is in a negative mode energetically and hormonally at service

Weight and/or especially fat content of sow

←— Lactation —→

So how you feed and manage the sow way back in pregnancy, and especially in lactation, can markedly influence re-breeding

**Figure 1.** The 'nose-dive' effect. How it could affect ovulation - returns to service and litter size

Sows in lower than normal condition at weaning can often give poorer subsequent litter performance as well as re-service problems.

Don't let the sow lose too much condition down through lactation and/or (new suggestion), lose less condition **overall** but suddenly lose it badly towards the end of lactation.  For each 10 kg weight loss in lactation, subsequent litter size can be reduced by 0.5 pigs/litter, and for each 1 kg less food eaten, 23 day weaning weight can reduce by 400-500 g per weaner, dependent on litter size and sow condition.

A great deal of future sow productivity depends on the stockperson's skill in defending the sow from the depredations of a voracious and, these days, larger litter.

The following checklist gives a list of factors which can help.

## CHECKLIST – HOW TO AVOID THE 'NOSE-DIVE'

✓     ***Body Condition Score**** so you can quickly detect when it is occurring and its degree of severity.

✓   Don't breed gilts too soon,  too light, too thin.

✓   Flush all gilts (new advice on feed intakes available – see Table 5).

✓   Buy breeds/strains with good appetites.

✓   Keep sows cool (21°C/70°F maximum) in lactation.

✓   Feed a special sow lactation diet (altered for hot weather). The gilt needs a special lactation feed, different to a sow lactator.

✓   Adequate water and accessibility especially in lactation.

✓   Water by bowl, not by bite drinker.

✓   Feed wet (by pipeline).

✓   Don't overfeed in pregnancy, especially 7 days before farrowing.

✓   Take the load off the sow by fostering/piglet swapping/ "weaning by weight/ not date" etc (but take care if PMWS is about).

✓   Feed 3 times a day, last main feed at night.  Maintain feed freshness.

✓   Avoid all stressors, especially discomfort.

✓   Be especially careful in hot weather.

*   Body Condition Scoring has its detractors among academics, due to its undoubted imprecision and subjectivity. However, many years of advisory work has convinced me that its value lies in encouraging stockpeople to really examine their sows by feeling as well as looking.  It is especially valuable in monitoring the start of the nosedive phenomenon (Figure 1) and in determining fat cover and fleshing post-weaning. Academics, while right to point out its disadvantages, should ease up on their public denunciations for fear of damaging a useful sharp-end stockmanship tool.

To help ease this "stockmanship versus academic" controversy, there is now a relatively inexpensive hand-held muscle/fat scanner, which will help the stockperson, new to body condition scoring by touch and by eye, relate his subjective opinions to what the scanner reveals. The scanner has other uses - for grading and for challenge feeding (see page 110).

# PRACTICAL ADVICE

•   Check carefully with your nutritionist that his diet contains sufficient protein/ amino acids and energy to satisfy the weight of progeny to be suckled daily. Litters are larger these days.

•   Many successful breeders follow the U.K. 'Stotfold' lactation feed scale. (Table 7). There is also a simpler American 'Feed-scoop' system (Table 8) which works

**Table 7. Lactation feed scale (Metric)**

| First 10 days (All sows/gilts) | | | Sow Identification |
|---|---|---|---|
| Day | Kg | Fed | Total Fed: |
| 1 | 2.5 | | Date Farrowed (Day 1) |
| 2 | 3.0 | | NOTES: *This dietary scale is now widely used in Europe.* |
| 3 | 3.5 | | *Liaise with a pig nutritionist to formulate a diet density which* |
| 4 | 4.0 | | *will satisfy the published daily intakes (cf Close & Cole* |
| 5 | 4.5 | | *'Nutrition of Sows and Boars', NUP 2000). Total litter weight* |
| 6 | 5.0 | | *at weaning can be on some farms over 30% higher than for average* |
| 7 | 5.5 | | *herds.* |
| 8 | 6.0 | | |
| 9 | 6.5 | | |
| 10 | 7.0 | | |

| Gilt<10 piglets Sow<9 piglets | | | Gilt 10 piglets Sow 9 piglets | | | Gilt 11 piglets Sow 10 piglets | | | Gilt 12 piglets Sow 11 piglets | | | Gilt 13 piglets Sow 12 piglets | | |
|---|---|---|---|---|---|---|---|---|---|---|---|---|---|---|
| Day | Kg | Fed | Day | Kg | Fed | Day | Kg | Fed | Day | Kg | Fed | Day | Kg | Fed |
| 11 | 7.0 | | 11 | 7.5 | | 11 | 7.5 | | 11 | 7.5 | | 11 | 7.5 | |
| 12 | 7.0 | | 12 | 7.5 | | 12 | 8.0 | | 12 | 8.0 | | 12 | 8.0 | |
| 13 | 7.5 | | 13 | 8.0 | | 13 | 8.5 | | 13 | 8.5 | | 13 | 8.5 | |
| 14 | 7.5 | | 14 | 8.0 | | 14 | 8.5 | | 14 | 9.0 | | 14 | 9.0 | |
| 15 | 8.0 | | 15 | 8.5 | | 15 | 9.0 | | 15 | 9.5 | | 15 | 9.5 | |
| 16 | 8.0 | | 16 | 8.5 | | 16 | 9.0 | | 16 | 9.5 | | 16 | 10.0 | |
| 17 | 8.5 | | 17 | 9.0 | | 17 | 9.5 | | 17 | 10.0 | | 17 | 10.5 | |
| 18 | 8.5 | | 18 | 9.0 | | 18 | 9.5 | | 18 | 10.0 | | 18 | 10.5 | |
| 19 | 9.0 | | 19 | 9.5 | | 19 | 10.0 | | 19 | 10.5 | | 19 | 11.0 | |
| 20 | 9.0 | | 20 | 9.5 | | 20 | 10.0 | | 20 | 10.5 | | 20 | 11.0 | |
| 21 | 9.5 | | 21 | 10.0 | | 21 | 10.5 | | 21 | 11.0 | | 21 | 11.5 | |
| 22 | 9.5 | | 22 | 10.0 | | 22 | 10.5 | | 22 | 11.0 | | 22 | 11.5 | |
| 23 | 9.5 | | 23 | 10.0 | | 23 | 10.5 | | 23 | 11.0 | | 23 | 11.5 | |
| 24 | 9.5 | | 24 | 10.0 | | 24 | 10.5 | | 24 | 11.0 | | 24 | 11.5 | |
| 25 | 9.5 | | 25 | 10.0 | | 25 | 10.5 | | 25 | 11.0 | | 25 | 11.5 | |
| 26 | 9.5 | | 26 | 10.0 | | 26 | 10.5 | | 26 | 11.0 | | 26 | 11.5 | |
| 27 | 9.5 | | 27 | 10.0 | | 27 | 10.5 | | 27 | 11.0 | | 27 | 11.5 | |

Developed by the UK Meat and Livestock Commission – Stotfold PDU

Notes on how to use the feed scale

(1) Assess piglet and sow condition on Day 10

(2) Select appropriate scale consistent with piglet number and rearing ability of sow (*e.g.* a highly productive sow with 10 piglets may require the feed scale for a sow with 11 piglets)

(3) Where deviations from the scale are appropriate (either up or down) record the amounts consumed in the 'Fed' column

(4) Cross off the days in the 'Day' column as lactation progresses – allowing relief stock persons to refer to and maintain correct feed intake levels

(5) Record alterations to piglet numbers and change to the appropriate scale

(6) Feed lactating animals at least twice per day

(7) Two diet feeding system is recommended; the lactating sow requiring higher energy and lysine levels than the pregnant sow

(8) Ensure an adequate water supply. Drinkers should flow at least 1.5 litres per minute

(9) Ensure correct room temperature. As sow feed intake increases, room temperature should reduce from 20° to 16°C. Maintain at 16°C for the last 10 days of lactation

(10) When day time temperatures are high, feed one third of the daily requirement am and two thirds pm.

well. An American feed scoop holds about 4 lbs (1.82 kg). Sows are fed 3 times a day and receive 0, 1 or 2 scoops at each meal. If there is feed left over from a previous meal then no new feed is added. If there is a small amount remaining, say 2 lb (about 0.9 kg), 1 scoop is added. If the sow's feeder is empty, then 2 scoops are fed. The exception to this pattern is 2 days after farrowing when less is fed to avoid udder complications/MMA etc. (See Table 8).

Caution: If more than one person is feeding the farrowing room over a period then a way of telling the next stockperson what the sow's appetite has been like at the previous 2 or 3 feedings is needed. This is done by using clothes pegs to an agreed code, or rotating the sow's record card likewise. Table 8 illustrates this simple system.

Table 8. "To appetite" feed method (Kansas State University)

| *Number of 4 lb scoops to feed at each feeding from day 0 to 2* | | | *Number of 4 lb scoops to feed at (4 lb = 1.82 kg)* | | | | |
|---|---|---|---|---|---|---|---|
| *Feed in feeder* | *Feeding* | | *Feed in feeder* | *Feeding* | | | *Totals (kg)* |
| | *AM* | *PM* | | *AM* | *Noon* | *PM* | |
| Empty | 1 | 1 | Empty | 2 | 2 | 2 | 10.9 |
| <2 lb | 0 | 0.5 | <2 lb | 1 | 1 | 2 | 7.3 |
| >2 lb | 0 | 0 | > 2 lb | 0 | 0 | 1 | 1.8 |

Source: Dritz and Tokachi (2010)

Sow feeding by the simplified scoop system used in the USA - in this case for dry sows needing a boost.

- Try to buy females with an ample appetite potential. Ask the breeder for evidence of likely appetite, then contact his customers for their opinion on how easy or otherwise it is to get sufficient daily food intake in lactation.

## CORRECT LIGHTING PATTERNS

Light, like water, is relatively cheap but its importance is poorly appreciated in the breeding unit. The mating area and the areas where the newly weaned sow or new-entrant gilt are held prior to service need really bright light – at least 350 lux (lumens). This is bright enough to read a newspaper easily. Bright summer sunshine in a green field is between 500-600 lux. Figure 2 suggests that in some breeding units white fluorescent tubes are needed, and suggests the spacing and height needed of 100w fluorescent tubes. Remember, the light needs to be taken in by the sow's eye so the tubes need to be over stalled sows' heads, or just forward of them, not over the back.

100 watt fluorescent strip lights, white
Aim for 16 watts per m² (1.6 watts/ft²)

Place the lights so that the majority of the light falls via the eyes

**Lighting pattern**

*On* maximum 16 hours/day
*Off* for 8 hours/day

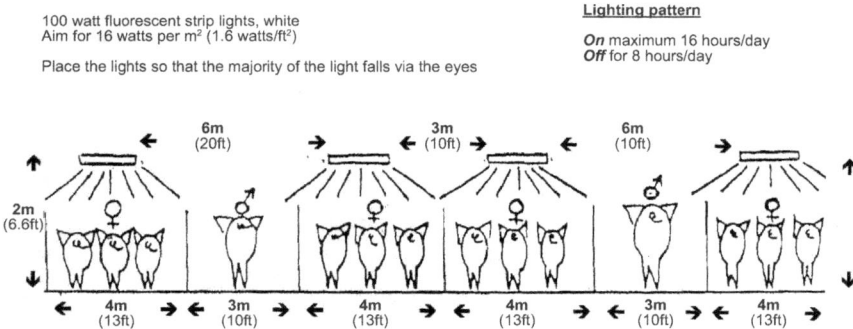

**Figure 2.** Lighting a mating/breeding house

## LIGHT PATTERNS

Most people – but not all academics – think that a 1/3 / 2/3rd pattern of off : on is best, say 8 hours of near darkness (10-12 lux) followed by a maximum of 16 hours of bright light (350 lux+), maybe 14 hours for gilts. I myself am quite convinced that this pattern is correct from following up too-dark breeding areas advice. A definition of lux (and also lumens, not the same) can be found in the Glossary. For how to measure light levels, see Figure 3.

## LITTER SCATTER, AN UNDER-USED LITTERSIZE CHECK ON MANY FARMS

Litter scatter is a good indicator of infertility disease arising or already present, problems with ovulation/implantation or poor boar service. These are all pointers to litter size problems.

Litters with <8 : Target 10%. Action level 15%.

Use a photographic light meter - now redundant & cheap. Choose any setting to cover daylight. e.g.

**1**

1 On a bright sunny day, point the reverse (receptor) face towards the sun (not directly at it).
This will register approx. 600 lux
MARK THIS POINT

2 On a starlit night, the needle will register approx. 25 lux
MARK THIS POINT

**2**

3 Now estimate and mark where 300-350 lux would be on the dial

4 Holding the receptor side of the meter away from the sow's eyes (about 200 mm away) check that the needle reaches the 300-350 lux mark you have made = correct light intensity

**3**

**4**

200 mm

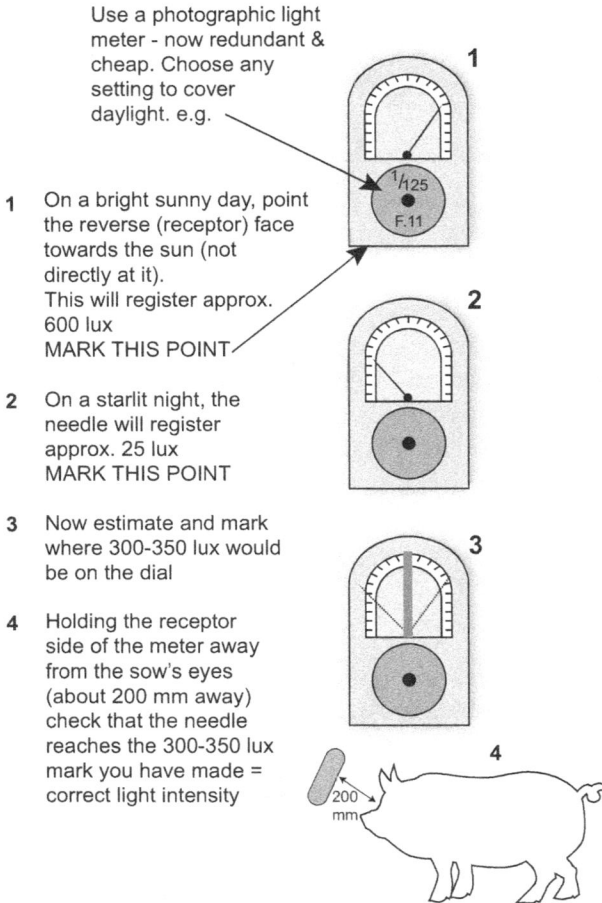

**Figure 3.** How to measure sufficient light

# HERD AGE PROFILE – DON'T GET CAUGHT OUT

Because many herds, due to non-specific infertility and in hot weather, leg problems, are forced to cull prematurely (Table 9), the ideal herd age profile can quickly distort (*see Culling section*). This can affect litter size by up to 2 pigs a litter for a period and 1 pig/litter is very common.

This can also allow virus disease to gain a hold because of reduced herd immunity from too few established 3 to 6 litter sows. This alone can reduce litter size substantially.

**Table 9.** Number of young females culled for poor reproductive/performance reasons

|  |  | *1980s* | *1990s* |
|---|---|---|---|
| Overall herd replacement rate |  | 37% | 45% |
| Reasons:– | Reproductive Failure | 32% | 48% |
|  | Health & Losses | 30% | 27% |
|  | Legs & Feet | 18% | 10% |
|  | Other | 20% | 15% |

Source : MLC Yearbooks (UK)

*Practical advice* : Keep an ongoing graphical account of your herd age profile. Have ample gilts available *suitably acclimatized* in a gilt pool; buying weaner gilts can assist here. Consult with a pig specialist veterinarian if your enforced culling is more than 33% due to non-specific infertility. Choose gilts with strong leg bone structure and good spring of pastern; this should not unduly affect a tendency towards coarse/ heavy bone structure in the finished progeny.

Remember – it is enforced early culling which can alter your herd age profile which eventually affects litter size.

## SHORT LACTATION LENGTH

The pressures of economics favouring the reduction of lactation length are well-known. European producers, at the time these notes are written, have stabilised at 23/24 days (even though a move to 28 days has been proposed), but the move towards 16 days seems to be preferred in the US, especially on larger and newer units. A fall-off of 0.5 pigs litter at least can be expected between 23 days and 16 days, but this could reduce as management and re-breeding skills improve among those permitted to wean early outside the Welfare limitations now in place in some countries. The Americans are now considering moving back to at least 19 days as their litter size and post-weaning problems have hit home.

## REPEATS (RETURNS TO SERVICE)

Check repeats like this : –

*Regular repeats (21 ± 3 days)*     Target 10% Action level 15%

*Irregular repeats (>24 days)*     Target 3%  Action level 6%

Attention to better pregnancy diagnosis and thus reducing repeats by 33% has improved litter size by 0.3 pigs/litter.

# DISEASE

If disease is an influence on litter size it will strike mainly post-service. Check that gilts are vaccinated against parvovirus and test for PRRS. With your vet's help determine if causes are infectious (especially stillborns, together with mummies and size of mummifieds) or non-infectious causes (*e.g.* born alive but suffocated/weak).

**The routine use of a pig specialist veterinarian is an important defence against low litter size.**

# BIOSECURITY

Follow a *proper* biosecurity protocol. Quite a few used today are now outdated in technique and products used. (*See Biosecurity section.*)

**Mycotoxins**. Various mycotoxins at very low levels may have a bearing on low litter size.

* Always include a mould inhibitor in the feed or stored grain.
* Consider also (and it is especially advisable after wet warm harvests) adding a modern mycotoxin absorbent (not clays).
* Sanitise (*i.e.* steam clean and dry out) bulk bins regularly, probably twice a year.
* Agitate with bin manufacturers for 'bulkhead door' access ports at mid-height and swing-away food auger boots. It is invariably too onerous and dangerous to gain access to a bulk bin via the top inspection hatch.
* Feed hoppers can be a source of mycotoxins.
* Damp or mouldy bedding is a source in grouped sow yards.
* Fodder maize is also a potential source.

*(All this is discussed in detail in the Mycotoxin section, page 203)*

# GENETICS

All body functions implicated in litter size are under genetic control to a greater or lesser extent. However as a large number of genes are involved in the physiological processes which determine litter size, it is unlikely that any two females would be identical, thus considerable variation is likely to exist between groups of females.

Secondly, different females respond differently to stresses, further complicating the situation, especially regarding ovulation and implantation.

Various experts have reviewed the complexities of genetic improvement of litter size in pigs. The difficulties revealed show :

- The low heritability of the trait (<0.1)

- The low repeatability of the trait (<0.15)

- A very large sample is required to measure differences.

- The influence of heterosis.

- The influence of environment and management at all stages of the female's life.

The author fully accepts the scientific wisdom thus expressed, but has evidence of at least three cases of a dramatic improvement in litter size (range +0.92 to 1.86 b/a) where a batch of gilts from a different breed was tried – an average improvement of 21%. In all three cases the indigenous breed was a high lean hybrid, while the replacements were a cross based heavily on female traits (LW).

The general expert opinion is that the boar has little or no influence on litter size. *However*, individual boars (or AI) within a breed can have a very significant influence on litter size if semen concentration or quality is low, and some boars can produce lethal genes which can result in some embryo deaths or chromosome abnormalities rendering them infertile. In future these defects could be screened out before boars and/or AI semen is used.

A careful examination of boar/semen use is important in any investigation into low litter size.

## A BOAR PERFORMANCE CHECKLIST

✓ No more than 15% of young boars under 9 months should be 'on duty'.

✓ Litter size varies between boars; keep a check on this.

✓ Check boar's success rate:–

For each boar, multiply the farrowing rate of his sows x the total born average litter size across 100 of his services or inseminations.

*Target* : 1000                    *Action / exploratory level* : 800

*e.g.*    Farrowing rate of his sows   85% ⎫   Across the last
         Litter size of his sows & gilts  9.1 ⎭   100 services

85 x 9.1 = 773.5   Below 800 = Action needed.

✓   For herds using natural service, check on use of 'favourites'. This is quite a common, if understandable, weakness of stockpeople. Use a check-board sited in a prominent position. (Figure 4). If put in a passageway it can be viewed sideways (at an angle) and any over or irregular/sporadic use of a boar can be quickly noticed.

✓   *Quality of service*. Hurried services can reduce litter size (Table 10).

**Table 10. Conception rate and litter size in relation to quality score at first service**

| Quality score | Duration of intromission (mins) | N°. of first services (%) | Conception rate (%) | Mean total borns |
|---|---|---|---|---|
| 1 | <1.5 | 7 | 86 | 7.67 |
| 2 | 1.5 – 3.0 | 28 | 75 | 10.11 |
| 3 | > 3 | 49 | 91.8 | 11.46 |
| 4 | > 3 | 16 | 75 | 11.50 |

Source : Bell R., et al (1994)

✓   *Multiple Services* :  have an effect on litter size. Two being better than one, and three being better than two (Tilton & Cole 1982) in this case the two latter on consecutive days. However in the last 20 years breeders have paid more attention to correct heat detection and the placement of services/inseminations with more even 12-hour spacing. In this case two *correctly timed* service inseminations will maximise litter size.

✓   Your heat detection timing.

✓   Your subsequent service timing.

✓   Your quality of service. Supervision & patience is essential with natural service.

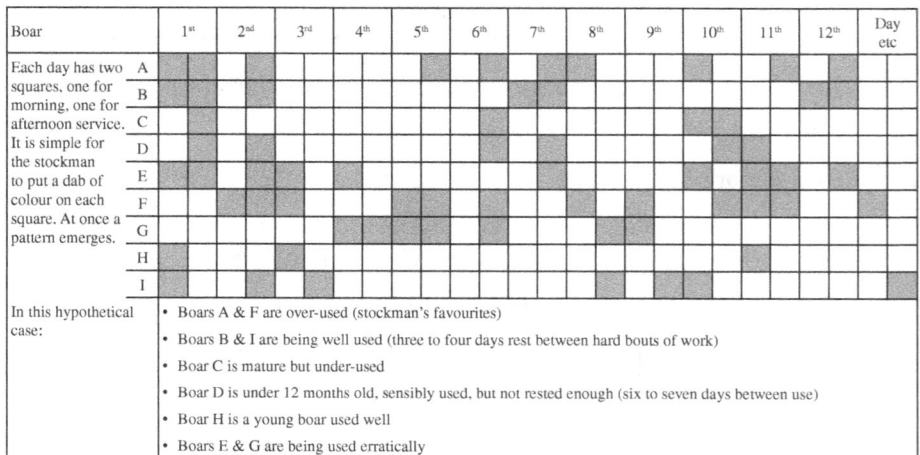

| Boar | | 1st | 2nd | 3rd | 4th | 5th | 6th | 7th | 8th | 9th | 10th | 11th | 12th | Day etc |
|---|---|---|---|---|---|---|---|---|---|---|---|---|---|---|
| Each day has two squares, one for morning, one for afternoon service. It is simple for the stockman to put a dab of colour on each square. At once a pattern emerges. | A | | | | | | | | | | | | | |
| | B | | | | | | | | | | | | | |
| | C | | | | | | | | | | | | | |
| | D | | | | | | | | | | | | | |
| | E | | | | | | | | | | | | | |
| | F | | | | | | | | | | | | | |
| | G | | | | | | | | | | | | | |
| | H | | | | | | | | | | | | | |
| | I | | | | | | | | | | | | | |
| In this hypothetical case: | | • Boars A & F are over-used (stockman's favourites) | | | | | | | | | | | | |
| | | • Boars B & I are being well used (three to four days rest between hard bouts of work) | | | | | | | | | | | | |
| | | • Boar C is mature but under-used | | | | | | | | | | | | |
| | | • Boar D is under 12 months old, sensibly used, but not rested enough (six to seven days between use) | | | | | | | | | | | | |
| | | • Boar H is a young boar used well | | | | | | | | | | | | |
| | | • Boars E & G are being used erratically | | | | | | | | | | | | |

**Figure 4.** How graphics can make complicated things clear - to reveal pattern of boar use, look at an angle laterally. Thus siting the board on a passageway is useful to encourage this sideways view.

# REFERENCES

Dritz, S. and Tokach, M. (2010) KSU Breeding Herd Recommendations For Swine 8-10.

'Litter Size'. On Our Farm (report) Farmers Weekly March 2000.

Tilton & Cole (1982) Effect of triple v double mating on sow productivity Anim Prod 34 279-282

# 3

# BIG LITTERS *AND* GOOD BIRTHWEIGHTS

Birthweights are – on the working farm – defined as the weight of those born alive. The definition of a 'born-alive' can be taken as a neonate which took at least one breath after being born. The 'bucket-test' to help in determining those born alive from those born dead which never drew breath is given at the end of this chapter.

For their own purposes research workers may need to record both born-alives and born-deads, but in this section we deal with the birthweights of born-alives only. For farmers, knowing their birthweights is very important.

## TARGETS

These are changing year on year, but 10% under 1.3 kg and 50% over 1.5 kg, leaving 40% between 1.3 and 1.5 kg gives a good start to the growth performance of the finished pigs.

## SO BIRTHWEIGHTS ARE IMPORTANT

We all know that - but how important? Recent surveys allow predictive models to suggest . . .

**More piglets survive.** For each 100g additional bodyweight at birth, preweaning mortality is likely to fall by 0.4%.

**Weaning weights are better.** Each 1g heavier at birth is likely to provide 2.34g at 18 day weaning and 2.70g at 26 day weaning. This can be even greater (Fig. 1) where a change in 0.5 kg birthweight was equivalent to a 23 day weaning weight of 1 kg.

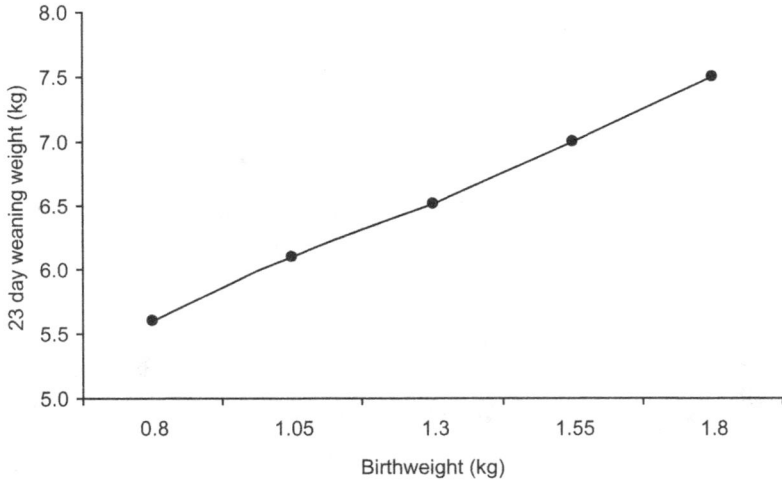

**Figure 1.** The relationship between birthweight and weaning weight.
Source: Sprent *et al.* (2000)

And my own measurements from clients' herds revealed . . .

**Litters are more even.** Very often a piglet weighing under 1.1 kg at birth will cancel out any profit at slaughter from one in the same litter weighing 1.70 kg.

• You should know your birthweights even though it adds a further task to a busy time. You cannot estimate birthweights by eye. 'Baby-scales' are cheap and effective. My clients tell me that the extra time taken is under 15 minutes per sow per year (+1.25% extra labour). This is recouped if 33% of the slaughter pigs get shipped 2 days sooner.

• Does this happen? It seems so as Table 1 shows that careful attention to birthweights improved days to slaughter by 2.7 days for the whole herd, paying back about four times the extra trouble taken.

**Table 1. Results after Birth Weight Audits, action being taken on several of the areas described below**

|  | *Before* | | | *After (+ 14 months)* | | |
|---|---|---|---|---|---|---|
|  | *Under 1.0 kg* | *1 - 1.3 kg* | *1.3 kg +* | *Under 1.0 kg* | *1 - 1.3 kg* | *1.3 kg +* |
|  | 13% | 45% | 42% | 9% | 28% | 63% |
| Av. days to slaughter (birth - 88kg) | 156 | 151 | 142 | 157 | 151 | 141 |
|  |  |  |  | Av. days saved to slaughter 2.7, whole herd. | | |

Source: Clients records

Figure 2 provides a guideline on how different birthweights can affect days to slaughter.

**Caution:** The graph is based on British conditions in the mid-2000s where in general the growing/finishing housing needed refurbishment on the farms monitored, due to stringent financial conditions in the industry at the time. Every pig industry should publish similar guideline graphs based on their own circumstances. I have worked in 31 different pig industries and a third of them are different enough to merit their own figures.

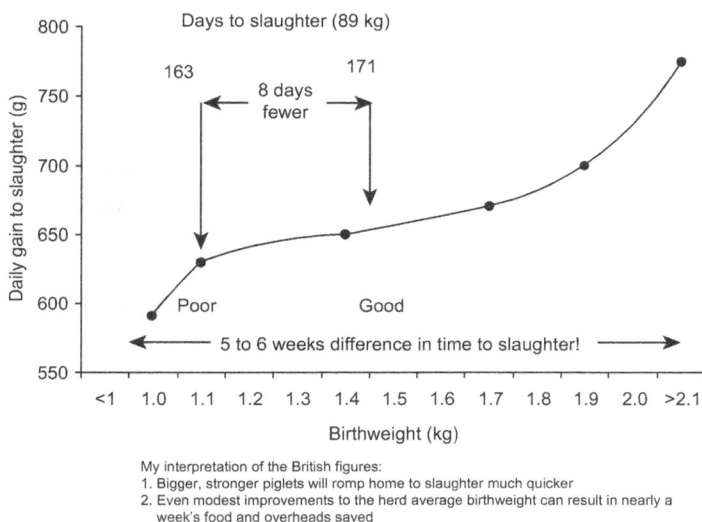

My interpretation of the British figures:
1. Bigger, stronger piglets will romp home to slaughter much quicker
2. Even modest improvements to the herd average birthweight can result in nearly a week's food and overheads saved

**Figure 2.** How better birthweights influence days-to-slaughter

- Evenness of birthweights is more important than averages. Too many piglets at, say, 1.1 kg or below can markedly affect average days to slaughter, especially if the litter number is high (12.5+). First, my records suggest that a 1.0 kg birthweight takes a week longer to get to slaughter than one at 1.45 kg and the pigs were up to another 2 kg lighter at 28 days.

- Variations in weaning weight are more important to profit than average weaning weight. 250 kg of extra weaning in a 5 kg weaner is worth a lot more to your profit than 250g extra in a 7 kg weaner. As that great pig researcher and teacher the late Dr Peter English said "Farm for the smallest pig in the litter".

- Next, 25% of my clients' birthweights are between 1.1 to 1.3 kg and these small (but not over-small) piglets took 2.7 days longer to reach slaughter than the 30% between 1.5 and 1.65 kg.

Of very small neonates (800 to 900 g), 62% died, with many being kicked to judge from their injuries. Farms with over 5% under 900 g neonates had a 2% higher pre-weaning mortality than those with only 2%. Target for as few as 10% under 1.20 kg, and as many as 50% over 1.45 kg. That's a good and achievable birthweight scenario, I find, profitwise.

# THE PROBLEM OF 'AVERAGE BIRTHWEIGHT'

Average birthweight *conceals much larger differences in the numbers of smaller and larger piglets* born over a production/costing period.

## Good Birthweights = More Lean Meat Sold

More MTF (Saleable Lean Meat per Tonne of Feed, *See Business Section*) is quite dramatically improved. One recent trial of mine showed piglets of only 1.05 kg birthweight provided 211 kg of saleable lean meat for each tonne of food fed to slaughter, while those of 1.41 kg birthweight provided 270 kg. 67 kg more meat on the same food and conditions!

Is this important? Seems so. Table 2 gives some figures published by a well-managed European commercial/demonstration farm, and another very similar unit, nearby.

Table 2. Actual birthweights / mortality ratios from two farms

| Birthweight category (kg) | Distribution of born-alive birthweights (%) | | Pre-weaning mortality (%) | |
|---|---|---|---|---|
| | Farm A | Farm B | Farm A | Farm B |
| < 0.5 | 0.5 | 1.8 | 80 | 78.2 |
| 0.5 – 0.74 | 2.2 | 1.4 | 62.4 | 63.1 |
| 0.75 – 0.99 | 6.2 | 11.8 | 24.7 | 25.2 |
| 1.00 – 1.24 | 16.5 | 20.9 | 13.4 | 13.0 |
| 1.25 – 1.49 | 24.1 | 29.1 | 6.6 | 6.2 |
| 1.50 – 1.74 | 27.9 | 24.3 | 3.7 | 3.5 |
| 1.75 – 1.99 | 15.1 | 6.4 | 2.5 | 2.6 |
| > 2.00 | 6.9 | 3.8 | 1.7 | 1.7 |
| *Average pigs born alive* | *11.7* | *11.1* | | |
| *Average birthweights (kg)* | *1.48* | *1.37* | | |

Comment : Despite a seemingly small difference in average birthweights (under 8%) Farm B had 0.6 fewer pigs born alive; more than three times the piglets born alive under 0.5 kg, which involves hard work to keep them alive; nearly twice the percentage of pigs under 1 kg (the ones that got crushed/chilled/scour); and half the percentage over 1.75 kg (the ones which get to weaning 2 kg heavier or 5 days quicker).

I've been around looking at new born litters for some 45 years now, and I still find it difficult to single out litters which recorded *average* (and good) birthweights, say, of 1.4 and 1.5 kg – as in Table 1 for example. The temptation is to say "Big enough – everything's OK".

Now this difficulty of mine may well be due to my own incompetence, sure, but I don't think so! Knowing his *spread* of birthweights alerts the stockperson to a problem – either there already or developing. So, back to "Is this important?"

Table 1 was from a breeder/feeder client of mine with not especially good birthweights (which was why I was called in) and shows the advantage achieved from improving the birthweights all up the scale – overall nearly 3 days faster to slaughter at 88 kg.

Do certain antibacterials increase birthweights by being fed to the sow 14 days before farrowing and continuing on until weaning, when weaning weights were also better? It seems so, but in certain countries this is not legally permitted. There are other products, however, which could escape the anti-bacterial cachet and I'm sure we will hear more on this subject in the future.

## TWO SUGGESTIONS WORTH EXPLORING

1.  I wonder if the global problem of pre-weaning mortality, still stubbornly over 10% of born-alives when it is consistently under 6% on some farms, is linked to the fact that stockpeople regard weighing pigs at birth to be a pain. Well, it *is*, but I find fully half of those farms which get well down into single figures now make time to do this. They know their lower birthweight litters and do something to improve them, or tighten up on culling.

2.  And as mentioned in the 'Mixing Pigs' section, some expert breeder-feeders are taking birthweights into consideration as well as evenness for size when batching and matching weaners. Research is needed on this suggestion – that pigs of similar *birthweights* tend to do better **as a group** to slaughter – the spread of close-out time is shortened by 3 to 4 days per group. An interesting hypothesis and important too, as batch production depends on minimal close-out variation to achieve one major advantage of the concept of batch shipping.

## A BIRTHWEIGHT CHECKLIST

### Start early

Success has come from convincing the stockperson that what is done *very early in the sow's reproductive cycle* can influence birthweights on the commercial farm.

✓    Good implantation at 12-24 days from service is vital

✓    Provide rest and quiet at this time

✓    Freedom from stress and sexual excitement is beneficial post-final service

✓    Try not to mix sows at this time

✓    A relatively high nutrient intake between weaning and service seems to be beneficial. For sows with a birthweight problem, try feeding 1.5 kg/day more than you currently allow them. Also, as routine, feed sows a lactation diet at this time, i.e. before and across insemination. Both may help follicle release synchrony – more follicles are released closer together in time – and/or it enables the womb surface (endometrium) to regain receptivity sooner. This is covered in the Litter Size section.

✓    Don't feed too good a diet in pregnancy. See your nutritionist for a low lysine (0.55% total) pregnancy food – a bit lower than currently favoured. However, just increase feed allowance (1.8 to maybe 3.0 kg) of this diet as pregnancy progresses rather than keep flat as the text-books advise for sows in reasonable condition.

✓    Never let a sow nose-dive in lactation. Here we are looking at the effect on the next litter of what was done in the ***preceding*** breeding cycle.

✓    Don't deliberately feed more and suddenly just before farrowing ***if birthweights are low***. This could be acceptable for other reasons.

✓    Don't worry too much about big litters affecting birthweights - with the new 'hyperprolifics' you can have big litters and big newborns, up to about 14 total borns anyway.

✓    Several multisite farmers in the US, after changing from farrow-to-finish monosite, do not report lower birthweights next litter when weaning at 16 days old (but their litter size seems to be one pig less than in developed pig industries in Europe).

✓    Don't use prostaglandins too early, as foetuses could be growing up to 60 g/day just before farrowing.

✓    Latest research could be indicating that transferring to organic trace elements rather than inorganic (from rocks/soil) will be the future of trace element nutrition in the breeding animal.

✓    Genetics are not likely to be involved. Maybe only if the super hyperprolific strains are much in evidence.

# SMALL PIGLETS CAN DAMAGE PROFIT

Having done many cost-calculation exercises for farmers in my lifetime, I find time and time again that the extra cost of raising the smallest surviving pig in the litter tends to cancel out most or all of the profit of the biggest pig in the litter. Surprisingly, while the average to slightly smaller pigs in the litter don't convert food to slaughter much worse, or at all worse, compared to the larger piglets which get there faster, I find the very small piglets definitely do convert worse by 0.3 or more. One of the reasons why, in the broiler-like batch-production conditions of SEW/multisite in the USA, these 'very-smalls' may be sacrificed at birth. They are just not "profit potential"; at best only break-even. And why I said earlier that with birthweights, averages can be misleading and we need to study individual weight sectors to see which remedies are best.

# PERSUASION DIFFICULT!

The problem has been to persuade farmers to go the extra mile and start weighing born-alives as routine.

If you take a modest average birthweight of 1.25 kg, which is a typical figure presented to me by problem farms, then an average birthweight 0.25 kg lower is likely to reduce saleable meat/tonne of feed (MTF) at 100 kg slaughter by 31 kg, and 0.25 kg above it will increase MTF by almost 36 kg. This is from speedier, more efficient growth from those that live, not from lower mortality. In this latter respect, however (the value of reduced mortality) turn back to Table 2. In birthweight terms Farm A had 0.6 more weaners/litter from their lower number of 'smalls' and larger number of 'heavies'. Thus Farm A sold about 5.4% more finished pigs out of the yearly sow and boar's food share of, shall we say 1.4 tonnes. Let's also assume the farm sells, out of each sow, 22 x 100 kg live pigs year, or at 75% KO, 1650 kg saleable meat. 5.4% more saleable meat is another 89 kg; and spread over 1.4 tonnes this is another 64 kg of meat sold off each tonne of breeding food sow per year. This is added to the benefits of faster growth already quoted. So what's one kg of saleable meat worth to you? Each extra kg of monetary income therefore reduces the sow food cost/tonne by that same figure. Work it out. Quite an eye-opener isn't it?

# A BONUS WORKS WONDERS !

This is why these clients have started recording birthweights despite all the extra hassle involved. Sure, the farrowing house stockpeople still have the extra work to do, but explained this way, some owners have agreed to a 50% bonus based on an improved MTF over the current achievement. The bonus is fixed for 3 years and then reviewed again, up or down.

The breeding section heads use my checklist to explore the suggested avenues of improvement. Even so, some hard thinking had to be done by the manager to ensure the time was made available *and that other jobs around farrowing didn't suffer*. Three of my clients work this system. The bonus has meant that their stockpeople's take-home wage has risen by about 10%, so wage costs accordingly have increased 1.4% but as far as we can tell, increased productivity probably linked to the concept has improved by 30 kg MTF/tonne fed – although this was about half what was expected. *Nevertheless this is enough to pay for the bonus three times over*.

# THE BUCKET-TEST FOR TRUE BORN-DEADS

This helps distinguish between a true born-dead and a neonate which was born alive, but died soon afterwards. When trying to establish the possible causes of low litter size, the factors which led to pre-farrowing and post-farrowing deaths are different, and need different remedies.

Fill a bucket with water and lower the eviscerated lungs into it gently. A true born-dead's lungs will sink relatively rapidly, while those from a born-alive will sink more slowly, if at all.

This is because even if the neonate took just one breath some of it will remain in the lungs, while a true born-dead never drew breath at all.

Once you have observed a few of both, you will easily detect the difference.

# REFERENCES

Deen, J., Dion, N., Wolff, T. (2006) Predictions of piglet birthweight. Procs. IPVS Conference.

Sprent, M., Varley, M.A. and Cole, D. (2000) BSAS Meeting. 'Effect of birthweight on subsequent performance of pigs.

# AVOIDING POST-WEANING PROBLEMS

## THE POST-WEANING CHECK

The slow-up in growth rate seen immediately after weaning. Correctly defined as the period in days the newly-weaned pig takes to recover the degree of daily gain achieved in the last 24 hours on the sow.

## THE EXTENT OF THE PROBLEM

The post weaning check can be a matter of hours only to as much as 18 to 24 days, with 7 to 9 days being commonplace on typical farms and half that again on the poorer-run units. We shall see what effect this has on profitability later in the chapter.

In keeping with high pre-weaning mortality, poor growth rates to slaughter in relation to what can be achieved with today's good genetic material, high 'empty' or non-productive days (NPDs), and low sow productive life (SPL), the post-weaning check is a fifth area where a disappointing lack of progress at farm level has been seen across the last 30 years.

### Primarily – maybe exclusively – a nutritional problem

In the author's opinion this is mostly due to stockpeople and owners failing to appreciate that the problem has been – and to a disappointing extent still is – due to *incorrect nutrition* over the crucial transitional period from the time when the piglet is on the sow to its inability to process solid food satisfactorily immediately after weaning.

### Problem 1 – failure to invest sufficiently in the design of the post-weaning food

Much progress has been made by nutritionists on the design of diets to make the dietary transition as easy as possible for the piglet's digestive system. The trouble is that these specialised foods can be expensive – around three times more than what farmers have been used to or are offered for sale from manufacturers over-keen to get their business. There is no escaping the fact that to lower the immediate cost of post-weaning diets compromises the specialised diet design needed to avoid indigestion. The growth check problem then emerges, especially when the piglets are weaned 'early' *i.e.* 16-21 days. It is a tough discipline the seller of baby pig food has to adhere to.

## Why are these new post-weaning feeds so expensive?

- They need a high essential amino-acid content but tied to a relatively low crude protein content, so as to avoid indigestion when pigs are weaned. Forcing down protein but maintaining high amino-acid levels paradoxically costs money.
- Certain common and cheaper ingredients which aggravate the gut wall are excluded. As are other raw materials which contain anti-nutrient factors (ANFs).
- Several cereals and soya need to be heat treated to improve digestibility.
- Added enzymes help in counteracting ANFs. Fermenting some ingredients, too.
- Pellets have to be carefully made (not overheated) by slow and careful processing. The machinery is expensive.
- Some immunoglobulins may need to be added, or new materials used which are able to bolster immunity.
- Organic trace elements (not the cheaper inorganic sources) are used as well as specially protected vitamins.
- Manufacture and warehousing constraints to preserve freshness call for smaller production runs which inflate costs.

This is further explained in the *Creep Feeding Section.*

## Problem 2 – dirty feed receptacles

The second contributory factor is the farmer's failure to provide these expensive, high-quality transitional feeds in a clean enough manner. Troughs get fouled too quickly and too often. The contaminants further compromise the animal's delicately balanced digestive system at a challenging time digestively and immunologically, when both these systems (along with its thermoregulatory system) are underdeveloped and need all the help science and stockmanship can give them.

The following pages are an attempt to persuade those who are responsible for raising weaners to . . .

- 　　　Pay sufficient for the right food and
- 　　　Feed it properly.

Keeping post-weaning troughs clean pays off handsomely:

| | Days to 108 kg | Savings/pig | Minutes worked/pig | Mortality |
|---|---|---|---|---|
| Trough meticulously cleaned | 156 | €5.26 | 18 | 4.2% |
| Casual trough hygiene | 168 | - | 7 | 7.0% |

Spending 10 minutes more per weaner in the nursery cost €1.20/pig more, but saved over €5/pig

Source: Client's records (2008)

## What does the post-weaning check to growth cost ?

My survey of a substantial sample of British breeders revealed that 96% thought that some form of post-weaning interruption to growth was bound to occur and that, on their own farm . . .

88% thought it lasted 2 days or more
52% thought it lasted 4 days or more
30% thought it lasted 7 days or more
16% thought it lasted 10 days or more

What is quite serious is that, even in a sophisticated pig industry like ours in the U.K. so many of those questioned were resigned to a 10 day check at weaning as being 'normal'.

In fact the best of us limit it to 2-3 days.

At Dean's Grove farm, even 20 years ago, we measured growth rate and often we estimated we had got it down to a 2 day check. We had a few pigs growing at 923g/day at 25 kg – which is close to the magic 1000g (1 kg/day) by 25 kg which the nutritionist/geneticist says is possible and is a figure which many producers disbelieve. I can remember lifting (with difficulty!) one or two 9½ week (66 day) monsters out of the nursery weighing 33.5 kg. These must have achieved a daily gain of 485g/day from birth.

This is almost double what many people obtain.
And thereby hangs a tale. .

*A minimal post-weaning check has a maximal effect on growth rate to slaughter, and thus the amount of saleable lean meat (MTF) produced for each tonne of grow-out food purchased*

This is confirmed by some figures from the UK (Table 1) where a reduction of 9 days in the check (12 days down to 3 days) saved the producer just under £20 per tonne of all food fed to slaughter, or 14%.

**Table 1. Losing typical growth impulsion at weaning raises the cost of all growing food by over 10%**

| Liveweight | Length of check | Days to 94kg | Daily gain | First graders | Carcase lean | Lean per tonne of food |
|---|---|---|---|---|---|---|
| 5.8 kg | 12 days | 156 | 567g | 72% | 52.3% | 166 kg |
| 5.8 kg | 3 days | 142 | 621g | 86% | 53.1% | 182 kg |

16 kg more LEAN for every tonne food

Worth, at 150p/kg* retail meat = **£24.00**

And 14 days fewer overheads at 24p/day = **£3.36**

**Total £27.36 more income pe tonne of feed**

* UK / Dec 2009                     Source :  Based on A1 Feeds data

Expressed another way, 16 kg more meat/tonne feed (M.T.F.) at 5 pigs to the tonne of feed is worth £5.47/pig or 7% more income.

**Remember**, the average weaned pig checks for 7-10 days, thus every day the pig slows down over 3 days adds another £2.28 to the cost of a tonne of growing/finishing feed he will eat to slaughter.  On a world pig feed price this is about 1.4% feed price increase for each day's slowdown in post-weaning growth.

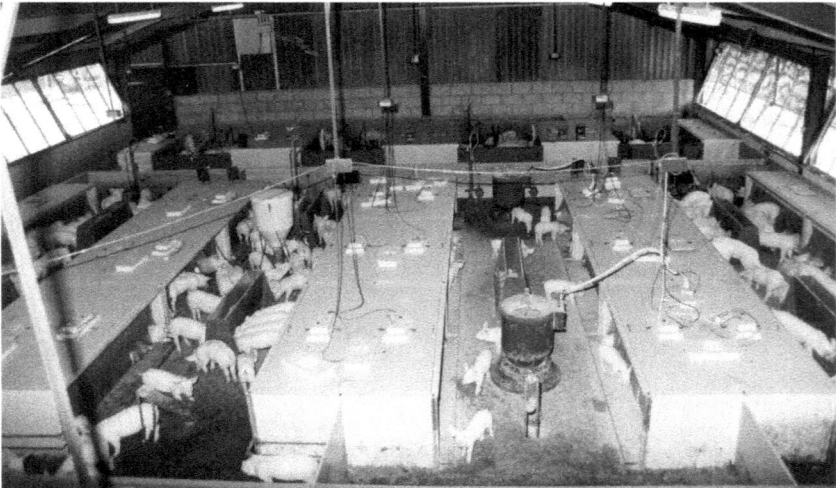

A modern nursery in Europe using a redundant cattle barn. Note the temperature regulated boxes and ample feeders/ambulatory space.

# RESEARCHERS RARELY GO FAR ENOUGH IN BABY PIG EXPERIMENTS

I have read many worthy research trials which indicate statistically significant differences in performance *to weaning or to the end of the nursery period* – and then stop there!

Some even have gone on to give the financial benefit of the product or technique used (as distinct from the performance improvement) for that portion of the growth curve. However, smart farmers say "Yes, but the treatment barely paid back, despite the improved physical performance demonstrated ... *at the conclusion of the nursery trial!*

Had the trial been continued to slaughter, however, the cost-effectiveness picture could have changed markedly for the better even if the eventual physical performance improvement percentage could have slipped a bit by then.

*I believe no young pig trial is satisfactorily concluded until the econometrics have been assessed on both groups of pigs raised under similar conditions to slaughter. Most of us sell slaughter pigs, not end-of-nursery pigs!*

One may not be able to be as statistically certain because of subsequent variables, but any negative or 'not-worth-it' *economic* conclusions early in the growth curve could be altered by slaughter weight especially if the treatment tested had a positive effect on the post-weaning check.

I understand the reasons (extra money, lack of facilities and staff needed) which hamper research departments carrying on young pig trials to slaughter in this way, but it is a weak point in their current approach which at least needs consideration at the trial design stage. After all it is the improved profit which comes from better physical performance which matters. Generally we don't sell weaner pigs, but we do (or someone does) sell the finished animal – and the finished pig incurs a lot more food cost/day than does the weaner.

# WHY IT PAYS TO SPEND, SPEND, SPEND ON THE LINK FEED

The interim specially-designed pre-starter feed has been called the Link Feed – a better description than 'pre-starter' which could also refer to a creep feed. Many breeders refuse to pay the £700-£850 asked for even a moderately well-designed Link Feed (I have seen costs of £1,200/tonne for a really top class feed) when they are paying £350/tonne "without too many problems".

In world pricing terms that is two to three times more.

"No feed can be worth *that*," they exclaim in disbelief.

Let's look at the situation coolly and dispassionately in terms of payback.

First, that statement "without too many problems".   The problems referred to are digestive ones, such as scouring stall-out or inappetance.  But the real problem is the underachievement at slaughter which goes unrecognised, like this:-

The Americans are adding data to those I've quoted in Table 1.  The University of Minnesota has quoted a financial loss of 10 cents per lb gain from 11.5 lbs to 50 lbs (5.2 – 22.7 kg).  Note that this loss had already occurred by 23 kg (51 lb).  Table 2 cites the University of Georgia where 8 days were lost to slaughter on a conventional post-weaner diets.  And while the food cost in the 7 days post-weaning was double on the more expensive diet, this was recouped three-fold by slaughter weight.

**Table 2. Do post-weaner link feeds pay?  Extrapolated from American data**

|  | Conventional post weaner diet | High digestible diet |
|---|---|---|
| Weaner growth rate/day 0-7 days post weaning | 100 g | 200 g |
| Days to 105 kg | 171 | 163 |
| Relative cost of food eaten in 7 days | 100 | 199 |
| Relative savings in costs to slaughter (food & overheads) | - | 513 |
| Relative value of highly digestible food | (513-199 = 314) | 314 |

While the post-weaning food cost twice as much, the net income at slaughter was a third more

1.    Link feeds pay best in the first 5 to 10 days after weaning

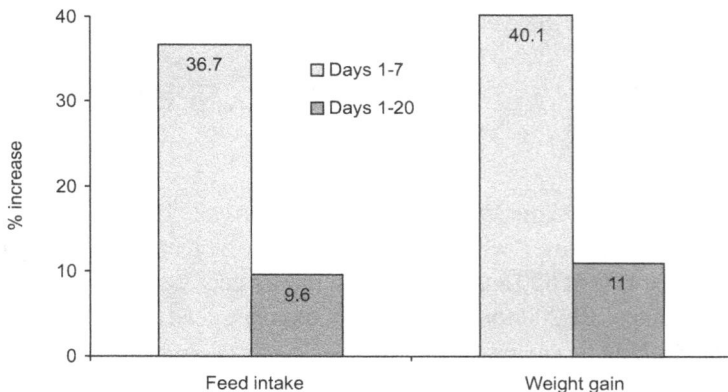

2.    A little link feed after weaning shortens time to slaughter

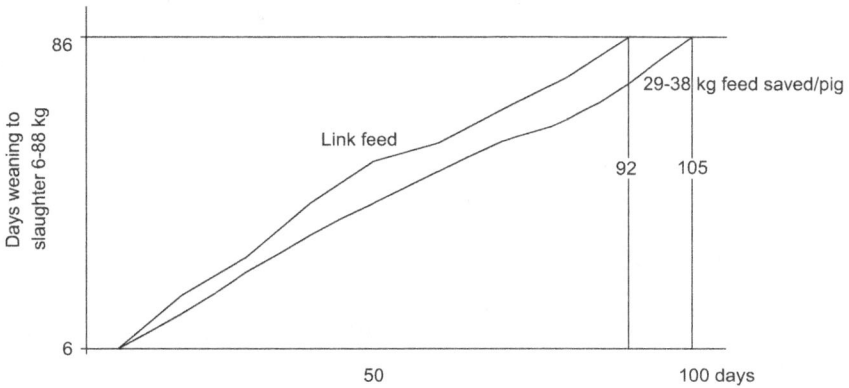

**Figure 1.** Improvement of higher cost link feed over standard starter feed
(British Data)

Table 3 shows some data collected when I was working in Iowa and Minnesota just as the Link Feed concept was being tried out – hesitantly due to the ×3 increase in the cost per tonne asked for the whole feed! The table shows what each farm could have afforded to pay for the Link feed before the extra investment was eroded, based solely on food saved, not on reduced "overheads" as well. Including overheads would have given a 2 to 2.5:1 payback on extra feed cost. Incidentally, the standard of nursery feeder hygiene was not good on farms 2&3, reinforcing my point about trough cleanliness.

**Table 3. Breakeven costs and paybacks from feeding Link Feeds on 4 US farms**

|                                   | *Farm 1* | *Farm 2* | *Farm 3* | *Farm 4* |
|-----------------------------------|----------|----------|----------|----------|
| Days used post-weaning            | 5        | 10       | 7        | 12       |
| Breakeven cost per ton ($)        | 2,631    | 1,840    | 2,135    | 3,100    |
| Price actually paid per ton ($)   | 1200     | 1200     | 1500     | 1350     |
| Pay-off ratio                     | 2.2:1    | 1.5:1    | 1.4:1    | 2.3:1    |

Clients' Records : 2000-2005

# A POORLY UNDERSTOOD EQUATION

Some farmers do not seem to realise how little the post-weaner pig eats in the 7 days post-weaning when the Link feed does so much good.

•    A typical pig weaned at 21 days eats about 3.25 kg to 7.5 kg food in the 5 to 14 day post-weaning period. This means that one tonne of Link Feed will feed about 300 down to 133 weaners. Let's say 250 to 100 weaners to be on the 'safe' or pessimistic side. From this, take an average of 175 weaners.

- But because of the post-weaning check ('stall-out' in the U.S.A.) on low grade (conventional) post-weaning feeds, each of these weaners is costing 5 to 8 days in extra food by slaughter – 10.5 to 17 kg of food. At 16p/kg this is £1.68 to £2.72/pig. Let's take £2.20/pig as a fair mean, on food costs alone – no saved overheads. A conventional post-weaning diet will cost from £350 - £400/tonne.

- So if a special high-cost Link feed avoids this waste of finisher food you can afford to pay 175 pigs x 2 = £385/tonne *more*, even at the most pessimistic, customer-friendly scenario, before the extra cost is eroded. About £430/t with saved overheads incorporated.

- Thus you can afford to pay around double the cost of a conventional starter/ grower feed – even using pessimistically-weighted figures – before any advantages are eroded.

## DO THE SUMS

All I ask is for you to do your own sums based on this approach. You can always ignore my assumptions and substitute your own. Also, you should convince yourself to do a nursery trial with an expensive Link feed against your current choice and use the benefits as your base performance matrix. My experience is that only very rarely will a good Link feed fail to succeed econometrically **at slaughter** if your nursery stockmanship and housing is good.

The interesting exercise is how far up the price scale asked/tonne do you need to go? It does not seem to be a question of 'the cleaner and better the nursery environment the lower the price is required'. Rather the converse; very good nurseries seem to do much better, the better the design of the post-weaning feed they use. This suggests that the geneticists are right in constantly telling us that at the sharp end we are nowhere near exploiting the genetic potential – in this case growth rate – already locked into the genetics we have purchased.

And it reinforces my plea that post-weaning research is never finished until the pig is shipped at slaughter.

"It's never over until the fat lady sings!"

# WHAT HAPPENS IN THE GUT AT WEANING?

I hope I've convinced you in economic terms to pay more for a well-designed Link feed.

But what do I mean by 'well-designed'? First, we must understand what happens digestively when we wean a piglet. Nature never intended for it to be removed suddenly from its dam – it evolved a gradual process taking at least 16 weeks, and more commonly 20 weeks, which allowed the gut to become accustomed to digesting solid food little by little. The bacterial and chemical pathways had time to adjust and change from milk to dealing with plant roots, acorns, mast, grass and weeds, apples and the soil bacteria and fungi ingested with it. And there was always time for a quick milk suckle to help level out any inconsistencies even late in the process.

Apart from feral pigs, and to a much lesser extent modern outdoor pigs, all that has gone. By weaning abruptly from 17 to 32 days we put an impossible strain on the piglet's digestion. If we don't help it counteract the suddenness of the changes, then the post-weaning check inevitably occurs. This is what happens . . .

Look at Figure 2. This is a simplified, diagrammatic representation of a very complicated – but extraordinarily elegant – chain of events best understood like this:–

- In the 3-week-old weaner the stomach is both a reservoir and a pre-digestive mixing tank holding about 0.2 litres – say a wineglassful.

- Milk from the sow arrives every 35 to 45 minutes or so in carefully measured amounts. The sow does this in response to the suckling stimulus by releasing milk from the udder cisterns, and switching it off again after around 17 to 30 seconds. However long and vigorously the piglet suckles it only gets its hourly 'ration' of about 150-200 cc.

- The stomach of the unweaned piglet is not very elastic and can only hold a certain volume of contents – as we've seen, about 200 cc or 0.2 litre.

- Cells in the stomach walls liberate both digestive enzymes and hydrochloric acid to start pre-digesting proteins (proteases, etc) and carbohydrates (amylases, etc.) especially. The acid helps disable pathogenic bacteria which are involuntarily eaten along with the food. The contents – sow's milk in the unweaned piglet – already contain nutrients in the right form for this to happen easily and within the 35 minutes or so needed for these pre-digestive (enzymes) and sanitation (acids) processes to take place.

- After this time the stomach contents are then passed into a short pipe-like channel, the duodenum, where fats are pre-digested. It also holds about 0.2 litre.

- At the third gut movement (each one of these instigated by the call to suckle) the duodenal contents, now largely ready prepared for absorption, enter the small intestine. They are also relatively free by now of potentially damaging organisms the piglet may have eaten as it scampers about and investigates life around it.

- The food (sow's milk) is now properly predigested and made safe for absorption in the small intestine.

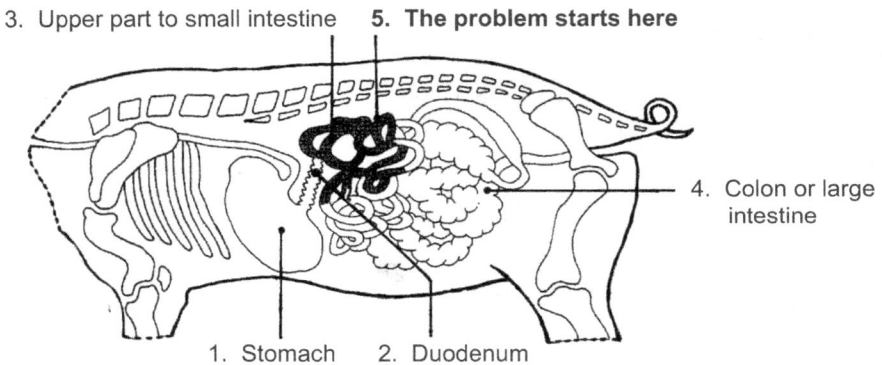

3. Upper part to small intestine    **5. The problem starts here**

4. Colon or large intestine

1. Stomach    2. Duodenum

**Figure 2.**   Digestive system of a 5.5-6.5 kg, 3 week old weaner.

Understanding what will happen digestively when the piglet is weaned helps a great deal in solving food, food intake and growth check problems. This is what happens.

1.   **Stomach** Holds 0.2 litre, does not expand. Food needs to remain here for nearly 45 minutes so as to be infused with acid to knock out harmful organisms and be washed with enzymes which get starches and proteins ready for digestion later on. Engorgement does not give sufficient time for either to happen. The stomach then refills the duodenum and is itself replenished by fresh sow's milk.

2.   **Duodenum** 2" thick 9" long, also holds a little under 0.2 litre. Cells in wall wash stomach contents, once they arrive, with fatsplitting enzymes so that fats in the food are made ready for digestion and absorption in the next part of the digestive tract. Too much food from the stomach, too soon, causes the digesta to be pushed on only partially prepared.

3.   **Upper part of small intestine** 5 to 7 yards long, convoluted with a huge surface area equivalent to half a football pitch, due to thousands of millions of tiny villi, or little microscopic fingers which absorb food. Billions of surface cells absorb

the predigested nutrients. Insufficient predigestion – no absorption. This part of the tract cannot cope with poorly processed food.

4. **Colon or large intestine** 2 yards long, 1" thick. Water and fibre absorption. Not involved to much extent in the post-weaning check to growth.

5. **The problem.** Piglet is weaned. Stops regular ingestion once every 45 minutes. Gets hungry… then overeats (engorges). Food not long enough in stomach or duodenum, so is insufficiently "sanitised" or pre-digested in the small intestine. Blockage occurs, bacteria breed causing villi to truncate (shorten and wither). Water is liberated and piglet scours / dehydrates

6. **The solution is to:**
   - To accept that piglets *will* engorge after weaning;
   - And *will* therefore overload the small intestine with food;
   - So provide a post-weaning prestarter food which is so "pre-digested" that **overloading will not harm absorption**.
   - Only restrict feed such a diet for 12-36 hours, and then only marginally.
   - However, some 'SUPER-LINK' feeds, more akin to creep feeds, are so digestible that little restriction is needed, if any. Take advice from the manufacturer and be very careful not to over-order and then store these feeds properly.
   - Then blend into a normal grower food once they have been on the link feed for 7-10 days.
   - Have electrolye solution available and plenty of clean fresh water.
   - Alternatively wean later (26-30 days) and accustom the piglet to sufficient daily creep feed.

Crude, indigestible fibre *is* indigestible and sow's milk is virtually fibre-free.


## WHAT OFTEN HAPPENS WHEN WE REMOVE THE SOW AT WEANING ?

- The piglet gets a little hungry as is quite normal after 45 minutes to an hour, and looks for a feed. But mother is nowhere to be seen.

- After an hour or two the stomach is empty, the duodenum is empty and even the fore-end of the small intestine has moved its contents further down to other adsorption sites and additional processing by beneficial bacteria further on down the gut.

- "Yes some idiot has put down this solid but quite pleasant-smelling creep pellet/ meal," thinks the piglet. "But it is not wet, it is not warm, it is gritty, it doesn't taste or feel like milk and I suspect it contains more of that fibre stuff than is good for me. I'll pass it by in the hope that mother will appear soon."

• By now three or four hours have passed. The piglet is ravenous. Moreover some of its bolder or hungrier penmates are beginning to eat the solid food provided. "Perhaps I could try a bit too," it thinks. It does, and while the solid food is a poor substitute for the real thing, it eventually **overeats** to remove its hunger pangs. This is called 'engorging'.

• But its stomach is inflexible. It cannot handle the volume of solid food which the piglet throws down to it, and there are only two ways the ingesta can go – either back up again and the piglet is sick, or through the more natural route into the duodenum and on further to the intestine – which is calling for replenishment and thereby activating the hunger response.

• So the ingested solid food does not remain sufficiently long in the stomach or duodenum for proteins, carbohydrates and fats to be prepared for absorption in the small intestine. Neither has it been sufficiently washed by acid to eliminate the hostile bacteria which nature has made susceptible to a natural, high acid level.

• The food arrives too soon in the small intestine with the wrong chemical signatures for absorption, and also loaded with damaging bacteria.

## WHAT HAPPENS THEN ?

• The ingesta forms a traffic jam – a blockage – in the forefront of the small intestine. It cannot be sufficiently absorbed, so it stays there. It is a serious form of indigestion. But it is now an ideal breeding ground for the bacteria which have free-loaded in with it. These quickly proliferate and their toxins aggravate the delicate absorptive structures – the villi – which are covered in cells which recognise and absorb *properly digested* nutrients but refuse to accept those not predigested sufficiently.

• The bacteria cause the villi to reduce their length defensively (called truncation) so the huge absorptive area (about half the size of a football pitch in each piglet!) can be reduced to no more than the penalty area. Nutrient processing is drastically reduced.

• This villous reduction process stimulates cells (crypt cells) at the base of the villi to exude water. This liquifies the ingesta and stimulates bowel movements to flush the blockage down the gut to be voided. This is what scouring (diarrhoea) is – it is a lavatory-flushing operation to help cleanse the gut of potentially lethal material.

• This is why post-weaner pigs are prone to scour. It is a defensive mechanism - all that they can do to try to put things right.

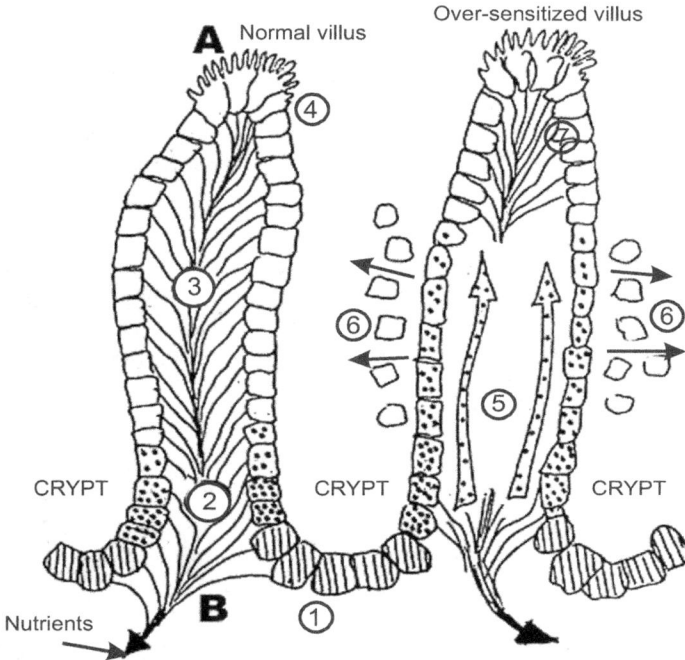

**Figure 3.** Microsection of villi.

| | |
|---|---|
| **A** | Inside of intestine containing food |
| **B** | Mucous membrane of gut wall |

} A to B = ½ width of a pin shank

1. Cells which secrete enzymes and in the scour of scour - water.

2. Replacement (germinative) cells which are non-absorptive at this stage. Within 3-5 days these become...

3. Absorptive cells which absorb amino-acids, sugars, water and minerals from food. (An ideal ratio is 1 germinative to 5 absorptive).

4. Microvilli on villus tip further increase absorptive surface area.

5. Germinative cells are speeded up if certain foods are given too early.

6. Healthy absorptive cells are pushed off the villus surface.

7. Absorptive area is drastically reduced - piglet cannot digest sufficient food - bacteria multiply, invade and scour occurs.

# INSUFFICIENT WATER

• Trouble is the 6-7 kg piglet only has a limited amount of water in its bloodstream and body cells to 'flush the lavatory'. When this runs out, the blood thickens unless the water can be quickly replaced. Blood both conveys nutrient energy to the muscles (arterial blood sugars) and then removes toxins via the venous system to the organs (liver, kidneys) which can then process them for excretion as urine and also in the faeces.

With thickened blood the piglet is starved of muscle energy (gets sluggish) and also gets cold (shivering). It already starts to poison itself with the accumulating toxins (feels very ill). Thus an **immediate source of specially treated water** helps the piglet avoid these traumas. Treated in the form of added electrolytes which are simple minerals allowing the crypt cells to *insorb* water at the same time as *exsorbing* it. In other words it can now take on water from the forepart of the gut while continuing to expel it by the scouring process at the other end.

• This is why as soon as looseness is noted, an electrolyte solution should be provided either as an additive to the normal water supply (in the early stages) or as a replacement for it when scouring is acute. The important deciding factor is never to affect the piglet's ability to drink clean water. Allowing an electrolyte container to run dry is disastrous and accustoming them to a new source of water needs to be borne in mind. So some farms, especially in hot, dry climates, provide an electrolytic solution as routine after weaning.

Here is an electrolyte formula recommended many years ago by a specialist pig veterinarian. (Table 4) However several commercial products exist and are less trouble to make-up.

**Table 4. Home made electrolyte solution**

| To 2 litres (3½ pints) of water, add … | |
| --- | --- |
| Pure Dextrose BP | 45g |
| Sodium Chloride (salt) | 8.5g |
| Citric Acid | 0.5g |
| Glycine BP | 6.0g |
| Potassium Citrate | 120 mg |
| Potassium Dihydrogen Phosphate | 400 mg |
| *Scouring pigs* – full strength for 2 days for all pigs in room | |
| *Post Weaning Depression* – half strength for 10 days | |

# HELPING THE WEANER THROUGH THIS DIGESTIVE IMPASSE

Now we know what happens in the gut of the newly-weaned pig, we can do something about it.

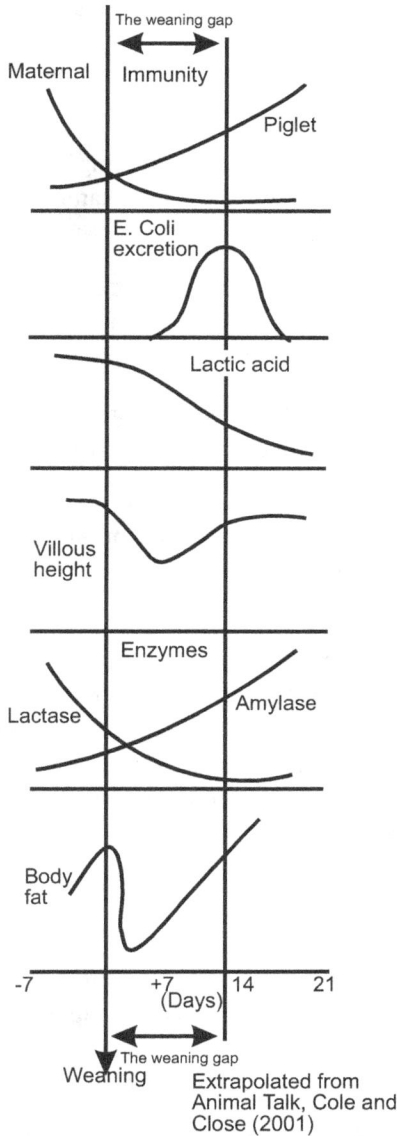

The weaning gap

Maternal    Immunity

Piglet

E. Coli
excretion

Lactic acid

Villous
height

Enzymes

Lactase    Amylase

Body
fat

-7        +7      14      21
          (Days)

The weaning gap
Weaning    Extrapolated from
           Animal Talk, Cole and
           Close (2001)

**Figure 4.** Why 7 days after weaning causes so much trouble.

1.  We can pre-digest the solid food to such an extent that it doesn't need to remain in either stomach or duodenum for the necessary 35-40 minutes for normal processing. Some ingredient raw materials are already largely preconditioned/ pre-digested and are essential at this time.

2.  We can additionally add the essential enzymes (preferably from natural sources) needed to pre-digest proteins, carbohydrates and fats, as well as helping the piglet deal with ANFs.

3. We can add minerals and trace elements in a far more absorbable organic form.

4. We can condition and control, often by heat treatment, what fibre must be present so as to make it more easily digested.

5. We can add extra acid to the food (or drinking water – moderately acidified water is surprisingly palatable) to help pre-sanitize the gut contents as they arrive in the stomach.

6. We can make the food exceptionally palatable (texture, flavour and smell) to dissuade holding off eating solid food and lessen the resulting overeating (engorgement). *Freshness* is the best palatant.

7. Judicious creep feeding can accustom the piglet to solid food while on the sow. This will be of vital importance if, on welfare grounds, weaning is not allowed before 28 days.

8. We can, if we wish to adopt the system, pre-ferment the diet or individual cereals/soya by adopting FLF (Fermented Liquid Feeding).

The upshot is that all these things – the raw material feeds, the additives, the processing and the precautions taken in manufacture and storage shelf-life *are much more expensive than with conventional foods*. By suggesting you should use a link feed at three times your current cost does not mean the manufacturer is ripping you off. On the contrary, these are usually genuine on-costs and any reputable firm will be glad to explain things to you and answer your questions however doubtful you may feel.

## FEEDING METHOD MUST BE CORRECT

Generally speaking the younger and lighter are the piglets you wean, the better Link feed they need. This does not always follow, as the *way you feed them* is of great importance to success. This is best described by a check-list.

## THREE CHECKLISTS FOR GOOD POST-WEANING FEEDING PRACTICE

**A.    FIRST CHECKLIST – THE FOOD ITSELF :**

✓    What degree of post-weaning check do you suffer?  See Target Growth Rate section (page 337).

✓ Have you considered a specially-designed Link feed? The degree of sophistication – and thus the cost – will largely depend on the magnitude of check your pigs suffer.

✓ Have you discussed a suitable Link feed with a nutritionist experienced in the design of these diets? Also, some pig specialist veterinarians and other consultants are a useful source of advice on how your housing and management will influence the quality of diet you need.

✓ Do not be overconcerned about the cost per tonne – this will repay itself in improved performance and thus dietary cost savings, *by slaughter*. Find a good Link feed and stick to it.

✓ If you sell end-of-nursery weaners, make sure the buyer appreciates the trouble taken and extra expense you have incurred in helping your pigs to reach his slaughterweight quicker. You are entitled to a premium for this to offset your higher feed costs. **To give you an idea, the investment of one monetary unit before 30 kg is worth at least 2.5 monetary units at slaughter.** Use this as a negotiating factor on your weaner sale price/ slaughter pig price.

✓ Nearly all farms cannot make their own Link feed. They haven't the plant to make it, and many vital ingredients are only available in bulk lots, or unobtainable outside the feed trade due to restricted supplies. Leave it to the experts.

✓ But you should ask the supplier about his turnover of stocks both of raw materials and finished goods, especially in summer/hot weather. These should preferably be days, not weeks.

✓ In this respect, never hold more than 14 days' Link feed yourself. Order frequently and often; accept small-load charges reluctantly if you have to, and store the food in a dry, cool space. An old ice-cream or frozen goods container is excellent. Never store any bags in the nursery. Because of the small quantities required (5 x 200 pig nurseries totalling 1000 weaners from 6-12 kg will eat – at 350 g/day growth rate and with a food conversion of 1.2:1 – over 10 days on the Link feed only a little over 4 tonnes of food). So bags could be considered rather than bulk. For one 200 pig nursery, the amount needed will be around 1 tonne only – a 1.5 tonne fortnightly order, maximum. But beware, some nurseries can require up to double this.

✓ Water adequacy, cleanliness and accessibility is essential to adequate uptake of a Link feed, which by its nature tends to be thirst-making. This itself is no problem if water supply and management is good – indeed it will increase feed intake which the design of a good Link feed encourages, without digestive kickbacks.

✓    The water problem is made worse by piglets drinking less once weaned. At this stage liquid intake can fall dramatically from as much as 800 ml/day on the sow to only 200 ml from a waterer (Figure 5) Until the weaner learns how to get all its needs from water, feed intake drops, its ability to digest food reduces and performance suffers.

This may be the fault of drinker design.  The Japanese (Zen-noh) have an excellent tongue or leaf drinker in aluminium or bright metal specially for 5-12 kg weaners – easy to maintain and keep clean.

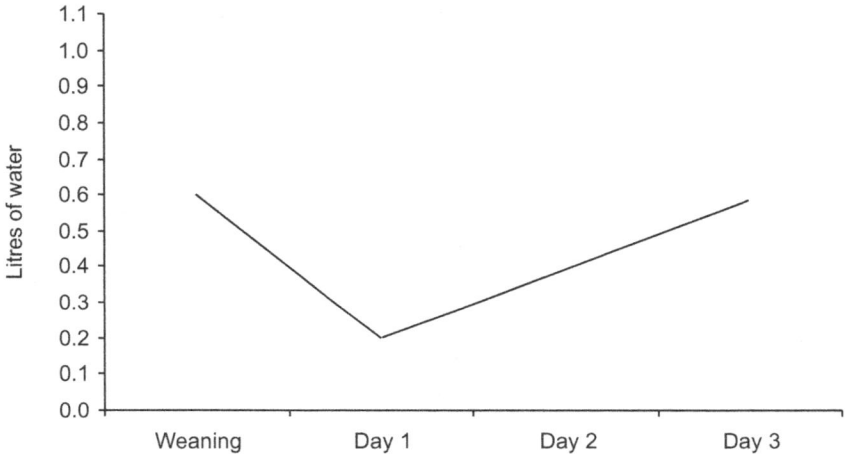

**Figure 5.**  Effect of weaning on water intake. Source: Tibble (1992)

✓    Learn how to 'drive' electrolyte provision.  While electrolytes can be put in the feed, the water route is far preferable.  Auto-dispensing devices are available, but many smaller nurseries use dedicated canisters.

## Post weaning nutrition

By following the advice given in the chapter on creep feeding you should have, in Table 5, weaners at the 28 day weight given as a starting point for continued target growth rate through and beyond the nursery period.

Table 5. Target growth rates. Published by BPEX (2009)

| Age in days | Age in weeks | Liveweight (kg) |
|---|---|---|
| 28 | 4 | 8 |
| 35 | 5 | 10 |
| 42 | 6 | 12 |
| 49 | 7 | 14 |
| 56 | 8 | 17 |

*BPEX = British Pig Executive, the leading pig advisory body in Europe today.

## Tremendous growth potential nowhere near being reached

One correspondent found it difficult to accept the target liveweights in my table as 'surely not possible' as he struggled to reach 8 kg at 35 days 'despite good conditions and management'. For him and other doubters I reply that most pig producers do not fully realise what tremendous growth performance lies buried in today's advanced genetics. Mavromichalis (2009), a leading specialist baby pig nutritionist, reports that pigs fed on reconstituted cows milk alone under experimental conditions from 10 to 50 days of age, **at 30 days pigs weighed 15 kg** (as around 9 kg in my target table) and **at 50 days, 32 kg** (compared to 16 kg in my target table) a remarkable 100% more!

Now these growth rates really are astonishing and put my Table 5, which seems optimistic to some people, into perspective. It supports my claim that everyone can reach my targets now and go beyond them in the future.

Even so, producers are right to ask to see what nutrient specifications are needed in practical post-weaning and nursery foods. Those which are considered advisable today under 2010 manufacturing and cost constraints are given in Tables 6 and 7.

**Table 6. Example of recommended dietary energy and lysine requirements for nursery diets.**

| Bodyweight (kg) | DE/mg/kg | g lysine MJ/DE | Total lysine |
|:---:|:---:|:---:|:---:|
| 3-5 | 16.5 | 1.05 | 1.73 |
| 5-8 | 16.0 | 1.00 | 1.60 |
| 8-12 | 15.5 | 0.95 | 1.47 |
| 12-18 | 15.5 | 0.90 | 1.40 |
| 18-25 | 15.0 | 0.85 | 1.38 |
| 25-35 | 15.0 | 0.80 | 1.20 |

**Table 7. Ideal amino acid profiles averaged from USA (NRC), UK (ARC), France (ITP) and Netherlands (CVB) published tables 2001-2006, compared to sows milk.**

| | Sows milk* | Advised ratios to one another for both total and true ileal digestible proportions |
|:---|:---:|:---:|
| Lysine | 100 | 100 |
| Methionine | 33 | 26-30 |
| Meth+ cystine | 56 | 57 |
| Threonine | 55 | 62 |
| Tryptophane | 16 | 18 |
| Isoleucine | 55 | 55 |
| Valine | 73 | 69 |

*True ileal digestible basis

Many producers still prefer to consider total lysine as a guideline as to diet value. As can be seen from Table 6, the ideal amino acid profile is established from the lysine estimation. Then for adequate protein accretion in the weaner's body, 120 mg true ileal digestible lysine is needed per gram of protein. If you or your feed supplier do prefer to work in total lysine terms, this is 145 mg total lysine per gram of protein based on a true ileal digestibility of lysine at 82% for most practical nursery diets.

### 'Overage' of lysine usually adopted by the better feed manufacturer

Overage is to include more than the levels recommended in Table 6, as a safety precaution. Under good ordering and storage conditions on-farm, as a safety margin the feed mill may add 5% more lysine to their pellets and no extra for meal. Under poorer storage and less frequent ordering times 10% can be added to pellets and 5% to meal. The manufacturer does this in order to protect himself and his product's reputation. The salesperson is encouraged to report back to the feed mill on these on-farm aspects, so for this reason alone it pays to order baby pig foods twice a month and store them cool and dry. The overages for certain vitamins are even higher.

### Lower protein nursery diets?

There is current interest in the reduction of dietary protein in nursery pigs fed lower protein diets (of less than 20%) which have been shown to be less prone to *E. coli* infection leading to scouring post-weaning. Also from the environmental aspect, low protein diets have also reduced nitrogen excretion by 30-50%. In general, for each 1% reduction in protein, nitrogen excretion is reduced by approximately 8%. Protein reduction in nursery feeds is permissible if the amino acid profile is maintained along the levels of those in Table 7, if necessary by adding synthetic amino acids if cost permits. These are fully digestible and you need not have worries on that score.

### Energy

Pigs are able to maintain correct energy intake by adjusting their daily feed intake, thus appetite is dictated by both genetic capability and pig size. Growing pigs of high genetic merit selected mainly for fast lean growth may therefore have a reduced capability for eating enough to support that genetically-manipulated lean growth. Dietary energy concentrations below 15.5 MJ/mg are expected to reduce energy intake in most modern pigs weighing less than 15 kg, while a minimum of 14 MJ/kg is suggested for pigs over 15 kg. Each MJ of ME energy reduction is likely to reduce energy intake by at least 1.5 MJ.

The energy concentrations in Table 6 should be used as a guideline only, as ingredient selection and availability of added fat determines the upper levels of dietary energy.

Also genetic capability for appetite can be different between breeds, lines with different breeds and even between lines of the same breeding company, so it is important to seek advice from the breeding company you have chosen and to ascertain that they have a recognised *pig* nutritionist and to ask who he is, as I have found some generalist nutrition advisors to be behind the times.

## Lactose

I mentioned in the section on creep feeding, when dealing with tight control of raw materials in the diets of young pigs, there has been a tendency to overuse lactose, and I have been asked to clarify the situation. We have known for 60 years that baby pig feeds supplemented with milk products rich in lactose such as dried whey and skim milk improves performance. But over this time the price of such materials, especially if dried, has escalated.

Recent studies have shown that dietary lactose can be rapidly lowered in the initial two weeks following weaning. And in pigs over 12 to 15 kg bodyweight, there are no real benefits from feeding lactose - even from the scour-prevention aspect. Table 8 gives the latest advice on lactose. If you are exceeding these levels of lactose etc, you will not be doing the pigs any harm - only wasting your money, which would be better spent on including other higher cost ingredients such as organic rather than inorganic trace elements.

Table 8.  Lactose (from simple sugars such as lactose - which is cheapest - dextrose, fructose and sucrose) in nursery diets.

| Bodyweight (kg) | Minimum[1] | Optimum[2] | Maximum[3] |
|---|---|---|---|
| 3-5 | 20 | 30 | 50 |
| 5-8 | 15 | 20 | 30 |
| 6-12 | 5 | 10 | 15 |
| 12-18 | 0 | 2 | 5 |
| 18-35 | 0 | 0 | 0 |

[1]Minimum for acceptable growth in low cost production systems
[2]Optimum concentration for balanced ingredient cost and growth performance
[3]Maximum concentration for accelerated growth performance
Source: Mavromichalis (2009)

## B.   SECOND CHECKLIST – THE WAY YOU FEED THE LINK FEED

✓    Troughs and hoppers must be spotlessly clean and dry – an essential part of any AIAO process (*see Biosecurity section*).

✓  Allow *at least* 25% more trough space than the conventional allowance (70 mm, 2¾ inches, per pig at 5 kg) until the pens have settled down. In other words calculate the space needed – but remember, the weaners will have come from feeding all at one time on the sow. I myself prefer allowing each piglet their shoulder width plus at least 25%. A temporary extra trough or hopper, suitably 'barred' to deter nosing the food to one end and stepping right into it, is essential.

✓  The floor under any trough should be a solid 'comfort board' arrangement, even if it is used as a temporary slab, cover or board, or even a permanent tray as is popular in 'big pens', (nurseries on solid floors) to contain peat or shavings.

✓  The trough should be opposite to the drinker/elimination area etc. but as far away from it as possible.

**How much food to allow in the first few hours ?**

This depends on several things – the design of the food; the weight and fleshing of the weaners; trough adequacy, group size and stocking density; stress levels on arrival; creepfeed consumption before the day of weaning; and watering facility.

The ideal is to feed ad-lib and while this has been known to be successful, it is not usually undertaken in piglets weaned early (under 18 days, say 4 kg) but is more likely to be feasible in pens containing weaners of over 6 kg (21 days) and quite possible with weaners of 7 kg and over (24 days and over).

*First* : Contact your manufacturer. He will know how weaners under differing conditions, when fed his food, respond to various feed allowances and timings in the first critical 2 days. So seek his advice.

*Second :* Check that all the environmental desiderata (see Checklist C) are in place.

Here is a pelleted feeding plan I have used successfully for the past 12 years (see Table 9). It is especially valuable for light-weight batches which have had to be removed from the farrowing house, and/or if creep feeding has not been adequate, and/or the farmer is reluctant to pay enough for a really good Link Feed.

The idea is to avoid postweaning scouring by not pushing such vulnerable piglets too hard.

As baby pig nutrition progresses and the acceptance of the expensive link feed concept becomes universal, then immediate ad-lib feeding after 17 day weaning will become standard practice. It is starting to happen now on the best specialist nursery units. Until then a cautious 'trial and error'

**Table 9. Post-Weaning Feeding – A suggested schedule for a hot nursery where the weaning skills are only average (this table can be largely ignored by experienced stockpersons using the latest feeds).**

| Post Weaning | Time | Weaning at 17 to 21 days – pigs at 5 kg (11 lb) or under Expected consumption | Weaning at 4 wks (Pigs 6 to 6.5 kg) Expected consumption |
|---|---|---|---|
| Day 1 | 10 am – wean. Do not feed for 2 hours. 12 noon – Place about ¼" (6 mm) of food in base of hopper or ½ round trough. 2 pm – Inspect. If clean add same quantity. 4 pm – Inspect. If clean add same quantity. 6 pm – Inspect. If clean add same quantity. Last thing: Inspect, tidy up, add ½" (12 mm) food. Leave light on over hopper. | 60 to 70 g/pig over the day (2.1-2.5 oz). Certainly not more than 0.7 kg per 10 pigs (25oz/1.5 lb). | You can probably allow about 33% more on Day 1. |
| Day 2 | 8 am – Inspect. Tidy up, add similar quantity i.e. 12 mm approx. 11 am – Inspect. If clean, add similar quantity. 3 pm – as for 11 am 7 pm – Add appreciable quantity, enough to last the night on the basis of looseness-free consumption up to now. Leave light on. | 70 to 90 g/pig (2.5-3.2 oz) over the day. Not more than 0.9 kg (2 lb) over 10 pigs. | Careful! Some pigs will stand 100g (3.5 oz) but others won't; keep to a lower level if so. |
| Day 3 | Check and inspect. Check food eaten, looseness. If looseness is apparent, you are overdoing quantities, or the feed is not digestible enough. If OK feed to appetite or x2 or x3 times daily, as you see fit. | 100 to 120 g/pig (3.5-4.25 oz) over the day. Do not exceed 1.2 kg/10 pigs (2.7 lb). | To appetite x3 day |
| Day 4 | Ad-lib under x3/day supervision | Ad lib | Ad lib |

- Spreading the food down the trough is essential at all times.
- Batches will vary in acceptance, thus each pen may have to be treated individually. Some can be ad-libbed from Day 2 evening onwards.

- You will find more variation in 4 to 5 week weaned pigs than in 21-22 day weaners in the amount they can eat in days 2 and 3 without looseness.
- Watering needs checking/cleaning at every feed.

**REMEMBER:** This is a cautious feeding table which should avoid digestive overload on many 'average' farms. If you are weaning later; have got a good creep feed intake by weaning (500g/day+, Varley, SCA, 2002); have correct trough cleanliness; the environment right and a well-designed link feed, you can increase the quantities offered substantially and quickly within 4 hours of weaning.

approach similar to that outlined above is often needed, because both the suitability of the foods the farmer has chosen and the degree of investment and care in the housing and management of the weaners is, I find, still very variable.

## C.   THIRD CHECKLIST - THE ENVIRONMENTAL ISSUES INVOLVED

✓   All weaners should be weaned within an AIAO regime.

✓   As well as the piglet having an undeveloped digestive system at 4 to 6 kg liveweight, its thermo-regulation and immune defence systems are also rudimentary. We must do all we can (as we have with the feed) to compensate for their lack of development.

✓   **Hygiene**. Check that the feed troughs are clean, disinfected and dry before first use, and then kept clean and 'sweet' thereafter. Stale accretions must be removed several times a day.

✓   Check that the in contact surfaces are warm *and dry* before entry. Up to twelve hour pre-heating is wise.

✓   **Temperatures**. Above-back temperatures should be 29°C (84°F) for well-fleshed 3.5 kg weaners, 28°C (82°F) for 4.5 kg weaners and 27°C (81°F) for 6 kg and over weaners. For 'thinnies' allow 1°C (2°F) warmer. With ample dry strawed pens, 'thinnies' temperatures up to 4°C (8°F) warmer may be permissible in still air conditions without affecting appetite. A 'thinnie' is a standard-sized weight-for age weaner but lacks fleshing.

✓   Supplementary heating is essential/advisable in cold/temperate climates.

In general airspeed over the newly-weaned piglet's back at thermoneutral temperatures should not exceed 0.15 m/sec (about 7 seconds to cross one metre or yard). Fans should begin to accelerate when the temperature is 0.5°C (1°F) above the correctly-set temperature and switch on when the temperature is 0.5°C below set temperature.

✓   A very common fault is chilling at night – even in the tropics.

✓   Draughts disturb airflow patterns. Check for draughts at night – use your wetted arm or back of hand and, or better, use a small smoke pencil.

# REDUCING STRESS AFTER WEANING

Chilling and draughts raise stress (anxiety, worry) and generate low-level hormone reactions which dampen down both appetite and digestive competence.

So learn how to use door and window tape sealers and/or simple air deflectors.

- When setting a temperature, allow for the smallest pig in the batch. The others will do no worse for being a little warmer.

- Always check the lying pattern of the pigs both at the warmest and coldest time of the day. This will mean the occasional night-time inspection in cold or windy weather. Don't switch lights on, take a torch and move quietly to detect resting patterns and satisfactory breathing.

- Never assume in your nursery that the temperature corresponds to that set on the control panel. Check, check, check! Call the electrician if you suspect an error. Over 1 in 5 nurseries I visit have got it wrong by 2°C or more. This is enough to cause low-level stress, slow growth rate and raise FCR (Table 10).

**Table 10. Fluctuating temperatures cost money**

Effect of temperature variation in the first 2 weeks after weaning (6 kg)

|  | *Variable*<br>*More than ± 2°C from*<br>*set temperature* | *Steady*<br>*Within ± 2°C of*<br>*set temperature* |
|---|---|---|
| Daily feed intake | 443g | 404g |
| Average daily gain | 306g | 344g |
| FCR | 1.45 | 1.17 |
| Extrapolated extra weight at<br>9 weeks at +47g/day overall | | +2.33 kg |

Source: NAC Pig Unit (1989). These figures are fairly elderly now, but are frequently found on average farms across the world today.

- Reduce house temperature progressively so that at 11 kg the air temperature is 24°C (75°F). This is hotter than most people expect for this weight-for-age, but a weaner loses a lot of fat cover after weaning if a growth rate check occurs (Figure 6).

- Trough hygiene is vital. To a certain extent clearing up the food allocation will 'polish' a trough – but don't assume so. The best nursery stockpeople use a cleaning stick for trough corners, a garden trowel for removing stale uneaten food and a bucket sponge/swab and cloth for drying out corners. This sort of attention pays in the first 3 days, at least – it is not 'unnecessary'! We wash up food receptacles for baby humans – you must do the equivalent for baby pigs. ***Our present standards are far too low!***

- Once on to *ad lib* feeding, trough and feed 'sweetness' (*i.e.* freshness) is vital.

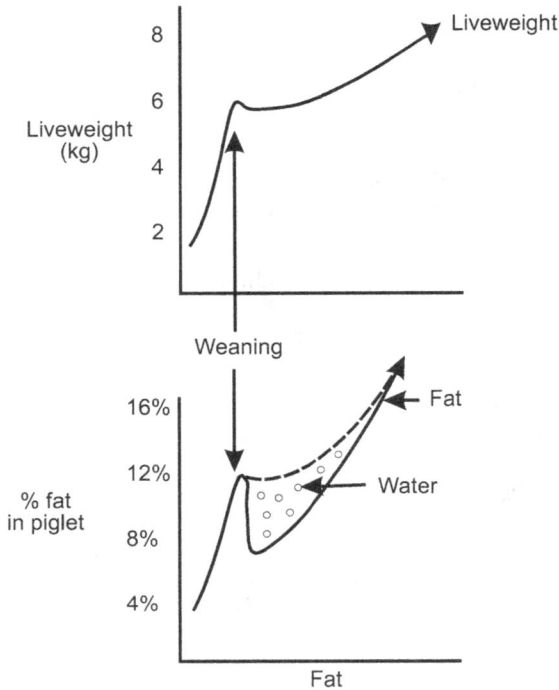

- Piglet recovers its weight gain but loses fat reserves immediately after weaning.
- Some or much of this is replaced by water in the tissues.
- This shows we must keep the piglets **warmer** after weaning, be very careful about **draughts and chilling at night**, and ensure the **water supply is accessible**.

**Figure 6.** The fat situation after weaning.

Too many *ad lib* troughs/hoppers are dirty and single-space feeders especially so. As well as food spoilage pathogens, mycotoxin poisons are a danger to young pigs, especially.

- Newly weaned piglets must not be without any food for more than 2 hours. This means constant monitoring and supervision – a reason why dedicated nurseries in a three-site arrangement are so successful – the staff have time to attend to routine inspection.

## MORNING OR EVENING WEANING ?

Weaning the pigs in the quiet of the evening or the bustle of the day? 8 pm and 8 am have been tried.

American work suggests that the pigs settle together more quickly with a night's rest to come, and by 28 days feed intake was 5% higher than those weaned during morning hours and their weight gain 6% better. Little food was offered to both groups in the first 24 hours, however. Is the lack of weaner supervision in the first 8 to 10 hours of darkness compared to daytime weaning a drawback? It doesn't seem so. However, weaned sow movement is made more awkward and most have to stay put, without their litter, until the morning of the next day. Plenty of time for them to fret with little to take their minds off things as would be the case when they are moved to the service house straightaway. Would stress have affected conception rate? The American work doesn't say.

Me? I still favour morning weaning as I feel I need to be there during those first critical hours!

# IN CONCLUSION

## What top nursery managers tell me

I have been privileged to talk to several of the best nursery managers in the business - with the experience and skills to coax the level of performance out of their young growers shown in Table 5. The following points were made in addition to their adherence to the usual ground rules of weaner production which are well published.

✓    They cultivate their opposite number in the breeding unit to make sure he/she appreciates their need for regular and even deliveries.

✓    They make sure the supplier has a strict biosecurity system in place to ensure the health status of the batches they receive is stable. Note the key word 'stable'.

✓    And that the supplier has/employs a specialist pig veterinarian who liaises with their own pig vet so that the trauma of transfer is made as smooth as possible from the disease angle by both vets monitoring their own herd's likely disease profile.

✓    Age is important. Each batch has an age bracket attached. Uniform weights are desirable, of course, but pigs which are small **for their age** need special attention, such as keeping a link feed on longer, providing more trough space and a slightly higher temperature, compared to those pigs which are small because they are younger. Bigger pigs for their age, also healthy, which are nevertheless what Prof. Whittemore calls 'thinnies', may also need similar treatment until they have fleshed out. Several ask the supplier to tattoo the day of the year (1-365) on each weaner so that the weight for age could be monitored as they grow.

✓   Some of those I interviewed asked their supplier for details of which weaners came from gilts litters, which may need housing together and separately from other incoming batches (see *Parity Segregation* Section for the reasons behind this request).

✓   None of them co-mingled - all weaners came from the same source.

✓   Few of them were keen on sorting arrivals. Only the smallest pigs (with requests to the supplier to improve uniformity!) which were then always managed separately.

✓   Several kept 15% more space available for pigs which turned 'peaky' on arrival or subsequently, this in addition to the conventional hospital isolation pens for sick animals. Emaciated pigs are immediately euthanized.

## Staffing

✓   Despite careful training and 'buddying' of new employees - 'double-tasking' alongside an experienced stockperson for 2-3 weeks - a tight eye is kept on each stockperson's routine tasks (injecting, trough replenishment and sanitation during AIAO etc) as familiarity and routine can cause things to slip, even among dedicated staff.

✓   Part of this training is to try to teach what I have long called the 'doability' of groups of weaners. 'Doability' is recognising which pens of pigs are growing well, or seem to be hanging back despite being perfectly healthy. An then to check the possible reasons for this - temperature/overcrowding/feeding/ventilation etc - for which checklists are provided starting from this one . . .

# WEANERS TALK TO YOU ALL THE TIME . . .

✓   Appearance - alert or depressed

✓   Body condition - normal or thin

✓   Abdominal shape - round or gaunt

✓   Skin - sleek/'polished' or dry

✓   Hair standing proud/gingery in white breeds

✓   Appetite - feeding at the feeder or hanging back

✓   Dehydration - sunken eyes

✓   Lying position - supine or semi-sternum

✓ Even, quiet breathing - listen last thing at night

✓ Wrong-mucking

✓ Huddling

**Action**

✓ Detailed referral checklists (environment/nutrition/sanitation etc) along with an action plan (When you notice this . . .) are pinned to the rest-room walls.

✓ Briefing meetings are held each morning and 'where have we got to/any suggestions' meetings for everybody are held weekly.

## What contributes to a post-weaning stallout?

From a holistic (far reaching/exhaustive) survey of some 180 research and farm trials over the past 8 years, the following contributors to the post-weaning check to growth can be recorded. They are individual measurements and not in any way additive.

| Faults | Influence on the post-weaning check to growth |
|---|---|
| Overstocking by 15% | 2 to 3 days |
| No creep feed | 2 days |
| Not using a modern link feed formula | 3 days |
| Poor quality pellets (too hard/too dusty) | 1 day |
| Unskilled mixing of weaners | 2 days |
| Too cold (-3°C below LCT) | 3 days |
| Too hot (+2°C above ECT) | 2 days |
| Water stress | 2 days |
| Inadequate trough space | 1 to 3 days |
| Dirty troughs | 2 days |
| Poor feeder throat adjustment | 2 days |
| Mycotoxin presence | 2 days |
| Poor flooring | 2 days |

It is quite common to see 3 to 5 of these errors evident on one farm at one time. this raises the post-weaning check from between 2 to 3 days achieved by the best of my clients, to 9 to 10 days. This 7 to 9 day extra stallout at weaning will always be magnified to slaughter, usually by a factor of two and often as much as three times more.

So a quick getaway after weaning is a major influence on profitability.

Finally, Fermented Liquid Feeding (FLF) can put a new dimension into the weaning procedure. This is covered in the Fermented Liquid Feeding section.

# REFERENCES

Cole. D.J.A. and Close, W.H. 'Bridging the Weaning Gap' Pig Talk 8 No 5 May 2001.

Farrell, C., Templeton, C. Procs. London (Ont.) Swine Conference. 2007.

Proceedings 'Understanding Heating & Ventilation' Course NAC Pig Unit, Stoneleigh, UK

Tibble, S 'Clearing the Confusion' 'Feeds & Feeding' May/June 1992 pps 9-11.

Mavromichalis, I. 'How Much Lactose for Piglet Diets?' Pig International, Oct 2006, p 23.

Varley, M. 'Piget Nutrition: The Next Five Years'. Procs. J.S.R. Genetics Annual Technical Conference, Nottingham, UK. Sept 2002.

# 5

# IMMUNITY - EVERYONE'S BLIND SPOT

The condition of being immune, or non-susceptible to the invasive or pathogenic effects of micro-organisms, viruses and cancers. The mechanisms of immunity invoke the body's ability to detect and combat substances within it which it interprets as foreign to its wellbeing. When such substances enter the body, automatic complex chemical reactions are commenced to defend the body's cells and tissues.

When I was a farm student many years ago, we had very few vaccines and no in-feed antibiotics. As a result we tried to know all about what we then called, in our ignorance, 'natural resistance'. As a possible result disease levels were surprisingly rather less on pig farms than they are today. Yes, we were less intensive, and that must have helped.

## A CHECKLIST OF 50 YEARS AGO.
## HOW TO MAINTAIN 'NATURAL RESISTANCE'

✓ We knew for example that our farm could be too dirty – and too clean; thus the solution was to try to get the balance right.

✓ We knew that we needed a good proportion of mature sows in the herd – that way disease was less prevalent.

✓ We knew that we needed good strong weaners – not to wean too soon or at too light a weight. We ignored the new fashion for 21 day weaning and weaned only when all the piglets were well over 11 lbs or 5 kg – more like 24-26 days. (Small pigs we back-fostered until they were 5 kg.) So we weaned by *weight*, not by *date*.

✓ While we weren't all-in/all-out, we knew that meticulous cleaning of the farrowing and weaner follow-on pens was essential.

✓    We quarantined all new stock and practised our own method of feedback to in-pig sows (afterbirth and a six month roll-over mixture of minced-up piglet guts kept frozen in the fridge. Today, however, this may not necessarily be the right thing to do).

✓    We didn't serve our gilts too early – in those days over 115 kg seemed adequate, but not today – they need to be even heavier and grown less quickly to 135 kg or heavier at 240 days of age.

✓    Stalls had yet to arrive. We kept sows as far as possible in groups on bedding and could afford plenty of space for them.

✓    We didn't use the vet very often as a result. (There were no pig specialist veterinarians in our area.)

Now I'm not saying the modern pig breeder needs to follow that advice to the letter, as some of it was misguided or expensive and parts of it dangerous under today's conditions. But re-reading my students' notes of the 1950's one thing stands out clearly compared with today's average producer, **we knew – instinctively – about immunity.** We had to!

# IMPROVE YOUR KNOWLEDGE ABOUT IMMUNITY

We have to raise our awareness of how to stimulate natural immunity in all our pigs, but especially the sow and the baby pig up to about 20 kg. Why ?

*First* : Because several 'new' viruses appearing on our farms are at present quite good at shrugging off existing preventive vaccines – and for others of them there aren't any, yet. But even if one new virus is protected against – another seems to appear.

*Second* : The old-favourite, drugs – especially the in-feed ones – which are useful to control the secondary infections which the viruses allow in, are being increasingly constrained by bureaucrats and the buying public mainly on grounds of antibiotic resistance, and latterly 'food scares'.

*Third* : We are stressing our pigs more in the race for productivity at all costs. Stress neutralises or inhibits immunity.

*Fourth* : We have been shown how much food energy and other nutrients the pig uses to rebuild his damaged natural immunity (Table 1) or to set up the necessary defences should the challenge be high.

**Table 1. By having to cope with a high disease challenge, genetically improved[†] nursery pigs 6.3-27.2 kg eat less, grow slower and eventually have a poorer quality carcass.**

|  | *Immune stimulus required* | | |
|  | *Low* | *High* | *Difference** |
|---|---|---|---|
| VFI, kg/day | 0.97 | 0.86 | 12.8% more |
| ADG, g | 677 | 477 | 42% more |
| FCR | 1.44 | 1.81 | 25% better |
| Protein gain (g/day) | 105 | 65[†] | 62% more |
| Fat gain (g/day) | 68 | 63 | 8% more |

* In favour of low immunity needs;  VFI: Voluntary Feed Intake
[†] The leaner the genotype the more the protein gain is damaged
**Note:** Both sets of pigs could be considered "healthy". A high disease challenge is described typically as a 'pig sick' building and a low disease challenge environment as 'all-in/all-out/multisite' scenario, properly disinfected.
Source: Stahly et al (1995)

This subject of immunity is so important to lowering our costs and in defending our profits that farmers need to understand it thoroughly. Failure to do so, and not act on what such understanding reveals must mean the costs of keeping their pigs healthy will rise substantially.

But it is not easy….

# A CONFUSING PARADOX - UNDERSTANDING HIGH AND LOW IMMUNE SHIELDS

Table 1 clearly shows that in the growing and finishing pig the immune shield needs to be as low as possible. The newcomer to the subject finds this odd. "Surely it is best to have as high an immune shield as possible to fight off disease?"

Ideally yes, but if a high immune defence is called for (because the growing pig is challenged by a large number of pathogens) and the pig automatically diverts food nutrients away from growth into building that high barrier so as to continue to keep itself healthy – then this costs money.

How? Look again at Table 1. Notice the high drain on **protein gain** due to the pig's need to remain healthy by defending itself against a vigorous pathogen assault. Less protein being available for growth means less meat. Less meat means less profit.

Sure, a solution would be to increase the protein intake to satisfy both the needs of a high immune defence and supply enough to maintain growth. But again this costs money.

**It is far cheaper to lower the chance of pathogen attack in the first place**.
From clients' experiences of what they did and didn't do - probably 2.3 times cheaper to keep the pig in a cleaner and less stressful environment than losing out on the level of lean meat formation signalled in Table 1 and some 3 times cheaper than having to pay through the nose for a special higher nutrient density diet. Then you don't need either solution.

That is why I have placed the two chapters, on combatting Stress – stress erodes immunity, and Biosecurity – which if good makes a low immune barrier possible - next in sequence. They are all part of the immune story which farmers must understand from now on. Not understanding it is costing a lot of money.

# BUT FOR SOWS IT IS DIFFERENT

On the other hand, sows need a good strong and high immune shield as quickly as possible.

Why 'quickly'? Because the gilt is stressed in having to produce a lot of babies so soon in life and nature takes time to bring her up to full immune strength - around her second or third litter anyway.

Why 'high and strong'? Because the sow lives far longer than the slaughter pig which is gone in 20 weeks or so, while the sow hopefully lasts 6-8 times longer.

Yes, she too needs more and better food to sustain a good output, especially down through lactation, and to keep healthy across her lifetime. But this extra cost is affordable being spread across her Weaning Capacity of, say, 500 kg of weaners potential income in her lifetime.

A similar table for the sow to that of the growing pig in Table 1 is given in Table 7. showing how high and low immune thresholds affect her.

That difference does not mean that, for the sows, we can afford to let up on the same disciplines of "lowering stress and tightening biosecurity" as are vital for the growers.

But we need a high immune barrier and for gilts and sows, and we can afford it.

So…in a nutshell:-

- For growing pigs a Low immune shield is good

- For sows a High immune shield is good.

- For both of them, good biosecurity** and lowering stress is very helpful in reducing costs

**Biosecurity in this case goes way beyond just keeping things cleaner and visitors away from the farm etc. but total biosecurity – see my definition at the start of the biosecurity chapter.

## Where does vaccination come in?

Vaccination is merely - but an increasingly important - way, if supervised by a veterinarian who knows your farm – of ensuring a nice high immune barrier in sows and `filling in gaps` in any nice low barrier in growers which might be leapt over by a specific nasty disease in the locality or which may have slipped into your farm through the biosecurity barrier.

As the Americans discovered several years ago when their vets got 'needle happy' over vaccination by sticking just everything in, this seemed to reduce natural immune protection in breeding herds, which started coming down with all sorts of simpler things which had rarely occurred before. I give an example in Table 2 but don't follow it (!) as it was specific to the disease picture of the time in that particular pig industry.

Things are better now, but it reinforces my belief that any vaccination and natural immune protection protocols should be **left to your veterinarian to supervise** and that any routine vaccination programme as advised by manufacturers should nevertheless periodically be referred to him for an opinion. Some of them on his advice may even not be needed, or not be needed for a while as the immunity they have bestowed over time is now adequate.

Yes, vaccinate when it is needed but take professional advice.

Vaccines are bricks in the immune structure, filling in gaps and sometimes so critical that the whole building might collapse without them. The structure itself is built from naturally-acquired immunity in the broadest sense,and may or may not need vaccines to help out. As vaccination knowledge increases, and more vaccines appear, vaccination used properly will become increasingly important in establishing a sound immune defence.

# ACTION PLAN TO IMPROVE YOUR PIG HERD IMMUNITY

Here is my advice on what every pig farmer needs to do to establish a better immune status in his herd. Time is short and you need to act now. Do it at once – the viruses are not going to wait.

1.  ***Study the subject.*** Go to every meeting you can on the subject of immunity. Read and file articles. Talk to your veterinarian about your own circumstances. Within one year you must be as knowledgeable on the subject of immunity as you are today, for example, on mating procedure at which you are indeed expert.

2.  ***Contact a specialist pig veterinarian.*** Compared to the 1960's a good local pig vet is often available. Many of us have them on our doorsteps. So use them – it is one big advantage Europeans possess over the 'low-cost' pig producers in the Far East, for example, where pig veterinarians are scarce or a long way away.

    Show him (or her) round the farm, give him time to think (and maybe do a few tests) and then have a 'what-to-do session' with him involving gilt pools, batch farrowing and even parity segregation (q.v.).

    His or her action plan may or may not involve remodelling expense – it very much depends on a lot of things, including how you both decide on your present and future exposure to the 'new' virus diseases. ***My experience is that the remodelling needed is often – maybe usually – far less costly than the theorists have proposed in print.*** (Table 5, line 4) So don't panic. For example, some disease-breaking ideas need not be onerous, but they will involve an altered and meticulously-followed routine. If you understand how immunity functions you will convince/discipline yourself to do – and spend – what is necessary.

3.  ***Adopt a more disciplined approach.*** We are very much on a tightrope situation with regard to the present virus diseases and the 'killer-secondaries' they let in. It is very easy to fall off a tightrope, but if you are trained to it and become practised and never lose concentration, then it is relatively safe. But you and your staff have to do exactly what you are trained to do in disease control, with no deviation or omissions. This is particularly true of cleaning and disinfection.

    Your veterinarian is the keystone in the disciplinary structure. You must allow him, *i.e.* pay for him, to ***disease-profile your herd*** on a regular (6 to 8 week) basis and set up what the Americans – now well versed in this – call a 'protocol' – a clearly set-out programme of pre-vaccination, medication and management which may change month to month according to how the disease challenges rise and fall in your herd.

4.  *With your vet – and with the help of other advisers – ag. engineer, nutritionist, geneticist, general consultant – there is a need to analyse what is stressing*

*the pigs, and reduce it.* Stress lowers immunity to disease. We know this with our human ailments in our relatively comfortable life at home. Do a stress audit. There are so many stress-inducing things we do to pigs which lowers their immunity these days, from 15% overstocking, to culling too early, to allowing in mycotoxin poisons, and a whole group of other stressors. Identify and ameliorate.

# HELPING THE GROWING PIG BY LOWERING THE IMMUNE CHALLENGE – A CHECKLIST

**How can a producer avoid high levels of chronic immune stimulation and hence maximise productivity at least cost?**

✓ Reduce the need for immune stimulation from other pigs. Older pigs are a major source of disease challenge to younger pigs, so segregate by age.

✓ Adopt an all-in, all-out policy wherever possible.

✓ Thoroughly clean and disinfect weaner, grower and finisher accommodation between every batch. This includes correct pre-cleaning with detergents (not just plain water), fogging enclosed air spaces and sanitising the water system.

✓ Attend to under-slat areas as well as slurry pits.

✓ Reduce dust levels in pig houses. Dust particles are virus 'taxis', and inflame the problem.

✓ Where continuous production has to be practised, institute short production breaks either by selling young pigs or following the 'partial depopulation' idea. Utilise these breaks to clean and disinfect thoroughly.

✓ Adopt tight on-farm biosecurity, especially from vehicles delivering supplies and removing stock. Biosecurity involves at least 30 other things apart from showering-in/showering-out which is what many people think 'biosecurity' means. Study the subject in depth; many of you need to catch up with the latest advice and transfer to the latest products. (*See Biosecurity section.*)

✓ Avoid stressing the pigs. Do a stress audit. (*See Stress section*)

✓ Don't overcrowd/overstock.

✓ Have plenty of 2nd to 5th litter sows in the herd.

✓ Follow a new-stock induction programme agreed with your veterinarian. This is ***not*** the same as (also essential) quarantine, which is preliminary total isolation. Induction is planned progressive merging, ***not*** isolation.

✓   Be careful about vaccinating 'as routine'. The ideal situation is for a pig veterinarian to disease-profile your herd and advise on what natural immune stimuli are needed, backed up if needs be by specific vaccination. However, there is already suspicion that some American 'needle-happy' farms are overloading their pigs' ability to acquire a robust immune defence. Table 2 gives a typical American vaccination protocol for a breeding herd in the recent past.

**Table 2. Suggested breeding herd immunisation schedule (USA 2001)**

| *Time/Age* | *Immunisations/Treatments* |
|---|---|
| **Gilts/Sows** | |
| 6½ months | Leptospirosis, erysipelas, parvovirus, PRV. Feed fresh anure from boars/sows; repeat one week later. |
| 7 ½ months | Repeat vaccinations |
| 6 weeks before farrowing | E coli bacterin, AR, TGE, rotavirus, PRV |
| 2 weeks before farrowing | E coli bacterin, clostridium toxoid, mycoplasma, rotavirus, TGE, AR |
| 3-5 weeks after farrowing | Leptospirosis, parvovirus, erysipelas, PRV |
| **Boars** | |
| First 30 days in isolation | Blood test for brucellosis, lepto, parvovirus, APP, TGE, PRV |
| Every 30 days in isolation | Erysipelas, leptospirosis, parvovirus |
| Every 6 months | Revaccinate for PRV, leptospirosis, erysipelas, parvovirus |
| **Pigs** | |
| Day 1 | Clostridium antitoxin |
| Day 3-7 | AR, TGE |
| Day 7 | Mycoplasma |
| Week 3-4 | Revaccinate for AR, mycoplasma |
| Weaning +20 days | Erysipelas, APP |
| 10-12 weeks | PRV; revaccinate for erysipelas, APP |

Source: Pork Industry Handbook, PIH-68

TGE = Transmissible gastroenteritis, AR = atrophic rhinitis (bordetella/pasturella), leptospirosis = 6-strain leptospirosis, PRV = pseudorabies virus (called Aujeszky's elsewhere), APP = Actinobacillus (Haemophilus) pleuropneumonia

Quite a workload! And quite a load on the pig's response system! My advice is to see which of these or others is definitely necessary or just advisable from your own veterinarian's experience of your conditions. So consult him – often.

# COLOSTRUM

There can be very few producers who are not aware of the importance of colostrum and the part it plays in immunity.

The piglet is born with insufficient antibodies to defend itself from hostile bacteria and particularly the viruses it meets as soon as it is born. It must obtain protection from these pathogens from its mother's foremilk containing immunoglobulins, known as colostrum.

These are primarily…. ( Ig = Imunoglobulin)

| | Protection provided | Against | Proportion in colostrum(%) |
|---|---|---|---|
| IgA | Internal linings - gut, throat, lungs, etc. | Bacteria | 17 |
| IgG | Whole body, via bloodstream | Bacteria | 76 |
| IgM | Starts the piglets immune response system | Predominantly viruses | 7 |

IgA is the only immuno-protein to be present in subsequent sow`s milk, but at a much lower level.

Immunoglobulins have a large molecule size and the cells lining the newborn`s gut able to absorb it, start to close their absorbtive areas within 6 hours of farrowing and may be fully closed within 12-16 hours.

This means that…

• All piglets should drink sufficient colostrum within 8 to 12 hours at least.

• The piglets born last are at a severe disadvantage. Not only will they be weaker but they are in danger of 'missing the immunity bus'. The earlier-borns can enjoy as much as 30 mg/ml of antibodies while those last to arrive and suckling poorly might only achieve 4 to 6 mg/ml (Varley 1989)

• With much larger litters being born these days, the subject of speeding up farrowing is growing in importance – see later. The more protracted the farrowing the weaker are the last to be born . "The piglets start to die as soon as the farrowing process commences" one expert has observed.

Attended farrowing is another technique which helps the last to be born and thus weaker piglets to obtain a sufficient intake of colostrum. as they can be placed on a teat as soon as they arrive or by split suckling a little later on. As colostrum is very rich in quickly-absorbed energy, this also defends against chilling and resultant overlaying.

## How much colostrum?

The sow should produce between 1200 to 1900 g/day. Total intake by the newborns is variable but is probably 200 - 450g/piglet (range 100-1400g/day within the first 24 hours of life) so with today's big litters this equation suggests that there may not all that much of a surplus - the weaker piglets being the first to suffer   Minimum intakes should be be at least 60g (ml) within 6hrs of birth and at least 100g within 16 hours.(BPEX 2008) dependent on the quality of the colostrum.

Avoid teeth clipping during this time as it is bound to deter suckling.

# COLOSTRUM QUALITY

Many producers do not realize that this can vary in immunoglobulin presence. This variation is due to the age of the sow (poorer before two litters and then again after six) and her previous exposure to pathogen challenge. This latter is another reason why the veterinarian should be used to disease-profile the herd regularly and which I have already described in several contexts elsewhere in this book. Reverting to my simplistic analogy, she may need to have the 'holes' in her 'immunity building' filled in by specific vaccination 'bricks'. By using a wide range of vaccines available to him, how many 'bricks' and where to put them in the structure, the vet. can determine from his tests and from his knowledge of what virus diseases are currently prevalent in the area, what to recommend to suit the farm at that point in time

This is a correct way to use the veterinarian - in a fire prevention role – not as a fire brigade as so many farmers tend to employ him. The vet is a key part of any farm's immune status

## Gilts and their colostrum

The quality of the gilt's colostrum is very probably below that of her colostrum as a more mature sow, which makes supervision and identification of what stimulatory challenges she is likely to have encountered before she is bred – or not, as the case may be, most important. Only the veterinarian is capable of deciding which vaccination 'bricks' are needed to be inserted into her as-yet incomplete immune structure, and when and where to make it sound. This will then go on to influence the quality of her colostrum when she farrows for the first time and could even have an effect on her second, especially if she  has been over-stressed up until then.

Good nutrition also has an effect on colostral quality. The sow needs to be properly fed, especially leading up to farrowing and I have been recommending to farmers whose sows look a little below par in the last 14 days before farrowing that the gestation diet could be changed to a sow lactation diet.

In my experience, probably thanks to better-designed lactation feeds, this has not resulted in udder troubles such as MMA and/or inappetance post-farrowing, which breeders have been concerned about in the past.

Maybe this strategy could be adopted as routine? An aspect of sow nutrition worthy of research by the feed trade? Possibly even providing an immediate pre-farrowing diet with more of those nutrients (about 8 so far) believed to help both colostrum quality and quantity. Yet another fruitful area for more research.

Along these lines, too, in the section on gilt feeding I have already recommended that gilts need special diets leading up to first farrowing which in part contain these same ingredients which could assist building a better immune structure more quickly and to save waiting for natural means to do it one or two litters later.

## Colostrum substitutes

I have covered the skilled technique of how to collect colostrum from existing sows, how to store it and then stomach-tube orphaned and colostrum-deprived piglets, in some detail in another book (see References).

Alternatively colostrum substitutes, mostly derived from bovine colostrum, are available which follow the Ig pattern of the sow material very well. I have not used them myself, preferring to milk a sow – any newly-farrowed and docile sow from the same herd will do, but never from a different herd or separate section of a large unit – as I was trained to do this and got quite skilled at the procedure, only managing to lose three piglets in around 500 dosed.

Research by Dr.Jim Pettigrew in 1994 showed the value of one American colostrum substitute, survival rate rising 11.2% from 79% to 90% and weaning weight at 19 days by 318g against untreated litters. The cost of the product at the time was £0.17 per piglet with, I calculate an REO (including labour of administration - about the same again) of some 15:1 on survival rate alone. Good value.

If you are unable to or unwilling to go to the bother of collecting your own, then a colostrum substitute could be a useful stand-by for those weaker neonates who may not have obtained their 100ml or so of the real thing. Substitutes seem less readily available than they were 15 years ago, but with larger litters these days they could well make a comeback.

# SUPERVISED FARROWING

Very much a part of the immunity story as it is the one factor which will ensure – with proper attention - that those 3% of piglets which do not survive due to insufficient colostrum intake, will then do so.

In addition to this, another 2% are born anoxic and immediately suffocate. With someone there at farrowing most of these later arrivals can be saved too.

For decades now we have struggled to fall below the 10% mortality of born alives to weaning statistic, and there exists the means to get down to the 5 to 6% which the best producers manage to do

I can vouch for this as 8 out of ten of my clients in the past achieved this level of success. All of those that have, have attended all their farrowings by ensuring the sows farrow during working hours

The last time I collected data from my clients was several years back, and Table 3 gives an indication of the performance differences between being present for as many farrowings as possible As you know, this is done by using prostaglandin analogues, if needs be supported by oxytocin injections (check any restrictions in your own country) so as to ensure a large majority, probably 95%, of sows farrow in working hours.

**Table 3. Attended versus non-attended farrowings. 3 before-and-after trials.**

|  | *Attended* | *Non-attended* | *Attended* | *Non-attended* | *Attended* | *Non-attended* |
|---|---|---|---|---|---|---|
| Born alive | 10.66 | 10.00 | 10.81 | 10.67 | 10.01 | 10.12 |
| Weaned | 9.91 | 9.12 | 10.10 | 9.64 | 9.83 | 9.01 |
| Mortality % | 7.00 | 8.88 | 6.60 | 9.70 | 4.30 | 11.00 |
| Extra pigs sold per 100 sows/year | **185** |  | **108** |  | **190** |  |

The average extra income from the 'attended' strategies was £84 per litter while the cost/ litter including labour was c. £25/ litter - an REO of 3.36:1.

Because the stockperson is there, not only are piglets saved which would otherwise not survive, but many of the later and weaker arrivals are immediately put on to a teat, and/or split-suckled, or even stomach-tubed to ensure <u>every</u> neonate gets an <u>adequate</u> amount of colostral immunoglobulins, which often does not happen.

The results suggest that attended farrowing increased the number of pigs sold per sow per year by well over 6% on average, which doesn't sound a lot, but on a farm achieving 25 pigs sold per sow per year, moves it up comfortably over the 26 mark.

## WHEN DO PIGLETS DIE?

Table 4 was taken at the same time, averaging the records from 11 countries when I questioned the section-heads on 18 farms - 5 of the farms with the whole of their farrowing teams present - not on how many sucklers died or what they died of (as I suspected I might not get worthwhile replies from this line of questioning) but their impressions of <u>when</u> they died.

**Table 4. When do pigs die? Average losses:** *(11 countries: 18 farms interviewed. Av.10.8 born alive).*

| | | | *Hours/days* | | | | |
|---|---|---|---|---|---|---|---|
| *From birth* | *0-12hrs* | *12-24hrs* | *24-48hrs* | *3-7 days* | *8-14 day* | *14-21 days* | *Overall* |
| Losses (%) | 37 | 32 | 12 | 8 | 6 | 5 | 13.5% |
| Pigs died | 0.53 | 0.46 | 0.17 | 0.11 | 0.09 | 0.07 | 1.43 pigs |
| | <--One pig--> | | <---------------Half a pig--------------> | | | | |
| | in the first day | | in the next 20 days | | | | |

Comment: If we are going to get unstuck from the world-wide 12 to 13 percent mortality-to- weaning figure, which has hardly improved at all over the past 20 years, then saving half a pig more per litter within the first 24 hours would dramatically improve the situation. The outcome in Table 4 should be viewed with caution as they were people's opinions and only three results came from the records. None of the farms used prostaglandins, several being not allowed to do so.

Nevertheless the percentage mortality of born-alives to weaning is not unusual at 13.5% - virtually 1.5 pigs lost in every litter. With today's larger litters this is nearer two pigs lost/litter these days and I wonder how many more could be saved if adequate colostral intake was achieved by attended farrowing?

## SPEEDING UP FARROWING

With the arrival of much larger litters – up to 40% greater than 15 years ago – it can be expected that farrowing time would be extended by a quarter to a third. The Danes are optimistically forecasting litter sizes of 15.5 by 2016 (Pedersen, 2007). This is more than enough to exhaust many sows – and especially a gilt, so that the last-born are anoxic and often do not survive for long.

Not only does this strengthen the case for being present at farrowing to assist those born last but also interest has heightened around the possibility of shortening the whole farrowing process if at all possible.

Several years ago I was witness to the use of the drug neostigmine, a nerve stimulant to encourage smooth muscle contractions (different from the hormone oxytocin which awakens the farrowing process).  Neostigmine was injected after the birth of the 4th to 5th piglet to address the rising stillbirth problem at the time. This reduced the birth interval by about 9% but the stockpeople didn't like doing it and as the emphasis in those days was on reducing stillbirths, so its use has drifted away as opinion turned to gas heaters in the farrowing barns being suspected as  the main culprit.

In 2007 a commercial product 'Parturaid' (SCA) was administered by paste into the mouth of a sow just before farrowing,  and it has shortened farrowing time by some 20 minutes, which was certainly useful. But again the attendants baulked at it and it is not widely used, which I think is a pity in view of the high neonatal mortalities linked to very large litters - 1.6 born-deads are now  increasingly reported from litters of 13.5 total-borns. A lot of this loss must be down to suffocation.

> *"A piglet starts to die as soon as the birth process commences"*

> (English, 2001).

## Practical measures to consider

So back to basics until science can try again.

*Temperature:* Not too hot, not too cold. 21°C is ideal. Many farrowing houses I visit are too hot, 24°C.

*Sows in good condition.* This goes without saying of course, but the jury is out on whether the sow should have an energy boost 2 days before predicted farrowing time as this might affect MMA. My experience is that if the farrowing area is kept very clean over the farrowing event then this is much less likely.

And that some sows can take the energy boost (from say, a kg of creep feed/day is one way it has been done) but others will not and the attendant that knows his/her individuals will not be too troubled by it

Have a cautious try, as the energy boost does seem to reduce mortalities in the larger litter neonates.

*Muscle tone:* We all know that exercise improves suppleness. So too with sows especially the older ones. This must be another of the benefits associated with group housing especially if generously-sized preferably straw) yards can be afforded.

*Disturbance level:* Some farrowing sheds on the large units can be very noisy with piglets squealing and careless stockmen banging things together. Go about as quietly as you can. Distraction can hold up the birthing process.

*Moulds and mycotoxins:* Not proven but I suspect there is a link with trouble-free farrowing, as farms I have visited with an obviously high degree of mould presence had their farrowing troubles alleviated when we jumped on the resultant mycotoxins.

A 2% reduction in stillborn pigs due to as speedy a birth process as nature allows, is worth 46 more pigs sold per 100 sows per year. With these larger litters stillborns are <u>rising</u>, not falling at that level.

## DISEASE PROFILING

Getting a pig specialist veterinary practice to disease-profile your herd is a worthwhile investment. We did this at our Dean's Grove Farm in the 1980s and got superb performance. Figures from clients in the USA confirm the value of the idea as Table 5 shows.

**Table 5. Before-and-After Results from Using a Pig Specialist Veterinarian to Disease-Profile 3 farms, with extra vaccination & Re-Modelling expenses costed in. (US$ Per Sow)**

|  | *Before* | | | *After* | | |
|---|---|---|---|---|---|---|
| *Farm* | *A* | *B* | *C* | *A* | *B* | *C* |
| Estimated cost of disease per year* | 284 | 186 | 300 | 80 | 96 | 109 |
| Cost of veterinarian | 8 | 3 | 12 | 30 | 27 | 31 |
| Cost of vaccines & medication† | 26 | 18 | 30 | 18 | 20 | 21 |
| Cost of remodelling (over 7 years)** | – | – | – | 27 | 45 | 33 |
| Total Disease Costs (US$) | 318 | 207 | 342 | 155 | 188 | 194 |
| Difference (Improvement %) | – | – | – | 51% | 9% | 43% |

\* Disease costs *estimated* from items like the effect of post weaning scour and check to growth on potential performance; respiratory disorders, ileitis, abortions, infectious infertility, *etc.*
† Note that the cost of planned preventive medication was *lower* than for reactive curative medicine.
\*\* Not including parity segregation
Source *:* Clients' records and one veterinary practice

# WHAT DOES INADEQUATE IMMUNITY COST?

There must be a hundred thousand answers to this! It is impossible to quantify in general terms any more than there is an answer to "How much will I save if I don't get disease?" or "How much will I lose if I don't fertilise my fields?"

## Immediate detectable losses

No reader needs reminding that the cost of inadequate immunity lets in very serious diseases some of which, especially the viruses causing PRRS, PMWS/PDNS; Swine Fever (Hog Cholera), Swine 'Flu, Coronaviruses, etc are deadly to profits. The damage these pathogens have done in eroding my clients' profits – even when the pig price was good – have varied from 40% to 100%, sometimes lasting as long as 18 months, and sometimes with carry-over losses extending to six months after the disease had seemed to have gone. To these losses must be added the costs of vaccinations and vet/med attention. Such routine preventive costs alone can be 8% of production costs, excluding labour, with protective in-feed medication adding another 1.5%.

## Insidious/hidden costs

What many producers fail to realise is the penalty which the pig's body puts on performance when its immune system has to respond to a high degree of challenge. From time to time useful research has appeared to quantify this and I illustrate two examples, both from the pioneering work, on the interaction between immune demand and nutrition which Iowa State University carried out in the mid to late 1990s.

Table 1 summarised what can happen when a young growing pig has to activate its immune system to a high degree compared to one which has no need to do this to anything like the same extent. Notice how protein gain – the primary objective of any of us as meat producers – is severely reduced, and what is more, we are the very people in animal farming leading the trend to purchase high lean-gain genetics!

I have attempted to put a cost to this fall-off in nursery performance in Table 6.

It is important that we try to quantify what this can cost the wean-to-finish producer because this underachievement *from perfectly healthy-looking growers* (the activated immune system has seen to that) is far higher in lost performance than the cost of providing a low immunity challenge environment. This is where many producers are falling down today by not realising that...

**Table 6.** **How failing to match dietary quality to the current disease status can affect economic performance in MTF, PPTE and REO terms***

| *Immune activation* | *Lysine needed per day 7-102 kg* | | *Advantages from altering diet density (average + 2.81g/lysine/day) for high health pigs* | | |
|---|---|---|---|---|---|
| | *(g)* | *Extra feed needed (kg)* | *MTF (kg)* | *PPTE* | *REO* |
| High ('Low health' pigs) | 5.7 to 16.1 | +25.26 | 286 | – | – |
| Low ('High health' pigs) | 7.9 to 19.3 | – | 321** | +£33.25/t ** | 5.3:1 |

\* The New Terminology.......
   MTF = Saleable Meat per Tonne of Feed
   PPTE = Price Per Tonne Equivalent – a figure relating the MTF improvement to an equivalent reduction in feed/cost/tonne.
REO = Return on Extra Outlay. Extra outlay in this case is £6.25 to provide the better diet, thus REO is £33.25÷£6.25=5.32:1
** MTF includes a 0.91% improvement in yield. Base calculations from Williams (1995) and Stahly (1996)

*What this table shows*
Failing to match dietary quality to immune status can be equivalent to a 24% price rise in the cost/tonne of all feed from 7 to 102 kg.

• It is cheaper to provide a low challenge environment in the first place than to make the growing pigs protect themselves by (over) stimulating their immune system.

• And it is cheaper than loading the feed with protective drugs – often needing more over a period of time to achieve the same effect.

Not matching the diet to the current immune status can be equivalent to your paying a quarter as much again for all the food you need from weaning to slaughter (Table 6). That pays for a lot of disinfection, better housing, and veterinary monitoring/guidance.

# THE SITUATION IN BREEDING SOWS

Iowa State workers have also shown that continuous activation of the immune response during an 18 day lactation reduced sow feed intake by 0.5 to 1 kg day. This resulted in a reduction of litter weight gain of 0.32 kg/day probably from poorer quality milk (Table 7). This, at slaughter could itself cost 9 kg MTF or a 6.5% increase in the cost/tonne of all grower feed from 7 to 25 kg.

**Table 7. Impact of immune system activation on lactating sow and litter performance**

| | Immune system activation | |
| | Low | High |
| --- | --- | --- |
| **Sow traits** | | |
| Feed intake – kg/day | 5.36 | 4.80 |
| Body weight change – kg/day | 0.74 | 0.69 |
| Backfat change – mm/day | 0.19 | 0.24 |
| | | |
| **Litter traits** | | |
| Number of pigs weaned | 12.6 | 12.6 |
| Litter weight gain – kg/day | 2.60 | 2.28 |
| Estimated weaning weight – kg/pig | 5.53 | 4.93 |
| | | |
| **Milk & milk component yield** | | |
| Immunoglobulin G (mg/ml) | 4.3 | 5.4 |
| Immunoglobulin A (mg/ml) | 12.4 | 17.8 |
| Yield – kg/day | 11.5 | 10.1 |
| Energy (Mcal/day) | 14.4 | 12.7 |
| Protein – g/day | 683 | 612 |
| Fat – g/day | 726 | 675 |

Source: Sauber *et al.*, 1999

This is why any attempt to keep the sow more comfortable, cleaner and to reduce the strain on her system in lactation is so cost-effective, not only to herself – but to her progeny right through to slaughter.

And that if a higher degree of challenge is imposed in lactation, a better lactation diet is needed.

Standards of biosecurity and cleanliness to reduce the need for the pig to activate a high immune barrier are given in the Biosecurity section.

# IMMUNITY IN THE YOUNG SOW

The breeding herd is exposed to debilitating disease some 5 to 7 times longer than the grower/finisher, and the young sow therefore needs a good solid immune barrier as soon as possible and for as long as possible (Figure 1).

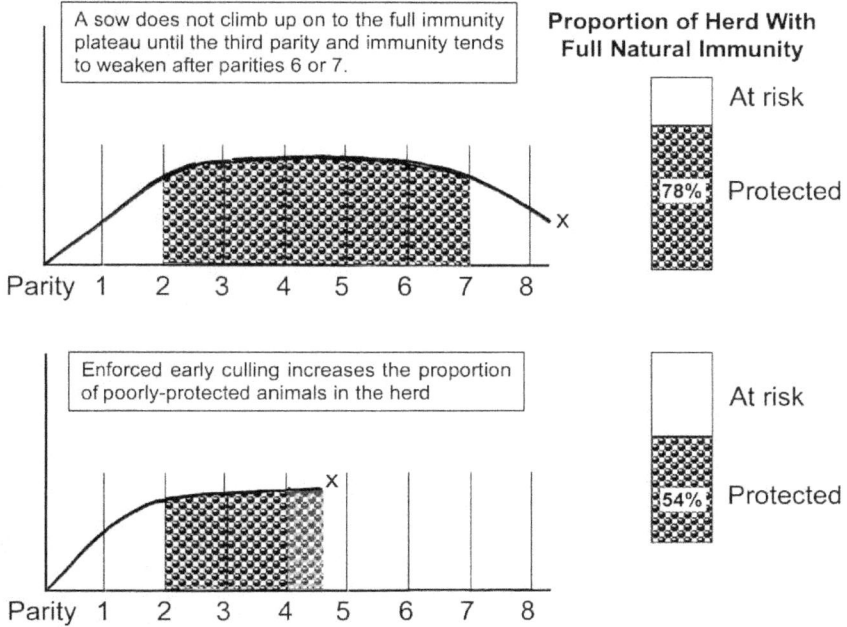

A sow does not climb up on to the full immunity plateau until the third parity and immunity tends to weaken after parities 6 or 7.

**Proportion of Herd With Full Natural Immunity**

At risk

78% Protected

Parity 1 2 3 4 5 6 7 8

Enforced early culling increases the proportion of poorly-protected animals in the herd

At risk

54% Protected

Parity 1 2 3 4 5 6 7 8

**Figure 1.** The danger of a short herd life. If the immune status of the gilt/young sow needs to be high to protect her future productivity - so the feed they need has to be of high quality to sustain this extra immune demand.

## IMMUNITY AND HEARD AGE PROFILE

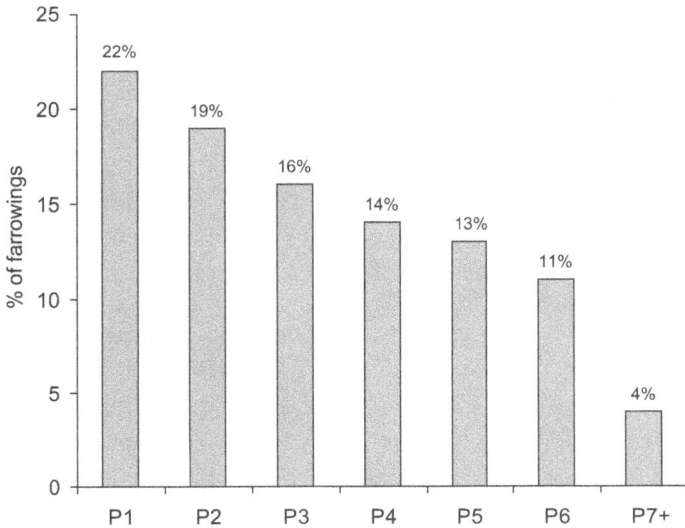

**Figure 2.** A commonly agreed ideal herd age profile at the time of writing, called the 'Resting Lion" shape.  Source: P.I.C. 2010.

## A look into the future

However, we are now in the age of the hyperprolific gilt and sow, when properly managed and fed, a sow herd could productively and economically well last longer, in which case the ideal shape would be longer (8+ parities) and lower in profile (across a range of 19% to 4%). This would provide a weaning capacity (*See Business Section*, page 252) of between 600-650 kg from today's target of 500 kg.

## Target herd age profile

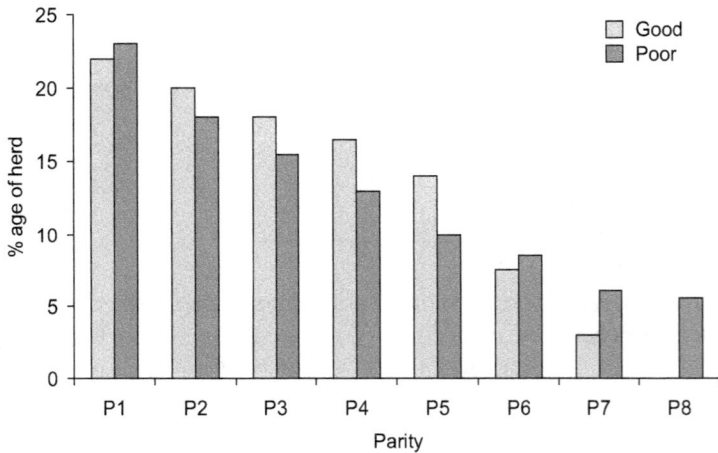

**Figure 3.** Two different herd profiles. Source: BPEX 2009.

Based on a 7 day parity profile, the light bars in Figure 3 represent a much more profitable profile than the one with the darker bars. Both the profiles have a similar annual replacement rate of 42%-44%. However, the darker bar herd has lost nearly 30% of its gilts by parity 3 while the light bar herd only 16%.

Thus the light bar herd has nearly half of its sows in the peak immunity parities 3 to 5 while the dark bar herd only has just over one third and is vulnerable to disease. The 'resting lion' shape is the one to keep in mind - not the 'ski run' shape of the darker bars. Try to have 45%-47% of the sows in parities 3 to 5 and/ or 54% in parities 3-6. The gilts start off at about 20%-22% of the total sows in the herd but some of the best breeders keep this input figure nearer to 18% as their sow productive life is longer and the need for replacements is therefore lower. Gilt numbers can be ideally anywhere between 17% and 21% (dependent on farrowing index and replacement rate). If down towards 17% the Lion does not have much of a 'head' if at all and his back is straighter! Either profile is productively efficient.

However, from the trial cited earlier, maybe our lion of the future could have a rather lower but much longer profile?  No loss of productivity, but much lower replacement costs?

I hope it does, because it fits into the correct economic theory of SLC or producing the Same at Less Cost.  SLC doesn't flood the market, and pockets the cost savings as profit.

## HOW TO KEEP THE LION'S BACK HIGH, LEVEL AND LONG

- Have a gilt pool.

- Have a sufficiently long induction period.

- Use the vet to advise on induction protocol measures.

- No serving gilts before 240 days.

- Use gilt developer, gestation and lactation diets.

- If large gilt and second litters – take the burden off the mother by a variety of options – fostering, split suckling, early latest-formula creep feed, rescue decks, etc.

- Meticulously graph out your herd age profiles every month. (This is so important yet only 10% of the farms visited did it, and none did it every month)

- Follow the 'do's and dont's; of sow culling, (see Gadd 'Pig Production Problems' (publ. NUP 2003, pps.68-71).

To my mind there are several important reasons why the immune status of our breeding animals is often too low: -

## A CHECKLIST: STIMULATING GRADUAL AND NATURAL IMMUNITY IN THE BREEDING HERD.

✓   We don't give the gilt a long enough acclimatization period from entry on to the premises to fully merging her with the herd.  5 weeks is minimal, 6 to 7 weeks may be necessary to combat these "new" viruses (PDNS, PMWS, PRRS, circo- & coronaviruses etc).

✓   As to a minimal length of time in order to save on costs, ask your vet, who should be monitoring the disease profile of your herd, and also knows the prevalence of the viruses in your area and can liaise with the vendor's veterinarian if you are buying-in replacements.

✓   We are going about the challenge protocols in the acclimatisation period too casually, often using the same old techniques (afterbirth, fence-line culls etc) when a specific planned and varied programme is needed as the pathogen population changes, including vaccination.

✓   Ask the vet again to advise on what challenge procedures to adopt, for how long and when in the induction period. Normally this will be in the first 14 days; the next month being a 'rest & recuperation' period.

✓   We are growing gilts too fast between purchase or selection at 90-95 kg and first service at 135 kg. The range is dependent on lean gain genotype. Slow them down! Let immunity acquisition catch up with the modern gilt's precocity. Let her hormone system catch up with her ability to grow fast. She looks like a sexy 21 year old woman but with the hormone and immune development of a 14 year old schoolgirl – or should I say 12 years old in modern 21$^{st}$ century society!

✓   Consult a nutritionist about a Gilt Developer Diet to grow them no more than 550 g/day at 100 kg rising to 750 g/day at 8½ months old (135 kg). Gilts grow very fast these days and a few take some holding back. This feed needs to be high in certain nutrients but fed under control.

✓   Feed and manage the pregnant gilt and first-litter sow differently to your standard, established sows. She is a totally different, developing animal, and quite apart from her nutritional needs, failure to do so could compromise her subsequent immune status.

## IN SUMMARY

Bone up on this whole subject – developing the gilt to service, the latest flushing technique, first pregnancy feeding and management, helping her cope with that strange, stressful first farrowing, and using a special first litter sow lactator feed. Examples are given in the Further Reading Section, particularly in Close & Cole 'Nutrition of Sows & Boars' 2000, the cutting edge textbook on pig breeding nutrition and very clearly written/easily understood.

## IMMUNITY IS COMPROMISED BY TOO-RAPID SOW TURNOVER

During speaking tours across the world I am worried about my audiences' knowledge gap between what is known and published about the gilts physiological needs from initial selection at 5 months – to the end of that first litter 8 months later – and the listener's acceptance of what is needed.

There seems to be a disturbing trend towards faster and faster sow replacement rate. I don't think the breeding companies are entirely to blame as some say; it must be the unacceptably-high young sow culling rates mostly due to early-in-life infertility.

# 40% TO 45% REPLACEMENT RATE TOO HIGH?

It is precisely because we are being forced to cull so many sows prematurely for reproductive failure at the end of the second or third parity that we have ended up with herds which are only partially immunized to disease – in some cases under 50% protected (Figure 1, page 103). Moreover, these young sows may be unable to produce sufficient of the correct antibodies to protect their piglets from eventually succumbing to diseases like PDNS/PMWS and PRRS. PRRS is a real headache worldwide, and won't seem to go away. How much of this is due to a too rapid turnover of the sow herd?

Replacement percentages should be in the upper 30s to low 40s, preferably the former. My clients who have a good long herd life (averaging over 5 parities) seem to have fewer disease problems, with vet/med costs half to two-thirds lower than most. Look again at Figure 1 – the key to their success lies there.

# BUT WHAT IS THE COST?

They all have longer induction periods under veterinary control and while the cost of this is substantial in extra housing and feed needed, raising the cost of the first litter by about 15% (range 12.8% to 17.1% from my clients' records) the payback from a higher and longer herd life seems to far outweigh the time, trouble and money invested early on into the gilt and that first parity sow.

Table 8 is from a selection of my UK clients. It shows that if the extra cost of getting a gilt properly prepared and looked after is, say 15% more, the longer productive life likely to result from this early investment actually makes the gilt 50% cheaper per pig sold, so the REO is 50÷15 or 3.33:1 on this basis – more than a 3:1 return for the extra cost and hassle involved.

Table 8.  **Why sow longevity is important – your gilt investment is halved**

| Cost of getting sow to first litter | Sow lasting 3 plus litters (40 pigs) (per pig sold) | Sow lasting 6 litters (70 pigs) (per pig sold) |
|---|---|---|
| £300 | £7.50 | £4.29 |
| Plus empty-day lag in replacement female at 4th parity 25 days at £3.10/day ÷ 40 piglets = | £1.94 | Nil |
| Total per pig | £9.44 | £4.29 |
| | | (at least 50% less) |

UK Costings (2010)

# WHAT HAPPENS WHEN A VIRUS STRIKES

Simplifying a complex procedure, readers will forgive me if I draw an analogy with a modern battle, but it is so important that immunity is understood.  When a virus invades, a firefight commences.  A variety of soldiers are called up to deal with the pathogen enemy, which has already invaded healthy cells in the animal's tissue, reproduced inside them and emerged to take over other cell territory.

## The reconnaissance

As the virus bursts forth from its bridgehead, **Helper T cells**, already on watch in the body, identify them to HQ as **antigens** or foreign invaders.  This is an alarm alert.

## Mobilisation

1.  **Forward defence screen**

    However **natural killer cells** are also an on-watch force which do not need the antigen alarm.  They immediately go ahead, seeking out and killing some virus and cancer cells.  But they soon need to be reinforced.

2.  **Rapid reaction troops**

    Responding to the antigen alarm from the Helper T cells, **Macrophages** (white blood cells) are a rapid reaction force who in peaceful times live in the bone marrow 'barracks'.  (These bone marrow barracks in wartime – *i.e.* when the disease organism invades – become training and replacement camps.) Macrophages deal with some, not all, of the invaders, especially bacteria, fungi, cells invaded by viruses and cancer cells.  Thus they identify the enemy and liaise with the Helper T cells to mobilise the assistance of B cells.

### 3.    Primary defence force

*B cells* are the heavily armed troops.  Their armament being *antibodies*.
Specific antibodies for specific antigens – in the same way that an army uses
different weapons to deal with different challenges – anti-aircraft, anti-tank,
mines, machine guns, etc.  It can take 14 days for full mobilisation to happen
(*i.e.* with vaccination) but under attack the rapid reaction troops hold the line
as a much more solid defensive build-up develops.  So during mobilisation, as
well as the macrophage defence troops, one type of Helper T cell goes for the
invaders, destroying virus-infected cells, while another variety organises the
correct armament (the right antibodies) needed by the B cell heavy-duty troops.

## The battle commences

After the build-up in response to attack by disease (during which time all the troops
involved need all the help you - the civilian population/government edicts - can
give them by reducing stress, keeping things clean, not overstocking and managing
warmth and ventilation well) while the heavy duty B cells start tearing into the
antigen invaders.  Each B cell 'regiment' recognises and reacts to only one specific
antigen by destroying the body cells harbouring the enemy antigen (such as a virus)
or neutralising the virus itself.

## Battle over

### 4.    Logistics and intelligence corps

Another form of Helper T cells have been in reserve, called *Suppressor T
cells*.  These detect that the battle has been won and stand most of the troops
down.  Without it the various troops, by now in full fighting mood, could begin
attacking healthy body cells, too – not just those occupied by the enemy.  If
you remember, the heavily and specifically armed B cells recognise and react
to only one form of virus enemy, and *Memory B cells* and some T cells stay
in the body waiting for any re-invasion of the antigen 'enemy' *i.e.* if there is
re-exposure to the same antigen, or invader.

# DON'T CONFUSE ANTIGEN WITH ANTIBODY

*The Antigen* is the invader, and there are many different types: *unwanted*, *i.e.* viruses,
bacteria, fungi, cancers etc, as well as the various families within each group.  And
*planned*, as in a vaccine.

A vaccine antigen is a 'teaser' invader to alert, stimulate and mobilise just sufficient defensive troops to fight off an unwanted invader should it materialise.

*The Antibody*. These are protein structures (principally IgA, IgM, IgE and IgG) which fight the foreign invading agents. In our analogy, the weapons and ammunition the defending troops can call on.

## CAN NUTRITION HELP WITH IMMUNITY?

The pioneer work of Stahly, Williams, Cook, Sauber, Zimmerman and others has already been mentioned, *e.g.* on pages 87 and 101. It is important to discuss the nutrient density of your diets with a nutritionist who understands the work of these pioneer researchers. Feed design may be affected both by appetite (intake) and immune status differences, and a nutritionist experienced in designing diets to match – as far as present knowledge allows – both these variables is a useful ally for the future.

We immediately run into a problem. Not insurmountable even if it presents, at first sight, practical difficulties.

The problem lies with diet design. Farmers, always pragmatic realists, ask "Fine, but these pigs I'm looking at, how do I know – and more important how do you know – where they are on the immunity activation ladder?"

Good question ! An absolutely vital question, in fact, because getting it wrong could raise, in UK terms, at least 15% of the cost of producing a pig from imprecise nutrition alone.

## THREE OPTIONS

As I see it these are three possible solutions.

1.  **Serology**. Here one uses the vet to blood-sample the herd to try to establish a disease profile. Snags are that even the cutting edge of serology cannot identify certain diseases – so what happens if the current challenge happens to be mostly from one of these ? Secondly, it is expensive and could be time-consuming. Serology, as knowledge advances, could help more in future. But what can we do now?

2.  **Challenge or test feeding**. This concept takes 50 typical growers, feeds them on what the nutritionist calls a 'non-limiting diet' and periodically, say every 14 days, monitors growth, FCR and lean gain (by using a deep muscle scanner). In this way, along with carcase data at slaughter, the nutritionist has a good idea

of the grow/finish herd's lean accretion curve and can design a farm-specific diet or diets to satisfy it. Done twice a year as routine, or if the disease picture changes markedly (Table 9).

**Table 9. Challenge or test feeding concept to match nutrient-intake to immune status**

*Method*

| | |
|---|---|
| 1. | 50 representative pigs 25 kg – 105 kg |
| 2. | Fed non-limiting diet |
| 3. | Weighed every 14 days |
| 4. | Ultrasonic test every 14 days |
| 5. | Results sent to nutritionist |
| 6. | Lysine accretion curve calculated |
| 7. | Least cost diet designed to match it – Farm Specific Diet (FSD) |
| 8. | Done x 2 per year, Summer/Winter or if the disease picture alters abruptly. |

Snags:– In the past the scanner has been expensive so it has been really the province of a feed manufacturer who can organise it and whose clients have a computerised wet feeding facility. Much cheaper scanners are on the horizon so the concept moves up a gear towards being farm-feasible. Why wet feeding? Because with this equipment any variety of diet can be made on-farm from only two (or three) basic formulae. This cuts down a custom mix inventory drastically; in fact one feed compounder known to me has, across two years, reduced his normal pig diet list by 50% despite increasing his custom mix clientele by 60% – and his pig business by 300%! He also dispensed with several feed reps as he had no need for selling on price – the predicted lean gain curve dictated the price (Table 10).

**Table 10. Financial & physical performance benefits from using lean gain feeding vs conventional formulation methods**

| | *Conventionally-formulated grower/finisher for all clients* | *Diets designed specifically for the lean gain and appetite potentials of the genotype used* | |
|---|---|---|---|
| ***Physical Performance*** | | | |
| Deadweight FCR | 2.97 | 2.87 | 3.36% better |
| Av. daily lwt gain – g | 786 | 846 | 7.63% better |
| Av. daily saleable carcase gain* – g | 581 | 660 | 13.6% better |
| P2 Backfat – mm | 12.1 | 11.6 | |
| ***Financial Performance*** | | | |
| Av. cost/tonne of feed | 100 | 107 | 7% more |
| Margin over feed cost | 100 | 116 | 16% more |
| Nett return | 100 | 121 | 21% more |

\* assuming pigs started with 80% saleable carcase wt at 25 kg

Source : Farm Trials (2000-2004)

Is this the shape of the future? Could be. With pig units getting larger, the feed manufacturing trade is moving steadily towards farm-specific diets, so to take a further step forward to encompass challenge feeding is not a substantial one.

3.    **Measuring growth rate**. At present – and it is early days yet – there could be a possible correlation (linkage) between growth rate and immune activation. Measuring growth rate accurately is something the producer can do if he sets his mind to it, so the idea looks workable, is farmer-friendly – and doesn't cost much!

Snags? Of course, there are other things besides immune stimulation which can easily affect daily gain. Temperature, stress, feeder management, overcrowding, water, wet/dry feeding and so on. What we need is confirmation from research that this potentially simple and workable guideline is indeed a viable option.

# MATCHING DIETS TO IMMUNE RESPONSE – A NUTRITIONIST'S HEADACHE

Most commercial nutritionists at the present time view the subject as a "nightmare", to quote a leading European formulator. When pigs encounter pathogenic challenge, cytokines (a type of protein chemical messenger) are released which reprogram the animal's metabolism to divert nutrients away from growth, especially lean growth, in order to ensure the immune process is prioritised. Cytokines alter nutrient intake and utilisation which – first headache for the nutritionist – need to be compensated for in order to lessen the damage to productivity.

At the same time – second headache – metabolic changes are occurring which both increase and decrease nutrient requirements. Fever places demands on energy and while the consequences of fever – reduced activity and more sleep – lessens it; a reduction in growth rate lowers the demand still further. On top of this, appetite reduces when immune response is high, even if the animal feels healthy enough.

I quote Paul Toplis, a leading European commercial pig nutritionist responsible for making sense of it all:–

> *"Appetite changes can be unpredictable. For example, if a healthy growing pig with an appetite of 1.5 kg a day requires 15g of lysine then the diet specification for lysine should be set at 1.0% (10g/kg). Now if this pig encounters an immune challenge its lysine requirement might fall to 14, 13 or perhaps 12g per day and the feed intake might fall to 1.4, 1.2 or 1.0 kg per day, giving the nutritionist nine possible diet specifications to work with."*

Toplis (1999)

# BUT DOES IT MATTER ?

An understandable question from the producer. Yes, it could well do. Taking the variables Toplis quotes and the reduction in performance quoted in the Iowa State results, even for the less extreme differences, under current UK prices at March 2010 if you *underachieve* immune demand intake this could reduce saleable meat sold per tonne of feed fed by slaughter by 11.6 kg, and also incur 12 days longer to slaughter in overhead costs almost as costly again now that good housing is so expensive. But if you *overachieve* the immune demand this might cause you to pay an unnecessary 8% more for your food as the pig won't use all of it, and will just excrete the overage – another extra cost in more slurry disposal.

# CAN FEED ADDITIVES HELP WITH IMMUNITY ?

## Zinc

Zinc supplementation has long been recommended by the medical profession, particularly for the older human patient, to help bolster their immune defences. In animals it is known that zinc plays a critical role in both reproduction and immune-competence, but unfortunately there are no clear guidelines as to the optimum requirement for the latter. The levels are likely to be considerably higher than the requirement for growth. In terms of the immunocompetence of the animal, zinc has a positive effect on both the immune response to pathogens and the prevention of disease by maintaining healthy epithelial tissue (epithelial = tissues involving many varieties of cells, in zinc's case those deterring or delaying invasion by pathogens).

So if the zinc needs are higher to assist immunocompetence – how much higher? I don't think we know yet, not fully. What could be an important lead ('breakthrough' is a too dramatic term to use yet) is the way proteinated or 'bioplexed' trace elements are better used by the animal. Let me try to explain it in layman's language at the risk of over-simplifying a complex metabolic pathway – academics please bear with me!

## The Bioplex Concept

In the case of zinc, the mineral is linked to an amino-acid, in this case methionine, which 'tows' it through the point of absorption (for methionine) in the intestine, which also happens to be where zinc is absorbed. Result, more zinc is absorbed so less is needed in the diet. Such linkages care called bioplexes. Therefore, less is excreted as unused by the pig and pasture contamination and watercourse pollution due to small but prolonged soil build-up is reduced substantially (amino acid-linked trace minerals are also called 'proteinates').

Is the zinc, when more of the bioplexed form is absorbed, better used for both production and immune status? It seems so, although rather more evidence has accumulated to date on the productivity side than on the immunocompetence area, which is not surprising as it is much more difficult to measure.

The experts still seem to be a little bit undecided whether all zinc should be derived from the bioplexed form, or whether just some of it, or even a good proportion of it. Meanwhile follow their advice, which at the time of writing seems to be moving towards total replacement away from conventional inorganic sources.

What we are certain of is that more **nutrients** are needed when disease starts to challenge. For example in poultry, some 1% of all nutritional needs are used to maintain a normal immune level, while this rises to 7% when disease activates the bird's immune defences. American work hints that this could be even more in the case of the pig, especially high lean gain genotypes. Anyway, using a bioplexed form of zinc either in whole or in part seems to be a good idea. It looks – at the dose rates advised – as if it won't harm anything i.e. through over-availability, and it could do a lot of good. REOs of between 5:1 and 22:1 for either bioplexed, organic zinc, iron or copper have been obtained as the inclusion costs/tonne are not excessive.

## Oligosaccharides - the new subject of Glycomics

Oligosaccharides are simple sugars derived from brewing by-products (fructo-oligosaccharides or FOS) or yeast manufacture (mannan-oligosaccharides or MOS).

Both these additives have come into prominence now that antibiotic growth promoters are being increasingly banned. Being natural and safe by-products (sugars), they provide a useful and cost-effective alternative.

Originally they were thought to work mainly by competitive exclusion of gut pathogens on the gut wall. They do, but in the case of MOS, other more complex mechanisms seem to be at work.

Again at the risk of oversimplification, the latest evidence so far seems to be that Biomos – the bestselling source of MOS – is getting results which are surprisingly good with some disorders and could be unlikely to do so solely due to its proven 'capturing' effect of pathogenic bacteria and holding them fast until gut peristalsis (peristalsis = wavelike movement on down the gut) removes them out of harm's way to the outside of the digestive tract in the faeces. Something else may be at work – could this be a strengthening of immunocompetence?

MOS appears to enhance immune function in a variety of ways ...

- Oregon State University reported a 25% increase in secretory IgA.

- Researchers have found that MOS enhances macrophage response.

- Other workers find that in germ-free pigs MOS influences both humoral and cellular (B cells and T cells) immune systems, although the levels measured seem to be widely different. (For definitions of these technical terms, see Table 11).

**Table 11. So you don't get lost ... Some immunological terms**

---

**Humoral immunity = B cells, lymphocytes**
> Memory cells which remain behind after an infection, recognize the reappearance of the pathogen and quickly call up the correct defences again.

**Cellular immunity = T cells**
> These stand guard against pathogen challenge, are limited to body cells in various tissues susceptible to pathogen ingress.

**Systemic or mucosal immunity**
> Local humoral or cellular antibodies ideally present when body surfaces are exposed to the outside – nose, throat, gut, outer reproductive tract.

**Active immunity**
> After exposure to infection, stimulated antibodies remain in the sow which are transferred to the offspring via colostrum for a while in the form of antibodies IgA, IgG, IgE, IgM etc. The dam is **active** in passing on the immunity.

**Passive immunity**
> The piglets accept the antibodies (i.e. are **passive**) and this lasts as long as the maternal antibodies survive. As no memory cells (lymphocytes) are provided or formed so the immunity is not permanent.

**Acquired immunity**
> After a pig recovers from disease or vaccination it develops acquired immunity.

**Antigens**
> Foreign material which triggers the body's defence mechanism – pathogens or vaccines.

**Antibodies**
> Protein structures (IgA, IgM, etc) which fight hostile antigens and unless overwhelmed, prevent disease

**Phagocytes**
> Cells which ingest and so destroy pathogens.

**Macrophages (white bloodcells)**
> Large immobile cells, usually originating in bone marrow, which become actively mobile when stimulated by inflammation, immune reactors and microbial products.

**Cytokines**
> Messenger proteins which control macrophages and lymphocytes.

**Immunosuppression**

When an immune system is not working properly because of dirty conditions, overcrowding, a poor diet, pre-existing disease, stress and mycotoxin presence, etc. some of the newer viruses seem to carry their own immunosuppressive capabilities, which is giving them a head start at present.

**Titre (titer)**

A numerical measure or test of a pig's immunity. An antibody titer measures how much antibody is in the pig's blood. Expressed as 1 followed by a number. For example if one volume of blood was diluted with 64 volumes of saline solution and antibodies were still detectable the titer is 1:64. The higher the number after 1 the more antibodies are present and stronger the immunity present.

**Serology**

The expression of antigen:antibody reactions by laboratory test.

**Inflammation**

A localised protective response caused by injury, destruction of tissues or injected poisons (e.g. insect bites) to block off, or destroy or dilute the injurious agent and protect the affected tissue.

---

So… Biomos seems to facilitate the complex interactions of all these disease-fighting substances. This is called immunomodulation (the effect on immune response). Much still remains to be discovered, but already research suggests Biomos helps resist infection from E coli, campylobacter and salmonella, so the beneficial effects of this useful alternative to antibiotic growth enhancers can now, with confidence from a growing number of such trials, be added to the original benefits from the inhibiting effect of 'pathogen-capture' under which banner Biomos was first announced more than 5 years ago.

# REFERENCES

Close W H & Cole D J A *'Nutrition of Sows & Boars'* Nottingham University Press (2000) p 359.

Gadd, J (2005) Stomach Tubing. In: *What the Textbooks Don't Tell You.* Nottingham University Press pps159-162.

Molitor T (1992) 'Immunization' Nat. Hog Farmer Blueprint Series, Spring 1992 p 28.

Sauber T E, T S Stahly and B J Nonnecke,(1999) *J Anim. Sci* **77**: 1985-1993.

Stahly T S and D R Cook (1996) ISU Swine Research Report, Iowa State University Ames, ASL-R 1373 pps 38-41.

Stahly T S, S G Swenson, D R Zimmerman and N H Williams (1994a) ISU Swine Research Report, Iowa State University Ames, AS-629, pps 3-5.

Toplis P (1999) 'Interactions Between Health & Nutrition' Procs JSR Genetics 10th Annual Technical Conference p2

Williams N H and T S Stahly (1994) ISU Swine Research Report, Iowa State University Ames, AS-633, pps 31-34.

Williams N H, T S Stahly and D R Zimmerman (1994) *J Anim. Sci.* **72** (Suppl 2):57.

# 6

# STRESS EXPLAINED AND HOW IT AFFECTS PROFIT

**Stress :** The total of all the biological reactions of an organism (in our case pigs) to any adverse stimuli. These can be physical, mental or emotional – those which disturb the smooth functioning (stability) of the pig's metabolism.

**Stressor :** Any individual factor or action (collectively called stimuli) which disturbs this stability. There is a long list of known stressors, and there may be more to be discovered due to physiological intereactions.

**Strain :** The effect on the animal's physiology of a stressor e.g. affecting various organs, neural pathways etc. Stress and stressors are factors <u>outside</u> the pig which create strain <u>inside</u> it.

It is a reflection of how little we know about stress that this section is not one of the longest in this book. It should be, as I am convinced that stress is certainly one of the most important subjects for the pig producer to be aware of, not only affecting the welfare of his pigs (and ourselves as their guardians) but his profits, too.

## STRESS AND STRAIN

Why define the difference? Because first, we need to keep in our minds the factors which could cause stress in pigs and thus lower the likelihood of resultant strain affecting productivity. And second, recognise the signs of strain in the pig and alleviate them should they appear. But there is another definition we need to deal with . . .

## STIMULATION

Stimulation? Now that's a beneficial word - so what has it got to do with the other two which are just the opposite? **Because farmers confuse stimulation with strain**. Stimulation (of appetite, sex, rooting and exploring, nesting, boar presence and suckling) - all of these and more are natural and essential processes leading towards better performance.

For example, that awful racket created by gilts approaching puberty - is that anticipatory excitement, or (I find all too often) their protest at bullying and aggression brought on by overcrowded conditions? Again - growing pigs - that painful peak-decibel clamour before feeding! Is it joyful anticipation of food arriving or a protest about recurring hunger which has been present for too long? The sound is different. Again, away from feeding time, the vocalisation of the hunger 'grizzle' - a lower register and more muted, is different from the higher, more urgent sound of thirst. Quite a few times when walking through a barn with the attendant they have remarked "They are a bit hungry - we haven't fed them yet" "No", I replied "that's the thirst sound, go in and check the drinker(s)".

**Stimulation or strain?** Sometimes it is impossible to tell. Those stalled dry sows chewing the bars - is it frustration (strain) or stimulation? Could it only be something to pass the time just as we do of an evening watching television? The corticosteroid levels (one approximate test of strain) of a bar-chewer can be no higher than one not doing it.

## Now for a bit of science

Simplified, as stress is a very technical subject, nevertheless, it will help you (and your pigs!) if you understand at least some of what happens.

Different types of stressors cause different types of physiological (within the body) reactions, but they all show similar biological measures separated into two main categories (Figure 1).

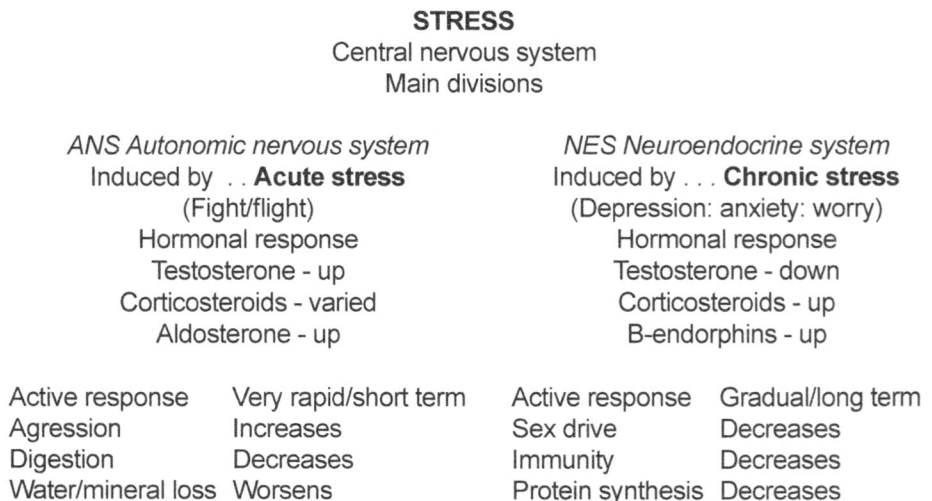

**STRESS**
Central nervous system
Main divisions

| *ANS Autonomic nervous system* | | *NES Neuroendocrine system* | |
|---|---|---|---|
| Induced by . . **Acute stress** | | Induced by . . . **Chronic stress** | |
| (Fight/flight) | | (Depression: anxiety: worry) | |
| Hormonal response | | Hormonal response | |
| Testosterone - up | | Testosterone - down | |
| Corticosteroids - varied | | Corticosteroids - up | |
| Aldosterone - up | | B-endorphins - up | |
| | | | |
| Active response | Very rapid/short term | Active response | Gradual/long term |
| Agression | Increases | Sex drive | Decreases |
| Digestion | Decreases | Immunity | Decreases |
| Water/mineral loss | Worsens | Protein synthesis | Decreases |

**Figure 1.** Extrapolated from Airey (1991)

Information about a stressor reaches the brain which then tells the body how to respond within a millisecond - so fast that it is one of the wonders of nature especially as there are hundreds of sources of stress for pigs and the number of different responses which they evoke are probably just as many.

The two main categories waiting in the brain's 'computer' are 'flight/fear' and 'depression, anxiety, worry'. Two very different areas of strain to which the brain has to respond. It does this either by sending signals down the autonomic (autonomic = not subject to the body's control) nervous system, called ANS. Alternatively it releases hormones from the neuro-endocrine system, NES which is largely under the body's control. Figure 1 is a very simplified description of what happens.

## What does the ANS do?

When the stressor presents a sudden threat the brain quickly activates the ANS. The immediate effect is to increase the availability of energy, increase heart and respiratory rate and put digestion on hold - all by hormonal activity. The pig can then either fight or flee. Once the threat is deemed over the activating hormones quite quickly reduce and disappear, so the effect can be of short duration - unless the fear reflex is re-activated of course.

## What does the NES do?

However, if the pig thinks that it cannot respond positively to the situation, such as an uncomfortable floor or too many companions squashed in with it, the brain activates the NES. This affects the release of other hormones involved with more long-term (what the scientist calls 'chronic') resistance to stress.

The serious aspect of NES activation is that it affects the organs in the body which deal with growth, especially protein formation; as well as the body's immune system which defends it against disease; and yet again those metabolic pathways which control reproduction. These are the most important strain effects caused by stress and why the two terms, stress and strain, need to be different.

## NES activation is by far the main problem

It can last a long time and is not nearly so evident as in the 'fight/flee' response of the ANS.

**Growth.** In the event of the NES being activated, protein synthesis (formation) is damaged and water mobilisation interrupted. Because meat is mainly protein and water, food conversion is quickly worsened as less meat is formed from the food and water consumed. FCR goes up and stays up until the stressor/s is/are removed. Those stressors responsible are primarily and all too commonly - temperature, discomfort, overcrowding, disease challenge and a host of others right down to uncaring and overworked stockpeople, see Table 1.

**Immunity.** I am sure from long experience of comparing the level of stressors and disease incidence on farms in over 30 countries, that stress does lower the immune shield and is also the one effect of NES activation least recognised by farmers. One important pathway is that the corticosteroid and endorphin hormones released by NES (see Figure 1) reduce the number of protective white blood cells which do the vital job of engulfing pathogens - it is as simple as that, I feel. No need for involved scientific explanations.

**Reproduction.** Also vulnerable to a wide variety of stressors. The phases which are controlled by NES activation, such as ovulation, and the implantation of the embryo are particularly vulnerable to the strains caused by stress of incorrect nutrition, lack of rest and quiet after service, and unfavourable conditions at farrowing.

Indeed, the main activation of NES seems to occur when females are getting pregnant and when giving birth. The gestation period gives far less trouble and could be a reason why the bar-chewing in stalled sows which worries people so much may not be all that important a stressor after all. Potentially far more serious is aggression around EFS (Electronic Feeding System) stations as grouped dry sows wait to enter them. This frustration of the more timid sows and grumpiness of those who want them out of their way can activate the ANS as well as providing a permanent worry for the less-dominant sow getting hungrier with the feeling she is never going to get a look-in, thus activating her NES as well.

Also why grouped dry sows on solid floors with no/minimal bedding need a quite different layout to those bedded in deep straw. We are learning a lot about keeping sows in groups (to meet the Welfare requirements here in Europe increasingly demanded by many of our customers for pig meat) and how best to keep them placid and contented. Keeping sows in groups is not easy as a great of it involves avoiding combative stress, however being able to use/afford ample bedding is a considerable palliative.

## SO WHAT DOES STRESS COST IN PERFORMANCE TERMS?

There is very little information available. I only wish I had collected before-and-after evidence on stress on the many farms I have visited (as I have done for more easy-to-measure subjects) where the results of stress alleviation might have been measured. Producers who have been persuaded to follow my stress audit checklist have certainly noticed improved performance, and in Table 1 I give a very approximate *estimate* of what this might be – from 35 years of the audit's use.

**Table 1. Conjectural figures postulated for what stressors of all varieties may cost a competent breeder/feeder pig producer in, say, the top third of his nation's performance tables\***

| | |
|---|---|
| *Breeding unit* | 3 fewer pigs weaned/sow year (Comprised of 6 more annual empty days; 2% more pre-weaning mortality, 8% greater sow replacement rate, and 10% fewer born alives)<br>**Plus** 12% more vet/med. costs, 2% more housing costs and 1% more labour costs |
| *Growing Finishing unit* | Food conversion worse by 0.15<br>Daily gain 7-100 kg lower by 30g<br>M.T.F. lower by 19 kg |

Plus general overheads increased by 3%

\* This is purely the author's assessment of improved performance when the producer was persuaded to concentrate on stress relief alone as a permanent part of his daily management routine. Even so it should be viewed with caution until research confirms these suggestions.

There are further costings on the return for extra time and care in reducing stress at the end of this chapter.

At farm level there are three areas we need to consider:

1.  *Anticipation*  What stressors are we likely to see in and around the pigs?

2.  *Observation*  What responses are the animals making to these adverse stimuli?

3.  *Action*  What action can we or must we or should we carry out within the bounds of cost and feasibility so as to alleviate the stress?

Since I trawled the literature several years ago to help compile Table 1, I have arranged a bit of trial work myself and found some further published information.

## STRESS AND OVERSTOCKING

Many - and I am tempted to say - most pigs are overcrowded on nearly every farm I visit. I see it everywhere on my farm call-outs, and so convinced I was some years ago of the damage it was doing to profit, that with the co-ordination of three producers we did three trials to see what might happen. This work will be found in the stocking density section, but for convenience I repeat it here.

We compared deliberately-overstocked pens of growers by 15% (two pigs in pens of 14 when there should be 12) with the current stocking density advice and recorded performance (Table 2).

**Table 2. Likely costs incurred by overstocking a nursery and finishing house by 15%.**

| | Pigs 6-35 kg | | Pigs 36-100 kg | |
|---|---|---|---|---|
| | *Correct density* | *+15%* | *Correct density* | *+15%* |
| Daily gain (g) | 518 | 480 | 844 | 848 |
| Days in pen | 56 | 60 | 77 | 77 |
| Overhead costs at 24p/day (£) | 13.44 | 14.40 | 18.46 | 18.46 |
| FCR | 2.02 | 2.12 | 2.42 | 2.63 |
| Total food eaten in period (kg) | 58.6 | 61.5 | 156.7 | 171.0 |
| Total food costs, p/kg (£) | 11.13 | 11.69 | 27.53 | 29.93 |
| Extra costs per pig (£) | | 1.85 | plus 2.4 | Total 4.20 |

Savings in 15% less housing costs per pig (at £8.20/pig) was a saving of £1.23/pig
Thus final cost was £4.20 less £1.23 = £2.97/pig, giving an REO of 3.4:1

**Conclusion:** The average payback from deliberately destocking to guideline levels on all three farms was well over 3 to 1. Even with healthy pigs as on these farms it just does not pay to overstock. REO = Return on Extra Outlay.

## Overcrowding gilts

So much for growers. The modern gilt, that animal which holds out so much promise as the lynch-pin of future productivity, is often overcrowded prepubertally and before service, made worse by poor matching of animals within the group when bigger, faster-grown gilts bully the submissives. This raises ANS in the bullies and both NES and ANS in the bullied, both of which stress pathways can affect ovulation and conception **in both categories** I am told (the bullied much worse as is to be expected). To what extent?

Regrettably, I have no neat and tidy trial figures similar to report as in Table 1 - only from 12 farms who were persuaded to lessen the obvious overcrowding in their rather narrow gilt pens holding about 6 to 8 gilts from an average of 1.5 to 1.8m² per animal, then changing to gilt pools with a more generous 2.8 to 3.0m²/animal in more square pens of 10-15's. Returns were down by a mean of 13.6% (range 0-21%) birthweights up by 200g and on five units, litter size better by as much as 1.8 pigs on an original litter size of 10.1. These results impressed the producers, but how much of this was due to the more generous space or to the 'gilt pool' effect, where the novelty leads to better supervision I do not know.

Anyway, as so many of you do - **do not overcrowd your gilts.** They are delicate future breeding machines, not finishers destined for the knife!

# ADVISED GILT SPACE ALLOWANCES SEEM TOO LOW!

While there are clear spatial guidelines for gilts laid down in the Welfare Codes, the currently advised 1.64m²/gilt is in my view too low. I would advise 2.6m² and maybe even more for these large ladies we are now holding until first service at 135 kg. Look what happened among those 12 farms when the space constriction was relaxed. Yes it cost a good deal more, but over a normal gilt housing amortization period of say, 12 years, not nearly so much as the costs in Table 1.

## STRESS AND MIXING PIGS

Any form of mixing is potentially stressful - but how much so? The ANS is often massively activated.

**Table 3. Effect of splitting litters after weaning**

|  | *Growth (g/day) 20 days after weaning* |
|---|---|
| Litters mixed after weaning | 240 |
| Kept together as litter groups | 350 |

Varley (2001)

**Table 4. Skill in batching and matching at weaning**

|  | *Days to slaughter* | *MTF kg* | *Weight range* |
|---|---|---|---|
| Litters mixed 'any old how' at weaning | 150 | 237 | 6.1-101 kg |
| Litters carefully batched and matched | 148 | 248 | 6.3-100 kg |

MTF = Saleable Meat sold per Tonne of Feed Based on killed out percent
Author's records (2008)

**Comment:** An interesting result. Amateurish mixing didn't seem to affect overall live growth rate much but significantly reduced the amount of lean meat sold. The extra MTF was worth an equivalent to all food from weaning to slaughter **being 10.2% cheaper.** Worth getting those stockpeople trained in the skills of batching and matching properly - whether you mix litters or not. Also, always relate growth rate and FCR to MTF, as MTF teaches **you** a lot about value for money, as in this case!

### Mixing just before shipping

We all know this should not be done, but producers succumb when a bulge in production occurs and space is urgently needed for the next input. What damage it can do was picked up from this example where the farmer very kindly - and generously - tried it out for me.

**Table 5. Enforced mixing of pigs from 10 days before shipping suddenly slows growth rate considerably**

Pigs varying in weight by 10.8 kg (av. wt. 82.1 kg) growing at an average of 760g/day were mixed from 4 pens into one pen of 15 pigs (with adequate stocking density) until average shipping weight of 92.2 kg was reached. These were compared to pigs of similar weight which remained in their pen groups until shipping.

|  | *Mixed pens* | *Unmixed pens* |  |
| --- | --- | --- | --- |
| Av. daily gain, g | 696 | 805 | (+13.5%) |
| Av. feed intake per day kg | 2.05 | 2.21 | (+4.87%) |
| Av. FCR 25.1 to 92.1 kg) | 2.94:1 | 2.73:1 | (7.7% better) |
| MTF (25.1 - 92 kg) | 259 | 278 | |

Client's records (2003)

**Comment:** This shows clearly how enormously costly mixing pigs before shipping can be. In this case just 13 days before shipping weight, 19 kg less saleable meat per tonne fed was forfeited due to the ANS stressors affecting protein formation and water/mineral balance, both major components of lean meat. Far cheaper to put up with the lower income from sending-on surplus pigs (called 'topping') as underweights so as to free up the space - about 2.8 times cheaper in this example as 2010 prices.

# STRESS AFFECTS IMPLANTATION

Implantation is another major stress-susceptible area. The female needs rest and quiet during the period when the fertilised eggs implant themselves on the womb wall. Failure to provide these restful surroundings, very common in stalled sows where the recently-served are all mixed up with those awaiting service and those later in gestation. Newly-served gilts are especially vulnerable, litter size reduced by 0.2 to 1.8 born-alives, birthweights 200g lighter and more uneven litters showing in such noisy, workaday circumstances.

# TWO THEORIES

This damage to productivity is thought to be due to stressors affecting the regeneration of the womb wall (the endometrium) after farrowing, especially the first time - the 'gilts' litter. One concept is that due to certain stressors parts of it become less receptive early on to the embryos trying to attach, so they migrate to areas which are not so affected and establish a 'hold' there. they therefore tend to 'clump' (my description) and the weakest cannot find a place at all so are lost (result - a smaller litter). Or if they can get a space, they are crowded out by those which arrived earlier (result - more smaller newborns/uneven litters).

This situation is less favoured by other workers who are more concerned about the speed of release of the newly forming embryos to engage in the implantation process. Stress can delay some embryos from arriving, so these latecomers also find enough room to expand difficult to secure - ending up as runts. The earlier arrivals have 'bagged' the best camp sites!

I don't mind too much which theory is correct as it is stress which causes the trouble in both cases. Figures 2 and 3 illustrate the two situations.

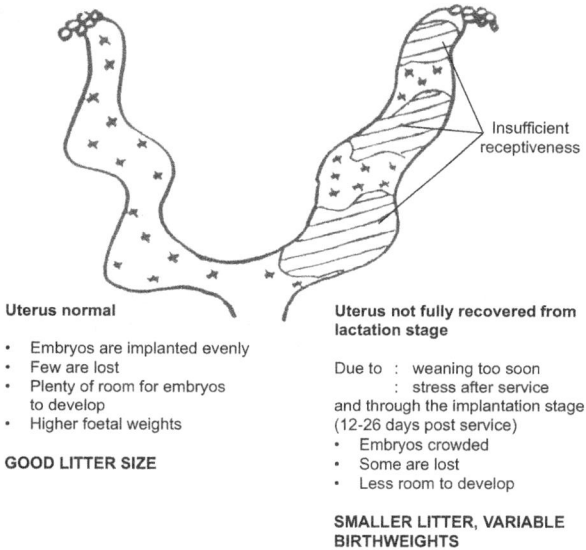

Insufficient receptiveness

**Uterus normal**

- Embryos are implanted evenly
- Few are lost
- Plenty of room for embryos to develop
- Higher foetal weights

**GOOD LITTER SIZE**

**Uterus not fully recovered from lactation stage**

Due to : weaning too soon
: stress after service
and through the implantation stage
(12-26 days post service)
- Embryos crowded
- Some are lost
- Less room to develop

**SMALLER LITTER, VARIABLE BIRTHWEIGHTS**

**Figure 2.** The current theory, may be flawed?
Uterine condition may affect birthweights.

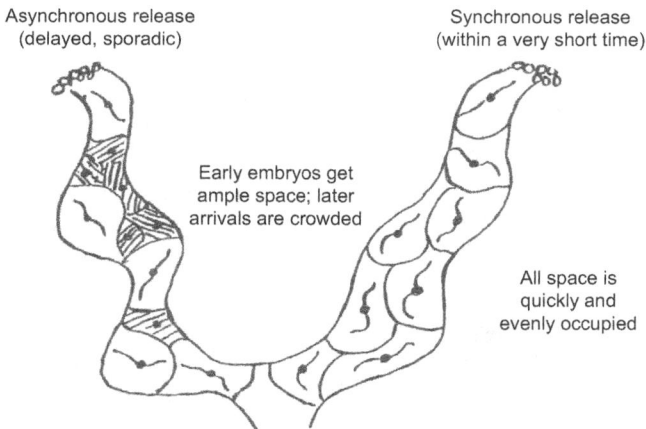

Asynchronous release
(delayed, sporadic)

Synchronous release
(within a very short time)

Early embryos get
ample space; later
arrivals are crowded

All space is
quickly and
evenly occupied

**Figure 3.** The latest proposition.

# FARROWING AND STRESS

The third major area affecting costs. We all know that making the sow as comfortable as possible is important at this time and appreciate what is needed, but the area below is not so well-known - or practiced.

## Being there at farrowing

I've long been a convert, having been there on night shift in my early days. After all, we learned that we lost one piglet from born-alives in the first day from farrowing (53% of these in the first 12 hours, no less) then only half a piglet more in the next 20 days to weaning. If only I could get that one piglet lost in the first 24 hours down to half a piglet, we were not far off the magic 6% target lost to weaning. The night shift did it and the extra income from over 1000 more pigs sold from our 1100 sows in a year not only paid for the overtime but got me a handsome bonus as well.

Table 6 shows the advantages when later in my career I promoted the idea to some of my clients. Figures in the table taken from the records.

**Table 6. Attended versus non-attended farrowing. 3 weaner-producer trials.**

|  | Attended | | | Non-attended | | |
|---|---|---|---|---|---|---|
| Trial No: | 1 | 2 | 3 | 1 | 2 | 3 |
| No. of litters | 78 | 176 | 130 | 40 | 87 | 126 |
| Estimated attended farrowings (%) | 85 | 80 | 100 | 15 | 20 | 0 |
| Total born per litter | 11.30 | 11.72 | 11.21 | 10.50 | 11.51 | 11.38 |
| Born alive per litter | 10.66 | 10.81 | 10.21 | 10.00 | 10.67 | 10.12 |
| Birthweights (kg) | 1.40 | 1.26 | 1.20 | 1.45 | 1.31 | 1.10 |
| No. weaned per litter | 9.91 | 10.10 | 9.82 | 9.12 | 9.64 | 9.01 |
| Mortality to weaning at 23 days (=/- 1.5 days) % | 7.00 | 6.80 | 6.20 | 8.80 | 9.70 | 11.00 |

After adopting the technique - extra weaners sold per 100 sows per year

| (Figures supplied by the farmers) | 185 | 108 | 190 | - | - | - |
|---|---|---|---|---|---|---|

**Comment:** Trials 1 and 2 were before and after trials, but trial 3 is interesting because it was a concurrent trail using two shifts of the same staff on a rota basis, so feeding, environmental temperatures, etc were exactly the same. Attended farrowings not only give you more weaners to sell but also reduce stress in those sows in difficulty (or just slow to farrow). I wish I'd recorded sow performance in the next parity as well, but the figures supplied a year later by the participants give a clue to what happened.

# THE STRESS AUDIT

Doing a stress audit periodically is a good idea.

**Table 7.  A checklist of natural stressors under today's conditions**

|  |  | ANS | NES |
|---|---|---|---|
| The small pig | Birth | ✓ |  |
|  | Pathogens |  | ✓✳ |
|  | Establishing itself/competition | ✓✳ | ✓✳ |
|  | Temperature/cold |  | ✓✳ |
|  | Thirst |  | ✓✳ |
|  | Weaning/re-establishing itself | ✓ |  |
| The farrowed sow | Parturition | ✓ |  |
|  | Lactation/water availability |  | ✓✳ |
|  | Temperature/hot | ✓✳ |  |
|  | Comfort |  | ✓✳ |
|  | Weaning | ✓ | ✓ |
|  | Parasites |  | ✓✳ |
| The pregnant sow | Ovulation        Helped by the |  | ✓✳ |
|  | Implantation    'Feel Good Factor' |  | ✓✳ |
|  | Confinement/comfort/boredom |  | ✓✳ |
|  | Competition (groups) | ✓ | ✓✳ |
|  | Gutfill/fibre |  | ✓✳ |
|  | Temperature/cold |  | ✓✳ |
|  | Parasites |  | ✓✳ |
|  | Legs / floors |  | ✓ |
| The gilt | Onset of puberty |  | ✓ |
|  | Competition/bullying | ✓✳ |  |
|  | Poor light |  | ✓✳ |
|  | Space |  | ✓✳ |
| The boar | Temperature / hot and cold |  | ✓✳ |
|  | Lack of Exercise |  | ✓✳ |
|  | Frustration |  | ✓ |
|  | Gutfill |  | ✓✳ |
|  | Boredom |  | ✓✳ |
| The grower/finisher | Temperature/variation/diurnal |  | ✓✳ |
|  | Space |  | ✓✳ |
|  | Pathogens | ✓✳ |  |
|  | Food and water access |  | ✓✳ |
|  | Sleep adequacy |  | ✓ |
|  | Boredom/satisfying activity needs |  | ✓✳ |
|  | Transport/lairage/handling | ✓✳ |  |

✳=Stressor under your control

Table 7 can be used as a basis for a stress audit.

# A FEW EXPERIENCES FROM 40 YEARS OF STRESS AUDITING

- *Do not wear white overalls*. Wear green or dark blue. Keep quiet, move slowly.

- *Do it quietly!* Observe the pigs under their normal behaviour patterns. Open nursery doors an inch, and listen before switching on the light and/or entering. Listen to the breathing, for restlessness, wheezing, 'snittering' (light, irritant sneezing). The same in the farrowing house and in the grower houses.

- *Observe them unawares.* Still with the door slightly open, switch on the light if needed and look for the resting pattern, huddling (piling) and where they are lying relative to air placement. Do not enter until you have looked at as many areas as possible before the pigs disturb themselves. A good time to detect problems is last thing at night with a torch.

- *Observe them stirred up.* Enter the room and immediately move quietly along its length breathing the atmosphere for gases and checking temperature. At the same time look for stiffness/lameness/reluctance to move and listen for laboured respiration/coughing.

- *In nurseries and farrowing rooms*, look for piglets which are lying awkwardly/ lifting their undersides in the semi-sternum posture. This is advance warning of digestive upsets.

- *In lidded kennels etc*, try to get a quiet peep inside but ensure all occupants are eventually 'banged-out'. Scrutinise those last to leave carefully.

- *Use the farrowing crate to palpate the sow's udders*, feeling for unusual conditions/discomfort.

- *Check hoppers and troughs*, for stale food, contamination, cleanliness. Check evidence of wastage. Are pigs 'nosing' the food in the troughs? A sign of nutritional dissatisfaction from hunger to unpalatability.

- *Get as many stalled sows* to stand as you can. Condition score those that do and check legs/sores/rubbing (see Lameness Checklist). Check water adequacy *i.e.* colour of urine. Check for discharges.

- *For group-housed sows*, get them to move past you and study gait and alertness. Check their water supply. Check too much queuing disturbance at ESF feeders.

- *It is a good idea* to do an audit with a neighbour present occasionally and you do it for him likewise.

There are many more tips than this of course, embodied in Table 7, but the above will give you a flavour of what a stress audit does.

**Make time (without distractions as it needs good observation - you need to CONCENTRATE) to do a stress audit every 2 months.**

# MEASURING AND MONITORING VITAL SIGNS

A stress audit cannot be carried out to its full potential unless you are able to check the physical constraints which you have imposed on the animals in your care. These are:-

- **Temperature.** *(Common weakness)*. Thermometers should be properly positioned as close to the pigs height without being damaged. At least two are needed per house located at a 'neutral' spot on the floor plan *i.e.* out of extremes of ventilation pattern. While auto-recording is best as this gives day/night variations, individual max/min thermometers are adequate as long as they are:–

  Clean; consulted frequently; calibrated to no more than a 1°C error against a BSI-rated instrument.

- **Check air movement.** *(Major weakness)* : Part of a stress audit involves checking that fan speeds, placement and operation are as designed/ intended. *A ventilation engineer should do this at least once a year*. It is remarkable how often a specialist will pick up individual major errors causing a 0.2 worsening of FCR for growing pigs in the error zone – which easily pays for his visit.

Things you can do between such visits are to have smoke tubes/phials to 'see' air movement, and to wet the back of your hand or bared arm to detect cold draughts. This idea is particularly useful to discover cold downfall draughts on to the backs of sleeping pigs close up to a wall especially at night. This can be simply and cheaply counteracted by nailing a 3 to 4 cm triangular wooden batten to deflect the air falling down close to the wall surface into the rising current of warm air rising from the sleeping pigs and so 'lose' the draughts by natural means.

There is no substitute for knowing how air moves in a piggery and the importance of correct air placement.

- **Check stocking density** *(Common weakness)* : Check, check, check that you are within the safe spatial guidelines *(see Stocking Density section)*. Stocking density quickly gets out of control, and the stress audit draws this to the attention for every pen. Fully half the farms I visit are infringing stocking density recommendations somewhere or other – and as you can see from the Stocking Density section it does matter!

- **Attend to water** *(Common weakness)* : Again, the audit must measure that not only water flow rate is up to standard, but the *ease of access* is adequate.

  Examples are: – only one drinker in a pen; no use of height-adjustable fixings; bite drinkers in a farrowing crate, not troughs/bowls; drinkers set too low; no separate, additional water-only points in a wet-feed system or wet/dry troughed pen; siting a drinker in a corner; dry sows in yards watered from bite drinkers, not a trough.  All are common errors which raise stress.

- **Check that any manipulable materials e.g., straw, compost etc are still adequate and not fouled.** Also that they are still using any toys (balls, piping etc) as they soon get bored with them. Change the plaything if so.

- **Monitor flooring and bedding** (*Major weakness*) : I have always been a 'bedding' man, certainly for sows and young pigs.  However 80% of the farms I visit outside Sweden and Britain use little or no bedding for economic or logistical reasons. This means that correct floor design is paramount. You need to check that many in-contact areas are not too small for tiny feet in the interests of cleanliness. If so, provide a temporary solid comfort-board for tender-footed weaners so that they can at least get on to a solid area as a respite.

  Maintenance of floor quality (slipperiness, gaps/holes, roughness) must all be checked during the stress audit – reference to the Lameness Section will provide more information.

  Deep-bedded (straw) yards are bound to increase in future.  Keeping them 'sweet' and free from dead, coagulated, over-fouled areas in the yards, particularly corners and pen-fronts in hot weather, needs physical aeration which is one of the hardest physical tasks I've ever done! Nevertheless there are mechanical devices to do this, free-moving aerators suspended from a monorail for sawdust can take the effort out and so encourage more frequent attention which is essential – and save on bedding as well.

- **Review conditions at implantation** (*Major weakness*): Far too many sows are stressed during the 7-28 day period post-service. They need rest and quiet, freedom from aggression or discouragement from aggressing others. They need adequate gutfill even though the nutritionist cautions against too much feed despite some of them being out of condition. Skilful use of supplementary hay, edible straw and especially dried sugar beet pulp can make a huge difference to a feeling of well-being.

They need to be kept out of draughts, warm (18°C) but the air needs to be free of gases, *e.g.* the marker gas ammonia present at under 12-15 ppm. This can easily be measured by an appropriate chemical discolouration tube.

## CAN STRESS BE MEASURED?

To a certain extent. Corticosteroids, endorphins, heart rate and blood pressure. But from your point of view, why bother? Leave these to the researchers who admit anyway that a more integrated approach is needed and that they have much to learn. For the producer, the pigs` well-being is the best source of measurement and they have many ways of telling you.

# SO - DO PIGS TALK TO YOU?

Yes, all the time! As well as vocally (hunger – a sort of low register 'grizzle' and thirst - a more urgent, demanding highly-pitched clamour) we need to listen to their body language and actions (behaviour) as well as watch for their bodily functions (bright yellow urine, constipation), the smell of scour just before it appears, a 'staring' coat, a 'knobbly' udder, any vaginal discharges and finally, abnormal behaviour. Lots of ways of communication to interpret!

So we need to keep vigilant, using our five senses (sight, sound, hearing, smell, and touch) to understand their complicated language and be ready to pick up the messages quickly.

**Too cold** – easy! Huddling (piling). A 'stary' coat – raised hairs.  However sows which may be cold in stalls are more difficult to detect (check for night-time draughts especially in end-stalls near outside doors) but easy enough when they are in groups.

**Too hot** - wrong mucking This is such a common  method of pigs talking to us that it is worth a few words here.

Many growing/finishing pens get dirty in hot weather. This is the pig's natural reaction to urinate in the `wrong` place i.e.. where it normally rests, a place to be kept dry and comfortable.  It deliberately wets this resting place in order to cool itself by the evaporation of moisture from the skin surface The remedy is to attend to the ventilation over the normal voiding area where you want it to wet, both in its adequacy and direction (see Ventilation Section) and if this action is insufficient due to excessive heat, then to start wetting the area artificially (ibid).

One vital exercise when the messing is in some pens and not in others, is to review carefully the differences in ventilation and air placement between the dirty pens and the clean ones, especially at the warmest time of day. This will tell you what is wrong.

When walking the barns I find myself doing this exercise for stockpeople many times during the summer. They all agree to having their eyes opened when I have explained on the spot how the placement of the air is all wrong and what are the solutions. It is one of the quickest ways to learn about ventilation, and it all stems from listening to what the pigs are telling you – positively shouting in this case!

When **Tailbiting,** they are talking to you again. Long experience has suggested to me that they are looking for something to do. Yes, I know that overcrowding and the tendency towards a stuffy atmosphere is a primary cause, but in cases where they were definitely not overcrowded they were still tailbiting.

Pigs are sentient beings - inquisitive and explorative - so giving them something to do has certainly helped, as we discovered with our own pigs 40 years ago. We chucked in sods of earth, paper bags from our feed mill and later, 1-metre lengths of plastic piping which they liked to push around. This is called 'environmental enrichment' - the scientists' euphemism for just giving them something to do! That it took at least 30 years for the penny to drop, as it has at last, has always amazed me.

When **Scouring** is about to commence, the pig will sometimes lie in the semi-sternum position with its lower body lifted off the ground supported by the forelegs, possibly because with a disturbed innards it feels better if it lifts the belly off the ground? Seen like this, it is time to spraymark it and start medication and not wait for scouring to appear.

Sometimes to you can notice the acrid smell of incipient scour before it actually occurs – I'm sure I have nipped an outbreak in the bud by taking advance action just in case.

**Constipation** is an involuntary means of communication. Get into the habit of noticing dung consistency – it should deform slightly as it hits the floor and be easy to compress with a light touch of the boot, if recently voided. Constipation could be caused by lack of water or digestible fibre. Another such visible sign is a distinct yellow urine in sows – lack of water again, and a slowdown in excreting body toxins which is what urine does.

**Bar chewing?** Boredom, raising stress hormone levels or just something to do? I've alleviated but never really stopped it, by giving stalled sow culprits a tad more food for a week or so with a fibre additive or a gut-filler like a little dried sugar beet pulp.

**Sluggards?** Pigs slow to 'bang out' from a hut or covered pen, or slow to come to feed are also talking to you, and need to be watched like a hawk. Something could be wrong or starting to develop outside simple sleepiness.

But this leads me on to an aspect of stockmanship which is most important and in my opinion distinguishes an experienced stockperson from the average– that of **deviation from normal behaviour.** This is important. 'normal' in this case is what you have experienced with them over time. If they seem to be behaving differently then something is either up or is shortly going to be – time to check that everything, especially things mechanical and electrical, are in working order. This is where monitoring equipment such as sensors, gauges and meters are important.

Water consumption is one measurement which can predict the onset of disease and is another form of pigs talking to you involuntarily. The environment equipment-monitoring firm Farmex have done sterling work on this recently.

# DO YOUR PIGS ACTUALLY LIKE YOU?

Not such a silly question! Researchers Paul Hemsworth and Harold Gonyou especially, but also others in the behavioural field like Temple Grandin (on movement which can cause a lot of ANS stress) have spent much of their lives studying stockperson/ pig interrelationships. Hemsworth especially has shown great ingenuity in trying to measure this scientifically and its effects on pig performance. He has quantified how 'Pleasant' treatment (gentle manner, soothing voice, slow approach) contrasts with `Adverse` handling (noisy, rapid approach, abrupt movements, use of sticks or electric goads – now rightly banned in some countries) and how this effects performance of both growing pigs and breeding females.

In growers the fall-off in daily gain seems to be 8%, the average from many trials, and the stress hormone corticosteroids raised by as much as 30%. Where breeding gilts are concerned the conception rate can be as much as 50% to 66% lower between the two, with corticosteroids up by 40%. Intriguingly some of Hemsworth's measurements are made from how long it takes a pig to approach a stockperson standing still in a pen with whom they are familiar, or are fearful of, or who is new to them. The range varies from twice as long to 14 times, with some fearful groups not approaching at all

This response seems to be particularly important in a one-on-one situation such as around service time. Especially is this so with gilts as the above research suggests.

## KEY POINTS

So, to capitalize on the effect of your behaviour on the pig, the following seem to be the main points emerging from the behavioural work so far:-

✓ Human behaviour does seem to have a significant effect on the productivity of both growers and breeding stock and this has been quantified.

✓ Stockpeople who maximize the number of physical actions of a positive nature (pats, strokes, tone of voice, gentle movement among, and allowing the pigs to sniff them regularly) are likely to improve their productivity considerably – both the pigs and their own!.

✓ Stockpersons should regularly monitor the level of fear in their animals by the individual responses to the 'approach test', which anyone can do.

✓ In situations where the level of fear is high or is increasing, the attendant should re-assess his/her behaviour when near to the pigs.

✓ Approaching, handling (i.e.. moving) and attending females gently at breeding - especially gilts around service – and also farrowing, is beneficial in terms of formerly, conception rate and latterly more piglets weaned.

**Comment:** Most stockpeople are sympathetic and caring around farrowing and after, being well aware of the trauma involved. The problem comes at the, I think, more critical service stage where the very routineness of getting the female bred and often the heavy workload at this time makes for hurried and unsympathetic handling." Oh, for Goodness' sake, get a move on!" is a natural reaction. The rise from such rushed attention in both ANS and NES stressors **just at the very time that the female starts to ovulate, accepts insemination and needs to implant the results** – all delicate hormonal balances which the stressors can interrupt - means that the stockperson needs to handle and look after the sow most diligently and in a gentle, 'friendly' manner over these critical 5 weeks or so. Especially the gilt who is new to it all.

## Cost effective?

On the assumption at the extra time put into not rushing things and taking the necessary effort to get the pigs to become really familiar with you - and even get to 'like' you – labour cost could rise by 3%, then the benefit from one person responsible for shipping 1000 finishers/ year achieving 8% faster throughput, and each 50 sows producing half a piglet more per man will **overtop this extra cost of the time and trouble <u>sevenfold,</u>** based on current economics.

A much happier family – less stressed, too - from owner, employee down to the pigs themselves.

## Other things . . .

- Talking to them? It helps, I'm sure, establish empathy.

- Playing music? The familiarity of background sounds must reassure animals when quarters are changed and pigs are mixed.

  **Caution!** I recently toured some piggeries where the stockpeople played <u>loud</u> rock music to growers. I presume for their own benefit rather than for the pigs! The animals were more than unusually excitable, nervous and scampered away from me in a rush as I walked down between the pens. ANS activation or something like it? I remonstrated with the owner who was with me and noticed on my next visit, a year later, that the pigs then present were far more placid - loud music having been banned.

- Sticking to a time routine? All animals habituate and following their expectations as to when things ought to happen should lower stress.

- Lighting – periods of distinct light and darkness must help sleep patterns – as it does with us.

- Giving them toys? Sure, why not. They must get bored stiff. This is becoming mandatory in some countries' welfare legislation.

Of course it is dangerous to become anthropomorphic and assume pigs respond to what we as humans may prefer, but I find *good* stockpeople do use anthropomorphism more than we care to admit.

It helps to lower stress – on both sides.

# MANIPULABLE MATERIALS

As pig welfare increases in importance both from the producer's profit viewpoint and from public concern, keeping the pigs busy and thus happier will come to the fore.

## Variety

Pigs get bored after a while with the same stimuli, so novelty is important. Varying things to keep them occupied is as important as the choice of stimuli as long as they are changed periodically..There is no reason why a couple of distractions/interesting items should not be before them at the same time.

Things like waste cardboard and paper (beware staples) feed bags, and tougher items like wood (not chemically-treated), logs (beware pine resins) rope (beware tar) alkatene piping, tough rubber sheets and traffic cones (beware the Law!)

## Composition

Behavioural research shows that pigs generally prefer soft things to nose around and push, or alternatively chew. Straw is the best example - but not near slats. Alkathene piping is good for nosing and attempted chewing and so lasts quite a time. Sawdust (not pine but whitewood) and mushroom compost are used but soon get fouled.

## Hanging items

Are good – but not chains! I've watched these over periods and they only slap other pigs in the face. Better are plastic 'Bite Rites', from which flexible rods protrude and the pigs spend time with heads raised trying to get hold of these evasive objects, which is not easy for them so it keeps them busy! Well-anchored (!) thick knotted ropes from vessels are good, but avoid tarred examples.

## Food enrichment

A few largish pellets mixed into the straw bedding keeps them happy for a while each day. Novel foods such as herbal mineral blocks, salt licks (ensure plenty of water is available) root vegetables, grass (short-lived, no thistles) and even soft hedge trimmings (beware thorns) are all possibilities. It does not matter too much that these could be of short duration – it is the variability along with other longer-lasting `toys` that seems to do the trick. A little dried sugar beet pulp in a stalled dry sow diet provides a feeling of satiation as it swells sevenfold in the gut – so don`t use too much.

## Avoid…

Chains, tyres (due to wire reinforcement), insanitary items and those quickly-fouled.

# 7

# MANAGING TODAY'S HYPERPROLIFIC GILTS

Selecting a breeding female which will give a good first and second litter with minimum intervals between, and go on to produce at least four to five good litters thereafter.

## *TARGETS*

Good choice of gilts can set the standard for a future replacement rate of under 48%. 70 offspring might be produced in a gilt's lifetime (22-23 of these in the first 2 parities) and empty days kept down to 30 per year, every year of her life.

A whole section has been devoted to the gilt. Yes - deliberately!

* **The gilt is the most important animal in your herd.** She is your future profitability as a breeder and is the bedrock for a long productive life.

* **She is not only the most vulnerable, but also potentially the most dangerous animal in your breeding herd**. She has a partly developed immune system – full protection comes later – and during this initial period of her life she and her progeny are potential disease-shedders to the rest of your herd.

* **She is the most improved animal in your herd**. The geneticist has built an impressive degree of hyperprolificacy into the modern gilt. Litters of 13 – 14 are now likely under the right conditions of feeding and management and can be continued into later life.

Or not! Unless management and nutrition of the first (and second) litter sow in particular keep abreast of this new level of potential productivity, hyperprolificacy now bred into the young female can soon kick back into what I call the 'Shattered Sow' syndrome in the second litter and even on into the third.

**Many of the answers exploiting the geneticist's undoubted success in establishing the hyperprolific gilt will be found in this section.**

# CHOOSING A GILT

When I was young I worked for what was at that time England's largest pig farm. By today's standards it was small – 1200 sows, but in those days it was massive.

Our gilt replacement rate was about 8 per week and one of the jobs I had to do was select batches of 10 or 12 on Mondays.

After two years of this I got reasonably good at it – and developed a system which has stood me in good stead ever since. I made some pin money too, as local farmers paid me to do the same for them, or do a run-through of gilts which had been selected for them by the breeding company.

A few years ago I did my final job of this nature (I don't bend down so easily these days, and those teats/udder lines are so important!). So here's my own check-list for you. Generally I'm told that I rarely chose a bad one. This must have been helped by my need to use a guide to keep my attention on the points to look for, in sequence and in a deliberate, objective way.

## Choosing a good gilt

### First rule

*Don't try to carry everything in your head or in your mind's eye!* You've got a superb computer between your ears, but it is still not good enough to compare a pen of gilts objectively. So write it down! If you think it - ink it!

### Second rule

*Use a progress chart.* I illustrate my own personal route map (Figure 1), but you can design your own. The diagram shows how my eye travels around the gilt, noting details as I go. A planned progression is important – or you'll miss something important. My own scoring system is 1 – 10 with *anything* below 6 not being selected.

That computer between your ears is a fickle instrument. It tends to pick up what you are looking for – what you are concerned about (narrow chest, poor hams, legs and movement may all be characteristics you know need to be improved in your herd's gene mix). So you'll very likely pick these up – but miss others! Those last two teat pairs that aren't going to develop; that crossed toe; that tendency to be nervous. And so on. A progress chart slows you down and makes it less likely that you'll miss things.

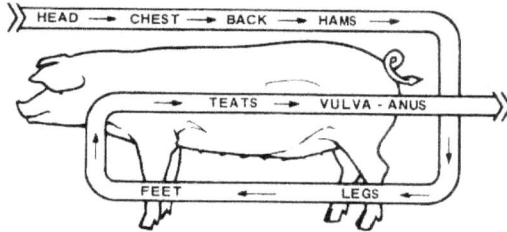

**Figure 1.** Chart for examining a gilt.

## Third rule

*Examine them one at a time.* Do a pen of ten or so at a time but concentrate on one, get the stockmen to move her about a bit, then finish and select or reject her. Do not jump from one to another, *e.g.* "Is *this one's* ham better than that one's". It is a great temptation and you will get confused if you do this.

## Fourth rule

*Favour the docile gilt.* This is my own personal rule and it is not scientific by any means, but in the UK with our wide range of sow temperament it has rarely let me down. Ignore it if you want to – I just find it useful. It is this:-

I try to *choose a gilt which is quiet and amiable* and quite unconcerned by me, a total stranger, giving her a visual check close-to and prodding her around gently, even clapping my hand close to her ear to startle her. *These girls almost invariably breed well!* Of course, if there's something wrong visually or in locomotion, of if she's low-rated on the index, out she goes. I wouldn't be at all surprised if, one day, some bright young PhD proved to us that prolificacy is positively correlated to an easy-going temperament. I find it so after selecting thousands of gilts.

## Fifth rule

*Do not select from a group of less than 10 gilts* (and 20 is better, if not always possible). This is because you need to select from as many replacements as you can so as to give yourself the best chance of improving your genetic traits over a period.

## Sixth rule

*Always get a gilt to trot.* If she's an easy mover her legs will almost certainly stand the loads to come, fed properly, of course, and on the correct floor surface. Horse breeders know exactly what I mean by a 'fluid mover', and while pigs are not nearly so elegant you need to get the same feel about a gilt. Strangely, a well-known horse breeder and I both judged a group of 30 gilts on the move and selected the same dud to an animal – there were 5 of them and we compared notes and agreed exactly. Suspect gilts with short strides and a stiff gait. Conversely, long strides and a swaying back end (future back trouble?). Just drive them on a bit so that they do a fast walk – you need to see how they *move*. And do it last, on the way to the draft pen as it disturbs them.

Check for swollen joints, tendons and ligaments, weak pasterns front and rear, and extra straight and stiff hocks. Legs should have a good spring and cushion, but not to an extreme. The Legs Checklist Section in my previous book "Pig Production Problems, p. 372" deals with this in much more detail.

# SOME OBSERVATIONS

## Narrow chests

Countries like Britain can be cold and damp, so respiratory troubles are a problem to us. Unscientifically again, I associate a narrow chest (*i.e.* poor spring of rib) with more pneumonia, often in the offspring too, and this applies to boars as well as gilts. Also I *never* select an animal with a dip behind the shoulder blade, real trouble here: little stamina and maybe slow growth in the offspring. If the farm is a good warm one, maybe; if the index is good and maybe if animals are a bit scarce, I *might* select such a gilt – otherwise I reject them.

## Teat troubles

You have to be very careful with blind, inverted and immature teats (button teats). I can't teach you this, it's experience of knowing when one cistern is not going to make it, now, or later. One tip is to feel the udder – if it is rough internally rather than silky, check for blind teats and inverted nipples very carefully. Also eight teats forward of the navel is a good sign, but increasingly rare – breeders please note! Discard if fewer than 7 functional teats are present on both sides. Check for poor, asymmetrical teat placement, discard if so as some future glands could be substandard. Initial screening for teat number, spacing and quality can be made at birth, weaning or in the nursery well before gilt selection.

## Vulva size

One major company rejects a gilt with a blind or inverted teat in the first two pairs. A small vulva coupled to a small pelvic spread is also trouble ahead; they tend to make suspect breeders at farrowing.

## Feet and legs

Extremely important, of course. Downgrade a limb structure which is extra vertically 'set' *i.e.* not sprung, when viewed from the side. This is very important in hocks and pasterns – if extra straight, score low on the chart. Duroc and Hampshire lines need careful examination in this respect. Feet and legs should have good springing to cushion the effect of hard floors, but some genotypes – often Landrace and Welsh – show extreme springing and thus hind leg weakness when older.

Downgrade gilts with swollen joints or inflamed tendons etc, and downgrade inverse toes if they differ by more than 1.2 cm on the same foot on a 70 kg + animal.

## A CHECKLIST FOR BUYING GILTS

✓    **Check your prospective multiplier:** In most cases you will be receiving your stock from a pure or cross-bred multiplier, not the nucleus or nucleus multiplier the seedstock firms like to talk about. Check him out – ask for names of other breeders who have bought his stock and telephone them. More important, search out others who have not been recommended and ask *their* opinion. Once the seedstock house has been agreed as your supplier *go and see him* (or in this electronic age some suppliers have useful CDs. However, there is nothing like a 'live' visit if it can be arranged) and learn how he prepares 'your' pigs for entry into your conditions.

✓    **Ask for the vendor's conditions-of sale document:** These, like the multipliers, vary enormously. Read them carefully. If you don't like parts of them, say so and threaten to go elsewhere unless the condition is modified in writing or explained satisfactorily. Check closely what they say (or don't say) about animals which do not breed satisfactorily.

✓    **Check the general differences between the lines:** Differences are appearing in the genetic strains coming from various seedstock houses *i.e.* conformation, appetite, type of finishing food advised, docility and mothering qualities, leg strengths and hyperprolificacy. There are also quite major differences within the breeding companies own line structure, so check that you are getting what your market or system of production needs, not what they think you want, or maybe is convenient for them to sell to you. For the progeny of outdoor sows, concentrate particularly on *proof* of fast, lean growth, as despite what is said, this still can be a weak area compared to (their) white indoor breeds.

✓    **Get to know the vendor's salesman/pig specialist:** Once you have established a relationship based on trust (this takes time) he or she will be a key factor in ensuring you get the right stock for your system of production *on time, checked personally by him/her and old/well grown enough!* Part of the process of getting to know the vendor and his salesperson is to know the questions to ask them.

# SUGGESTIONS FOR ORDERING GILTS

When writing this book, I contacted three breeding companies about how they preferred you should place orders for gilts with them, on my assumption that by following their system you were likely to get the best service.

*JSR Genetics gave the clearest, and to me the most interesting suggestions, and I quote " We supply at 100 kg which means 26 weeks of age, on our recommendation that service will be at 240 days.(see page 156)* **We strongly suggest** *that producers order their animals by their target week of service rather than their weight to ensure they are serving at the correct age.*

*Some producers may take in multiple batches aged 3 weeks apart at one time, e.g. 20 gilts at 26 weeks (100 kg) 20 gilts at 23 weeks (85 kg) and 20 gilts at 20 weeks (70kg). Again this is based on their target service week and not their weight."*

# CHECKLIST: QUESTIONS TO ASK A BREEDING COMPANY

Check carefully their reply and what their answers are – it helps as the choice is wide, as it is these days.

| | | |
|---|---|---|
| ✓ Genetics | | I want to be assured that your likely performance of your seedstock can be expected to be better than average, so…. |
| | 1. | Damlines. What is your claimed current performance for numbers born alive, and ideally, for numbers weaned? |
| | | Sirelines. Growth (from when to when), fatness, FCR (permissible as they can measure it accurately – but MTF is a good cross-check see page 242), plus some carcase and meat quality measures are useful. |
| | 2. | How have these changed over the last 5 or 10 years? This will give you an idea of past phenotypic (see Glossary) progress. |
| | 3. | Next, please tell me your estimate of current genetic progress. This should be in the region of 10g/day growth rate (25 -105 kg) per annum and 0.2 pigs born alive. |
| | | Treat claims well above these figures with caution - even these days it is most unlikely to be possible. |

4. What is the size of your nucleus populations- the larger the better.

5. Please outline your testing regime. A clear explanation here, (not `blinding you with science/statistics`) gives confidence.

6. What analyses do you do on this, and how often?

7. Do you use any additional technology to aid your progress? Look for things such as molecular techniques, advanced scanning techniques (tomography) to appear in the answer.

   A good salesman can be most interesting to listen to in these areas. These questions tend to identify the serious players in the subject and gives an idea of how well-trained are their salesmen, i.e likely to be front-end individuals you can trust.

8. What is the rate of inbreeding in the nucleus populations?.This figure should be 0.5-1.0% per annum. Any larger then it becomes a problem; any smaller then they may not be pushing the genetic improvement hard enough

9. If I am to receive your gilts from a multiplier then what selection intensity does he work to?

10. Could I speak to a producer with a similar set-up to myself who is using your stock? What level of performance does he/she achieve?

✓ Health 11. Can my vet speak to your (preferably independent) vet about the health of your nucleus/ multiplying farm?

12. What was your last date for monitoring x disease? You choose those diseases you are most concerned about.

13. What has been the length of time between your last monitoring/ check/vet visit and the date my stock would be leaving your farm? Initially in the tie-up between you and the breeding firm this should be as short as possible especially for the first animals purchased, but once you become more familiar with the source this precaution is usually relaxed to allow more freedom for the pigs to be supplied.

    The sales person may need to get this information if he/she has not prepared it in advance.

14. Are any drug treatment administered as routine? It is as well to know.

15. Diseases which a responsible breeding company should monitor are in Table 1, which cover UK conditions at the time of writing. Some other industries could have other priorities.

**Table 1.** A breeding company should be monitoring the following diseases and disorders as routine.

| Disease | Type of check |
| --- | --- |
| APP | Clinical, and at slaughter |
| EP | Clinical and at slaughter |
| Enteric viruses (TGE,ED,VWD) | Clinical |
| Erysipelas | Clinical and vaccination. |
| Internal parasites | Clinical |
| Mange | Clinical |
| Parvovirus | Clinical and vaccination |
| PRRS | Clinical and serological |
| PMWS | Clinical |
| PDNS | Clinical |
| Progressive Atrophic rhinitis | Clinical and swab |
| Salmonella cholerae-suis | Clinical |
| Strep.Suis meningitis | Clinical |
| Swine dysentery | Clinical |
| Swine influenza | Clinical |
| Swine pox | Clinical |

Plus of course, clinical checks for all the Notifiable Diseases in the current legislation.

Source: JSR Genetics.

# QUESTIONING SALESPEOPLE

The questions in the foregoing checklist are interesting to ask. Salesmen may have been instructed not to denigrate the opposition. If so, accept it as part of selling ethics. But you never know - by judicious questioning I have obtained some useful and important cross-confirmed data on various genetic lines which diplomatically I will not publish here, and there is no reason why you cannot do the same.

For example, data from 7 or 8 sources have revealed advantages between various commercial blood lines in the areas of:-

- Meat per Tonne of Food (MTF) on the same carcase yield.
- Leg strength and Appetite under hot conditions.
- Feed Protein needs of the slaughter pig in the last month before shipping. The progeny of some lines might be taken on to 120 kg without excess fat.
- Loading and haulage stress.
- Docility.
- Presence of marbling genes *e.g.* 0.7% marbling fat v 1.3%.
- Killing out percentage *e.g.* ± 1% under identical conditions.

This has enabled me to recommend certain breed lines which are more likely to be suited to the specific farm conditions I've encountered. I know this works because in most cases the follow-up resulted in comments like "Since we tried (or changed to ) breed 'X' the problem has been **much** better." Remember, no one breeding company's pigs are necessarily 'the best'. I am frequently asked "which do you consider the best breed?" The best one is the **right one** for your conditions and to compensate for/remove your commercial weaknesses.

# GETTING TO THE TRUTH

Of course getting the relevant "classified" information out of people is difficult, and in a commercial situation most lay people regard it as impossible. Trade secrets are just that. Secret!

But an old journalist's trick is to 'float the negative'. You need to know the subject matter pretty well, and insert an assumption, statement or claim into the discussion which is just sufficiently and deliberately wrong for the victim to at once correct it with the right figure from his kindly or professional instinct to put you straight. There are a variety of conversational subterfuges like this, and the rest I'm keeping to myself, although if you go into a good bookshop and read up on modern interrogation methods you'll get the hang of it! Meanwhile – beware of journalists!

# WEANER GILTS – A NEW TREND
## (Junior gilts in North America)

This is an relatively new development. Many commercial breeders are now buying their replacement breeding females – not at 90 - 100 kg but at 25 - 30 kg. I forecast that many bought-in gilts will be purchased as early as this in Europe within the next few years – that is on the professional/efficient units. Already some are being purchased at 60 kg (18-19 weeks)

## Cheaper and better

The reasons are not hard to see. The economic and performance evidence is now coming through from the pioneers of the system who started about 12 years ago, as it is not until the fourth year beyond repopulation that all progeny are derived from sows bought-in as weaner gilts.

## Cheaper cost

In Europe the cost of a selected maiden gilt at 100 kg bought from a breeding company is about £220. Of course the price of a 32 kg weaner gilt from the same source is not going to be as low as the value of a 32 kg home-reared female destined for meat, but prices have varied recently from £90 to £110, and one at 60 kg £200. All these are list prices and can be negotiable among European breeding companies. Table 2 gives a typical breakdown of comparative costs.

The Newsham breeding company, now merged with J.S.R. Genetics, quoted savings of £20 - £25 at 95 kg (Brisby, 1998) which was then a 12-15% saving on their average maiden gilt price.

**Table 2.** Typical cost of weaner gilts in the UK

|  | *(£)* |
|---|---|
| Weaner gilt at 35 kg (median price) | 110.00 |
| Feed at £185/tonne 35-100 kg (FCR 2.8:1) | 33.67 |
| Water, bedding, vet & vaccination | 12.00 |
| Interest on cost of gilt and feed | 3.50 |
| Combined purchase and production cost | 159.17 |

Assuming 4 out of 5 gilts are selected at this stage the cost to 100 kg is .

| | |
|---|---|
| 5 x £159.17 | 795.85 |
| Less sale of non selected gilt, say . . . | 75.00 |
| | 720.85 |
| Net cost per gilt selected | £188.21 |

A saving of £32.02 per gilt or 14.5% on a £220.63 maiden gilt price

## Better performance

A comparison of 49 herds using standard gilts and 16 herds buying weaner gilts (called 'junior' gilts in the USA) showed a 5.9% advantage in farrowing rate, 0.07 more litters per sow per year, 17 fewer empty days per sow per year, 0.5 more pigs born alive/litter, 0.28 more pigs reared/litter and 1.39 more pigs weaned per sow per year on 60 kg less food required per sow per year.

Why is this? The rationale behind buying breeding stock replacements at an earlier age and lighter weight is to allow a longer and more effective acclimatisation period prior to full introduction to the breeding herd. At least six weeks (and with certain low level diseases present, 8 weeks) is now advised when buying in maiden gilts at 100 kg. This delay is expensive in itself, and these extra costs alone would make a properly acclimatised maiden gilt kept longer before full introduction to the herd under the new recommendations, even more expensive. The extra costs are at least a further 5% per gilt to add to the 12 to 20% savings likely from buying 'junior' gilts at 30 kg.

## And what of the 60 kg gilt?

Producers should negotiate a realistic price based on their cost of rearing the animal over 40 kg liveweight to 100 kg. At the time of writing, and having done the sums on several clients' farms, this should not be less than 15% of the price asked for a maiden gilt. A major breeding company who does supply weaner gilts, reports that even under their skilled management, they can expect a 28% drop-out between 35 kg and 100 kg. If the commercial producer finds himself about this failure level, he must be careful to do the sums vis-a-vis weaner gilt cost price, plus cost of production to 100 kg and drop-out rate, set against likelihood of better performance from these animals as sows (*See Better Performance* page 147)

## Disease lower?

A much longer acclimatization period should result in less disturbance to the current health status of the herd. The weaner gilt herd owners interviewed felt that breeding herd health was better and there were fewer re-occurring health problems. We must wait for further evidence on overall disease incidence but sow mortality was lower, 4.0% compared to 4.3%. However mortality from born alives was higher on the junior gilt herds – 12.66 v 11.18 per litter. The absolute mortality figure per litter (A.M.F.) was 1.19 piglets (maidens) v 1.41 piglets (juniors), but the juniors piglets were weaned 2.5 days later.

## Much more weaner weight produced per tonne of food

A very important difference hidden in the published figures was the amount of saleable weaner weight produced per tonne of sow and piglet feed. At 116.5 kg (maidens) against 142.2 kg (juniors) this is a 22% improvement. *Under European economics (for 2010) this is equivalent to a 9% reduction per tonne in the price of all breeding and piglet food.*

## Comparisons to 36-38 kg

Did the considerable advantages of the weaner gilt system at weaning continue up the Acceleration phase of lean growth, which usually starts to ease off around 35 - 40 kg? Yes, it did!

Daily gain (7 to 37 kg) was 585 g/day (juniors) as against 548 g/day (maidens) or 1.28 v 1.20 lb. There was a marked difference in FCR; 1.8 (juniors) to 2.23 (maidens). This in itself would suggest the junior-sourced pigs could cope better with disease challenges at this critical stage of growth. Because of this large food conversion advantage, the liveweight produced per tonne of feed used through this stage was heavily in favor of the junior-gilt sourced herds – 698 kg v 559 kg, a difference of 139 kg or 25%! Even more dramatic – the figures on the PIC costings of the time revealed a reduction of 70% on the cost/kg gain to this weight.

If these results can be maintained by typical breeders it is no surprise to find that my forecast of big savings from buying junior gilts will be correct.

# INTRODUCING THE GILT

This section is written for the commercial producer buying-in replacement females at 100kg. liveweight at around 170-180days of age. And for the breeder who wishes to receive them sooner at 60 kg or so.

So why a lengthy chapter on the gilt in a book about problems? This is because, due to genetic progress resulting in new-found and very welcome hyperprolificacy, the latest gilts on offer are in danger of being unable to maintain this initial high level of productivity into later life

Unless something is done about it. Which is the reason for this chapter.

# INDUCTION: ACCLIMATISATION: INTEGRATION

35% to 42% and more - which is becoming commonplace - of sows are replaced each year. More than this upper level of new intakes risks jeopardising the immune status of the herd. Most replacements come from multiplication units allied to seedstock houses who provide the foundation genes from their nucleus units.

Multiplication farms will have a different pattern of viruses, bacteria and moulds to your own herd and will have been managed, housed and fed differently – maybe only slightly differently, but still different to your own herd. These factors combine to provide the gilts before they reach you with a certain degree of protection (acquired immunity) to disease on the farm of their origin, **not necessarily on your farm.**

## Quarantine

Disease is an emotive word. Replacement gilts from a reliable multiplication source are unlikely to arrive 'diseased' as such - but they will be `different`. The protective antigens they have developed will have responded to the microbes they have encountered and not necessarily to those present on your farm. This difference may or may not be sufficient to cause trouble once they arrive, hence the need to provide strict quarantine for at least three days in case a latent problem should appear caused by the changes involved and also to enable the stockperson to check physical factors like leg movement, etc.

When certain diseases and their severity are present in a locality, a strict isolation period may require to be longer than 3 days, so it is essential you get advice from your veterinarian on how long to provide what is necessary, the degree of isolation advisable and how those staff looking after the quarantined intakes should be organized.

One occasionally sees in the literature confusion between quarantine and acclimatisation/induction - suggesting that the two are synonymous. They are quite different procedures, so don't confuse them.

## The process of induction

Once quarantine is over and the new intakes are thought to be in the clear, they need to be introduced gradually to your microbial and fungal populations.   This is to allow them to re-prime their immune defences against the organisms your own herd possesses as well as allowing your existing sows to upgrade their own immune patterns to deal with those microbes the new intakes bring with them. TIME is needed. The period the new intake needs to re-adjust their immune shield takes a while – usually several weeks

## Induction - how long?

These days longer than most breeders provide! 6 weeks is minimal and 8 weeks is advisable Why? Viruses in particular have changed over the years and are still doing so. 15 to 20 years ago many of the problem viruses were influenced by a swift immune response in the host animal attacked (Fig 2a) and recovery  was comparatively quick (Fig 2b) Today's virulent viruses are tougher organisms and it takes the infected animal longer to respond to their presence.
This is why we are seeing so many of these viruses – PRRS is a good example - having a longer 'tail' to the debilitation they cause. (Figs 2c and 2d)

This slower immune response can allow secondary infections, from a variety of bacteria for example, to gain a foothold while the animal continues to rebuild its whole immune defence system. This is why many of these viral outbreaks seem to go on so long – it may not be the presence of the original virus causing the prolonged debilitation, but the secondary infections which have invaded while the host`s immune system was struggling to cope.

What can we learn from this? With these new viruses around with the likelihood of a longer recovery period, a longer induction period is also needed to allow the gilt to build up a sufficiently robust immune shield to cope with them.

Longer – thus stronger!

Previously, viral diseases had this sort of immune response to infection (a),<br>and this sort of performance deterioration (b)

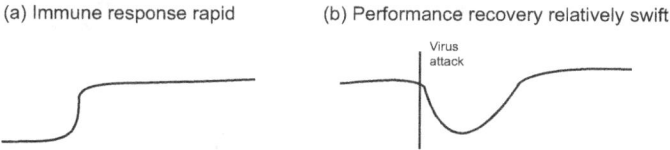

(a) Immune response rapid     (b) Performance recovery relatively swift

Now, the main reproductive reducing diseases demand longer recovery tails which may be due partly to a longer, harder climb up to full immunity (c) and/or being more favourable to secondary bacterial attack (P.R.R.S. and P.E.D. in particular)

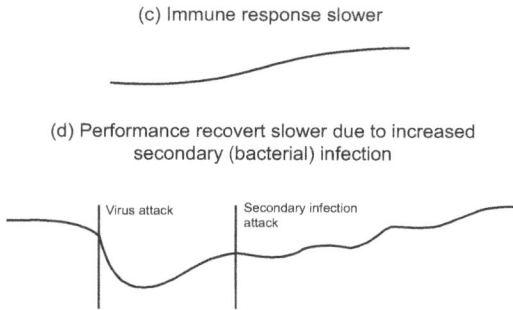

(c) Immune response slower

(d) Performance recovert slower due to increased secondary (bacterial) infection

**Figure 2.**

Remember this when you start fussing about the cost of having to wait patiently for those new intakes to be bred.

The next development the veterinarians advise to build a strong barricade against these new viruses, is to have a two-stage process within the induction phase....

## Two-week challenge period

This is when the newly arrived gilt has her immune system stimulated by deliberate and carefully-planned microbial challenges.

It is essential that this is carried out under veterinary supervision – not on a daily basis by any means - but the vet. needs to monitor the disease profile of your herd periodically (let him advise you when). The microbes on your unit probably change slightly or materially month-on-month. The veterinarian can track these changes, or sometimes lack of change, by carrying out various tests. He will also be aware of what diseases are prevalent in the area which will influence his advice what measures are advisable to further protect your unit – a measure, by the way, often omitted from guidance on Farm Biosecurity we read so much about these days.

This does cause comment that " This just gives the vet a blank cheque to run up a large vet/ med bill" a suspicious attitude which deflects breeders from what is really in their best interests.

But think – the right way to benefit from veterinary advice is in **preventive measures,** not in the far greater costs of curative medicine when disease strikes, let alone suffering the cost of a disease outbreak itself.

Table 3 gives a good example of how using the vet in a disease-profiling role pays handsomely.

**Table 3. Before-and-after results from using a pig specialist veterinarian to disease-profile 3 farms, with extra vaccination & re-modelling expenses costed in  (US$ Per Sow)**

| Farm | *Before* | | | *After* | | |
|---|---|---|---|---|---|---|
| | *A* | *B* | *C* | *A* | *B* | *C* |
| Estimated cost of disease per year* | 284 | 186 | 300 | 80 | 96 | 109 |
| Cost of veterinarian | 8 | 3 | 12 | 30 | 27 | 31 |
| Cost of vaccines & medication† | 26 | 18 | 30 | 18 | 20 | 21 |
| Cost of remodelling (over 7 years) | – | – | – | 27 | 45 | 33 |
| Total Disease Costs (US$) | 318 | 207 | 342 | 155 | 188 | 194 |
| Difference (Improvement %) | – | – | – | 51% | 9% | 43% |

* Disease costs estimated from items like the effect of post weaning scour and check to growth on potential performance; respiratory disorders, ileitis, abortions, infectious infertility, etc.
†Note that the cost of planned preventive medication was lower than for reactive curative medicine.
Source : Clients' records and one veterinary practice

There are a wide number of 'immunity stimulating' measures used in the challenge period, several of which are familiar to you already – fenceline contact of cull animals/ older baconers, putting a little faecal matter in the gilt pen, 'feedback'(scour matter, afterbirth, piglets intestines or gut contents, all of which can be dangerous, see Table 4) and of course vaccination.

**Table 4. Feedback can be dangerous in cases of the following common diseases**

| | |
|---|---|
| * Swine dysentery | * Pyelonephritis |
| *Clostridial dysentery | *Eperythrozoonosis |
| *Erysipelas | *PRRS |
| *Leptospirosis | *Toxoplasmosis |
| *Metritis | |

My point is that you should not take the decision of what 'challenges' to implement on your own even if what you have been doing as routine for ages "seems to work". You don`t know it is 'working' as your sows could be suffering from sub-clinical (not visibly apparent) disease which is dragging down performance, and which you accept as normal.

So please employ a pig veterinarian to design a correct challenge protocol for you, and let him advise you when it should be changed, what to, and why.

## 4 to 6 weeks fortification period

After the challenges the gilts now need as much rest and quiet as possible to allow the measures to 'take' and strengthen into a sound immune shield. Of course the gilts are young and excitable and maturing sexually - towards the end of this fortification phase they will need stimulating by flushing and brighter lighting anyway into a firm final oestrus, their third at least

100 watt fluorescent strip lights, white
Aim for 16 watts per m² (1.6 watts/ft²)

Place the lights so that the majority of the light falls via the eyes

*Lighting pattern*

On maximum 16 hours/day
Off for 8 hours/day*

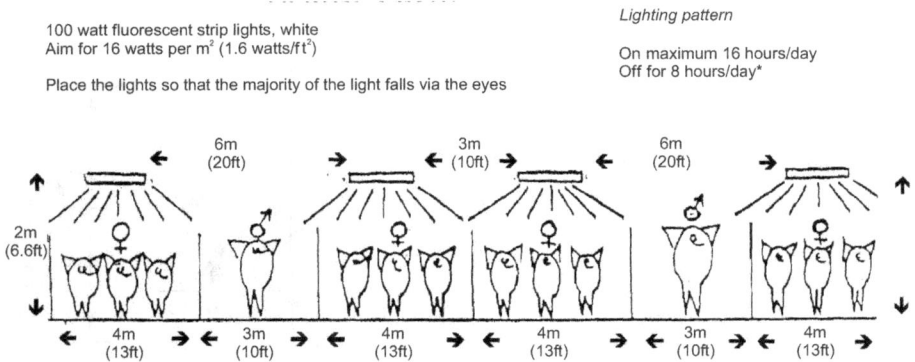

*It is suggested lighting for gilts could be slightly longer 17 hr/7 hr ratio

**Figure 3.** Lighting a mating/breeding house.

Overcrowding stress will not only dull this stimulation phase but may interfere/slow down acquisition of immunity. Overcrowding gilts is very common. Fleeing space is as important as m²/animal (I favour 3m²/gilt anyway) and square/near rectangular pens are far better than long narrow ones which are favoured to save on housing costs.

A good floor, preferably with bedding, helps provide the feel-good factor which a gilt needs at this time.

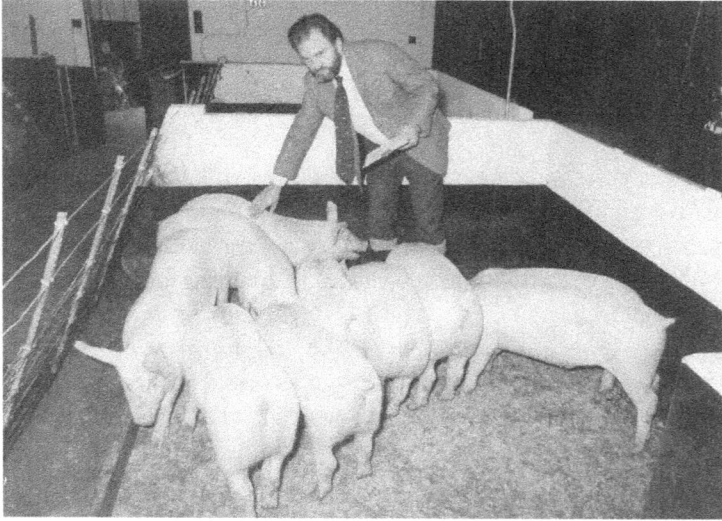

The author pre-selecting gilts at Deans Grove Farm in the 1970s.

Fig. 4 shows a gilt pen with a baffle board suspension which helps timid and/or smaller gilts get away from bullying.  In large yards straw bales are used.

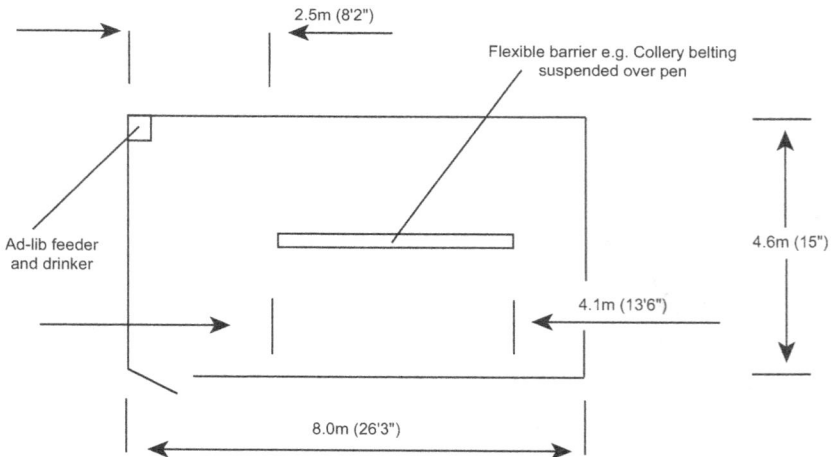

**Figure 4.** Introduction of gilts into groups.
Suggested layout for a mixing pen suitable for 10 gilts.

Old induction advice

W = purchase/brought on farm
Y = merge
Z = serve

Purchase

3 week induction

W

X
3 day isolation

Y Z

New induction advice (current viruses)

3d

14d

28-42d

Z = merge + serve

W

X
W-X 3 day isolation as before

Y
X-Y 2 week challenge period

Y-Z 4 to 5 week fortification period

Z

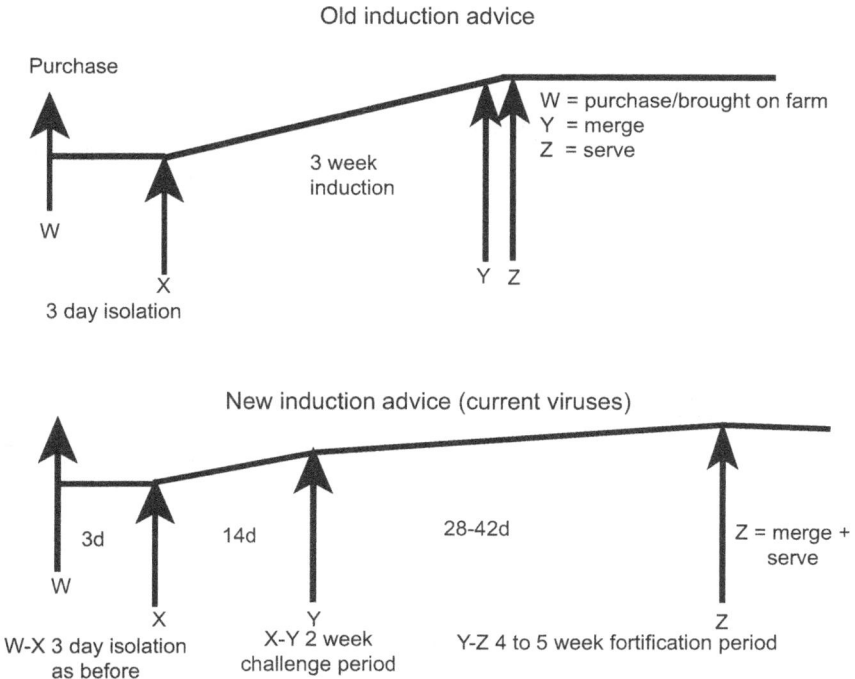

**Figure 5.** Infection can be materially discouraged by longer induction periods especially as no really effective vaccine is available for many viruses.

## Not too fast!

Lets go back to 'time' again – the latest genetically-superior gilt needs time for her reproductive hormones to catch up with her physical growth. She is physically precocious but not sexually so – that comes along very nicely later.

To take a human analogy, she has the body of a lovely young woman of 19 but the reproductive hormone development of a 10 year old schoolgirl! We must allow her endocrine system to catch up with the prodigious growth a modern gilt can achieve. Also, introducing her to the existing herd too soon risks high returns to service. A gilt will now grow at 1100 g/day. Too fast! We must hold her back a little to allow her hormones to catch up. Slow her down so that she reaches 135 kg (from 100kg purchase weight) at 240 days old (8½ months) some 9 to 10 weeks later. Table 5 suggests a weight–for-age progression. About 5kg per week is suggested.

This steady growth does two things. It allows her immune acquisition time to develop and for the reproductive hormones to mature.

**Table 5. Gilts:  suggested typical weights for age for modern high lean gain European breeds \***

| Gilt growth rate | Aim to achieve 100 kg in 170 - 180 days,  gilt growth-rate at 550 g/day, rising to 750 g/day  toward puberty | | |
|---|---|---|---|
| 100 kg | 180 days | 25th or 26th week | 6½ months+ old |
| 104 kg | 187 days | week 27 | |
| 108 kg | 194 days | week 28 | 7 months old |
| 112 kg | 201 days | week 29 | |
| 116 kg | 209 days | week 30 | |
| 121 kg | 216 days | week 31 | |
| 126 kg | 223 days | week 32 | 8 months old |
| 131 kg | 230 days | week 33 | |
| 136 kg | 240 days | week 34 | 8½ months |

\* Consult your seedstock supplier for actual targets.

# BUT WHY  SERVICE AT 240 DAYS?

And not the 220 days of old still followed in parts of N. America, and 220-230 days in Europe at the time of writing. The current advice here (and as I write it still seems to be carved on the tablets of stone of the advisory literature) is as on the left and the latest advice at the time of writing on the right......

| | *What many people are doing now* | *What is currently advised* |
|---|---|---|
| *Age* | 220-230 days | 240 days.... |
| *Bodyweight* | 130-140kg | ....which should then be 135-140kg |
| *P2 backfat* | 18-20 mm | Between 16-22mm. Note the wide range |
| *Sexual maturity* | 2nd or 3rd oestrus period | 3rd/even occasionally 4th oestrus |

Work by JSR Genetics (Figure 6) on a large number of samples shows that service at 240 days gives the highest born alives, closely followed by 260 days.

While service at 160 days, nearly 3 weeks later gives a similar result, this must result in a heavier sow in later life with its higher maintenance demand for food, as well as the extra cost of three weeks more food and overheads at the gilt stage. Figure 7 reproduces this work in scattergram form showing the very large number of recorded instances with the median line drawn through them.  I mark 'x' as the position where the most cost-effective age occurs - an economic trade- off between peak born alives and cost.

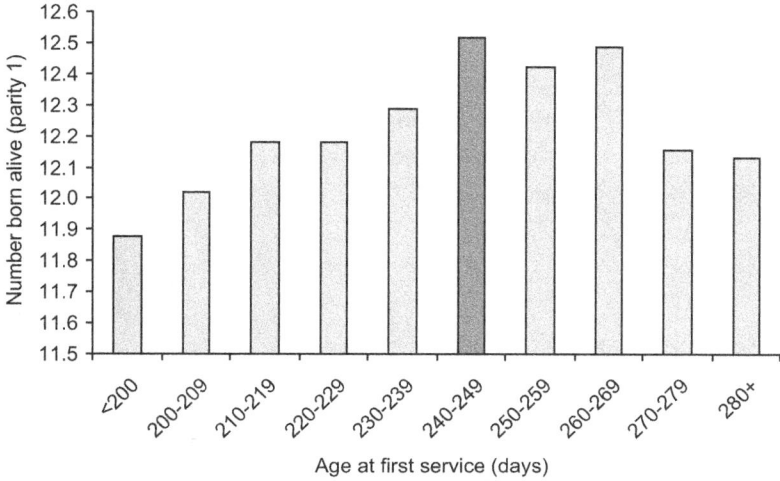

**Figure 6.** Gilts - numbers born alive to age at first service.

From this work 240 days is now considered the optimum age to serve. While a quite similar result is likely at 260 days, the extra cost of 20 days feed and overheads and the likely cost of a heavier sow in later parities with a higher maintenance demand for food renders the 240 days preferable.

Source: JSR Research (2009)

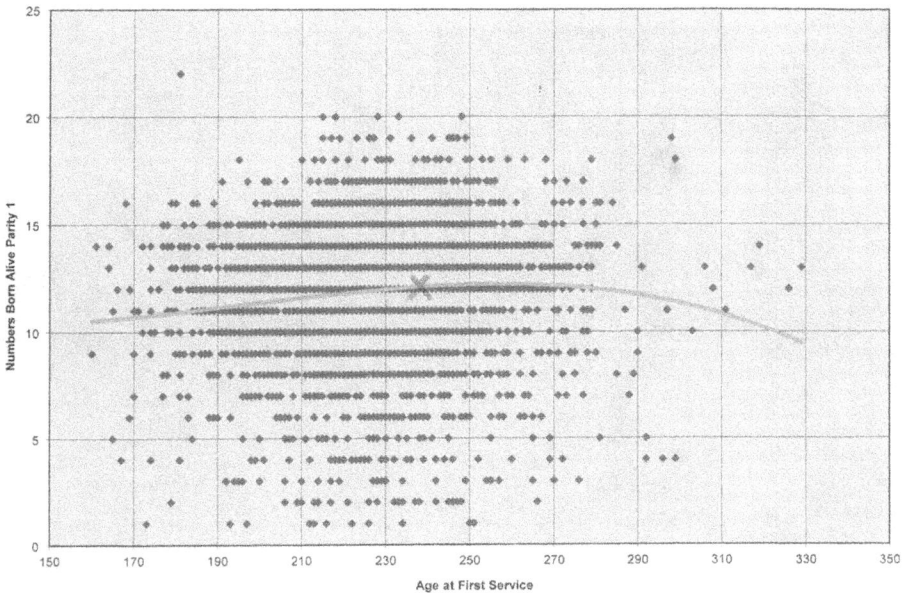

**Figure 7.** 240 days Scattergram)

And what of fat cover and bodyweight?

Figure 8 shows that the ratio between returns and and successful services with the gilts i.e. those in-pig) remains fairly constant when service was between 130-170 kg. Fig 9 reveals the same picture between weight at service and born alives across a comfortable fat depth range of 12-22mm P2.

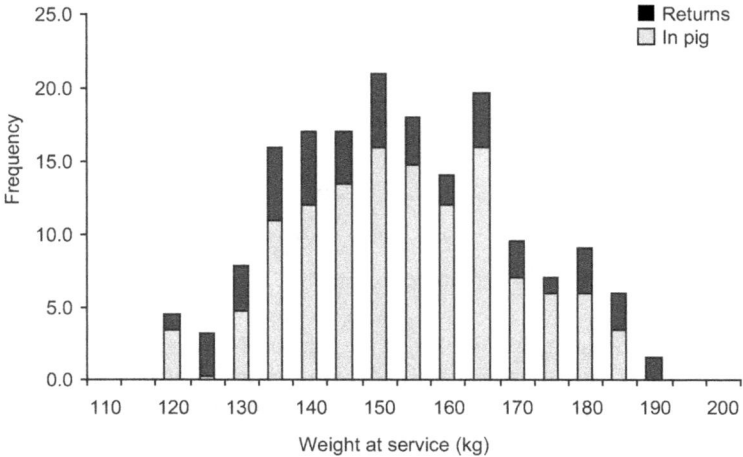

**Figure 8.** Relationship between bodyweight and outcome of first service. Notice that the proportion of successful in-pigs to returns is broadly similar between service at 130 to 170 kg bodyweight.

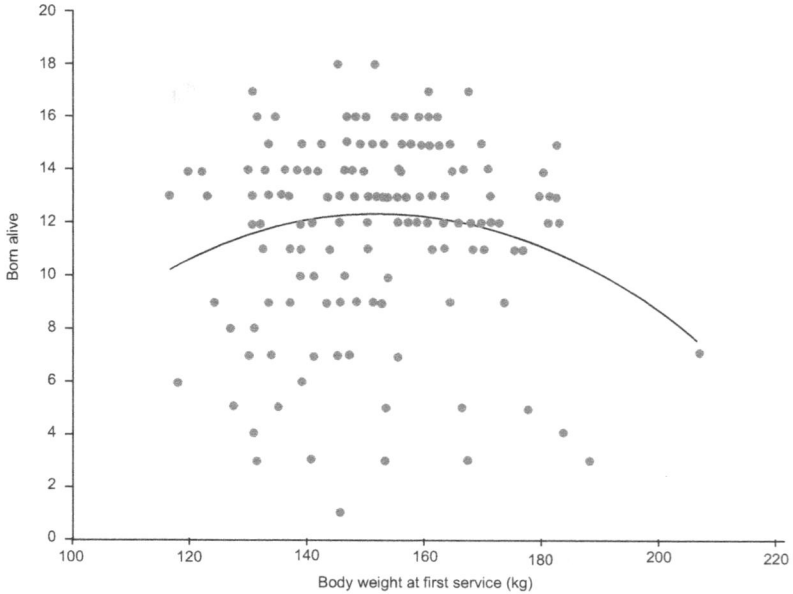

**Figure 9.** Relationship between body weight of the gilt at first service and numbers born alive of her first litter.

Notice from the predicted average that there is only a very small (<0.2) variation in numbers born alive between the wide range (32 kg) of body weights between 138-170 kg at first service. Conclusion - body weight at first service has little impact on numbers born alive. (Sources: JSR Genetics Research (2009))

Are these results particular to one line? It seems not as two other breeding companies at the time of writing support the concept of age at first service taking precedence.

**240 days provides the vital time to build up immunity**

If you look again at Table 5 the breeder now has at least 9 weeks to accommodate the steady acquisition of immunity before the gilt is merged with the existing herd.

Some breeders and veterinarians are preferring a longer lead-time with a much longer strict quarantine period than the precautionary few days advised at present for animals arriving at 100 kg or so. This is because some areas may have a good deal of virus infection present in the locality and buying in younger females at 60 kg or so is practiced so as to really ensure that they are 'clean' and to provide the vet with a more generous acclimatization period.

Again, this is why your vet should be masterminding the situation for you and that breeders should not go it alone. Seek advice from your breeding company too as they have experience with their own lines.

But - I have to say this from experience – be a little cautious of advice from the multiplying farms as understandably some are keen to get their females off their hands as soon as possible, even at a considerable discount,60 kg looks attractive to them and it may or may not be advisable so have a word with your veterinarian.

Don't get me wrong – I am not against the rather misnamed 'Weaner Gilt' concept as it can be useful.

# STIMULATING GILTS

Many of the units I visit are good at gilt stimulation. However from studying the major pig herd recording schemes, it is clear that most commercial piggeries operate their breeding herds well below the gilts and sows reproductive potential. If you are not achieving the following basic performance standards laid down by arguably the world's leading authority on gilt management, Dr Paul Hughes of Attwood University, Australia, then you should read these notes carefully.

| | |
|---|---|
| Cycling by 26 weeks of age | 50% |
| Cycling by 29 weeks of age | 85% |
| Non-cyclic by 32 weeks of age | 5% |
| Average age at first heat (weeks) | 27 |

# HOW DO WE REACH THESE TARGETS?

## 1.    Housing.

Try not to put more than 8 gilts together in one pen, which should be nearer to a square than long and narrow as this allows bullied gilts to get out of the way more easily. At least 2 m$^2$ per animal should be allowed - this is important. If your gilts are not particularly docile my own preference is to allow as much as 3m$^2$ as more `fleeing space` is needed. Pen shape and stocking density are important for gilts. I also suggest you take another look at figure 1 on page 410 which plan and dimensions work well, I find.

Do not house prepubertal gilts in <u>continuous</u> physical contact (including fenceline contact) with a mature boar, although short 20 minute periods of daily boar exposure are essential. The latest research suggests two such boar exposure periods each day are beneficial. Too much trouble? See my last paragraph for the paybacks that such, and similar attention to detail merits.

## 2.    Boar presence.

Regular boar contact undoubtedly provides the most potent natural stimulus for puberty attainment in young gilts. There is a general belief that all that is required is to put prepubertal gilts together with a boar regularly – but this is not good enough and there are several components to creating a powerful `boar effect` as follows.

### Maximising the boar effect – choosing the correct gilt stimulation 'window'

We want the gilt to start cycling reasonably soon even if she is not naturally- served or inseminated until the third oestrus. The gilt should be approximately 20 to 24 weeks (140-168 days) old, as response to boar stimulation reduces rapidly in gilts younger than this and also does not increase significantly in gilts older than 24 weeks old. These are pre-service stimulation episodes to help prime her reproductive system –service comes later at 240 days. When the time for service arrives, stockpeople tend to assume that the gilt has reached full puberty on the basis of vulval colour and swelling. This may or may not be so and waiting for another 24 hours could be beneficial in terms of conception rate and fewer returns.

## Maximising the boar effect  - choose the right boar

The stimulus value of the boar used is critical. Young boars less than 9 months old are of low or zero stimulus value. However not all boars 9 to 10 months old are of equal stimulus value. On average 20% of your boars will be of low stimulus value, and research suggests that using these will reduce gilts cycling within 21days by two-thirds! So the alert and patient stockperson has to assess each candidate boar for his sexuality and arousal/ excitement when in the presence of sows in estrus - and only choose candidates which definitely display this enthusiastic behaviour for gilt stimulation duties. It is also now thought to be preferable for gilts to be exposed to a herd sire which is being regularly used in matings, because his interest in the opposite sex may well be heightened.

There is an opinion among breeders that using any older, experienced boar will be sufficient – however, as in the older human male, he could well be sexually unexciting to the teenage female!

## Maximising the boar effect – touch rather than smell

Another misconception is that the smell of the boar`s pheromones is sufficient to stimulate cycling. Far better, as the pheromone(s) are not very volatile,  is that they need to be picked up physically by the female such as the 'snout in the mouth' behaviour, rather than being sensed in the air.

Fenceline contact is therefore advised, but even better – and in some herds dramatically better, but I do not really know why – is to house the stimulus boar at least a metre away, say the width of a feed passage, and then allow him under supervision to move in with the pen of gilts for 10 to 15 minutes, then removing to his pen one metre away. Could it be that the frustration of seeing but not touching stimulates both sexes?

## The vasectomised boar

Historically the use of a vasectomised boar was used to improve litter size and farrowing rate and thus make it easier to reach breeding target Whether the much larger gilt litter sizes now being achieved through the latest genetics will continue the practice remains to be seen, but those breeders with gilt litters averaging nine or ten could consider the concept.

Correct and skilful stimulation techniques are just as important with AI as with natural service, as it improves demonstrable standing heat which makes the stockperson's task easier. Also, pre-stimulation seems to speed up the insemination process as PIC have observed that these females needed 1.6 minutes less to take in all the semen than others which received only limited stimulation.

## Maximising the boar effect - separation

The response to the gilt to all the above advice seems to depend on how close together the sexes have been housed previously. The common housing system of penning gilts next to boars continuously during the rearing period looks to be a mistake, by allowing the females to be come too accustomed to the male`s signals. In the end they get bored with it all.

In this respect gilts are different to weaned sows, because the sows do not have enough time to get used to the boar`s presence during the short interval between weaning and mating. As well as less stress and bullying (being given enough space) one further reason why I prefer much smaller groups of gilts together is that the time the boar is allowed in under supervision with 6 to 8 gilts can be limited to 10-15 minutes per day while groups of 15 or more need at least 20 minutes, so Prof. Hughes has suggested.

# BUT IS ALL THIS CARE WORTH IT?

It is obvious that time, skill and patience is needed by the stockpersons in charge of breeding and that owners must allow such personnel enough time to follow these patient procedures, and also provide the right stress-free environment for the gilts. Some researchers have correctly shown that by doing so, gilt litter size is raised by certainly one piglet/ litter and often two. However correct gilt stimulation can have, in my experience, a further major benefit, as in my work on several really good breeding units where most of this advice has been followed for 5 years or more, sow longevity has increased significantly by an average of 2 more litters per sow lifetime (e.g. 6.6 litters in place of 4.6) and therefore a massive 42% more pigs produced per sow lifetime (67 v. 47). Under EU costings today, this easily covers the cost of one more trained stockperson per 500 sow herd plus providing a bonus scheme as well for achieving targets for all the animal breeding attendants, not just those responsible for breeding the gilts.

# SHOULD YOU BREED YOUR OWN REPLACEMENTS?

Again this may seem a strange choice in a 'problem' book. I think it is a problem to which the passionate adherents of 'breeding my own' fail to fully realize. I have had many an argument with clients and so convinced are they of their own capability in this respect, mentioning what they think is a saving in replacement gilt costs of 17%-21%, and 'less risk of buying-in disease' (now unlikely but old worries die hard) I haven`t managed to persuade a single one of my clients who breed their own to reconsider.

Having said this, I fully appreciate that under careful genetic guidance (**independent pig geneticists are as rare as independent ventilation engineers!**) together with sufficient investment in the strict discipline needed, it is possible to breed your own to a reasonable standard by having your own grandparent herd. In central Europe at present, breeding your own is quite common, but as soon as I step into their growing/ finishing houses I do not see the type of meaty, blocky animals which are now very common in, say Denmark, the Netherlands, UK, Ireland and Brittany to name but a few national pig industries. Their records show these home-bred pigs are growing satisfactorily fast **but not growing enough meat in the right place** – as I say, this is at once obvious as soon as I enter the building. In the past 8 years I have only come across one in-house breeder who impressed me in this respect – his growing pigs were superb - how he does it I know not. He is a genius!

I had difficulty in persuading him to build an international business out of selling his gilts which did produce those meaty, blocky finishers, rather than just expanding his sow herd to achieve his rather parochial ambition of being the biggest breeder in the district!

My argument is this. I fail to understand how any farmer can match the amount of computer calculations needed (millions for just one line) and progeny performance tests (hundreds of carefully, statistically-measured trials) to compete with the line development that breeding companies do. The genetic lag from such comparatively small resources of the on-farm breeder is bound to ensure they are falling further and further behind what progress can be purchased today in the genetically-improved gilt. Even the biggest, most sophisticated breeding companies say that it takes from 3 to 5 years of intense genetic development work to market a new dam line. What chance has the ordinary farmer home-breeder against that sort of genetic improvement work-load?

## Investment and discipline needed

Unless a breed-your–own herd can manage its gilt production pyramid separately and at some distance from their slaughter pig section of the farm, I feel that they should go no further.

This needs discipline in establishing a high standard of recording, dedicated labour and sufficient investment in order to avoid poorer quality gilts than can be purchased at the moment. The relatively recent appearance of hyprerprolific gilts from the seedstock firms only goes to reinforce my argument.

At the end of the day, producers who prefer to breed their own replacements tend to underestimate the costs of it all <u>when done properly</u>. You may think that I am insulting the worthy efforts of some very skilled producers, which is not my intention at all. I am just being realistic, I guess.

The penny has dropped with AI semen, where the producer has willingly left genetic improvement to the seedstock houses and AI stations, and buys the very best male lines they can from them - even studying the technique of EBV selection to ensure they get the most cost-effective progeny they can. So why not apply the same line of thought to the gilt? As I see it – the problem is surely an illogical and outdated attitude. So I am a one-hundred per cent breeding company man. They pay me not a cent for saying this - a conviction which comes from a hard-nosed attitude to helping my clients make the best use of their money and skills, and not to anything the breeding companies might think of paying me as a writer to take this line. In fact –to their credit they never have.

## " WITH YOUR INTERNATIONAL EXPERIENCE, WHAT BREED DO YOU THINK IS BEST?"

I am often asked this. There isn't one!

The right breed, line within a breed, or breeding company to choose is the one which....

- Makes up for, fills in and corrects the deficiencies in the pigs you have. (You need continuous and diligent records to know what these may be).

- Is most likely to suit your market outlet.

- And your climate e.g. hot/wet, slats or bedded floors, or outdoors, etc.

- Has a track record of performance and can provide customer names to support their claims which you can telephone or visit.

- Will allow you to dialogue with and perhaps visit the multiplier from whom you will purchase their gilts.

- Will provide a veterinarian-supervised delivery programme.

- And a reliable and customer-caring after-sales service.

And only then...

- Who provides a negotiable and competitive purchase price for gilts, boars and semen.

I believe, despite their claims which I do not doubt, that there is not a great deal to choose between the better firms on the technical aspects where females are concerned – they are all very good. However I think there are considerable differences in the male lines - but that is another story.

# THE GILT POOL

The purpose of a gilt pool is to have an immediate and adequate supply of selected and preconditioned gilts between 90 kg and the producers preferred service weight, in order to fill vacancies caused by enforced removal of sows and planned culling of older sows. Not every breeder subscribes to the idea of the gilt pool, but after visiting about 50 units who have invested in the concept it is noticeable after studying the past pre-pool records of 10 of them that their herd age profile (see page 104) was better, their empty days lower and their weaning capacity (see page 252) higher.

With their costs in front of me this enabled me to see that they benefited by an average payback of 5:1.after the gilt pools – mostly converted buildings or redundant cattle courts - had been operational for about two years.

Where the farmer without a gilt pool stands to lose (not all do but many have) is the downtime lost when a portion of the herd is short of properly- prepared gilts ready for service.  In such circumstances overheads continue even if the sow inventory is deficient and the herd is not earning up to its invested capacity. What the producer then does is to hurry a bought-in gilt into service improperly prepared, or even slot in one of his own finisher females. Both reduce future performance at best - or worse, highten the risk of disease.

## How big a gilt pool?

Mathematically, a 500 sow herd farrowing weekly and achieving 11 per litter which is short of 14% replacement females (1 in 7) in correct condition (and age!) for service when an expensive farrowing crate space suddenly falls vacant, stands to lose 400 weaners of potential yearly output, some 9% productivity (22.6 weaners sow year at 21-day weaning). Target for a gilt pool proportion of at least 12% of the working herd size.

For those converting from continuous breeding to batch farrowing, a gilt pool is mandatory in order to assemble the groups (batches) at the correct time. This is explained in the chapter on Batch Farrowing.

# AN INTERESTING GILT POOL EXERCISE

Three years ago three Eastern European units transferred to a gilt pool system and I asked to view their pre-pool records.

**Table 6. Results from 3 farms who adopted a gilt-pool approach (replacing an order-them-when-you-need-them system) 2 from a seedstock house, 1 from their nucleus unit. 1 year's figures before; 1 year's figures after.**

|  | *Conventional period* | | | *Gilt pool period* | | |
|---|---|---|---|---|---|---|
|  | A | B | C | A | B | C |
| Av number sows | 370 | 120 | 800 | 368 | 126 | 830 |
| Sow replacement rate % | 38 | 37 | 42 | 39 | 36 | 43 |
| Av empty days/sow/year | 37 | 41 | 35 | 34 | 37 | 34 |
| Weaners per week (Corrected to hear size) | 164 | 48 | 336 | 163 | 50 | 349 |
| Drug costs/sow | €21.60 | €14.62 | €27.91 | €14.71 | €11.25 | €16.98 |

**Comment:** We were disappointed to get such modest increases in weaner output, probably occasioned by the relatively good Empty Days on each farm. However, two farms noticed much less sow disease trouble which was confirmed by the drug-medication costs being substantially lower in the two herds which relied totally on bought-in replacement gilts. The transfer to the gilt pool system also meant that replacement females were on-farm for a longer period before being bred, and thus had a more gradual exposure (37 days average, after 3 days quantine compared to 24 days previously) to the resident herd's disease profile.

Interestingly their weaner output only rose modestly compared to their previous results and at first I feared a negative result, but their SPL (Sow Productive Life) at the end of nearly 3 years extended from 3.9 litters to 4.86 litters, raising their weaning capacity from 293kg/sow to 363kg by the time they were culled. This meant that they earned an average of 24% more from each sow which went through the pool than they had in their pre-pool days.

## A hidden benefit

It also resulted in their needing 17% fewer replacements per year, which eased their cash flow, but it also lessened the distinct overcrowding in their gilt pool pens of eight generously-endowed ladies (see illustration) down to six, which must have improved performance somewhat due to less stress.

Interestingly too, this happened on all three farms – not surprising as overstocking is rife in Eastern Europe.

The gilt pool is also a useful gathering ground to assemble gilts with PG600 into a batch-farrowing system.

Like AI/AO (All-in/All-out) the gilt pool is essential today.

### But a gilt pool costs....!

Often heard. Yes it does add to costs. While there is an inevitable range of costs needed to set up a gilt pool on any breeding unit – for example building from new is some 8 times more demanding in set-up capital than converting a redundant cattle court – the clients records I have collected suggests that it cost them an average of 2% (range 1.4%-3.9%) extra COP (Cost of Production) across 12 years estimated life. Extra labour demand seems to be more constant at + 0.5% COP. A total of +3% COP. which is about £1.40 on a £50 weaner today.

And the likely payback?

With a producer 12%-15% short of ready-to-serve, **properly-prepared** gilts costing 9% of potential output/year, this is getting on for £7 to £8/weaner – a healthy return and one which would comfortably overtop even new-build on a 15 year amortization/borrowings.

## MANAGING THE GILT POOL

Once gilts are 180 days old, try to move, mix and start boar contact at the same time.

Then check for signs of oestrus once a day, spray-mark those on heat and record the date when heat is first noticed.

'Combined operations' like this when all this activity is done in as short a space of time as possible, often causes a group to come on heat together and the `weak-heaters` more demonstrably, which is useful.

The use of a progesterone inhibitor is also a valuable tool to synchronise heats if required, as it is of course, for batch farrowing (q.v.).

Keep the spray-mark colour-coded for these synchronous groups of gilts using a different colour for each of three weeks, especially for continuous breeding farms.

This also helps plan weekly groups for service whichever system you use and makes it easier to know which animals are where.

If the 'botheration' factors are too much for your staff, then transferring to a batch farrowing and weaning system will make life easier for them – see the relevant chapter page 459.

# FEEDING THE MODERN GILT

I have an uneasy feeling that progress on sow and gilt nutrition has not kept pace with the remarkable progress the geneticists have made in their own field. However things do seem to be catching up on the gilt feeding side.

I make this clear at the start of this section as the tables and the figures which follow are examples of what I have gleaned from a variety of highly thought-of researchers, pig nutritionists and breeding company advisers.

Of course as has always been the case with scientists, not too surprisingly their own researches don`t always marry up, especially across continents where ingredients/ raw material availability and costs, climate and markets differ. This year I attended two pig conferences in two different Continents and the advice on sow diets was quite different too!  But as the academics seem to be reasonably in tune with each other where hyperprolific gilts are concerned, I provide the next four examples as guidelines but not firm recommendations - because I am frequently being asked`"What is different about modern gilt feeds – and am I feeding the right diet?"

## Where to go for advice

You should consider consulting a pig nutritionist responsible for designing correct daily nutrient intakes for your stock under your conditions, and also cross-check with the consultant nutritionist whom the supplier of your chosen seedstock employs. Why go this extra mile? Because I find there are different performance attributes between different genetic lines on the market which may need to be considered when designing a diet.

These variations, however slight, need to be taken into consideration before your nutritionist can start to design a diet and feed programme specifically for your farm and marketing circumstances.

Also it is wise to have an input from an experienced environmental engineer as environmental conditions can impact on daily nutrient intakes and thereby influence the design of the diet and its daily feed scale. He can do this from the measurements/ monitoring he can arrange – a growing sophistication of how to use modern pig diet design accurately.

**Table 7. Gilt feeding to first service**

| *Existing guidelines* | | | | |
| *Weight (kg)* | *Diet* | *DE MJ/ kg.* | *g total lysine/kg* | *Feedscale/day* |
| --- | --- | --- | --- | --- |
| 25-60 | Grower 1 | 14.0 | 12.0 | 2.5 to 4 kg |
| 60-100 | Grower 2 | 13.5 | 6.0-8.0 | 2.5 to 4 kg |
| 100-136 | Grower 2 | 13.5 | 6.0-8.0 | 3.0 to 4 kg |
| *Proposed new guidelines* | | | | |
| 25-110 | Gilt Developer | 13.5 | 8.8 | 2.5 to 4 kg |
| 110-136 | Gilt developer as above* | | | 3.0 to 4 kg** |

\* Some genotypes may need a Gilt Lactation diet, see table 9, this is because some emerging gilt lines seem to have rather smaller appetites than heretofore.
\*\* At least one breeding company suggests up to 5 kg/day.
 Both these circumstances underline how important it is to consult the seedstock house first before necessarily accepting generalisations found in the press.

There seems to be more divergence of opinion here. Across the first pregnancy period my interpretation is given in Table 8.

**Table 8. Gilt feeding in first gestation**

| *Existing advice* | *Aim to provide per day* | | *Proposed future advice* | | |
| *Weight (kg)* | *MJ DE/day* | *t/lysine/day g* | *Weight (kg)* | *MJ DE/day* | *t lysine/day g* |
| --- | --- | --- | --- | --- | --- |
| 130-225 | 26 rising to 36 | 6.0-8.0 | 135-230 | 40* | 14.0 - 21.0 |

\* This depends on genotype, especially appetite and prolificacy traits. And environment as well, especially in hot countries where higher fat inclusion is needed in the energy fraction. Consult a nutritionist.

Bearing in mind that the condition of the gilt is paramount and that feeding should not be overdone in the implantation period, and that a boost is given 10 to 14 days before farrowing as the foetuses in potentially large first litters are really developing quickly right at the end of pregnancy, then a Gilt Developer diet is fed throughout the first pregnancy.

So what is different about a Gilt Developer diet? It provides for a high lipid formation as well as a high lean accretion rate, has higher levels of Ca and P, extra micronutrients which influence fertility, such as vitamin E and organic selenium, and include a full range of organic trace minerals, plus ingredients which are said to assist natural immunity such as a prebiotic e.g. an oligosaccharide, and finally a mycotoxin absorbent, as some of these can knock the sensitive gilt's breeding potential sideways.

Something of a change from either a late growers diet or just a dry sow diet!

## Over-muscled gilts?

At the time of writing, one research establishment has suggested that the higher levels of protein (c.f. Table 8) could result in 'over-muscled' first litter sows, which they say lengthens farrowing time and decreases born-alives.

I have not noticed this in the gilt genotypes now on the market, but it is something to bear in mind should further research confirm it.

## Gilt lactation diet

The upgrade currently advised at the time of writing is to increase the lysine level in relation to energy and - to what has been fairly common practice up to now when a **sow** lactation diet has been used - much more lysine and its supporting amino-acid chain. This is thought to improve the gilt's milk yield and counteract a possible tendency towards a lower appetite in (some of) these modern gilts and supports the large litter sizes and weights these hyperprolific gilts can produce.

This, in tandem with the help of skilled creep feeding and other measures to help lift the burden of rearing from the young mother, will help avoid the second litter fallaway and eventually prolong her productive life.

**Table 9. Gilt lactation diet**

|  | *Existing advice* | *Proposed future advice* |
|---|---|---|
| 13.0-13.5 MJ DE/kg | t/lysine 6.0 - 8.0 g/kg | 14.5 MJ/DE kg   t/lysine 12.0 g/kg |

**Sources:** for the foregoing tables taken from publications and writings of :- Close, Campbell, Challoner, Gill, Hardy, Mavromichalis, Moore, Nicols and Tokach.

# A BIRDS EYE VIEW OF GILT FEEDING

Figure 10 provides a guide to the feeding regime from gilt selection to second pregnancy which can form a basis for discussion with your nutritionist and your supplier of replacement gilts.

It accommodates three critical areas of feeding the gilt and the young sow...

1.   A sufficiently long run-in period to encourage immunity and allow the reproductive mechanism to catch up with the gilts ability to grow fast.

2.  The use of a gilt pool to ensure continuity of gilts in the right condition for breeding and also to accommodate the new technique of Parity Segregation (q.v., page 467).
3.  Follows the peaks and lows of advised daily intakes established from past research.

# RE-LEARNING A NEW SKILL - NURSE SOWS

I include this subject in the Chapter on gilts because it is another way of lifting the burden of raising a very large family off the dam, whether she is a gilt or a sow. At the time of writing nurse sows are beginning to make a comeback.

### So why use a nurse sow?

**The Poorer Piglets.** Breeders do not need reminding that it is the smaller, more vulnerable piglets which drag down productivity and profit. As much time, trouble and expense has already been expended in getting them born alive as has been spent on their stronger siblings. And there is more effort to come in keeping them alive and well, and then getting them off to a good start on the growth curve. Nurse sows allow the lighter piglets to catch up as their stronger siblings are moved to a foster-mother while the weaker/lighter piglets are left to continue to suckle their own mother.

**Over-milked sows.** Not only is the nurse sow one way of helping underprivileged sows to make you more money, but the nurse sow concept has come into it's own as being quite a modern idea in that it takes the pressure off sows which have to cope with those big-litter 'bulges' which are happily occurring more frequently these days, now that improved genetics, more precise nutrition and advanced management skills are affecting reproduction, sometimes to an embarrassing degree - providing more weaners than the grow-out accommodation to house them can provide.

**Early weaning.** While the North Americans have learned the hard lessons of very early weaning at 12 to 14 days, and now favour 17 to 19 days as an earliest weaning age, the problem of larger litters and the accompanying risk of lighter birthweights still remain on any farm and make borderline weaning weights more likely. A planned and deliberate nurse sow policy can help with this modern trend (especially in those countries which favour early weaning) as it gives some breathing space to productivity by giving those 'underweight piglets' from a weaning point of view a little more time, vigour and better immune protection to withstand the traumas of such early weaning.

### Objections

1.  "It is not feasible" some people tell me. "You are always telling us to get those sows re-bred promptly and here you are suggesting that I sacrifice some of my

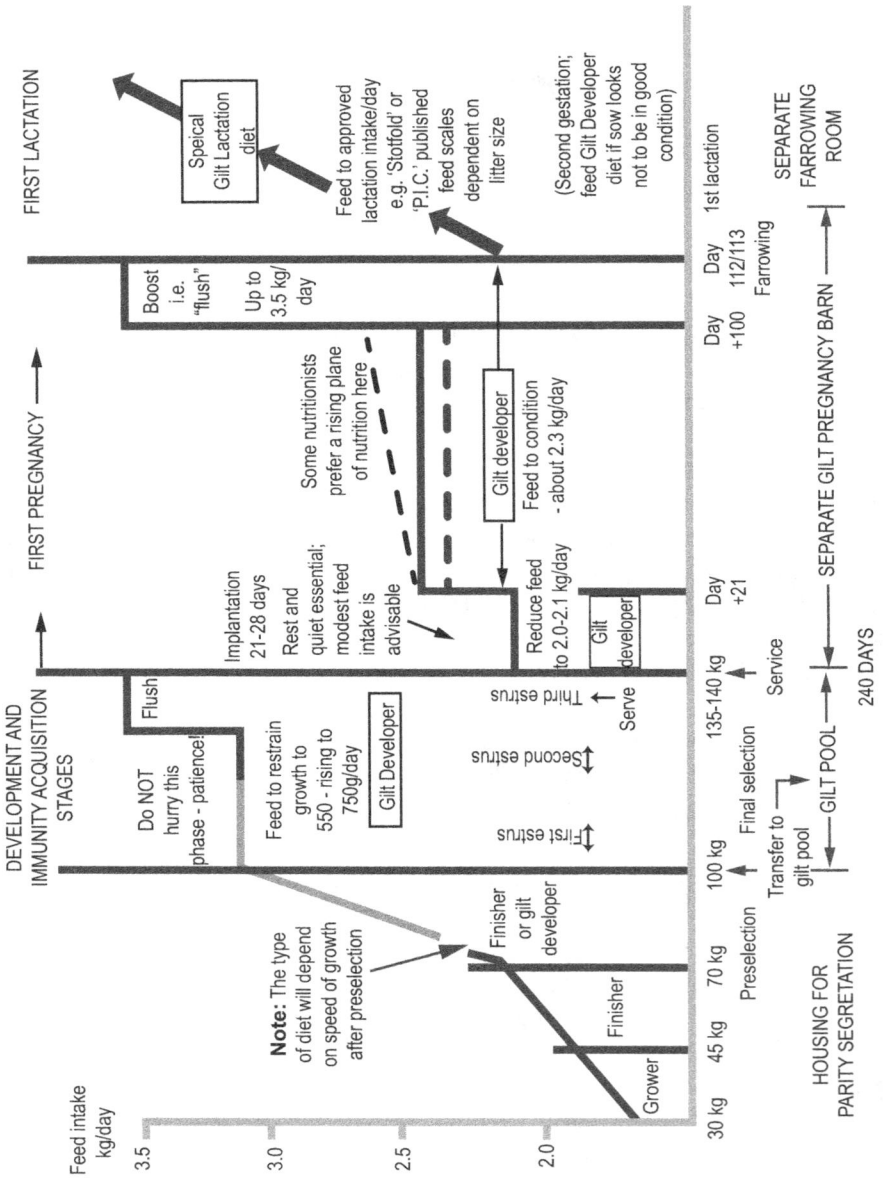

**Figure 10.** Latest advice on gilt feeding - 30 kg to first lactation.
(This regime can also be used with the new 'Parity Segregation' idea - gilts kept apart
from Parity 2 and all other older sows).

best milking sows to be out of production for 3 to 5 weeks when they should be making piglets"!

But look at it this way. Chosen well, a good nurse sow, for 3 weeks or so (or longer if you like to give her another batch) takes over, say, 10 surplus or underprivileged piglets (or alternatively some stronger piglets so as to give the weaker ones not transferred a better chance of a full suckle on their own mother) and so brings any lighter piglets up to a suitable weight and condition and weaning.

Many years ago I was told that "the smallest pig in the litter cancels out the profit made from the biggest one". This has been true enough if my records are anything to go by, so the more you have of these laggards at weaning, the more your potential profit is reduced. The nurse sow cuts the number of these underweight piglets down considerably.

If a nurse sow is kept and bred from conventionally like any normal sow she produces, if you are lucky, 11 weaners/litter or half a weaner a week over a full year. Good for her. But divert her into temporary nursing duty for 3 weeks (or longer if you like - by giving her a second batch) then she becomes a nurse sow rearing at least 10 surplus or underprivileged piglets, not only does she get those 10 weaker weaners up to speed, but in so doing **she has reared 3 piglets/ week during her term of duty, not half a pig/week.**

2.   "But she's plundered herself after suckling her own litter and then the equivalent of another litter too and possible a further one if she repeats the process" say the doubters. This rarely seems to happen if you've chosen the right sow and managed her correctly - a nurse sow needs to be a docile, prolific milker with a good appetite. Such sows, even though their milk yield may drop away slightly after a month since she herself farrowed, then up to a maximum 14 weeks of lactation. she won't lose enough condition to jeopardise rebreeding once her job of being a fast-food restaurant is done! After all, 14 weeks in milk was quite normal in the wild.

## Choosing the right nurse sow

Look in the records for evidence of **high litter numbers born**. A good benchmark is to look for evidence of how all their previous litters have done. A nurse sow needs to be a **good milker** with a **full range of active and well-placed teats**. They need to be **docile, placid** and provide evidence of a **good appetite in lactation**, especially in warm conditions (liquid feeding helps here).

**Method**

Two of them - the 'Double move' and the 'Direct' - I much prefer the Direct technique as it is the simplest. The double move involves an intermediate sow - and in all stockmanship tasks I've always found simplest works best. Let us forget it as an option.

In the Direct system the individuals to be transferred to the nurse sow are allowed at least 10 hours colostral suckling of their natural dam before being put on a newly-weaned nurse sow - as soon as you take her litter away then the new arrivals are placed with her. The technique works best, and is simplest to manage, if the stronger piglets are transferred as this allows the potential laggards of lighter weight to remain behind on their mother and gain much easier access to their dam's teats. Of course it goes without saying not to mix surplus 'heavy piglets' with any under-privileged 'light piglets' in the same group to be transferred. Keep the two categories separate in your mind, so plan ahead carefully. The policy of 'move the heavies/leave the lighter ones behind' is the best one.

**Transfer suggestions**

Acceptability is helped by the transfers being wiped over using a dampish but not wet cloth (no detergents/disinfectants - just plain water) which has just been used in the same way to way to wipe over the nurse sow's litter - not the nurse sow's udder as some do, this just disturbs her. This transfers the scent of her own litter to the new ones. She doesn't seem to notice that the piglets are smaller, but as a precaution keep the nurse sow occupied by giving her a special feed - we added a bit of creep feed or even added the occasional can of beer to her food. If you need to - give her a little bit of cereal straw or clean hay to play with afterwards.

If running an All-in/All-out system in the farrowing house - as you should be - the nurse sow will have to be moved to another room, or on the larger form to a special nurse sow house. This makes these distractions even more useful so as to occupy her mind.

So get her weaned a little sooner - I suggest 10 to 12 hours or so - than the rest of the batch of sows to be weaned so that the new sucklers are moved into a fully occupied farrowing room, then move them as a family to the new environment when the rest of the sows are weaned and moved to the breeding unit. Don't leave it until after the move.

## The future - 'Rescue Decks' to take over?

Having said all this - has the nurse sow a long-term future? We now have reconstituted milks which are every bit as good as sow's milk and are digestively stable if fed hygienically.

Now within the past few months, the arrival of specialised rearing and feeding 'boxes/ decks' have appeared which show promise as a means of rearing piglets after 4 or 5 hours colostral suckling on their dam. Too early to say, as they are not cheap, but it is claimed that one such deck can rear the hived-off progeny of some 24 sows, which brings the capital cost into perspective.

From the above notes, it is obvious that the current nurse sow system demands skill, judgement and a lot of dedication. Fortunately, I meet more and more 'professional' pig technicians - the new name for the stockperson of old - on the breeding farms of today who are quite capable of managing the nurse sow concept, and who enjoy doing it.

Rescue Decks will be another interesting development for the pig stockperson of the future.

# REFERENCES

I am grateful to the nine nutritionists listed in the text for their published views and for discussions with, and guidance from, several of them.

BPEX (UK) Yearbooks 2009 and 2010 for economic data.
Connor JF. 'Gilt developer units' Veterinary Practice manual, privately printed (Illinois c.2002)
Beckett, M. 'Cost of Weaner Gilts' Farmers Weekly, Feb 1994
Hughes, P 'Gilt Management to Maximize Lifetime Productivity' Austr. Pork Jnl. (March 2001).
National Hog farmer (USA) Blueprint 26 (1998) 'Gilt Pool Management' (Authors: Baas, Bee, Johnston, Levis, Grimes et al).
Univ. of Illinois: Knox et al (Aug. 2008) Gilt Management – DVD presentation.

I am also very grateful to Dr Grant Walling (JSR Genetics) for assistance with this chapter.

# 8

# PROPER BIOSECURITY -
# WHAT IT REALLY ENTAILS

Biosecurity involves everything which needs to be done to protect a farm, its livestock and its workers from disease.

Scope

Biosecurity is therefore a massive subject, encompassing not only hygiene (cleaning and disinfection), but also the protection of premises from disease ingress from other animals, (farm, domestic and verminous), birds, humans, insects, as well as from the weather, transport, air movement, waste food and liquids, the water supply, drainage, rivers and carcase disposal. Finally vaccination and farm location and layout as well as animal movement are all involved.

This section concentrates mainly on the measures needed to prevent infectious diseases getting into the farm; to deter them from increasing should they gain a foothold; and to eliminate those present wherever possible.

## *TARGETS*

The ideal is to reduce the level of pathogens to that which is low enough for the animals own defence mechanisms to cope with those remaining. No farm can ever be sterile, so the target is to obtain a working balance between disease challenge and effective defences against disease.

The level of natural immunity sufficient for the animal to do this varies due to the age and condition of the animal, to the degree of exposure to the organism responsible, and to beneficial or detrimental conditions under which the animal lives.

Some clues as to how 'clean' things should be have appeared from research and general pig farm experience. For example Waddilove, a pig veterinarian specialising in this area, suggests the targets in Table 1.

**Table 1.  Typical Total Viable Counts (TVC) of bacteria after pigs have been removed**

| State of house | TVC/sq cm |
| --- | --- |
| Immediately after pigs out | 50,000,000 |
| After plain washing | 20,000,000 |
| Hot wash and heavy duty detergent | 100.000 |
| Target after disinfection | 1,000 |

Source: Waddilove (1999)

Any pig veterinarian can carry out a TVC test(s) for you. Are you brave enough to ask him to do so? And so reveal your short-comings?

# DOES THE ORDINARY FARMER ACHIEVE THESE LEVELS?

I suspect – rarely!  On the few occasions that I have seen where swabs have been taken even after a clean-up 'blitz', the TVC before disinfection has been 3 million or more and in places many were at over a million *after* disinfection, with one or two over 5 million!  If the bacteria weren't being reduced sufficiently, then the viruses, which are more difficult to kill, were even less likely to have been controlled.

Ask most serious pig producers (as I have) and they say "But we clean and disinfect pretty well", "We have a set routine and stick to it", "Yes, we are now AIAO" (All-in/All-Out), "Don't disinfectants deal with everything?" and so on.

When I ask a few questions on what they actually do, great gaping holes appear between what we are told by the experts is necessary and what producers are still doing on their farms.  (Table 2)

**Table 2.  Survey revealed nine omissions or errors**

I've surveyed 119 pig farmers across 4 years.  Here are the results.

- Two thirds of them used no detergent in their pressure wash-down. Guys - this is **awful!**
- 52% of those that did, did not use a *hot* pressure wash, even so. Hot is best.
- Three quarters of those who used a detergent did not use a farm-specific detergent. Unwise - they weren't removing grease sufficiently.
- Only 2% troubled to monitor the effectiveness of their cleansing and disinfection procedure.

**Table 2. (contd)**

- Of the 9% who were sampled with swabs after disinfection, none at all had levels of viable bacteria remaining at or below the target level of 1,000 viable bacteria/cm$^2$. Dangerous.
- 50% of those swabbed had over 5 million/cm$^2$ viable bacteria from at least one swab. Very dangerous.
- 80% did not sanitise the water. Pipes can be slimy internally.
- 40% did not **regularly** combat vermin, most only did it "once a year or so".
- 90% "only fogged after a disease storm", not as routine. No one fogged their loft space. Viruses can survive up there.

Even if this is representative of pig farmers' attitudes, are they complacent? I don't think so. Complacency implies that farmers know what is correct but it is not done for reasons of time, or labour, or money or a failure to monitor things diligently.

To my mind more likely reasons are ….

- Failure to realise how much disease costs you in performance. Probably 0.3:1 food conversion from 7 – 100 kg and 4 fewer pigs sold per sow per year in the breeding herd. Yes, that much! And these are probably minima, I'm told.

- Failure to recognise that subclinical disease – the continuous effect of rumbling, low-level, largely invisible disease – **possibly costs you more**, over a period of say 2 years, than the outbreaks of clearly visible clinical disease we all worry about and take action on when they might happen over this period time.

- Failure to realise that modern pathogens are tougher, more resilient and more virulent than ever before and so need uprated detergents and disinfectants to combat them.

- Failing to clean properly before disinfection. Ever painted the outside of a house? I'm sure you have. Experience from previous disappointments tells you that it is the **preparation of the surfaces** which lead to a long-lasting effect, not so much the care in application or number of coats of paint subsequently applied. In the same way, pre-cleaning before disinfection has gone by default. We don't clean adequately so we end up not disinfecting properly, with disease being the disappointment.

We now have better virucidal (virus-killing) disinfectants, but they tend to be neutralised by organic matter and fat deposits.

These new viruses have stronger protective biofilms around them – they've changed so as to be better survivors. The 'old' disinfectants, like the phenols and quats, aren't so good at getting through this protection. Newer oxidative disinfectants (peracetic acid and also peroxygen) are more effective at the job, as well as being more biofriendly – one spin-off bonus is that pigs can even breathe one of them in at advised dilutions (fogging). Useful in pulmonary infections. One can also be used, with care, in the pig's drinking water. Consult your vet in both cases.

However, the new virucides do tend to be weakened by organic deposits on surfaces. In addition, fat and grease on the piggery surfaces can make the job of any disinfectant harder. The nutritionists are using more fats in lactating sow, baby pig and nursery diets these days, so there are protective grease layers deposited all over the in-contact surfaces.

*You must remove these barriers* to get a good kill of pathogens from the disinfective process – down to about 100,000 TVC (Total Viable Count) per cm² *before* disinfection (650,000 per sq. inch). To think that you can start with 50 million/cm² and only reduce this to 20 million/cm² if you just use a cold pressure wash before applying your disinfectant wash , you can see the problem you are giving any disinfectant! Waddilove moreover, puts it very succinctly.

> *"To understand the problem, think about the material you are trying to remove. It is often dried on, strongly adherent and greasy. Now think about washing up your own dinner plates after leaving them overnight after a greasy meal. With cold water this is nearly impossible, with hot water it is difficult, but add a detergent and it is much easier. So why, when they pressure wash, do most farmers use just water (often cold) and don't use a detergent?"*
>
> Waddilove (1999)

Yes - it costs to remove these barriers. For example, the correct heavy-duty detergents can cost up to 6 times more per m² than the best household (kitchen) disinfectants. but they are worth it to achieve a satisfactory pathogen kill under **farm** conditions.

# PRE-CLEANING PARAMOUNT

Over the past 15 years, as research into better disinfectants received prominence in the media, the importance of pre-cleaning has been sidelined. It is better now in the advisory literature, but some pig farmers are still too slow in implementing the advice to CLEAN THOROUGHLY FIRST.

# A PRE-CLEANING CHECKLIST

✓ Disconnect the electricity supply.

✓ Remove all moveable equipment.

✓ Open all inaccessible areas – fan trunking etc.

✓ Physically remove as much organic matter as possible from all in-contact surfaces.

✓ Flush out slatted storage bins and gullies.

✓ Use an approved degreasing detergent to loosen up the dirty surface

✓ Allow time (a minimum of 20-30 minutes) for it to soak in, but longer is better – half a day if possible.

✓ Pressure wash the complete building at 500 psi (max) with hot water (to remove all traces of grease). The temperature must be 70°C or higher.

✓ Inspect for thoroughness.

Next, make sure you are using a satisfactory detergent. A foaming version is preferable so you can see the areas covered.

# HOW TO CHOOSE A GOOD DETERGENT: A CHECKLIST

It must be farm approved. What does this entail?

✓ Capable of working well on all surfaces found on a pig farm. Unlike urban factories, there are many kinds of farm surfaces. Several of them are semi-porous (*e.g.* concrete, plastic and some metals). This variability makes it more important to use a product specifically designed for on-farm use. A heavy-duty formula is essential, stronger than those used in a catering establishment, for example. In my survey 18% of the farms used a well-known catering detergent 'because it was a lot cheaper'.

✓ Contamination in crevices and other poorly accessible places is more easily removed with a heavy-duty formula.

✓ Slats are more thoroughly cleaned. The build-up of dung on the surface facing between the slats is more easily dislodged. This is especially important with enteric organisms such as *E. Coli* and *Serpulina hyodysenteriae* (Swine Dysentery) and *Lawsonia intracellularis* (Ileitis).

✓    Good degreasing is vital. Just because a surface looks clean it does not mean it is clean of all pathogens. The presence of a greasy layer on the surface increases protection of micro-organisms by long chain fatty-acid molecules. A heavy-duty alkaline formula helps remove this protection. This is important as the newer, essential and better virucides don't work so well with fat protecting the organisms.

✓    Vital if time is limited – a heavy-duty detergent works quicker and faster.

✓    It mustn't interfere with the subsequent disinfectant's activity. This highlights the importance of using a fully integrated programme, such as the DuPont Pig Biosecurity Programme, when the products are specially chosen to be compatible, or in some cases help each other.

✓    Ideally it should be applied through existing equipment with minimal modifications.

✓    Foaming can be helpful. This increases the contact time and allows operatives to see where it has been applied. The foaming decreases the amount of water needed in the soaking and pressure-washing phases of cleaning. Reducing water reduces costs and problems with excess run-off to dispose of.

✓    It does not leave residues that can make the floor slippery and harbour micro-organisms. Especially, it should not leave cumulative residues.

✓    It should work in hard water situations.

✓    It should be non-toxic to pigs and operatives.

All these are what add costs to a good farm detergent.

# THE VALUE OF PROPER CLEANING DOWN / DETERGENT USE

Table 3 from Australia shows the performance improvement of growing/finishing pigs from proper precleaning.

**Table 3. Pre-cleaning itself boosts performance**

| Class of pigs | Cleaned buildings before disinfection | Uncleaned buildings before disinfection | % increase |
|---|---|---|---|
| Weaners | 572 | 500 | 14.4 |
| Growers | 736 | 692 | 6.3 |
| Finishers | 671 | 662 | 8.1 |
| Birth to market | 569 | 530 | 8.2 |

The buildings were AIAO (All-in; All-out) All weights in gms/day

Source: Cargill & Benhazi (1998)

# ECONOMICS

A 39g/day improved gain from 6 to 90 kg results in an improvement in saleable meat per tonne of feed used (MTF) of 24 kg. This is equivalent to a 15% reduction in cost/tonne of feed at a modest deadweight pig price. The extra cost of cleaning and the special biocide detergent used is equivalent to 5% of the cost of one tonne of feed. Therefore a payback or REO (Return to Extra Outlay) of 3:1 is achieved, even in times of very low pig returns.

# DISINFECTION

We should now have a clean and exposed surface with a TVC of 100,000/cm$^2$ or less bacteria. Such a bacterial threshold should also bring down viruses to a controllable level. The *surfaces* are now ready for disinfection, but we also have water tanks, lines and drinkers harbouring pathogens, and pockets of air, such as in lofts, which need attention to prevent recontamination, as well as inaccessible places among rafters. The problem with the older disinfectants has been that ...

- They are poorly effective against some of the newer viruses unless used at impracticable and costly concentrations.

- Due to toxicity they cannot be used in water lines or as space foggers.

- There is a wide range of correct dilution rates and coverage areas to deal with certain pathogens. Stockpeople risk getting them wrong and owners order up the wrong disinfectant basing their decision on price.

- Some are toxic or irritant. See Table 5.

# A DISINFECTION CHECKLIST

✓  There is a wide range of disinfectants on the market.

✓  You must choose one which is approved for the disease spectrum you and your veterinarian are likely to encounter.

✓  So either take advice from your veterinarian , or only buy from a well-known primary manufacturer who can advise you on which one to choose.

✓  Equally important as the choice of disinfectant is to follow the approved dilution rate which will differ according to the disease situation. Some manufacturers have a simple colour dipstick (*e.g.* DuPont) so that you can *quickly and easily* check that the product is correctly diluted for the purpose in mind.

✓    Also important are the instructions as to cover rate. Most people use a 'chisel' pressure washer at 200 psi or a spray nozzle. Managers should check the usage rate of the disinfectant purchased against the surface area which should have been covered in, say a month's use. In the examples reported to me during the serious Foot and Mouth outbreak in Britain, I suspected from cans used and discarded that the cleaning contractors had not fully followed the instructions either as to dilution or coverage rate. And if some of <u>them</u> didn't get it right . . . .!

✓    Check the time rate recommended for the disinfectant to act fully.

✓    Some disinfectants take longer to work in cold weather (Table 4). You may need a higher concentration, so check with the manufacturer in cold weather conditions

✓    If in doubt or you are unwilling to go into all these careful details, just use a peracetic acid disinfectant, the best known is Virkon or Virkon S. These are powerful oxidising agents and can rapidly kill most viruses as well as all bacteria; especially if you clean down well.

**Table 4.  Is your disinfectant effective in winter?**

| Dilution | Temperature | Able to stop bacterial growth | |
| --- | --- | --- | --- |
| | | *Formaldehyde* | *Modern formulated product at specified dilution* |
| 1% | 20°C | Yes | Yes |
| 2% | 10°C | No | Yes |
| 3% | 4°C | No | Yes |
| 4% | 0°C | No | Yes |

**Table 5.  Some properties of disinfectants**

| *Toxic* | *Irritant* | *Corrosive* | *Taint* |
| --- | --- | --- | --- |
| Chlorines | Chlorines | Phenols | Chlorines |
| Phenols | Phenols | | Formalin |
| Some iodophors | Formalin | | Iodophors |
| Formalin | | | |
| *Good speed of kill* | *Long persistence* | *Good virus control* | *Chloroxynol phenols* |
| Peracetic acid | Phenols | Peracetic acid | Chlorines |
| Peroxygen | | and peroxygen | Iodophors |
| | | are best, followed | |
| | | by iodophors | |

**Table 5. Contd.**

| | Action in presence of organic matter* | |
| Best | Moderate | Poorest |
| --- | --- | --- |
| Phenols | Peracetic Acid | Chlorine |
| Iodophors | Peroxygen | |
| Formalin | Quaternaries | |

\* To be safe, always preclean with a detergent

# AFTER DISINFECTION

## 'Resting' the Building

The writer has found that the common practice of resting the building before reoccupation (3 to 4 days winter; 2 days in summer) is generally not needed if the surfaces are quite dry before stock are put in.

Unfortunately this is sometimes not the case, and so a period of resting will help.
**So dry, dry, dry!**

An industrial kerosene space heater is very useful to achieve sufficient dryness in quite a short time, but check slatted areas and crevices in particular when the hot air currents can be deflected.

Again, as with cleaning down, check out the surfaces and, if possible, do a periodic swab test on close-contact areas. Your veterinarian can arrange this. The target threshold is 1000 TVC/cm$^2$ or 6500/sq in.

# SANITISING THE WATER LINES

The drinking water system is a potent source of viral reinfection (PRRS, PMWS especially) as well as digestive-disturbing bacteria like Balantidium coli. While QAC (Quaternary Ammonium Compounds) have been used for this, any organic matter – like slime – can inactivate them. Much better to use a peracetic acid like Virkon S.

# CHECKLIST FOR SANITIZING WATER LINES

✓ Make certain that the product you intend to use is recommended for this purpose.

✓   Check whether or not pigs can be present during sanitation. ***The dilution rate will be different for in-situ pigs***.

✓   Ensure the water system is not blocked, or if any drinker is leaking badly.

✓   Read the product instructions carefully.

✓   Dose the header tank. At terminal disinfection, (pigs absent) the strength can be much greater than if the water is santized with pigs present.

✓   For terminal disinfection know the volume of each header tank and the volume of the pipe-run it serves. Leave for at least 30 minutes, then allow to drain. Check with the disinfectant manufacturer's sales representative as I find this is often poorly calculated. See p. 187.

✓   For sanitation with pigs present, there is no such intense activation period, of course. Frequency of treatment depends on disease level (as does the dilution rate). See advice from your veterinarian or from the manufacturer.

✓   For sanitation with pigs present, know the volume of each header tank and use a ***dilution test strip*** as a double-check.

✓   Routine water treatment can depend on climate *e.g.* systems in the tropics may need more frequent attention, especially the header tanks.

✓   Do not mix water medication products with the water sanitizer.

✓   Keep your header tanks covered against dust, insects etc.

✓   Where pressure-washing the pens, ***attend to the underside of any drinker tongues***. The water sanitizer will insufficiently reach such an inaccessible surface which continually comes so close to a pig's mouth. Feel for any slipperiness/slime with your finger.

# SANITISING THOSE WATER LINES

Recently, just returned from working in Europe, I happened to be present on two farms which were replacing parts of their water lines, and stopped to take a good look at the inside of the discarded piping.

They were really fouled up and I certainly wouldn't want to drink water which had come through those pipes! Regular sanitising of water delivery lines has always been advised and this experience really confirmed how essential it is.

Fortunately, there are products which can do the job such as Virkon (DuPont) and CID 2000 (Cidlines).

Since my visits, when raising the subject with clients the question arose "But how do I measure the volume of water down through the pipe run so as to arrive at the manufacturer's correct dilution rate"? I had some equations from my college notes of 60 years back - based on gallons of course - and decided I had better get some more recent information.

Cidlines kindly provided the following equation based on litres.

$$r\,(cm) \times\ r\,(cm) \times 3.14 \times L \div 10$$
$$r = radius\ in\ cm;\ L = length\ of\ pipe\ in\ metres$$

Pipes are usually quoted in diameters, so as an example:

For a 1" (2.54cm) diameter pipe (= 1.27 cm radius, radius being half the diameter) and a line length of 100 metres counting in the drops as well, this would be:

$1.27 \times 1.27 \times 3.14 \times 100 \div 10 = 50.65$ litres, or 0.5 litres per metre
Any header-tank volume needs to be added onto this pipe volume.

Cidlines advise holding the treated water in situ for 4-6 hours before flushing through, and be sure to remove any of the resulting dark brown residues from any bowls or troughs.

## Another source of slime

Is from the undersides of leaf/tongue drinker bowls. I find stockpersons often fail to pressure-wash underneath the tongues as they need to be slightly lifted to do so. When I run my finger under a leaf in a newly sanitised nursery fitted with bowls it too frequently comes out smeared with dark slime. This can house an organism called Balantidium coli which I am informed is not a desperate pathogen, but nevertheless can cause tummy upset and inappetance in nursery pigs.

> *Remember: Water systems are often overlooked as a source of disease spread. Incorporate water sanitation into your routine.*

# AIR FOGGING

Dust and microglobules of water are pathogen 'taxis', carrying hostile organisms directly into the pigs mouth and lungs as well as acting as breeding surfaces especially in the upper reaches of a piggery, like ceilings and lofts. These microparticles can enter the smallest cracks, and changes in air pressure from outside or in ensures the movement and release of disease organisms *even after the structures/rooms below have been carefully cleaned and disinfected*.

## The disadvantages of previous fogging technique

Routine fogging with 500 ml liquid Formalin 40% poured on to 200g of potassium permanganate per 28m³ of air space which is closed up for 12 hours and not used for 8 hours after opening up again was developed by the poultry industry. This formula is effective and cheap but has serious drawbacks.

It is laborious  · It is dangerous  · The surfaces need to be wetted first and openings sealed  · A second person is advisable as a standby  · Livestock cannot be present  · Protective clothing *and a face mask* is essential  · Stockpeople dislike it  · The pot. permang., if bought in quantity, doesn't store well as it absorbs moisture from the atmosphere and then cakes.

## A major breakthrough

When oxygenating disinfectants appeared and could be used through an electric misting device most of these disadvantages were overcome. At the correct dilution rate, Virkon S, for example can be used to fog the atmosphere not only at terminal disinfection but *when the pigs are present*. It is thought that this in situ application as a breathable mist could well be beneficial where respiratory infection is present, even on a daily basis for a time. Seek veterinary advice on this.

With products of this sophistication, virtually all the disadvantages of the formalin method disappear, *but it is much more expensive*. Do your sums to explore this. I give some guidelines below.

## So should we fog as routine?

There is little concrete evidence of how easily or how much several pig viruses can spread via the internal aerosol (air droplet) route, though it is expected to be very likely. Routine fogging is likely to be a worthwhile method of killing airborne bacteria and viruses, and to sanitize the more inaccessible surfaces in a building where they can lodge.

At the time of writing fogging costs up to £2.50/sow/year in Britain. This cost pays for the products used for all surfaces and air space to be sanitized *each time* terminal disinfection is done (x11 times a year, farrowing house, x5 times nursery and x4 times finishing house) and also includes the cost of the misting equipment and all the labour involved.

As some viruses are such predators of profit these days, incurring costs varying from £500 per sow per year for Classical Swine Fever through Aujeszky's (£168) TGE

(£124) and EP (£100) and £75 for Infertility Viruses, the spending of about £2.50 per sow/year is put in perspective.

In the light of these potential costs, 21p/month per sow on fogging seems to me to be a worthwhile investment now that viruses are attacking us so remorselessly.

## LIMEWASH

Another old and well-tried technique. After solid floors have been cleaned and disinfected the in-contact surfaces are brushed with a mixture of hydrated lime to produce a thickness of thin salad cream, allow to dry hard.

## CHECKLIST - BUYING A DISINFECTANT

These notes can be used as a checklist for *any* disinfectant under contemplation before purchase.

Why are peracetic acid and peroxygen as well as glutaraldehyde-based disinfectants such a step forward?

✓ **Effectiveness**. The longest list yet of many hundreds of bacteria and viruses killed in test trials including virtually all the critical bacterial, viral and fungal families known to cause disease in pigs.

✓ **Stability**. Very stable which ensures long-term killing power.

✓ **Biofriendly**. Eventually break down into water, oxygen and carbon dioxide so do not remain a threat to the environment.

✓ **Organic challenge**. Work quite well in the presence of organic material.

✓ **Low temperatures**. Many disinfectants have to be increased in concentration to work well in winter conditions. Peracetic acid and peroxygen keep on killing in cold weather at their advised quite high dilution rates.

✓ **Note: Comparing the cost of disinfectants**: Always compare disinfectants on their power relative to the *approved dilution rate* and recommended *surface cover rates*. What may seem an expensive price per drum can often provide a cheaper cost per square metre treated.

✓ **Sewage and earthworms**. Even with strict European Union standards, they do not pose a threat to either sewage treatment or earthworms in soil, if used as directed.

✓ **Corrosion**. They are non-corrosive to metal, rubber or plastic.

✓     **Animal safe**. Some can be used for fogging and water treatment purposes with the pigs present.

✓     **Operator safe**. Maximum Exposure Levels can be up to 40 times more than other disinfectants – for example Glutaraldehyde.

Cement paint and phenolic disinfectant are another combination. Particularly popular in solid floored farrowing pens of the old-fashioned type, where even the crate rails were coated using a soft household brush. After 48 hours drying a hard but flesh-friendly antiseptic skin was left on treated farrowing pen surfaces. The technique is still popular in hot/dry countries where water is short and the atmosphere dry and warm.

However, wherever possible producers should use the modern technique described in this section as hydrated lime is not a match (even with some phenol disinfectant added at 30 ml to each 45 litres of limewash) for the modern cleaning and disinfectant agents.

## WHY ARE MODERN PERACETIC ACID/ PEROXYGEN DISINFECTANTS SO EFFECTIVE AGAINST VIRUSES IN PARTICULAR?

A virus has a protective *outer shield* consisting of …
- a fat layer
- a membrane
- a peplomer, or protein structure.

Inside this is a *capsid* or *inner shell* which envelopes the DNA-containing nucleus. This capsid consists of …
- capsomers; protein structures to hold and sustain the nucleus
- a membrane
- the nucleus itself.

The disinfectant has to break into both the shields and destroy or dysfunction the nucleus.

It has four weapons to do this …

1.    **A surfactant** to dissolve the outer fat layer and help attack the protein of the peplomer and capsomer layers.

2.    **Organic acids** which also attack these protein layers. Being organic, these are non-corrosive.

3.  **An oxidising agent** which cuts through the now-damaged protein structures and goes for the nucleus – even in cold conditions.

4.  **A buffering agent** (to increase acidity) which improves the biocidal effect, and also reduces the neutralising effect of hard water and organic material.

Thus these new sophisticated disinfectants are a composite of carefully-chosen chemicals each specifically damaging a part of the virus structure. Viruses are not only small organisms and thus difficult to target, but many are very tough and resistant (*i.e.* virulent).

## HOW THEY WORK. A SIMPLE SUMMARY

The surfactant *locates* the virus and *opens up* its outer surfaces to allow the nucleus to be *attacked* by the other two ingredients – the organic acids and the oxidizers – and the buffering agent *fends off the defensive action* of organic matter etc.

Simpler, cheaper disinfectants only complete some of the processes, and at a slower rate.

## DO YOU HAVE TO USE PERACETIC ACID/ PEROXYGEN DISINFECTANTS FOR EVERYTHING?

Not necessarily, although there is a tendency – laudable I think – to just use one or at the most two approved products for simplicity's sake and to avoid mistakes in mixing and application. There is also the adage '*Better safe than sorry!*'.

A generally agreed list of what to use/where could be:

**Virus infections**. Peracetic acid, peroxygen or iodophors and glutaraldehyde.

**Foot dips**. Iodine based, or if frequent replenishment is not likely – peracetic acid.

**Fogging**. Peracetic acid.

**Hands**. Quaternaries and soaps.

**Water sanitation**. Peracetic acid.

**Concrete surfaces**. Phenols. For very rough, broken *outdoor* surfaces, use an oil-based phenol.

**Loading ramps**. Peracetic acid or peroxygen (because you have no knowledge of what notifiable or transmissable diseases may be being brought on to your farm, so as broad a spectrum of control as possible is wise).

**Delivery and collection vehicles**. Peracetic acid as above, plus the advantages of minimal corrosion.

# WATER LINE SANITATION

Periodically, during my farm visits overseas, the staff were dismantling sections of the pigs' water lines, so I always take a quick look inside.

I certainly would not like to drink water from such pipes, which must have been rarely sanitized - if at all. Fortunately there are products which can do the job of removing the biofilm (a polite name for what I saw!) and you might like to have the formula for estimating the amount of water held in a run of pipeline - in order to get the recommended dilution right whose ever product you use.

One manufacturer has provided me with the equation given on page 187.

# COSTS AND PAYBACKS OF MODERN HYGIENE PRACTICE

Of course these better detergents and disinfectants cost more, even at the higher dilution levels their efficiency allows. Added to this are the extra tasks of things like pre-soaking time, hot pressure washing, fogging, water sanitation and thorough drying before re-occupation.

# IS IT ALL WORTH IT?

We need to examine typical results, both on-farm and research, and put some economic benefits to them. Then cost out the extra investment needed to update the countermeasures to satisfy the most up-to-date advice available.

Table 6 is based on 17 comparative trials (11 farm and 6 research) where on 16 of them the pigs were not suffering from any particular or obvious disease.

This is important, as good biosecurity should lessen the incidence of clinical disease outbreaks, where the benefit would be much greater – possibly up to 5 times as much.

# THE COSTS OF DOING IT PROPERLY

So if these are the possible benefits, what does a meticulous hygiene system cost?

**Table 6.** The benefits from better biosecurity techniques

| Reference | Trial type, basic details | Calculated value of finishers (Against controls or former practice) |
|---|---|---|
| Cargill & Benhazi (1998) | Cleaning AIAO buildings before disinfection with a detergent | + £3.37/pig |
| Overton (1995) | Salmonella outbreak controlled | + £4.17/pig |
| Jajubowski *et al.* (1998) | Using a peracetic acid disinfectant instead of NaOH | + £9.50/pig |
| Sala *et al.* (1998) | Full Antec programme v. iodine | + £2.27/pig |
| Sala *et al.* (1998) | Full Antec programme batch disinfection v. terminal disinfection only | + £6.17/pig |
| NCASHP Denmark (*nd*) | Partial v. total biosecurity programme | + £8.22/pig |
| Antec Trial (G&M, 1999) | Change to AIAO and updated disinfectant, result after 3rd batch | + £7.72/pig |
| Gadd (1994-1998) | Average of 10 clients uprated to full biosecurity protocols | + £6.08/pig |
| | Average : all results | £ 5.94/pig |

Assumptions:

Weights ranged from 6–90 to 30-100 kg
Food in last 14-21 days, range 2.2 to 2.25 kg/day
Finisher feed price £165/t (about twice the value of a finished pig in Europe today)
KO% standardised at 73%. £5.94 is about 5.3% of the value of a slaughter pig today.

I make a stab at it in Table 7 and go on to set this against what the trial work to date suggests you will gain in increased income/overall saved costs.

The additional cost of proper biosecurity is divided between the extra cost of the modern materials over what you use now (*i.e. virucides* rather than just bactericides); the costs of extra tasks now considered important (*i.e. hot* pressure washing using a *heavy duty* detergent; *fogging* the airspace and *sanitising* the drinking water) and the cost of the *extra* labour all this entails. Table 7 is an attempt to draw all this together for the first time. I have not seen this done before.

If from the trials cited in Table 6 the expected benefit from a complete biosecurity protocol is £5.94/pig then the REO (Return on Extra Outlay) is £5.94 ÷ £0.34 *or 17:1* (Table 7). When a good growth enhancer typically obtains 6:1 REO at best, this puts proper biosecurity into its true perspective – a very good bargain indeed.

**Table 7.  Proper biosecurity – the cost picture**

Figures based on the surface areas, labour and Approved materials needed to produce 100 finished pigs on a farrow-to –finish, largely solid or slatted-floored farm. Including all breeding unit surface areas needed for the 4.33 sows producing 23 finished pigs/sow/year.

| Expenditure on Materials | Correct biosecurity protocol | What is done (or often not done) now | Extra cost of correct action |
|---|---|---|---|
| Detergents | 12-19p/pig | Detergent not used | 12-19p/pig |
| Disinfectant | 2.3-3.2p/ pig | 1.0p/pig | 1.3-2.2p/pig |
| Airspace fogging | 1p/pig per treatment | Rarely done | Say 4p/pig* |
| Water sanitation | 2.7p/pig per treatment | Rarely done | Say 5.4/pig** |

* Based on 4 treatments/year (i.e..after every batch put through) = 4p per pig place/yr.
** Based on water lines sanitized once every 6 months.

| | *Cost of labour in the above plus depreciation cost of equipment.* | |
|---|---|---|
| Cleaning labour | No difference, cold pressure wash only | - |
| Hot pressure wash | Cost of steam jenny over 8 years £1.60/100 pigs | 1.6p/pig |
| Disinfection labour | No difference | - |
| Airspace fogging | £1.30/ 100 pigs | 1.3p/pig |
| Water sanitation | £0.65 p/ 100 pigs | 0.65p/pig |
| Cleaning loading bays etc | No difference | - |

**Totals: Extra cost approx. 26-34p/pig sold**

Source:  Assistance with materials prices, courtesy of Du Pont (UK).

Comment: These are **extra costs** for doing a proper job using the latest Approved materials at the makers dilution advice and surface coverage recommendations; not the total cost of sanitation which several sources, including veterinary practices, claim lie between  £1.00-£1.25 per finished pig.

Correct sanitation is NOT expensive as the following conclusions suggest.

# CONCLUSIONS FROM THIS ECONOMETRIC EXERCISE

•    Far from being expensive, proper biosecurity offers a very handsome payback.

•    Two jobs very rarely done when pigs vacate the building are water sanitation and air fogging to reach the areas difficult to access. Including labour these add up to under 12p/pig.

•    Farmers often baulk at the extra labour cost of doing these extra tasks.  In fact it is only 5.4% more.  Several producers and I who measured it carefully and obtained this average figure between us were really surprised how low it was.

- Farmers also complain about the higher cost of the correct, more powerful products per drum/bag, but it is still under 36p/pig (or 0.6% of production cost). Sure, it is over three times more expensive than using just a simple disinfectant alone – but as Table 7 suggests it should give you *over a 17 times return* on the extra cost.

- Farmers say that they just haven't got the labour to devote to the extra time needed. Then maybe they aren't correctly prioritising the work of their labour pool? A 17:1 return makes it just as important as most of the other vital labour tasks in potential value – farrowing, breeding, weaning.

In fact, moving up a gear to correct biosecurity is one of the most important routes you can take to more profit. I strongly recommend all pig producers reading this to reassess, with their veterinarian, exactly how much trouble they are presently taking over biosecurity, the adequacy of the products they currently use, and think it all through carefully.

You could be losing £16 on every £1 you are saving. Enough said?

## VEHICLES AND TRANSPORT

Past serious outbreaks of CSF and FMD in Britain have emphasised how dangerous a disease vector are the vehicles which visit your farm regularly.

I give below my own advance conclusions and recommendations as follows:

## A FUTURE CHECKLIST FOR FARM BIOSECURITY?

✓ Farms should have *perimeter* unloading and loading facilities for pigs, feed and general goods. No delivery/collection vehicle, including salesmen's cars or even the vet's vehicle, should be permitted past the farm boundary, or alternatively only allowed into a biosecure area with separate access on to hard standing with cleaning facilities attached. Figure 1 shows a suggested diagram for an ideal farm layout. Do as many of the suggestions to your own unit as you can.

✓ No pig collection or delivery vehicle should be accepted without a form/ certificate issued by the buying organisation /vendor signed by the buyer/ vendor's biosecurity supervisor that the vehicle has been properly cleaned and disinfected – inside, under and out – before calling at the farm. These organisations to train and appoint a biosecurity clearance officer as routine by law.

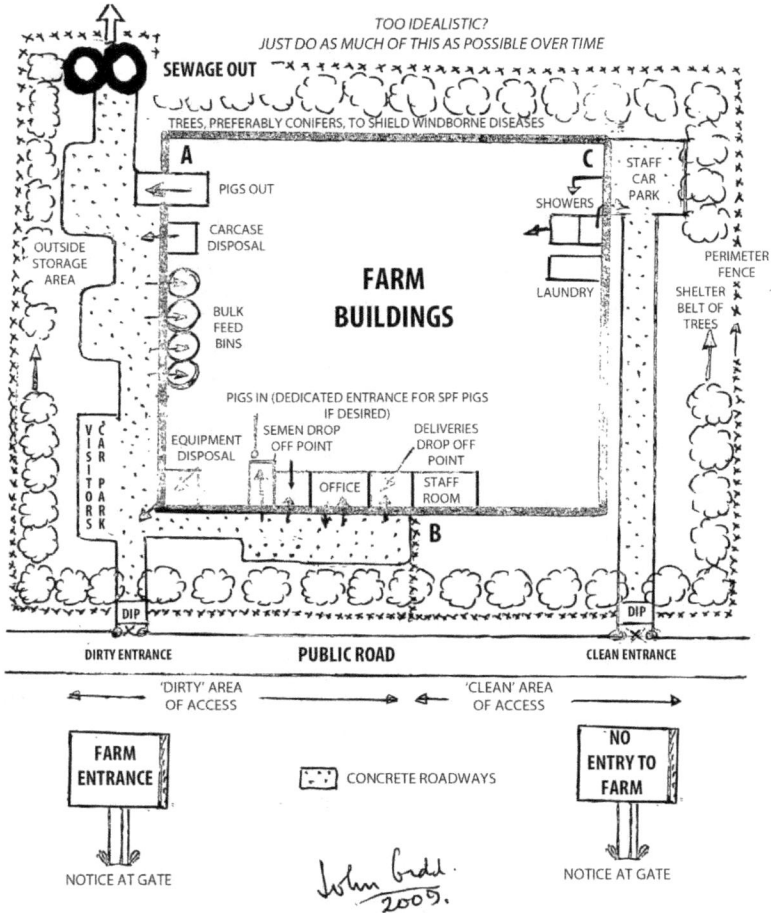

**Figure 1.** A biosecure layout minimising contamination
(modified from a PIC Company suggestion).

✓ In addition, each vehicle must have available a written load record which you can examine.

✓ No driver is allowed to assist in loading/unloading pigs.

✓ Anyone driving a transporter must not be allowed *by law* to keep pigs on any other premises. Legislation on this is overdue.

✓ All abattoirs, markets, feed mills and commercial pig-breeding companies must invest in satisfactory tunnel-type disinfecting bays with a supervisor attached to ensure thoroughness and avoid 'dodging'. Legislation may be needed to enforce this.

✓ Farm pig loading bays are essential, with wash-down facilities on tap. Drainage should be away from the farm.

✓ Casualties must be incinerated on site. In certain circumstances they can be left under licensed arrangements for collection *e.g.* rendering etc at an off-site venue, suitably protected from exposure and degradation, with the site disinfected after carcass removal.

✓ Wheel dips on routes to and from the farm are a wise precaution but must be adequate *i.e.* with automatic or manual spraying of wheel arches or undersides of all vehicles passing through. Such devices are now on the market.

✓ Place clear signs at perimeter exit/entrances, and provide your suppliers of goods and services with specific instructions on biosecurity-related measures you will expect them to follow.

✓ For bulk feed deliveries it is preferable to have your own off-loading blower hose, as many truck-mounted hoses are dragged across farmyards and rarely disinfected for fear of contaminating feed and for delaying deliveries.

**Remember** – all vehicles are a potentially serious disease risk. After our worst outbreak of FMD we in Britain realise this only too well. Think about minimising that risk where you are concerned.

## YOU AND YOUR STAFF - A PERSONAL DISEASE CHECKLIST

✓ Zoonoses are dangerous. You can easily catch some pig diseases.

✓ So wash your hands before you leave a main building. This is for your protection as much as the pigs. (The Japanese do this quite commonly and have a portable washbowl and soap/towel on a metal stand by the door.)

✓ Use a foot dip, replenished regularly, outside each main building.

✓ Wear one set of boots for outside (office and stores) premises, another for inside. The same with coveralls.

✓ If visiting a farm (like myself) leave your car off the farm boundary, *walk*, and carry your things with you! It's a pain, but I do it.

✓ Do not wear other people's boots.

✓ Shower and also wash clothes which have been on a piggery immediately after getting home.

✓ Wear disposable gloves when taking blood, castrating or doing post-mortems. Cover all cuts/grazes.

✓ Shower-in/shower-out (in my opinion) is probably over-rated, but continue until it is proved so, if ever. The quality of these facilities is generally far too low, anyway.

✓    Dust and mould spores can be very dangerous. Wear a proper, approved filter helmet when home-mixing or dealing with poor quality straw. Dust is much more dangerous to you than to the pigs. You live 15-150 times longer!

✓    After you have dosed a litter of young pigs against scour, your hands, and somewhere on your clothing may harbour as many as 10 billion (ten thousand million) of the organisms responsible. Only about 0.02% (or 2 million) bacteria ingested by you are enough to give you diarrhoea, especially if your current immune system is shaky.

So after treatment and flushing out the scoured areas in the pen with water, *wash your hands and change your overalls* before handling further piglets or eating rest-room food. Have special waterproof aprons to facilitate this. Stockpeople spread a lot of disease internally on a pig unit.

✓    Carbon monoxide from slurry pits stirred up before emptying is lethal – it nearly cost me my life, when I was a stockman. You cannot smell it. It knocks you out in a few seconds and kills you in a few more. Fortunately I fell down unconscious outside the piggery door, not inside it, otherwise this book wouldn't have been written.

✓    Likewise never enter a bulk bin. Cleaning must be done from the side through an inserted bulkhead door, with the top inspection hatch open. Manufacturers, PLEASE act on this!

**FINALLY - HAVE I ALARMED YOU? GOOD!**

# DOING A BIOSECURITY AUDIT:
## AN EXTENDED CHECKLIST/QUESTIONNAIRE

This list, while long, is not exhaustive. Nevertheless one or two items may make you or your suppliers sit up a bit!

✓    **Who visits the farm?** What precautions have you taken to keep them out/ sanitise them on entry and exit? This includes your veterinarian and his vehicle, bless him!

✓    **How do you dissuade bird vectors?** Netting over entry points, covering feed hoppers.

✓    **How do you control flies?** Have you a fly-control person?

✓    **How do you control rodents?** Have you a rodent 'king' in the same way?

✓    **How do you take in replacement stock?** Does your vet liaise with the vendor? Do you quarantine? Far away enough? For long enough? Inspect them frequently enough?  Have a disposal procedure agreed in advance for suspects/failures?  Buy from one source?  Never buy from markets?

✓ **Have you vehicle sanitising facilities**, both for you and for visiting vehicles if allowed entry?  Hard standings for visitors' cars?

✓ **Have you footbaths outside each main building?**  Replenished correctly and often enough.

✓ **Do you encourage/insist on handwashing/hand protection** with all surgical, farrowing and dosing tasks?

✓ **How clean** are your toilets/staff-rooms?

✓ **Have you put up correct guidelines in the staff-room** for inspection techniques/needle use?

✓ **And for AI procedure likewise?**

✓ **Have you made it clear** to all your suppliers that you insist that they follow a protocol when delivering to your farm?  Or at least discuss feasibility with them.

✓ And the same for your haulage-out requirements?

✓ **Have you asked them both** how, when and how often they sanitise their vehicles? Check up on this once a year with a visit.

✓ **Have you a locked perimeter fence and call bell/telephone?**

✓ **Do you have bulk bins on the perimeter?**  Do you use your own input hose?

✓ **Do you sanitise your bulk bins?**  It is scandalous that some bin manufacturers don't build in a bulkhead door on one side to get a ladder in. This should be mandatory by law.

✓ **If wet feeding** do you sanitise the tanks, line, troughs, periodically?

✓ **If not wet feeding** do you still sanitise your drinking water lines?

✓ **Do you load pigs out at a separate, specific place on the perimeter?**

✓ **Do you wash and sanitise this area after every batch is shipped?**

✓ **Do the washings drain away from the farm?**

✓ **Have you asked your vet** for a drug use record and guidance on usage/storage?

✓ **Do you sanitize your water header tanks and lines?**

✓ **Do you fog your buildings?**  Have you discussed this with your vet?

✓ **Do you do water tests/bacterial swabs periodically?**

✓ **Have you asked your feed supplier what salmonella tests he has done?**

✓ **Have you trained/made aware to your staff** that handling diseased animals needs *extra* hygiene and self protection/disinfection beyond their ordinary routine?

✓    **Have your staff got girl/boy friends on another farm?**

✓    **Do you ever borrow/share gear from/with other livestock farmers?** If so, how do you sanitise it before entry to the farm?

✓    **If outdoors, have you studied how to minimise disease spread** *e.g.* using the box of matches, field rotation/air and sunshine? Take advice.

✓    **Have you got chickens/sheep or goats in among the pig area?** Ban them!

✓    **How clean and tidy are the spaces between the houses?** The chaos I often see encourages vermin, the puddles encourage flies, spilt feed other predators – all of which bring in disease. *Make time* to have a blitz twice a year.

✓    **Have you a method of checking periodically that:-**

   • you are using the right detergent
      in the right way

   $\left.\begin{array}{l}\\\\\\\end{array}\right\}$ Dilution and cover rates are correct

   • you are using the right disinfectant
      in the right way

✓    Use dilution strips and have a check on what you've used from sales dockets covering the required total area every 3 or 4 months. The survey revealed that 3 farms were under-using the products by 35%!

   Get the salesman to monitor this for you. At 35% less sales it is worth his while keeping you straight!

✓    **Do you sanitise the slurry pits?** It is scandalous that many floor manufacturers/ house designers don't allow for a hinged access trapdoor in every pen to allow a lance to get in there as a minimum. We may need future legislation to make them do so.

✓    **The same for under slat areas.** These need cleaning too. Future designs of perforated floors should have hinged-upwards access. If not, removable hatch covers to allow the lance to get underneath the slats are essential.

✓    **Do you cover manure heaps against rains/leaching?** Where does the run-off go?

✓    **Do you inspect ad-lib hoppers for stale feed?**

✓    And wastage?

✓    **Finally**:

   Biosecurity also involves protective vaccination. You need to review your needs with your veterinarian twice a year. Pathogens change that quickly.

## HAS THIS LIST MADE YOU THINK? IF SO, GOOD!

# REFERENCES

Antec International - now DupPont International (1999) G&M Field trial report.

Cargill, C & Benhazi, T (1998) Proc. IPVS Conf. 3, 15.

Gadd (2000) 'The Revolution in Biosecurity' Vétoquinol Trans Canada Swine Seminars, March 10-15 2000.

Ibid (1999) 'Get it Clean Before You Disinfect' Antec International, Monograph

Ledoux, L. (Cidlines) (2010) Personal communication.

NCPBHP Denmark (1996) National Committee for Pig Breeding Health & Production Report     p 37

Sala et al (1998) Proc. IPVS Conf. 2 110 et seq.

Waddilove, J (1998) 'Disinfection on the Cheap?' Antec International, Monograph

Walburn, S. (DuPont) (2010) Personal communication.

# 9

# MYCOTOXINS - ANOTHER HIDDEN PROFIT THIEF

**Mycotoxin:** A toxic chemical produced by a mould.

## THE PROBLEM (MYCOTOXICOSIS)

The mould fungus can be innocuous or even dead but still leaves mycotoxin residues behind, commonly in stored and mixed feed, and especially in mouldy grain. Processing the grain, while it may destroy the moulds, still leaves mycotoxin residues behind. These need only be present at a few parts per billion in the feed (Table 1) to cause problems in pigs – infertility, anoestrus, prolapse, false pregnancies and embryo mortality; poor growth and vomiting. They can pass through sow's milk and remain behind in any slaughtered carcase. Damp straw is another common source of moulds.

I have over 40 articles and 8 surveys on the subject of mycotoxicosis filed away in my office, and must have read more than three times this number of articles and research papers on the subject. Trouble is, many of these worthy pieces are too technical for the working farmer and so tend not to get read, let alone understood.

Many articles repeat, almost to the point of boredom, the perceived opinion about mycotoxins - especially how they affect pigs visibly and directly. Of course this is important, but the insidious 'hidden' effects of mycotoxins are probably as significant to the producer's profitability and, I venture to suggest, even more so than the more familiar clinically identifiable effects of mycotoxin attack.

In this chapter I hope to provide something rather different for the working pig farmer and his manager in offering useful information I have collected about the problem for 40 years. I suppose I was one of the pioneers of some of the practical aspects of dealing with mycotoxin attacks in the late 1960s, purely because I encountered them in my advisory work on pig farms. We all knew so little about it in those days but realized that something important must be at work which was affecting susceptibility to disease. I discuss this early experience later on in this chapter which you may not have seen before. I add to it information from my own experience when encountering mycotoxins on farms across the world since then. Meanwhile I read all I could of what the experts were publishing and tried to put into practice what they recommended.

## IS MYCOTOXICOSIS A PROBLEM?

Yes, definitely, and it seems to be getting worse, especially regarding those 'hidden' aspects which must be affecting disease. We have heard a great deal about diseases like PRRS and PMWS. However I guess that the much wider range of diseases caused or aggravated by mycotoxin incidence probably costs you, the pig producer, even more than both of these serious and well-publicised viral diseases, but with 90% less publicity accorded to the part mycotoxins might play!

Why? Because much of the damage mycotoxins cause is insidious. The descriptions I have seen linked to mycotoxins such as 'silent killer', 'stealthy thieves' and 'the enemy from within' are fully justified, if only because the producer just isn't aware of how much these hidden predators are costing him. I provide cost-effective farm-based economic results of the counter-measures which can be taken against them later in the chapter.

## WHY DOES IT SEEM TO BE GETTING WORSE?

*   Changes in global climatic conditions which are conducive to the growth of moulds and subsequent mycotoxin production.

*   Increased use of by-products such as distillers grains (DDG) and alternative feed ingredients so as to reduce feed costs. These often have a high degree of mycotoxin contamination.

*   Improved analytical procedures which have increased mycotoxin detection in both conventional ingredients and complete feeds, as well as bedding. An example of this are new techniques in the laboratory which isolate so-called 'masked mycotoxins' which are bound to other molecules of glucose and some proteins. These cannot be detected by conventional analytical methods.

*   Increased intercontinental trade in raw materials grown from a mixture of crops grown under different climatic conditions which, when mixed with locally-grown crops, can result in a wide spectrum of mycotoxin presence made more acute in their effects on the pig by synergism, where two relatively harmless mycotoxins can combine to become toxic.

*   Badly-maintained harvesting equipment/conveyors and substandard storage facilities, all caused by periodic cash crises, can crack or erode the protective

grain pericarp before any drying process takes place and allows surface moulds to gain entry.

- Recent information confirms that when fungicides are not used at the correct dilution rates, this can increase moulds and therefore mycotoxin presence.

- Modern intensive agricultural cropping can lead to less crop rotation which can aggravate the prevalence of mould spores.

- Leaky feed and grain bins, containers with poor ventilation and excessive 'sweating' of the contents in unshaded bulk bins outside –all these can raise the moisture content above 14% (see Figure 1).

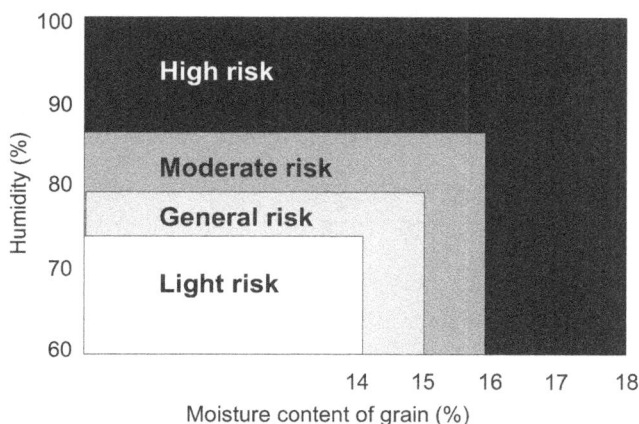

**Figure 1** Likelihood of mould contamination in stored grains

Devegowda in 2001 felt that mycotoxins were present in 27% of all pig feeds across the world. I am told that that the incidence in tropical pig industries is 'over 35%'.

## SO WHAT ARE MYCOTOXINS?

Myco = fungal. Toxins = poisons. Mycotoxins are a wide number of pervasive (widely-occurring) chemical agents produced by more than 300 different moulds. These residues vary from being harmless by themselves (but possibly made dangerous in concert with others in the feed or feed ingredients) to being highly toxic in tiny amounts – parts per **billion**. (Table 1) A few cases have been reported at parts per trillion, which is near to being unmeasurable!

I have been told that one part per billion is equivalent to one grain of sand in a bucketful - although I have not yet had the time to check this estimate!

While fusarium and aspergillus variants are the most common worldwide contaminants of pig feed, aflatoxin from aspergillus moulds are particularly prevalent in any tropical climate, affecting liver function which causes weight loss, loss of appetite and lowering immunity to a wide range of quite common diseases.

**Table 1.   Significant mycotoxins in pig feeds**

|  | *Main causal fungi* | *Major symptoms* | *Toxicity level (pigs)* |
|---|---|---|---|
| Aflatoxin | Aspergillus  spp. | Stunted growth<br>Liver damage<br>Feed refusal | 200-500 ppb |
| Zearalenone | Fusarium spp. | Estrogenic effects<br>(Eg. Abortion. Vaginitis,<br>Returns) Weak piglets | 200-300 ppb |
| Ochratoxin | Aspergillus spp. | Kidney damage<br>Tumors,  Scouring |  |
| Deoxyvalenol (DON) (Also known as Vomitoxin) | Giberella zea | Vomiting. Feed refusal.<br>'ISMT' | 1-10 ppm |
| T-2 toxin | Fusarium tricinium | Weak pigs.  Small litters | 100 ppb |
| Fumonisin | Fusarium spp. | Oedema.  Feed refusal | 5 ppm |
| Citrinin | Penicillin citrinum | Kidney damage | 500 ppm |
| Ergot | Claviceps purpurea | Necrosis  Kidney damage | 0.8% |

ppm = parts per million.   ppb = parts per billion.

There is no reliable treatment for mycotxicosis apart from the removal of suspect food or partial substitution of mycotoxin-free food to reduce the contamination down to, if possible, 0.1 mg/kg of feed (100ppb) which has helped. In practical farming terms this would be dilution with 90% of mycotoxin-free food. Even so, complete removal is the safest option. I do not advise dilution and 'feeding-off' contaminated food to dry sows in mid-pregnancy or to finishers in the later stages of growth, although I have witnessed this being done.

Certain feeds are more likely sources than others – although harvesting and subsequent storage conditions can influence the degree of contamination of any crop. Table 2 gives a broad outline of crops which can be affected by mycotoxin presence.

In tropical countries maize at high inclusion levels and groundnut are the primary suspect ingredients in a pig diet, with as much as 153 ppm in many feed samples of maize, groundnut with 200 ppm, (I would never, ever, feed groundnut to any pig as I've seen so many cases of trouble from doing this). Even as little as 0.5 ppm of a mycotoxin can trigger a wide range of disorders.

**Table 2. Suspect crops for significant mycotoxin levels**

| Crop | Aflatoxins | Zearalenone | Ochratoxin | DON, Fumonisin |
|---|---|---|---|---|
| Maize | + | + | + | + (High levels) |
| Groundnut/peanuts | + | + | + | + |
| Cottonseed | + | + | | |
| Copra | + | | + | |
| Soya | + | | | |
| Sorghum | + | + | + | |
| Cassava | + | + | | |
| Barley, wheat | + | + | + | + |
| Rice | + | + | + | |
| Oats | + | + | + | |
| Rye | + | + | + | |
| Millet | + | | | |

# HOW DO YOU KNOW IF YOU HAVE A MYCOTOXIN PROBLEM?

The following symptoms have all been associated with mycotoxins, but of course some of them could be due to other causes. If in doubt see your veterinarian.

### Breeding herd

Anoestrus, abortion, vulval swelling, vaginal prolapse, pseudo-pregnancy, increased weight loss in lactation, stillbirths, low viability piglets, splay legs, agalactia, udder oedema, reduced libido.

### Finishing and breeding herd

Reduced appetite, vomiting, rectal prolapse, liver and kidney damage, reduced feed intake, noticeably poor growth rate, scouring, respiratory oedema, skin irritation, increased water intake, immunosuppression (the 'Won't-go-away' disease syndrome). The very length, variety and frequency of these disorders supports my belief that mycotoxcosis is much more prevalent than pig farmers realize.

# THE IMMUNO-SUPRESSIVE EFFECT OF SOME MYCOTOXINS

## ISMT  - those 'Won't-go-away' diseases

Immuno-suppressive mycotoxicosis ( ISMT) is more of a problem than farmers **and quite a few of their advisers** realize.

I came across this many years ago almost by accicdent. In the 1970s I was plagued in my farm advisory work with what we called in our ignorance 'Won't go-away diseases'. Mostly they were disease-based but some were just sluggish growth and scouring in the young growers. Nothing we seemed to do long term seemed to help. Sure, things like Swine Dysentery, the pneumonias, Strep suis meningitis and E coli scour responded to in-feed treatment, and eventually the problems went away for a month or two – but back they came time after time.
Then I met Scottish pig vet. Sigurd Garden. He told me about mycotoxins and how he found the problems were more common after a damp harvest, after a wet spell, and when food was badly stored, especially on those farms who mixed their own feeds. He knew all about 'Won't-go-away' diseases, too.

*"John"*, he told me *" We're getting excellent results by persuading people to steam-clean their bins - preferably during hot weather so that  they quickly dry out.  Not just the 'crud' at the top but the sides and bottom exit trunks too. A really good clean, done twice yearly"*.

*"Trouble is that it is such an awful job. You can't get in there any sense – you need an inside ladder and a boy to descend it , as a grown man usually can't get his shoulders through the top hatch with the ladder present. You also need a safety rope for the lad and a special delivery pipe from the steam generator to go up-and-over. It is all too much bother and they just won't do it, but when they do, the 'Won't go away diseases' rarely come back"*.

*"What we need is to persuade every bulk bin manufacturer to put in a hinged bulkhead door in the side rather like those they have in ships, together with a swing- away exit-boot fixture. This will encourage farmers to clean their bins regularly and get rid of enough of the build-up of congealed food to stop it poisoning the pigs – or whatever is happening"* (Figure 2).

I followed this up with the (pretty reluctant) bin manufacturers and was told it would increase the price by 12% for the access door and another 10% for the swing-away boot. I haven't been very successful, have I, because such design features are still uncommon even after 40 years! Another example of people in our industry not

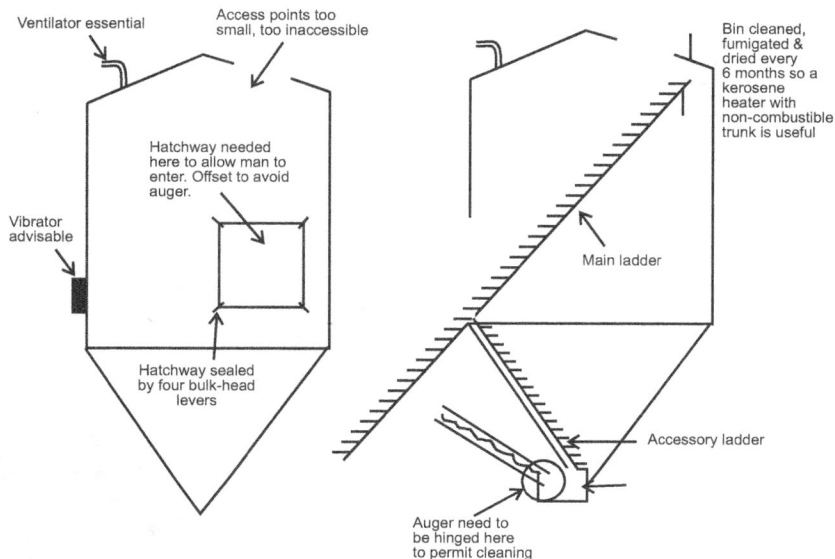

**Figure 2.** Bulk bin improvements/bulk bin hygiene

realizing what dangerous substances mycotoxins are and failing to take appropriate preventive action. If you examine the likely costs of even an occasional mycotoxin-induced disease outbreak at the end of this chapter, you will see how small this once-and–for-all extra price loading is (see footnote to Table 4) and what a false economy it is to ignore it.

**Bin manufacturers - will you <u>please</u> put in side bulkhead doors!**

## What could be the likely cost of ISMT today?

The indirect immunity problem could be a very large one. I have no idea how big but the indirect effect of slower growth and a higher disease incidence must, I guess, be reducing income/ raising costs by 20% on those farms who have (unknown to them) the problem of ISMT. I base this arbitrary assessment on over 20 'before-and-after' trials where the pigs were trundling along reasonably well with no obviously clinical symptoms of disease, yet when one or another (or more) of the precautionary measures outlined in this chapter were instituted, then there was a noticeable bounce upwards in improvement. Figure 4 cited by Dr Close gives one such example for young pigs where the effects are most dramatic, and Figure 5 (from the same source) for sows.

The evidence of the 20 trials reviewed above did not seem to involve the 'Won't go away' aspect of ISMT, in that clinical disorders were not apparent as shown in the tables below, but that the mycotoxins being brought under control by the countermeasures had been depressing performance all along – in some cases

In the 1990s bin cleaning - it if was done at all - was largely ineffective and very dangerous! (Pig International photo)

Praise be - a bin manufacturer who has seen the light. Side access door, simple swingaway exit boot and a succession of inspection windows.
(Uttley Ingham Ltd photo)

Typical bulk bins today with a tiny upper hatch and firmly bolted-on-draw-off boot. These are extremely difficult to clean properly (Author's photo)

If this farm could not be bothered to clean the outside of his bulk exit slide - what was the inside like! (Author's photo, 2009)

considerably below potential - however the producers had never noticed it. Hidden thiefs indeed!

## DOES BIN CLEANING WORK?

Tables 3 and 4 from my own clients` experience illustrate that it does.

**Table 3. Case histories of reducing or removing entirely the following problems after steam-cleaning bulk bins twice-yearly (Spring and Autumn).**

|  | *Positive result* | *Negative or inconclusive* |  |
|---|---|---|---|
| 1. Swine Dysentery | 3 | 1 |  |
| 2. Abortions in gilts | 2 | - |  |
| 3. Prolapses | 2 | 1 | (other measures tried) |
| 4. Mummified | 1 | - |  |
| 5. Returns to service | 2 | - |  |
| 6. Respiratory infection (non-specific) | 3 | 1 |  |

**Table 4. Case histories of before-and-after growth rates on 3 farms where bulk bins were steam-cleaned.**

|  | *Weight range kg* | *ADG before, g/day. (Av.of 12 months)* | *Disease situation* | *ADG after, g/day (Av.of 6 months)* |
|---|---|---|---|---|
| Farm 1. | 60-90 | 572 | Little | 652 |
| Farm 2. | 30-90 | 607 | SD and ileitis | 781 |
| Farm 3. | 25-86 | 616 | None detected | 697 |

Econometrics: Farm 3 bought new 10 tonne capacity bins (filled to 8 tonnes) and paid extra for bulkhead doors and recorded an extra 17 kg of saleable lean meat per tonne of food fed, despite their current high health status. This paid for the extra cost of putting a side-access door into each bin plus the extra labour cost of sanitation twice a year, some 12 times over - an REO (Return on Extra Outlay) of 12:1. Money well-spent!

## CAN LEVELS OF MYCOTOXINS BE TESTED?

Yes, but the current advice is that because of the very low levels (parts per billion) the test could be inconclusive. Representative sampling has to be very detailed and onerous due to isolated pockets of contamination, and most tests are expensive and do take time anyway. Most scientists would disagree with me over the need for testing – and I accept that - but I am practical man when it comes to my clients spending money and I would rather they used it used to reinforce the defences than be told what may be there or not.

But because of synergism, see below, is any farm test worthwhile? My own opinion is that producers, rather than worrying too much about testing, should take every precaution against moulds being introduced to and developing in the feed and then, despite taking such precautions, go on to assume that some dangerous mycotoxins will slip through the net and still be present in the feed. And so use in-feed absorbents **as routine** to lessen the risk  I do not think this routine precaution is a needless expense – see the payback figures later on in this chapter.

Whether you use some of your working capital to do some tests is up to you and your veterinarian.

## SYNERGISM

Synergism is defined as "The joint action of agents so that their combined effect is greater than the sum of their individual parts."

Much work is being done to show that synergism does exist between certain mycotoxins  A test might show that a dangerous mycotoxin is satisfactorily below the safety level.  Safety levels have been established in the laboratory **when the mycotoxin has been added and then its presence tested on its own**, and that the farmer or his vet. need not worry if the test reveals it to be below the published safety level.

However if another mycotoxin is present, maybe also below its own safety threshold, then the two together, due to the synergistic effect, can become harmful. Commonsense suggests to me that because of this possibility, on-farm identification testing  might not be so helpful after all?  Science will eventually tell us how likely this is, but the growing number of papers I've seen recently suggests that mycotoxin synergism could indeed be a problem.

World trading today in feed commodities ensures that parcels of contaminated food, probably containing different mycotoxins from raw materials grown in many different countries and conditions come into contact with each other well before the pig eats the food, so that synergism is more likely.

## SO WHAT CAN THE FARMER DO TO STOP MYCOTOXINS FROM GETTING ON TO HIS FARM, AND TO LESSEN THEIR IMPACT WHEN THEY DO?

### Keeping mycotoxins out

• Do everything you can to find out if deliveries of grain, vegetable proteins and finished food are declared free of mycotoxins or, as in EU legislation, are present below legally stipulated safety levels. Buy from such sources.

- Do not feed groundnut to pigs (personal opinion).

- Do not buy 'hot' or mouldy grain, feed or bedding.

- Train your staff to recognize musty feed and bedding by sight or smell and report it to you at once. Be prepared to discard such feed safely and empty out the bin completely.(Dependent on the degree of visible/smellable mould it is unwise to `feed it off` as is sometimes advised although I have seen it done without apparent effect at a 10:1 dilution or more to gestating sows in mid- pregnancy and to finishers at in the final 2 weeks before slaughter - never to other stock or to sows at other times. Note: Such dilution practice is illegal in some countries).

- www.hgca.com is a good source of internet advice on grain quality re insects and moulds.

# MASKING

Improved analytical techniques have brought this to our notice.

Some mycotoxins can be bound to innocuous molecules in the feed such as glucose and some proteins. These cannot be detected by more conventional and cheaper analytical methods. This is another reason (added to synergism, q.v.) to strengthen my opinion that routine protective measures are, at the present stage of our knowledge, a more pragmatic way for the producer to spend his money and energy than attempts to identify and quantify the level of contamination on his farm favoured by the specialists. Laudably this is probably done to persuade him to do something about mycotoxin presence – but my advice has always been for him to assume that his farm is under attack anyway and to take preventive action as routine

That this is cost-effective can be seen at the end of this chapter.

### Lessening the effect of mycotoxin attack

- Dry all grain to less than 12% moisture if possible, see Figures 1 and Table 6.

- Keep food as dry and as cool as possible.

- Don't over-order feed as it gets stale, as well as giving time for moulds to develop.

- Ventilate all storage areas well - especially important in a high humidity climate.

- Keep your feed troughs and self-feed hoppers far cleaner than you do. Last year I noticed stale, caked food in 35% of self-feed hoppers for dry sows - which must have been full of mycotoxins! When we established a regular cleaning routine on one such farm the overall disease level dropped immediately.

- Shade bulk bins to lower internal condensation.

- Clean your bulk bins, feed hoppers and wet feed lines regularly. I suggest twice a year during a warm spell to allow the bins to dry out naturally. The use of a kerosene garage/space heater with an 'up-and-over' hose is useful at other times, as it is for drying out nurseries properly after 'All-in/All-out' sanitation.

  Be especially careful when entering a bulk bin. Have a person nearby within call. This is why – for safety reasons alone - it is essential to have a side entrance hatch so that anyone feeling dizzy from breathing internal gases can quickly slide down the ladder into fresh air. Open the top hatch before starting.

- Never let wet feed mixing tanks build up a congealed layer of food above the fill level. This is hugely dangerous.

- If mycotoxins threaten, such as in a wet season, use a myco**stat** in the grain and/or feed so as to prevent the formation of moulds. Proprionic acid is commonly used as a first line of defence but has its drawbacks and I much prefer one of the proprietary products, which are more expensive but easier and safer to use. Costs are covered at the end of the article under 'Full Protocol'.

**Table 5. Comparison between proprionic acid and a combined mould inhibitor (Mold-Zap).**

| Treatment | Relative active agent content | % loss after 6 days | Corrosive effect |
|---|---|---|---|
| Proprionic acid | 100 | 67 | 100 |
| Mould inhibitor complex | 70 | 3 | 4 |

Source: Alltech (1998).

A **mould inhibitor** is a product used to prevent / lessen the formation of moulds in grain, protein supplements, final feed (or bedding) in the first place and is different from a **mycotoxin binder** which reduces or eliminates an most mycotoxins which manage get through into the feed.

The use of a mould inhibitor does not invalidate the need for a mycotoxin binder, and both used together are still cost-effective on many farms

## The drawbacks of clays

Certain clays, like bentonite which are available and cheap, have been used quite effectively as nutritional binders for the major mycotoxins in feed.

Trouble is, a lot is needed (a minimum of 4kg/tonne and up to 10 kg/tonne for some of them). Moreover, they can bind other useful nutrients such as vitamins and minerals and antibiotics.

They pass straight through the pig and so form a hard sediment in the slurry pits which is difficult to dislodge. Even at the lowesr effective usage level of 4kg/tonne, a 100 sow farrow-to-finish unit will accumulate 2.2 tonnes of clay binder in the slurry pits every year, which if left there too long will not be easily removed by the suction hose, if at all.

I know, as I once had the strenuous job of removing it by hand using a mattock!

Due to their source, clays can be contaminated with deleterious materials such as heavy metals, and dioxins and anti-nutrient factors (ANFs).

# DEALING WITH MYCOTOXINS WHICH 'SLIP THROUGH THE NET

Some will, whatever you do. It is, in my opinion based on experience from many farm visits' essential that you use, **as routine and not just when you suspect mycotoxins to be present,** a mycotoxin **absorbent.** These products vary according to what toxins they absorb but I find 'Mycosorb' (Alltech) based on glucomannans particularly suitable for the primary mycotoxins – aflatoxins, zearalenone and fumonisin.

## Glucomannans

These are natural nutrients in the food in the form of sugars,derived from yeast. These lock-up a much wider range of mycotoxins (clays mainly affect aflatoxin) at an inclusion rate of at least 8 times lower than with bentonite. Table 6 shows the broad spectrum attack possible with this type of product.

**Table 6. The mycotoxin binding capacity of glucomannan**

| | |
|---|---|
| Aflatoxins (B1+B2+G1+G2) | 85% |
| Fumonisin | 67% |
| T2 | 33% |
| Zearalenone | 66% |
| DON | 13% |
| Citrina | 18% |
| Ochratoxin | 12% |

Source: Trenholm (1997).

# IMMENSE ABSORBTIVE AREA

The second reason why I find Mycosorb so effective is its vast absorbtive capacity. The Irish Agricultural Institute (Teagasc) has defined Mycosorb's absorbtive area as follows...

•   One gramme of Mycosorb eaten by a pig provides a 'catchment area' to absorb mycotoxins of no less than 20 square metres.

•   This means that one growing pig's digestive system will be provided with 8 square metres of new mycotoxin-capturing absorbtive area every day when Mycosorb is added at Alltech's recommended inclusion level per tonne of feed.

•   This also means that each tonne of feed contains the equivalent surface area of 20,000 square metres (2 hectares!). It would take a whole army of mycotoxins within the food to get past that obstacle!

An absorbent based on glucomman is effective (Figure 3).

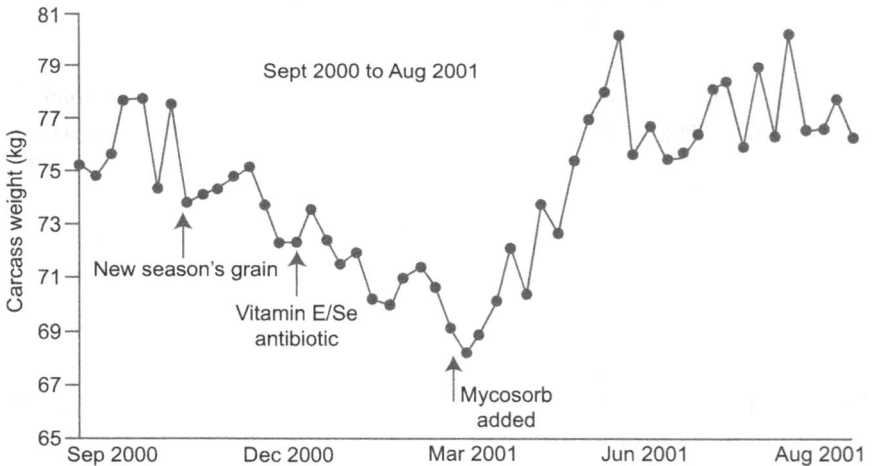

**Figure 3.** Effects of mycotoxins on carcass weight and effects of Mycosorb (Edwards, 2001, via Close Consultancy (2009)

# CHECKLIST – WHAT TO LOOK FOR IN A GOOD MYCOTOXIN ABSORBENT

1. **Binding capacity**

✓   There is considerable variation, so ask for the supplier's overall claims for the product as a starting base on which one to choose.

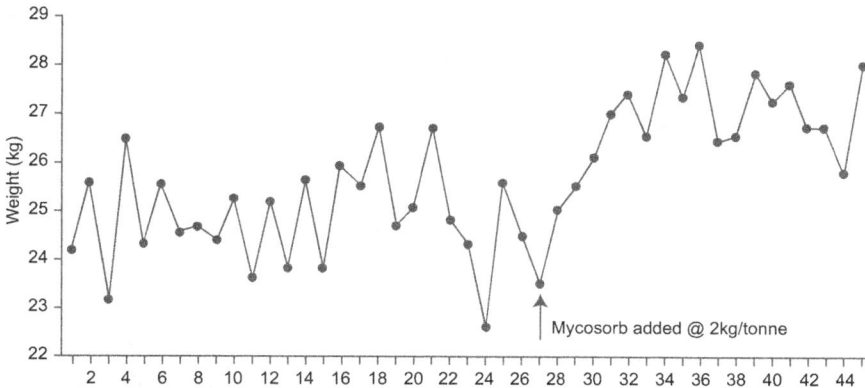

**Figure 4.** Nursery exit weights (at 68 days). Source: Close Consultancy (2009)

✓ This can depend on two factors (a) its ability to absorb mycotoxins present at **very high concentrations** so that the final concentration in the feed is below the threshold to cause toxicity.

✓ And (b) its ability to absorb **low concentrations** of mycotoxins – e.g. 10 to 40 ppb, so that even low levels which can still cause sub-clinical performance are controlled.

2. **Ability to absorb mycotoxins rapidly in the gut**.

✓ Mycotoxins can get into the bloodstream within 30 minutes and start to affect performance. An ideal binder must absorb the maximum amount of mycotoxins within this period.

3. **Stability over a wide pH range**.

✓ It is important that a binder can strongly absorb mycotoxins in the alkaline to acidic changes which occur all the way down the digestive tract. Clays are poor at this.

4. **Sustained degree of attachment**.

✓ The mycotoxin needs to remain attached to the absorbent throughout the digestive process and not lose hold of it, so that it is excreted safely out of harm's way.

5. **Low effective inclusion rates.**

✓ Some are effective at from as low as 0.05 to 0.2%. This means they can be as effective at 500g/tonne as clays at 4 kg/tonne, thus leaving room for more nutrients – important in baby pig feeds and sow and gilt lactation diets.

6. **Proven independent research data**.

✓ On live pigs i.e.. *'in vivo'* as distinct from *'in vitro'* in the laboratory. There should be as large a number of trials as possible to build up a reliable overall picture of the binder in question. Called an 'holistic approach' (Rosen 2006).

Putting these questions to manufacturers should help you come to the right decision on which product to chose – and then, and only then ask about inclusion cost per tonne.

Figures 4. and 5 are but two of the many pieces of confirmatory trials which show the value of using a proven mycotoxin absorbent.

Fig 4. confirms my own quite arresting findings (see cost-benefits below) where the young pig is concerned.

| Year 2003 | Control | Control |
|---|---|---|
| No. of sows | 4019 | 4254 |
| Litter size | 9.48 | 9.72 |
| Year 2004-2006 | Mycosorb | Control |
| No. of sows | 3815 | 4077 |
| Litter size | 10.75 | 10.82 |
| Difference | 1.27 | 1.10 |
| | +0.4 piglets/sow/year | |

**Figure 5.** Effect of Mycosorb on commercial sow production
(Henman, 2007 via Close Consultancy, 2009)

# THAT INCREDIBLE ABSORBTIVE AREA OF A LEADING MYCOTOXIN ABSORBENT!

## Question I'm often asked...

*How can a pig's gut possibly stretch to the size of a tennis court – that cannot be possible!"*

## Answer...

It doesn't need to 'stretch'. The combined absorbtive surface area of the millions of protruding villi in the small intestine is vast (the villi do the job of absorbing nutrient chemicals). Rather like hundreds and thousands of closely-packed haircombs attached to the gut wall - with their teeth being the villi sticking out into the digesta. The mycotoxin absorbent protects this huge absorbtive area by the very size of its 'moppimg-up' capability over and around the villi. As can be seen from the size of this protective shield quoted by the Irish workers, it has more than enough to do a good job of rendering a very large proportion of the mycotoxins harmless. The tiny amount that might still survive is almost always small enough to be dealt with by the pigs own immune system.

**What about the cost of an absorbent like Mycosorb?**

I find the important criterion in persuading pig farmers to help defend their pigs and eventually their profits from mycotoxins is to get them **to convince themselves** that adding an absorbent to the feed should be accepted – and afforded - as normally as adding a nutrient like vitamin E or salt. Price-wise it needs a new mind-set based on the acceptance of what mycotoxins are costing them in so many mostly hidden ways.

So let us look at this.

# THE ECONOMIC DAMAGE MYCOTOXINS CAUSE

I analysed the results of mycotoxicoses from 51 cases studied over 10 years and published them in 2005 (Gadd, Pig Progress **21**, 3 p.19).

Where the records were comprehensive enough, including veterinary diagnosis, a summary was… ( n = number of cases recorded)

**Small pigs**

Relatively mild outbreaks from 3 to 35 kg. lasted between 4 to 8 weeks and raised production cost by 8%.( n.= 8) Severe outbreaks lasted 4 to 5 weeks and raised production costs by 24% but on some farms gross margin **at slaughter** was reduced by 80%.(n = 15). Note this 'hidden' cost so often not appreciated because it doesn't become obvious until slaughter.

**Gilts**

Mycotoxins delayed entry into the herd causing whole herd empty days to rise by 15 to 34 days per litter. This raised production cost by 25% over a one-month breeding period. (n = 5). Another hidden cost caused by mycotoxicosis.

Note : A further long-term cost is difficult to assess at our present state of knowlege – if and how a mycotoxin attack in the young female affects her future breeding capability? Experts I have talked to suspect that this may occur and have kindly tried to explain why – but have managed to lose me after the first few sentences! (I gather that her reproductive organs are at risk of being damaged).

Any such effect has to be discounted from these equations.  But common sense on these grounds suggests that the producer should be especially vigilant over the mycological quality of the gilts feed, feeding management - and bedding, if used. On my farm visits I have 'smelt' gilts on musty bedding , and that is not good

## Sows

Mycotoxicosis of varying severity raised production cost by between 30% to 74% over periods lasting from 6 weeks to 6 months due to anoestrus, returns, abortions, mummifieds, prolapse, splaylegs and secondary infections.(n = 23).

## Cost/benefit analysis (51 farms, 1993-2004)

Outbreaks caused by various mycotoxins suggests they increased production costs of between 18% to 74% lasting from 4 weeks to 6 months or more.

The cost of a full year's prevention protocol (buying more expensive grain, drying if needed to 15%, better storage/ bin cleaning, cleaner hoppers/feed troughs, a mycostat and mycoabsorbent in the feed when needed), all these precautions averaged 9.5% extra outlay. Of this the feed additive treatment cost 2% to 3%, based on Alltech products and their costs in the UK at the time.

# …. AND THE PAYBACKS?

Dividing the benefits cited above by the costs…….

REO's ( returns on the extra costs)

*For the whole preventive protocol ….* between **1.9:1** (i.e.. 18%  divided by 9.5%) **and 7.8;1** (74% divided by 9.5%).

*Just for in-feed protection (based on Alltech products, 2010).*
**Pigs to 35 kg**.

The average rise in production cost <u>without</u> in-feed protection due to all mycotoxicoses was 21% or €8.82/pig. The protective  in-feed production cost varied from €0.38 to €0.43/pig., thus the…

### *REO would have been €8.82 ÷  €0.43 = 20.5:1*
 Notice the huge benefit from using protection at this early stage of growth, saving 20 times the investment. 20 times! Very few investments  provide that amount of financial yield.

*In AIV (Annual Investment Value) terms* (see Business Section, New Terminology) The paybacks are equally impressive, as the benefit from the additive is turned over some 6 times a year at least, nurseries being re-occupied about six times each year. (AIV 6 x €8.82/pig = 53) assuming the time in the nursery is 8 weeks plus 4 days cleaning down and disinfection between batches.

## Gilts

The cost of delayed entry into the herd and lower conception rate in the gilts parity (i.e.. parity 0) was €31 per sow (a 17.4% rise in production cost for the whole herd) and the cost of in-feed mycotoxin protection all the year round was €9.60 (a 2.3% rise in production cost).

*REO therefore 31 ÷ €9.6 = 3.25:1.*

Notice that if the gilts are slower entering the herd, how the whole herd's performance suffers.

## Multiparous sows

Total production costs rose between 30% to 74% (€126 to €311/sow) over an average period of 4 months, making the penalty as much as one third of the annual cost per sow, at €42 to €104/sow.

The cost of adding the in-feed protectants for a full year (not just 4 months) was €7.20/sow.

*REO therefore was 42 ÷ €7.20 = 5.8.1; to 104 ÷ €7.20 = 14.4:1*

Note: The above econometrics are expressed in Euros as the trial evidence was from several Continental farms and I needed to use their own prices and returns at the time to provide a consistent picture.

This is of little consequence as it is the **comparisons** between treatment and no-treatment which are important, and this should be similar in any currency.

## Conclusion

The on-farm data from a reasonable sample of farms and different outbreaks seems to fully justify, economically, the adoption of both a full preventive protocol and the use of modern mycotoxin control products in the feed.

# OVERALL SUMMARY AND CONCLUSIONS

- Mycotoxin effects on performance and disease incidence are underestimated by pig producers and some of their advisers.

- Their effect on disease incidence due to the suppression of immunity is increasingly being recognized by researchers, but less so by producers.

- Much better cleaning of bulk bins and feed receptacles is advisable, and improvements in bulk bin design to facilitate this are too slow in appearing on the market.

- After this is done there are excellent, carefully researched, well-proven and affordable countermeasures the producer can use, on the pragmatic assumption that for various reasons some toxins will get through, if only because such tiny levels of them can still affect performance.

- The paybacks look to be encouraging and farmers should consider making provision in their supplemental feed costs for the products and precautionary management measures described.

- While several authorities advise taking samples for identification and quantification – I question the need for this on the busy farm as such problems of synergism and masking ( q.v.) could distort the findings. Producers should concentrate their money and efforts on the protective measures outlined in this chapter, which are…

**Awareness.**  Monitoring and detection (visual and olfactory).

**Inhibition.**  (Reduction of mould presence).

**Incapacitation**  (Binding and inactivation of their mycotoxins).

# REFERENCES

I am grateful to Dr. Jules Taylor-Pickard for reading this chapter and making useful suggestions, and to Dr. William Close for use of information in figures 4 and 5.

Devegowda,G (2000) Mycotoxins : hidden killers. Is there a solution? Alltech Symposium Procs., 2000.

Connolly A.(2005) Prevalence of mycotoxins in Europe. Procs. Alltech Mycotoxin Seminar series, (booklet) p.145.

Diaz-Llano,G and T K Smith (2005) The efficacy of Mycosorb in preventing the deleterious effects of grains naturally contaminated with *Fusarium* mycotoxins on lactating sows.  Procs. Alltech 21st Symposium May 22-25. Poster session.

Fink-Gremmels (2005) Mycotoxicosis in Animal Health Alltech Mycotoxin Seminar series, pps.19-4

Gadd, J. (1975) Action now needed for those 'Won't Go Away Diseases' Pig Farming **10** 14-15.

Gadd ,J. (1998) Clean bulk bins lead to a clean bill of health. The Pig Pen **5**, 1, 1-4

Gadd ,J (2008). Mycotoxin Synergism – a new twist to the story. Pig Progress **24**, 7, 2.

Gadd, J.(2008) Economics of mycotoxin prevention and control.PigProgress **24**, 10, 11.

Meissonnier, G M et al. (2006) Aflatoxicosis in piglets… selective impairment of enzymes and the immune response. Procs Alltech symposium 22 (poster session).

Oswald I P et al. (2005). Mycotoxin effect on the pig immune system. Alltech Mycotoxin Seminar series pps.32-57.

Smith T K and G.Diaz-Llano (2009) Review of the effect of feed-borne mycotoxins on pig health and reproduction. *Sustainable Animal Production*. Eds. Aland and Madec. Wageningen Academic Publishers, The Netherlands. pps 261-272.

Trenholm H L,(1985) Toxic chemicals in swine feeds. Report, ARC Canada.1985

Varley, M. (2004) Where are we with the control of mycotoxins? Pig progress **20** 24-25.

Wyatt R D (2005) Mycotoxin interactions. *The Mycotoxin Blue Book* (Ed Diaz). Nottingham University Press (2005) 279-294.

# BUSINESS AND MANAGEMENT SECTION

# 10

# WHY WE NEED MORE COST-EFFECTIVE TERMINOLOGY

Much has appeared in the technical and lay media concerning problems associated with industrial pig production – pollution, disease control, training adequate labour and the perception of food safety and pig welfare by the consumer. This is true, especially from large new units.

## NEW TERMS FOR BUSINESS FARMERS

But much less remarked on by the press is the rising trend for pig farmers and managers to become increasingly business-orientated – to measure their progress by *profit*, not necessarily by *physical performance*. The day of the 'art' of keeping pigs by dedicated enthusiasts (apart from those worthy souls keeping the gene lines of rare breeds alive) rather than as vehicles for making money is over. Pig owner/producers of the latter half of the 20th century needed to see, touch and smell their charges daily; until recently they considered themselves as *pig* producers. Now after several storms of low pig prices they realise they might be in the business of producing *meat*, not pigs. As meat producers, not pig producers. Falling profits have accelerated this change in perception. In other words, they do indeed now see themselves as pigmeat producers. With more businessmen pig farmers around, their need to use terminology based on profit, not necessarily performance, is also essential.

This chapter makes a case therefore for new terminology based on profit, but which also embodies most of the performance-orientated terms which have served us so well as farmers in the past, and which, of course, the academic will continue to use with the precise measuring facilities available to him.

At the sharp end the two will continue to run together for a few years more, but as more businessmen producers realise how simple profit-related terms help them make better business decisions, the use of the New Terminology will become increasingly common.

The examples used to illustrate the use of the New Terminology draw heavily on matters considered in other sections of this book.

# RESEARCH AND SUPPLY TRADES NEED THEM TOO!

Both academics and especially feed and seedstock company technical staff will need to familiarise themselves with the new concepts. This is particularly important to feed and feed supplement company nutritionists as the New Terminology makes it easier for them to sell increasingly nutrient-dense and more expensive (per ton/tonne) feed in a pig production market place which is (sometimes to its own detriment) increasingly cost-reduction conscious.

The New Terminology also helps prioritise in the businessman farmer's mind how best to invest the 8 to 15% extra to his feed raw material costs which he can use to add value to his feed. Nutritional biotechnology products – like organic selenium, iron 'proteinate', the Bioplexes (especially zinc/methionine combination) mycotoxin absorbents and enzymes are all technical innovations which at first sight may seem to cost a substantial amount per kg in the bag or drum. Added enzymes especially are an exciting new area for pigs, too, especially in the future, and they are not cheap either.

## But which to use . . .?

Because of the relatively very low rate of use of all of these items, they can give quite dramatic *economic* returns of 10, 20 or even 60:1 across a year's use. *Such payback far outstrips their percentage improvements in physical performance terms*. This is why we need new measurement terms to highlight these hidden advantages, and keep pace with what they reveal in profit terms. In addition, because of this low rate of use both physical and economic space in the diet can be freed up for investment in other nutritional improvements – increasing amino-acid or energy intakes among increasingly appetite-challenged pigs, for example, and/or better/safer quality raw materials in the young pig.

# SCOPE OF THIS SECTION

I am certainly no economist or mathematician, but from working at the sharp end recognize that we need to move on in our world of business pig production and have available a new set of easy to calculate measurements which will get us faster to making a satisfactory profit.

On-farm consultants are successful when their advice is seen to generate more profit for their clients. Many years ago I began to realise that the physical measurements we all used were holding me back in persuading pig producers to take certain actions, and that an alternative set of terms was needed to reinforce my advice. To convince wary pig producers that a product or system which seemed expensive was in fact very cheap – far too cheap to ignore.

The complete New Terminology concept covers a wide range of disciplines *e.g.* EBV (Estimated Breeding Value) and WWSY (Weaner Weight/Sow/Year) in the case of breeding stock; PLR and ILR (Profit and Income to Life Ratios) in the financial, housing and equipment fields; and AMF (Absolute Mortality Figure) and others affecting disease costs in the veterinary area.

AIV (Annual Investment Value) is a valuable tool as it encourages the producer to look at his capital the same way as his bank manager. It is woefully under-used by pig farmers.

This chapter is mainly confined to those new terms useful in the economic assessment of the nutritional and feed supplement fields (examples are given in Table 1 but later on I also describe other terms mentioned in the book).

**Table 1.  The new terminology (as it affects nutrition & nutritional supplementation)**

| *1. Existing Terminology* | *2. The New Terminology* |
| --- | --- |
| Food Conversion Ratio (FCR) | Return to Extra Outlay Ratio (REO) |
| Average Daily Gain (ADG) | (Saleable) Meat/Tonne of Feed (MTF) |
| Cost/kg gain | Price Per Tonne Equivalent (PPTE) and Cost per % liveweight gain |
| Return on Capital Investment (ROC; ROI) | (Producing) More for the Same Cost (MSC) |
| | (Producing the) Same at Less Cost (SLC) |
| | Annual Investment Value (AIV) |
| Weaners/sow/year | Weaner capacity |
| % Replacements/year | Sow Productive Life (SPL) |
| % Mortality | A.M.F. Actual Mortality Figure |

# WHY THE NEED FOR NEW MEASUREMENTS ?

There is nothing radically wrong with the old terminology in column 1 above. After 70 years of use, it is certainly very familiar! Even so, it is not good enough for today's conditions. We can do better. We <u>need</u> to do better!

**Problem 1**    The existing terminology is based largely on *performance*. Today *profit* matters on pig farms far more than it ever did. You can have very good performance but still make less profit (Table 2).

**Problem 2**    The existing terminology also mainly covers *costs* of production (*e.g.* cost/kg gain). Again, a producer can have nice low costs but still suffer reduced *income* (Table 4) which can turn low costs on their head and result in less nett profit.

**Table 2. Profit rather than performance.**

The following very interesting results come from the records of over 40 good producers – either clients of mine or those supplied by several breeding companies keen to sell their breeding stock.  Sale liveweights averaged 105kg.

| (n) | Pigs weaned Sow/Yr (Pigs sold Sow/Yr) | Wt of saleable meat/sow/year (kg) | Wt of saleable meat/tonne feed (kg) | Relative nett profit per pig sold (%) |
|---|---|---|---|---|
| 13 | 30.1  (27.1) | 2846 | 447 | 110% |
| 8 | 27.2  (25.2) | 2646 | 428 | 108% |
| 23 | 25.0  (24.1) | 1808 | 482 | 119% |
| Typical | 21.8 (19.7) | 1635 | 390 | 100% |

Comment:

1. The producers at 25 pigs weaned per sow per year made the most profit, not necessarily the paragons at and over the `Magic 30`. What influenced the profit situation seemed to be that the top producers spent a great deal of their time and skills – admittedly absolutely first-class - on the breeding aspect of the business, so some of the growing finishing capital allocation/attentions to detail tended to let them down as a result. This was apparent with several top breeder-finishers.

2. It is interesting to see the increase in the amounts of saleable meat from both sows and feed-used-to-slaughter from those published only six short years ago. This is due to not only to the considerable rise in pigs weaned/sow/year among the better breeders but also to the heavier deadweights with better killing out percentages achieved these days.

3. Notice how meat produced per tonne of feed fed is a more reliable guide to profit than meat produced per sow per year. This is clearly shown in Table 3.

**Table 3.  Meat per tonne of growing/ finishing feed is a more reliable guide to economic performance than meat sold per sow per year**

| Meat sold per sow per year over and above 'typical' (in Table 2) | | | Meat sold per tonne of grower/ finisher food fed above 'typical' | |
|---|---|---|---|---|
| Weaned | kg | % | kg | % |
| 30.1 | +777 | +37.6 | +57 | +14.6 |
| 27.2 | +577 | +27.9 | +38 | +9.7 |
| 25.0 | +462 | +22.3 | +92 | +24.0 |

Comment: Compare the percentage improvement columns in both cases.
The feed-related results are very different to the sow-related figures.

**Table 4. Real-life example of adjacent farms in the same family business showing relative costs and sales income over a 3 year period**

|  | Low cost farm | Higher cost farm | Position of lower cost farm |
|---|---|---|---|
| Pigs produced/year | 5320 | 5610 | |
| Cost/kg gain (p) | 36 | 40 | 4p/kg savings |
| Deadweight meat sold year (1000 kg) | 345 | 387 | BUT . |
| Income per pig sold (£) | 66 | 80.7 | |
| Income per liveweight kg sold (p) | 112 | 117 | 5p/kg lower income |
| Nett margin per pig sold (£) | 8.02 | 8.68 | 66p/pig less margin |

***Comment*** : Lower cost/kg gain but much less profit per pig sold! Costs only give you half the story.

Source: Clients' records (1998)

Of course reducing costs is a good thing. But never to the extent that it affects income to a greater degree. The existing terminology doesn't necessarily identify or forewarn of such situations or trends, **which the new terminology does.**

Certainly, use both together. Scientists who need to assess physical performance accurately and use terms such as food conversion will prefer to stick with the old and familiar terms. But when dealing with pig farmers and the feed trade who supply them, their advisers will find an increasing move towards the new terms will help the feed compounder, the veterinarian and the pig producer.

# HOW THE NEW TERMINOLOGY HELPS THE FARMER

The new terminology clarifies the issues in the customer's mind in two ways.

1.  It sets the quantity of his primary ***input*** (food) against the total ***output***, meat (in our case, pork).

2.  Farmers are still very concerned with price per tonne of feed. While we can argue the validity of this attitude, salespeople can use it to help them sell successfully by using the new terminology to relate the value of the extra meat sold on a cost per tonne of food basis.

This is of great help to the farmer in arriving at a decision because pig producers all know their current feed cost per tonne and are equally familiar with the current price they get for their pigs, or more accurately saleable (*i.e.* deadweight/dressed carcase) meat.

**PPTE** The new terminology, in this case PPTE (Price Per Tonne Equivalent), presents performance data in a way which is very easy and swift for the farmer to do an econometric* calculation himself, because it shows the benefit if any, in terms of what it saves him on a cost per tonne basis. If he does this calculation himself he immediately **convinces** himself. As conversion to PPTE is a childishly simple calculation, he is very likely to do it.

*Econometric = the measurement of cost effectiveness.

# HOW TO CALCULATE PPTE

1.  Work out the MTF (saleable Meat produced per Tonne of Food used) of the new feed, growth enhancer, mycotoxin absorbent protein increase or whatever, which is on offer/what the research advises. (See page 242 on how to work out MTF). Supposing the MTF is 20 kg/tonne of feed better.

2.  What is your current pig price. Say it is £1.40/kg deadweight.

3.  Multiply the two together- 20 x 1.40 = 28.

4.  The monetary benefit is £28/tonne.Which means on the evidence supplied the change could provide £28 more income from each tonne of feed used.

    Or alternatively, a £28/tonne reduction in the current price paid for the feed.

5.  If the feed cost was, say, £180/tonne then the Price Per Tonne Equivalent is £180 - £ 28 = £152., or a price reduction of 18%.

    Many livestock farmers rightly or wrongly seem hooked on price per tonne, and PPTE accommodates their attitude, enabling them to assess any advantage in what they are being asked to do or purchase in their preferred price per tonne terms.

    I find it surprising that feed and feed supplement sales people are slow to use the term in this way.

# SO WHAT'S WRONG WITH THE EXISTING TERMINOLOGY ?

**FCR**  **FCR (Food Conversion Ratio)**

A useful yardstick *if it is measured accurately*. This is difficult to do on a busy working farm. Careful tests suggest that even if they attempt to keep track of it (under a third do) pig farmers still get it wrong by about 0.2:1, which is equivalent to a rise or fall in their price per tonne of feed of 15%! Most take a decision to buy feed or not on far less of a price difference than that.

While Table 5 is over 30 years old now, things don't seem to have improved much, as two investigations carried out last year revealed slightly greater discrepancies!

**Table 5. Farmer's calculated and actual FCR's taken from 5 complaints over poor performance (1975 - 1977)**

| Farm | Weight range (kg) | (lb) | Farmer's estimated FCR | Actual FCR based on careful measurements on-farm | Likely reason for error |
|------|------|------|------|------|------|
| 1 | 6 – 28 | 13 – 62 | 2.9 | 2.71 | Input food not weighed |
| 2 | 20 – 91 | 44 – 200 | 3.2 | 2.86 | Poor recording |
| 3 | 30 – 90 | 66 – 198 | 2.9 | 2.81 | Mistake over input batches |
| 4 | 25 – 86 | 55 – 190 | 2.6 | 2.92 | Guesswork (!) |
| 5 | 30 – 64 | 66 – 141 | 2.6 | 2.45 | Poor recording. |

Average error of 0.22 FCR over 48 kg (106 lb), an 8% error
Source : RHM Agriculture (Unpublished) 1977

## Average Daily Gain (ADG)

ADG

This is a measure of growth. But how much of the growth is lean meat, or low value offals, bone, fat etc? It may be good growth, *but is it the wrong sort of growth*? ADG doesn't tell you. This situation often occurs when growing pigs 'catch up' with unaffected pigs after an illness. On recovery they can grow faster but it tends not to be *lean* growth so much as low value gut offals and fat, often with a poor FCR despite the fast growth. ADG doesn't tell us this. **MTF does**. Economically ADG is really only useful in determining the contribution that fewer days to market and thus reduced overheads make to profit – often not insubstantial, it is true.

## Cost per % LWG

Cost kg/gain only measures cost, it doesn't marry it to income. Any businessman knows the two together are needed if profit is to be made. Costs can be cut so low that income falls to a greater degree so that profit is reduced. This is fully explained in Table 4.

## Weaning Capacity

WC

Sow productivity has traditionally been defined as the number of pigs weaned per sow per year. Since the mid 1990s - and especially over the past 10 years - the use of improved genetic selection techniques has led to a large increase in total numbers born which has made it possible for anybody to achieve 30 pigs weaned per sow per year. Table 6, published by one reputable Dutch seedstock firm in April 2009, is from a satisfactorily large sample of the customers, which reveals that in 2008, 101 of their breeders achieved 30.7 piglets **weaned** per sow per year. their live births averaged 13.8 with 12.3 pigs weaned per litter. There are other companies who tell me they are able to approach and maybe match this impressive performance. We are indeed in a new era of productivity and, as I have argued for 25 years now - need new measurement terms to keep pace with it.

Table 6. Technical performance results from one European breeding company.

| Number of farms | Average 2007 942 | | Average 2008 1006 | |
|---|---|---|---|---|
| | *Top 10%* | *Top 25%* | *Top 10%* | *Top 25%* |
| Average herd size | 412 | 391 | 506 | 465 |
| Weaned piglets per litter | 11.4 | 11.2 | 12.3 | 12.0 |
| Live births per litter | 12.8 | 13.1 | 13.8 | 13.6 |
| Weaned piglets per sow per year | 27.3 | 26.4 | 30.7 | 29.7 |

Source: Topigs (2009)

**Comment:** Total numbers born as a benchmark is no longer adequate in view of these large potential numbers born. Heavy selection pressure for total numbers born has resulted in some negative implications for piglet quality, growth rate through to market, feed conversion (thus saleable meat per tonne of food, MTF) and possibly carcase quality.

## A broader-based selection pressure

To combat this move to concentrating too single-mindedly on numbers born, geneticists are now focusing on a combination of traits so as to give a best economic outcome, rather than on one performance trait such as numbers born - however important it undoubtedly is in performance terms.

This is exactly what I have been saying for many years. We need new terms based on profit or income not just physical performance. For example, Food Conversion Ratio (FCR) must be replaced by Meat per Tonne of Food (MTF) as the former only measures <u>performance</u> while the latter includes figures which influence profit, such as killing out percent.

From a genetic perspective, balancing all the factors that contribute to improved Weaning Capacity means selecting for a range of factors relating not only to numbers born but also for piglet quality traits such as birthweight, born alives, numbers weaned, age at first mating and weaning to breeding interval.

## How do we arrive at this new term - Weaning Capacity?

Hypor - another forward-thinking European breeding company - has suggested that in their programme, which could be slightly different in other seedstock firms, those traits which influence (for them) a high level of weaning capacity are laid out in Table 7. While still according a high priority (44%) to numbers born, this programme still devotes 33% to piglet quality traits.

**Table 7. Proportion of traits influencing weaning capacity**

| *Piglet quality factors . . .* | |
| --- | --- |
| % born alive (i.e. influence of stillborns) | 14% |
| % weaned (i.e. losses) | 13% |
| Birthweight | 6% |
| | |
| *Other traits . . .* | |
| Meat percent | 5% |
| Daily gain | 5% |
| Weaning-mating interval | 3% |
| Age at first mating | 5% |
| Other (e.g. leg strength and conformation) | 5% |
| Number of pigs born | 44% |

Source: Extrapolated from Hypor (2009)

Thus, to achieve a weaning capacity there are three basic components, say Hypor:

1. The number of pigs weaned per litter.
2. Weaning weight.
3. Litters per sow lifetime.

## How to calculate weaning capacity

By multiplying these three figures together, a realistic benchmark figure for today's high performance situation would be:

12 pigs weaned per litter × 7.25 kg weaning weight (in this case 24 day weaning)
× 5.8 litters per sow lifetime = 505 kg weaning capacity

Weaning capacity clearly defines an individual sow's (or a whole herd's average) lifetime productivity and recognises the value of piglet quality and sow longevity, not just litter size or pigs weaned per sow per year as we favour today. Sow longevity (a short productive life) is a serious failing among breeders all over the world, which drags down the weaning capacity figure substantially. For example, even if the two initial performance figures are achieved, a breeder with a world average herd productive life of 3.6 litters drags the weaning capacity figure down to 313 kg - a 38% drop in productivity from a very expensive breeding machine.

# THE NEW TERMINOLOGY

**REO**
   **Return to Extra Outlay Ratio (REO)**
*Note: REO is not the same as ROI (Return on Investment - also now called ROC, Return on Capital). See below.*

REO is of great help to the feed trade, and should be to academics too. As we have seen, a farmer cannot possibly use all the additives / feed supplements on offer. Generally he is prepared to invest an increase in his cost per tonne (usually about 8 to 10%) to include a protective, growth-enhancing or nutrient-sparing additive. The question is which ones will give the best value for money? REO helps considerably because it indicates, from published trial work (usually expressed in the old 'performance-based' terminology) which of them might be the best value for money.

REO tells you, from each monetary unit invested per tonne of feed, how many monetary units are likely to be recouped also per tonne of feed. How much *return* you get for the *extra outlay - R.E.O.*

**ROC**
   Is this the same as ROC (Return on Capital = Return on Investment, ROI) ? No; REO involves the *smaller* sums spent on boosting profit or performance – like feed additives, minor improvements in housing or alterations to diet density – while ROI is used for more comprehensive investments - like new buildings, changing from stalls to group housing etc. Moreover REO is a measurement of the *extra* investment required, for example organic selenium costs more than Na selenite; non-GMO phytase more than dical, *etc.*, but the REO – the extra income likely from the extra investment needed – puts the requirement for extra working capital into perspective. In an urgent cost reduction situation, farmers object strongly to paying more for what they consider to be the "the same" additive or feed formula they have been used to. This is particularly true during times of marginal profits. REO helps to unblock this mindset, especially in the case of organic selenium which can cost up to 50 times more per unit of

selenium compared with bimodal selenite popular in the last 10 years. REO shows that, despite this, it is still extremely cheap for what it can do.

Table 8 illustrates REO taking organic selenium as a replacement for sodium selenite.

**Table 8. Economic paybacks from including seleno yeast (selplex) at the advised level in pig diets as a replacement for sodium selenite**

| Trial Source | Result | Physical Benefit* | ROI† | REO† |
|---|---|---|---|---|
| Janyck (1998) | Piglet growth rate | + 4.7 % | 7 : 1 | 17 : 1 |
| *Ibid.* | Litter size | + 6.7 % | 4.4 : 1 | 11 : 1 |
| Munoz (1996) | Drip loss, pork 72 hrs | - 12.0 % | 3.8 : 1 | 4 : 1 |
| Mahan (1998) | Sow litter performance | + 0.5 pig/litter | 20 : 1** | 25 : 1 |

Key
*       Over Na Selenite in feed
†       ROI = return on total investment/tonne (of including Se)
†       REO = return on *extra* investment/tonne (of Selplex compared to the cheaper cost of Na Selenite)
**      From extra meat sold from one year's sow's progeny at slaughter (estimate, computer model)

| | UK£ / tonne |
|---|---|
| Estimated cost of sodium selenite | <£ 0.05 |
| Cost of Sel-Plex organic selenium | £ 1.54 |

### Conclusion

While the *extra* cost of selenium yeast inclusion provides a 1% rise in raw material cost/tonne the REO paybacks could vary from 4 to over 20:1. Thus REOs put the apparent massive rise in inclusion cost for the 'same' 0.3ppm Se into perspective. In simple language – something costing 30 times more can give a return on the extra investment of between 5 and 25:1.

`REO`

## MSC and SLC

`MSC`

`SLC`

There are two main routes to making profit – produce More at the Same Cost (MSC) or produce the Same at Less Cost (SLC). Of course, there is a third way, which is to produce **More** at **Less** Cost, but this is only rarely achievable in practice – usually not the case in pig production!

Historically livestock production has concentrated on increasing productivity and trying to hold costs down, but all too often this has resulted in *over*-production causing the pig price to drop. This fall in income swallows up any extra profit from producing more – and the producer is no better off! So if we all adopt MSC and produce more at the same cost, we are on shaky ground in profit terms.

What is more sensible, especially today when producers are becoming more proficient at many livestock tasks, is to hold production at an adequate level, but *concentrate on reducing the costs of doing so*. This way over-production does not occur and as a result the pig price keeps up, meanwhile the reduced costs contribute substantially to the profit. So SLC or producing the Same (assuming adequate performance) at Less Cost seems much better.

## But How Good is 'Adequate Performance' ?

This is the key question. *How good does performance need to be to allow maximum attention to be given to reducing costs without damaging physical performance.*

Figure 1 gives two examples – the first involving sow productivity and the second slaughter pigs. The right hand area of each graph suggests the degree of importance the producer should devote to improving performance, and the left side to saving costs. As his performance improves, so the attention devoted to each sector moves from right to left.

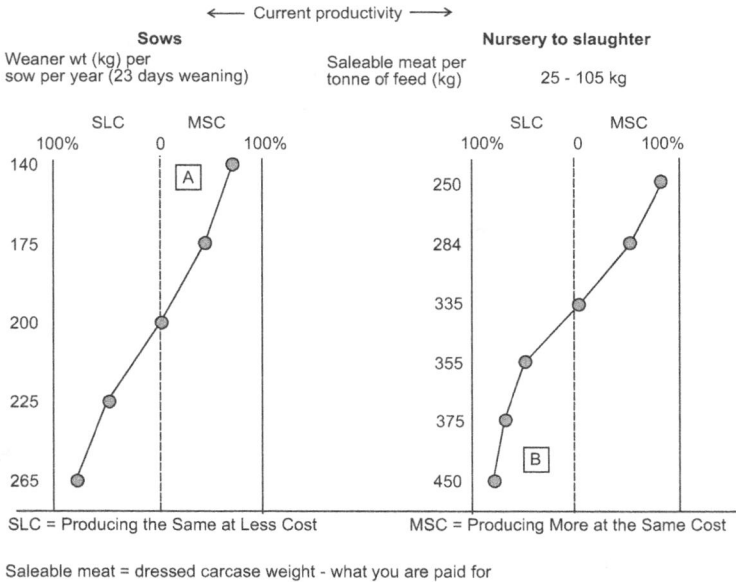

**Figure 1**. SLC or MSC - which gets priority?

This depends on current physical performance. I give above my estimate of the attention a producer should give to each dependent on his current physical performance (based on MLC Yearbook, 2009)

Similar sigmoid 'S' shaped curves can be plotted from data provided by computerized recording schemes (preferably with more than 300 sow litters and their progeny) using bottom 10%, bottom third, top third and top 10% and the median averages as locator points. A wide variety of performance criteria can be used, such as empty days, farrowing index, piglet mortality, saleable meat per m², *etc.*

Remember these are *guides* to how much management effort should be allocated to MSC or SLC and not hard-and-fast criteria, but they do help to answer the question "How good does my productivity need to be before I can ease up on improving productivity and really spend time (and money) on cost-reduction?"

## Instructions to farmers on how to use this figure

Calculate the amount of weaner weight produced per sow per year in kg; and the amount of saleable meat (kg dressed carcase weight) produced per tonne/ton of feed from exit from the nursery to slaughter.

Read off on the graphs how much attention (in %) you need to devote to SLC or MSC in proportion. For example, if you are at (**A**) you need to devote about 60% of your time to improving sow/weaner performance and perhaps 40% to saving costs without destroying what breeding performance you already have. At (**B**) however, the physical performance (FCR:ADG) is so good that it is unlikely to improve much further without incurring high costs. So while trying to maintain this performance the producer should devote about 85% of his time towards *reducing the costs of doing this.*

The key (left hand) performance scales vary between national pig industries, of course. U.S.A. would have very different reference scales to Thailand, for example, which is why they should be constructed from local national figures. If 5 reference points are taken for each scale, the sigmoid shape will nearly always emerge, some more pronounced than others as in the two examples given in Figure 1, which refere to UK conditions in 2010.

Remember – ROI/ROC involves *major* investments in new housing, a new farm or section of a business, etc. and REO relates to the *extra* investment to be committed to a current programme or strategy.  **ROI**  **ROC**

Generally speaking, products with the highest REOs are the ones to use first.  **REO**

## Using REOs in practice

Let's take an unspecified dietary enhancer. A variety of these are an excellent, if at present rather costly, way of making diets better.

From published trials several have an REO which gives an excellent return.    About 7.5:1. First class!

But would it be better to use one or more alternative supplements **which have a higher REO**?  They cost less to add, take up less physical space in a tonne of feed, and while individually they don't approach the payback of the original additive, cumulatively they might do so for less cost, sometimes much less cost.    And thus release more space in the feed, and more capital to be used elsewhere to improve the diet up to the 8% ceiling.  Table 9 shows the concept.

**Table 9.  How REO can be used to compare potential feed additives - total dietary cost £ 160/tonne. A good growth enhancer may cost up to 4% of the dietary cost (£6.40/tonne) and yields an REO of 7.5:1. £6.40 × 7.5 = £48/tonne). How would three alternative additives compare?**

|  | Rate of use  & cost per tonne | | | Expected return per tonne | | |
|---|---|---|---|---|---|---|
| Additive A | 1% | = | £1.60 | REO | 8:1 | = | £ 12.80 |
| Additive B | 2% | = | £ 3.20 | REO | 10:1 | = | £ 32.00 |
| Additive C | 0.5% | = | £ 0.80 | REO | 5:1 | = | £ 4.00 |
| Total | | | £ 5.60 | | | | £ 48.80 |

These figures are based on current European costings, but the same principle can be used with US Dollar costs, where the REO will come out at 3:1 overall.

**SLC**   These three additives have given us virtually the same return as the more expensive one used previously.  So . . . where's the benefit?  Why bother to change?

**MSC**   (1) **Either** SLC is improved (Producing the *S*ame at *L*ess *C*ost). We now have virtually the same return *for four fifths of the original capital invested*.  This reduces feed costs and improves cash flow because we've spent less.

(2) **Or** : MSC is improved (Produce *M*ore at the *S*ame *C*ost) – we can use the £0.80/tonne saved (12.5% of the cost of the major one-off additive used previously) for other dietary improvements *thus improving performance for no extra cost.*

**REO**   **REO is an excellent way of comparing products to make limited capital go further**

Please be assured, I am not against adding enzymes, growth enhancers, or increasing the amino-acid levels/energy, or the use of anti-mould protectants or any of the feed improvement products the producer can use cost-effectively these days.  But using REO figures *obtained from reliable published trial* work such as my example in Table 8 used by most good suppliers to support their sales claims enables the pig producer:

- To assess which of these products **are best value for money**.

- To use his valuable investment capital in the most effective way, which means achieving the shortest route to obtaining the most profit.

*Remember, REOs are worked out from the vendor's own claims. REO merely ranks these in order of cost-effectiveness. NB. Test the validity of the claims.*

## REO removes the apparent disadvantage of concentrated 'expensive' products

Many new feed additives are concentrated, low usage rate but expensive *per kg* products in themselves before addition to the feed. Nutritional biotechnology products are typical examples. Using the easy-to-explain REO concept puts their value in a true econometric (value-for-money) light and helps the customer prioritise the options on offer, and dissuades him from choosing something just because it appears cheap or cheaper than most on a cost per bag or per drum basis.

## How come some REOs are over 20:1?

Chromium is an example, but is sadly not universally permitted yet. It should be!

Any banker would sit up to attention if you proposed a scheme where he lends you a dollar and you benefit by as much as twenty dollars in a relatively short space of time. Is this really possible in pig production, you may well ask!

When a good antibiotic growth enhancer may have yielded what seems to be a modest 5:1 REO, how come some REOs can achieve the high 20's or more? The example of Bioplex Iron in the sow's feed is interesting. Recent trials have suggested excellent REOs of 12 to 18:1.

---

# AN REO CHECKLIST
# FEEDS AND FEED ADDITIVE PRODUCTS

---

✓   Find out the *inclusion rate* of the product you are asked to buy.

✓   From its unit price per kg work out its *inclusion cost*/tonne of feed. Include any mixing charges/extra labour costs, delivery charges, if these are additional to your current procedure.

✓   Work out the *performance benefits claimed* for the product, being very careful to request evidence/proof of performance and from where/how the evidence was obtained. If in doubt refer the data to an independent consultant/scientist.

✓   Convert the physical performance benefits into likely *economic benefits* – for example a Meat per Tonne of Food figure is a very useful one to set against inclusion cost/tonne feed or extra inclusion cost/tonne feed if the product is an alternative to an existing one. There are several others like WWSY (Extra Weaner Weight per Sow Year) you can use.

✓   Calculate the REO.

✓   Use the REO to compare it with other REOs from alternative products or systems you can use.

✓   Generally speaking the product likely to provide the highest REO is the one to use (first).

✓   But not always - do a follow-up check called AIV (Annual Investment Value). This is explained on page 249.

✓   Finally, an important refinement of the REO concept is to graft on to it an AIV which shows you how many times a year you would recoup the REO.

## MTF (saleable) Meat Per Tonne Of Feed

**MTF**   The new term MTF is as important as REO. This is because pig producers sell *meat* (pork and bacon) not pigs!

MTF relates the producer's total income against 58% to 66% of his total costs – food.

Using MTF as our primary yardstick means that we don't need to calculate FCR which as we have seen is difficult to collect and usually an inaccurate figure in the producer's hands. This is because meat is about 72% water and water is much cheaper than food! So the better the MTF figure the better *must* be the FCR – with no need to try to record FCR on the farm, as distinct from a research unit. Figure 2 on page 246 bears this out.

## MTF   MTF is easier to use

The MTF figure is much easier for farmers to record. This is because pig producers are paid – or should be paid – on dressed carcase weight (dcw) at a variable price per kg. So each week or month they know accurately what their 'meat' income is per kg of the pigs they have shipped for slaughter. As for food, they know from their feed invoices how much food the pigs have consumed on a running basis for that week or month, and also know how much they will be paying for it per tonne/ton. All these calculations can be done in the office, not on the farm.

# Price Per Tonne Equivalent (PPTE)

MTF is also useful because it can quickly be converted into an equivalent price per tonne figure (PPTE). As we saw with REO (Return on Extra Outlay) pig farmers can often be overly concerned with price per tonne. While it is important, price per tonne can be a fickle friend, but if the preoccupation with it exists with the producer, let's use it, not fight it!

If the MTF reveals a higher or lower figure for the period in question – for example 20 kg more MTF – the producer knows what the current price is for saleable meat, say £1.10/kg deadweight. 20 x £1.10 = £22, so in this case an improvement in MTF of 20 kg is equivalent to £22 per tonne cheaper food. Simple ! Work it out in your local prices to see how simple it is. PPTE has the advantage that it is easy and quick for the farmer to do his own calculations, thus he convinces himself the outcome is correct! And for the feed salesperson, if the customer is provided with the data and the method of working it out and he does the sums – the product is much easier to sell.

**Table 10. How to calculate an MTF figure**

---

(1)     Establish how many pigs are produced per tonne of feed
        *e.g.* FOOD EATEN 200 kg

$$\frac{1000}{200} = 5 \text{ pigs/ tonne}$$

(2)     Calculate saleable meat produced per pig across the growth period,
        say 30kg-105kg
        *e.g.* 75 kg liveweight put on × 75% killing out percent at the end of the growth
        period
                        = 56.25 kg (deadweight per pig)

(3)     MTF = 5 pigs x 56.25 kg    =   281 kg Meat per Tonne Feed

---

In some countries they have heavier pigs, with a higher dcw and eating more food. Thus the figures may look like, taking a US example .

(1)     Food eaten (from 45-265 lb) 725 lb

$$\frac{2000}{725} = 2.76 \text{ pigs/ ton}$$

(2)     220 lb put on x 78% KO percent = 172 lb dwt/pig

(3)     MTF = 2.76 pigs x 172 lb = 475 lb Meat per Ton Feed (215 kg)

---

This is particularly valuable in feed trials, something which both academics and many feed firms so far have chosen to ignore. Everyone involved in pig production should use PPTE/MTF and benefit from the advantages.

## What is a good MTF figure ?

MTF    Table 11 suggests what these could be on a world basis. Because of this note the modest FCRs quoted; even so they are fairly typical of average producers world-wide.

It seems that, at present, across the world a figure of 350-375 kg (30-105 kg) of saleable meat per tonne of feed is the one to achieve. However, in certain countries, and among certain producers, performances are higher than these world averages and 400 kg MTF is their current target with the top 10% producers achieving 450 kg MTF. All these refer to the 30-105 kg weight range.

**Table 11. Performances. World-wide 2009/2010 (30-105kg)**

|  | FCR x growth = range (75kg). | Food eaten per pig(kg). | Pigs per tonne of food eaten. | x Killing out % = | **MTF(kg)** |
|---|---|---|---|---|---|
| Poor | 3.3:1 | 247.5 | 4.04 | 74.0 | **299** |
| Typica | 3.0:1 | 225.0 | 4.44 | 74.5 | **331** |
| Good | 2.7:1 | 202.5 | 4.93 | 75.0 | **370** |
| Target | 2.5:1 | 187.5 | 5.33 | 75.5 | **402** |
| Exceptional | 2.2:1 | 165.0 | 6.06 | 76.0 | **448** |

Sources: Various national pig recording schemes and BPEX (UK) Yearbooks 2009/2010.

**Table 12. The effect of varying MTFs (30-105kg) on income**

Based on typical EU growing/finishing feed price of €175/Tonne and deadweight pig price of €1.20 /kg. Figures in Euros, winter 2009/2010.

|  | MTF (kg) | Income per tonne(€) | Income/tonne over feed cost/tonne (€) | |
|---|---|---|---|---|
| Poor | **299** | 359 | 297 | + 184 |
| Typical | **331** | 397 | 270 | + 222 |
| Good | **370** | 444 | 242 | + 264 |
| Target | **402** | 482 | 225 | + 307 |
| Exceptional | **464** | 557 | 215* | + 377* |

* From experience, feed cost for this type of skilled producer is often 8%/tonne higher due to the higher nutrient density needed for the very high lean-gain genes purchased.
<u>Comment</u>. The income per tonne from what is spent on feed is well over one-third lower (€85) for a typical producer not achieving today's MTF target of 400kg (30-105kg) per tonne of feed during a normal grow-out period from leaving the nursery. An MTF of 400kg (30-105kg) is perfectly feasible today with the advanced genetics and food quality now available to everybody.

## A word of caution

Just like the old term FCR, the value of MTF **depends on the weight range taken.** Pigs are much more efficient at turning food into meat early on in their lives than when close to slaughter. In Tables 11 & 12 the weight range taken is typically from 30-105 kg (end of nursery to slaughter).

FCR

MTF

Therefore a target MTF (world-wide) over this range is 400 kg.

However if the range is 7-70 kg (weaning to light slaughter) . . . the target MTF would be 357 kg.

And if 60 kg - 120 kg (the heavier grow-out period in U.S.A. and central Europe) the target MTF would be around 500 kg, or about 1100 lb in the US.

**Remember: MTFs will vary according to the RANGE of grow-out weights.**

## Does this mean that MTF is unwieldy / unusable ?

Not at all. We are using MTF to give a more profit-orientated method of assessing performance than FCR/ADG   As with this old terminology we are using MTF to demonstrate the improvement a feed or feed additive product can make over the controls / competition in a *comparative* manner. So as long as the weight range is the same there is no problem. As was always the case with FCR (but rarely done) we need to record over what weight range the figure refers to, so as to compare like with like.

ADG

So when using MTF in a production target manner, please can we start correctly and always qualify it with the weight range cited, *i.e.* . .
"A target of 400 kg MTF across the 30 - 105 kg weight range".

Written as "400 kg MTF (25-105  kg)."

## MTF replaces FCR as a more pragmatic measurement

Producers and some of their advisers - including feed and feed-supplement salespeople - are uncertain that substituting MTF for the old familiar FCR (quite apart from the fact that FCR is difficult to measure out on the farm - see page 381 - while MTF can be measured accurately <u>inside the farm office</u>) will not give the same sort of measurement as FCR.

In fact Figure 2,which is one of several farm tests I have taken the trouble to measure carefully (what a lot of work it involved!) shows that the two do follow each other closely, with perhaps a plus or minus 1% to 2% difference. This is a minor variation compared to the inaccuracy shown up by Table 5 on page 233 where FCR as measured by typical producers often varied by + or - 8 to 10% when I went back over the evidence.

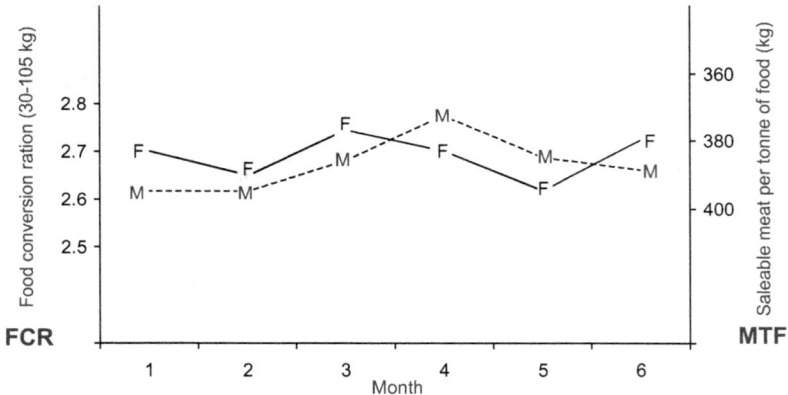

**Figure 2.** Close relationship between FCR and MTF.
Same farm, same pigs. PCR calculated on a 3 month rolling average.
MTF from the processor's returns. Source: Client's records 2008

| MTF |
|-----|
| PPTE |

## How the modern salesperson uses MTF & PPTE to sell products to farmers

At first MTF and PPTE may seem rather removed from the supplement-selling task as distinct from selling complete feeds. This is not true. At present, most trial results quoted in this area of meat production involve the company getting across better FCR's, ADG's or lower cost/kg gains of the products they sell. We've seen the disadvantages of . . .

**FCR :**   Yes, better, but at what cost ? Did it cost too much to get it lower?

**ADG :**   Yes, better, but how much of it was the right sort of growth (lean meat) ?

**Cost kg/gain :**   Yes, lower, but did the income suffer as a result ?

The intelligent salesperson takes the trial results given in the old performance-related terminology and grafts on to them the REO and MTF figures, because they mean more to the businessman-producer.  And if it doesn't mean more to him, then the salesman is in the advantageous position of explaining the benefits of these new profit-orientated ways of looking at the 'old' measurements.

PPTE  is a particularly effective weapon as it uses the prospect's weakness for cost per tonne.  Benefits like 0.1 better FCR, 4 days quicker to market, 2p lower cost/kg gain certainly mean something in the customer's mind, but .  using PPTE to add.
*"This is equivalent to a saving of £16 (or 10%) on every tonne of food you buy in future"* . . . is a far more impressive statement about the same improvements.
And using REO allows the continuation of a statement like . . .

*"And as our product with a likely REO of 4:1 only costs £3.20/tonne more (raising your cost/tonne figure by 2%), by using it you should be £12.80/tonne better off, equivalent to enjoying a 8% cheaper food in future if it is included."*

REO

So . . . both farmers and feed/feed additive salespeople should …

*   Look at the trial results supporting the product and the degree of expected significance.

*   Work out the REO, MTF and PPTE figures.  It is not difficult and can be done quickly.

*   Use them to make a greater impact while others struggle with the old and narrow terminology based on performance rather than profit.

As feed salesmen, just becoming familiar with and using REO, MTF and PPTE is sufficient to help them sell more, sell it better (at a higher price) and encourage them to ask for the order more quickly.  If feed and feed supplement companies do nothing else, get familiar with using these three "new terms" because the farmers and farm students reading this book will be doing so!

## MTF for sows

Is there an MTF figure for the sow?  If you like, but not involving weaners, as we don't, or very rarely, sell weaners for meat.

## Weaners

But we can keep a figure in our mind based on the liveweight of weaners produced per sow per year - as distinct from that of 'Weaning Capacity' which is based on a sow's lifetime's productivity.

Here, based on a sow eating 1.4 tonnes of food a year (including a little creep feed to help her out) and assuming she weans 24 piglets of 7kg in a year = 168kg, then a good target will be 120 kg of weaner (liveweight) output per tonne of sow food.

## Finishing pigs

Here an MTF(sow) would be calculated on the 1.4 tonnes she eats in a year to produce, say 23 finished pigs each of 105 kg liveweight (80kg deadweight/ dressed carcase weight) totalling 1314 kg saleable meat produced from 1400 kg food.

This is an MTF(sow) of 1840 kg.

An action level figure would be around a third lower at 876 kg - and I am still coming across plenty of those.

## Why cost/kg gain can be misleading

Cost per kg liveweight gain is as popular a measurement as Food Conversion Ratio - and can be just as misleading!

Dr. Phil Baynes (SCA Nutrition) is a level-headed pig nutritionist who always has his eye on econometrics – the measurement of cost-effectiveness. He claims that cost/ kg LWG can be misleading when taken in isolation when comparing feed trials, especially those involving post-weaning and nursery diets and their feed additives where there is much commercial competition. He goes on to give an example which I cite below.

SCA did two trials across the same time scale and under the same conditions comparing post weaning diets, one trial with the weighted average cost of three-stage diets fed being 45.5p kg (£455/tonne) and the other of three less sophisticated and cheaper diets costing, weighted average again, 33.2p/kg (£332/tonne)

The results were:-

**Trial 1** (Expensive diets) fed from 7-14.8 kg - weight gained 7.8 kg/pig
Feed cost/pig £4.55.
Cost/ kg gain therefore 58.3 p

**Trial 2** (Cheaper diets) fed from 7-13 kg – weight gained 6.04 kg/pig
Feed cost/pig £3.32
Cost/ kg gain therefore 55.0p

In cost/ kg gain terms the cheaper foods were the obvious choice as the pigs were 6% cheaper to produce – a 'no-brainer'!

But supposing we use our brains a little differently? The pigs in trial 1 were 2kg (12%) heavier. "Recalculating on a cost per percent/liveweight gain basis" says Baynes, "the pigs in trial 1 have grown some 111% over the start weight and those in trial 2 only 86%. We all know that better growth in the early days is of greatest benefit. Pigs in trial two will not make up the lost benefit." (I explain the reason for this on page 377)."If we recalculate on a percentage gain basis, the feed cost is about 4% in both trials, but the pigs in trial 1 are 2 kg heavier and have got at much better start in life from the more expensive feeds".

Before you take decisions on just cost/kg gain, follow it up with a cost per percentage liveweight increase figure as this could change your mind. Many a commercial early-feed trial has been won on a questionable cost/ kg gain figure. The same could apply to a grower/ finisher trial. So take care.

### AIV (Annual Investment Value) AIV

REO is a useful tool when comparing the price of feed additives and feeds based on trial significance. But a good high REO may take a long time to pay-back – and a lower one a much shorter time. So we need a further measurement of the return likely over, say, a year's investment. A year being the normal period of time over which a straightforward loan is granted.

For example, a product/additive giving a benefit in the creep feed may be turned over as many as 11 to 12 times a year; in the nursery 6 to 7 times a year; in the growing finishing stage down to 3 times a year, and in the sow only 2.4 times a year.

So **the time it takes to recover the expected REO** less the initial investment, less the cost of the interest on the capital funding – can be important.

Say a typical AIV based on an additive to growout diets has an REO of 10:1 by slaughter. Using Euro terms, turned over at 3 batches of pigs/year at a €3 per batch the (extra) inclusion cost/ton would be, at 3 batches/year €3 x 10 = €30 *less* the capital investment in use at the time plus the 30 cents or so interest needed to repay the €3 borrowed. This is €26.70 extra nett income per tonne per year. In this case while the REO is 10:1, the AIV is €26.70 ÷ €3 or 8.9 : 1.

## The higher the AIV the better is the investment

In contrast, take a creep feed additive which might give a much lower 2 : 1 REO at weaning for a higher €10/tonne extra inclusion cost. At first sight this looks a much poorer prospect in several ways compared to products with higher REOs. Indeed, this yields a modest €20/tonne return, but is **spread over a much greater number of pigs eating one tonne of creep feed**. Moreover €10 extra investment is turned over about 12 times a year with 24 day weaning (as is common in Europe). Our apparently lowly 2:1 REO in AIV terms requires an extra annual investment of €10/tonne + 10% interest, but the head start of 2: 1 REO it gives each litter is magnified by a factor of 12 in terms of slaughter pig output over a year's borrowing costs, *i.e.* €20 x 12 = €240, an AIV of €240 ÷ €11 = 21.8 : 1.

The moral is that a little investment in the correct nutritional area of the very young pig can give encouraging benefits at slaughter. We've always known this, and the AIV principle helps quantify it across a year's fiscal 'trading'.
**Advice:** Providing you are satisfied with the performance evidence claimed, go for the highest REO, but do an AIV check to check on how well you are using your investment capital.

**Remember:** Your bank manager looks at how he allocates his money like this . . . **how quickly will I get my money back** as well as **how much more will the borrower recoup**. AIV encourages you to look at your capital in the same way - how best to spend it.

## WWSY   WWSY (Weaner Weight per Sow Per Year)

1.   *Output/sow/year*   Pigs (usually weaners) reared per sow per year is the common current yardstick (*i.e.* 18 or 22 or 25 *etc.* 'Pigs per sow per year'). However, this takes no account of weaner weight per sow per year (Table 13) and can be misleading.

As Table 14 shows, **Weaner Weight per Sow per Year (WWSY)** is a better yardstick than the current Weaners Per Sow Per Year. Obviously the date/time of weaning will affect the weight produced per year, as it does the 'numbers of weaners' terminology used currently.

**Table 13. Two clients. Farm 2 won a national prize for productivity in his section in 1991, but farm 1 made more money!**

|  | Farm 1 430 sows | Farm 2 380 sows |  |
|---|---|---|---|
| Pigs weaned per sow and served gilt per year | 23.7 | 24.9 | + 5% |
| Average 3 week weight, kg | 6.7 | 5.4 |  |
| Liveweight output at weaning, per sow and served gilt/year, kg | 158.8 | 134.5 |  |
|  |  |  | -15.3% |
| Potential Value at £1.20/kg | £189.60 | £161.40 |  |

*Comment :* 5% more performance but 15.3% less income !
(UK figures)

Weight at weaning has a significant effect on economic performance, thus the suggested standards (Table 14) give target performances at 12, 21, 28 and 35 day weaning as well as the conventional figures.

**Table 14. Weight of weaners produced per sow per year, kg**

|  | Weaning age (days) | Poor | Typical | Good | Target | Exceptional |
|---|---|---|---|---|---|---|
| S.E.W | 10-12* | n/a | n/a | 103 | 103 | n/a |
| Conventional | 21 | 81 | 89 | 133 | 147 | 202 |
| Conventional later weaning | 28 | 97 | 109 | 184 | 217[†] | [ — ] |
| Swedish/Danish conditions | 35 | 115 | 133 | 261[†] | 299[†] | [ — ] |

\* Insufficient data for SEW technique, now largely abandoned, with the Americans reverting to 19 day weaning. [ ] Exceptional producers outside U.K. rarely wean over 24 days.
† These sorts of productivity place a considerable strain on the sow thus can be difficult to maintain consistently. All figures are corrected for farrowing index.

So after the WWSY figure an indication of weaning time, in days, is needed. For example, weaning at an average of 24 days should be expressed as . . .

158.8 kg WWSY [24]

**Secondly, the time of entry** of the replacement gilt to the herd should be standardised. Opinions differ on whether recording should start as soon as the gilt arrives on the farm and needs feeding and looking after, or on completion of first service. However as this can vary by several weeks and affect the weaner weight sow year (and the weaners per sow per year) by some 7 kg (or one weaner pig/yr) it is better to standardize on a 'start-date' for the input gilt *at the date/time she is put to first service*. While this ignores the bulk of the run-in or acclimatization period, it does ensure all farms are comparing like with like, as a new female replacement has to be offered for service at some time or other and this moment provides an equable start date in performance terms.

### WC   Weaning Capacity (WC)

Weaning Capacity is an important new term as it flags up the important drain on capital of replacement sow costs in new gilts, and also impacts on the disease defences of the herd. This needs a fuller explanation, See page 235.

*Piglet mortality to weaning.* Another vital figure, but with the advent of bigger litters in future due to genetic progress, misconceptions will increase. This is because the more piglets get born (alive) so the relative percentage mortality will increase. *Always relate percentage mortality to weaning to the born-alive figure.* Look at this, for example:-

*Looks bad* - **15% mortality** of 12 born-alives is 10.2 reared (12 - 1.8 = 10.2). While …

*Looks good* - **5% mortality** of 10.75 born-alives is also 10.2 reared (10.75 - 0.54 = 10.2)!
### Beware of percentages!

Certainly it is better to only lose half a pig than nearly 2 pigs per litter but the wide **percentage** differences in mortalities quoted among producers can be misleading. A better measurement is A.M.F.

### AMF   Absolute Mortality Figure (AMF)

Better than % Mortality is to give an Absolute Mortality Figure (AMF) – how many pigs died per litter of those born-alive. Not the % which died per litter, as **we don't know the litter size when a percentage is quoted**.

The clearest definition is to express it as. . .

AMF 1.2 of 12 b/a (b/a = Born alive).

Hopefully this could become familiar as AMF 1.2/12

This would be a 10% mortality of those born alive.

However AMF 0.8/12 is only 6.66% mortality and therefore good, while AMF 0.8/8 is 10%; again, not so good at all.

As a rule of thumb, using the percentage method, a target for the typical producer to aim for today is 10% *i.e.* in A.M.F. terms this is : for 8 born alive = 0.8 piglets lost; and for 12 b/a = 1.2 piglets lost. When it is expressed like this we at once can see that the problem in the first case *is the born-alives, not the mortality*! So always ask what the b/a figure is, as it is too important not to know this when measuring mortality.

And with the likelihood of 13 born alives being normal in future – possibly even 16 with advanced DNA selection techniques – expressing mortalities in this new way becomes even more important. For example, losing 0.8 piglets litter in a litter of 14 b/a looks high, but it is still only 5.7% pre weaning mortality – very good on most farms.


**ILR and PLR (Income to Life Ratio. Profit to Life Ratio)**

**ILR**

**PLR**

REO (Return to Extra Outlay ratio) enables the producer to compare paybacks from a wide variety of investment options all of which require extra capital. As we have seen in the REO section, using MTF or WWSY and Weaner Capacity to provide a comparison of performance improvement, generally the option providing the highest REO is the one to adopt first.

But if a high REO takes a long time to payback – years instead of months as can happen with equipment and housing alterations – then perhaps the option with the highest REO may not be the best economic bargain. ILR or PLR helps refine REOs in terms of the *time* taken to get the 'Return' part of the REO figure.

## Long term paybacks

So with ILRs and PLRs we are dealing more often with equipment REOs, which should have a long life (5-15 years), but with feed additives the payback date is relatively short and finite – at the end of the 160 day growth period, for example, or one sow's reproductive cycle, or her output over a year, or a product used only in the 6-9 weeks of the nursery period.

REO – or indeed ROI/ROC – along with ILR/PLR can be used to prioritise the expected effects of big and expensive refurbishment jobs (like ventilation renewal) or a host of smaller ones like replacing troughs, putting in height-adjustable drinkers, or comparable piglet heating systems (pads, lamps, under floor pipes, *etc.*) or even the comparative value of equipment used only occasionally, like hot weather showers, or on a small percentage throughput, like a sophisticated hospital/get-better pen.

As with feed additive paybacks, the results/benefits must come from acceptable trial results. This extra income (ILR) or (Nett) Profit (PLR) is divided by the time it takes to achieve it against the expected life of the equipment. Table 15 gives one example.

What does Table 15 reveal?

A study of Table 15 shows that a wide PLR is encouraging if capital is scarce, as with REO the wide ratio indicates the projects which are likely to repay soonest and best in profit terms.

**Table 15. How PLR (Profit to Life Ratio) helps assess capital improvements**

| | |
|---|---|
| **Situation** | Client has the option of investing the same amount of money either to refurbish his ventilation, or install new low-waste feed hoppers. From data supplied by the manufacturers of previous before-and-after results, alterations to the ventilation provided 10% more nett profit/pig and replacing the feed hoppers saved 4% feed (25-90 kg). |
| **Ventilation refurbishment** | 500 pigs. Cost £8/pig capital and interest over projected 8 year life before major renewal refurbishment is again needed. |
| | Nett profit improvement claimed as 10% (on £7/pig = 70p) |
| | Total benefit 70p x 3.5 pig places/year = £2.45 pig place / year. |
| | Payback therefore in 3.26 years leaving 4.74 years clear benefit. |
| | **Expected net gain in profit, after payback, over life of product.** £2.45 x 4.74 years x 500 pigs = £5758. |
| | PLR Extra profit per pig place over life of product £11.52. Investment cost £8/pig **PLR = 1.44 : 1** |
| **Installing low-waste hoppers** | 500 pigs. Cost £8 pig capital and interest over projected 12 year life before replacement is needed. |
| | Nett profit improvement is 4% less food wasted 25-90 kg (on 150 kg) |
| | 6 kg x 15p/kg = 90p/pig. |
| | Total benefit 90p x 3.5 pig places/year = £3.15 per pig place per year. |
| | Payback therefore 6½ months leaving 11.45 years clear benefit. **Expected net gain in profit, after payback, over life of product.** |
| | £3.15 x 11.45 years x 500 pigs = £18,034. |
| | PLR Extra profit/pig place of product £36.01  PLR = 4.5 : 1. |

**Table 15. Contd.**

| | |
|---|---|
| ***Comment*** | Both PLR or ILR (Income to Life Ratio) make you look at the *expected life* of the investment and the *expected financial benefits after payback*. |
| | So the low-waste hopper option is best (has a higher PLR) as long as the ventilation needing renewal is not causing respiratory disease. |
| | Costs and benefits claimed came from the manufacturers trial evidence in both cases. |

The payback (from the manufacturer's evidence) and the expected life of the installation relate both ratios to each other. If you are satisfied with the evidence, choose the project with the widest PLR or "highest=best".

Both PLR and ILR further refine this comparison and demonstrate *how quickly the capital required can be recovered* – ventilation 3.26 years with hopper replacement only 6½ months.

The recovered investment can then be kept as profit, or reinvested in other performance-improving (MSC) or cost-reducing (SLC) projects. As in the example cited, this rapid redeployment of (scarce) capital *can be more important, in value-for-money terms, than a straight REO figure.*

Always, especially in projects requiring long term capital, use PLR/ILR *in conjunction with REO*.

## Why ILR and not PLR ? (Income to Life Ratio v. Profit to Life Ratio)

ILR

PLR

Some businessmen pig producers prefer to use *income* rather than *profit* when calculating the effect of the life expected to achieve a payback on interest and capital invested.

This is because income is one, finite figure and is not subject to variation in interpretation as in 'profit'. Profit can be nett, gross, margin-over-food cost, profit per pig place, or per m$^2$ *etc.* Unless carefully defined this can cause confusion, and so some producers are more comfortable with ILR.

Again, the data from which the calculations are made can arrive at varying definitions of profit, or alternatively in straight, increased income per pig.

An ILR figure tends to be rather easier to calculate and compare between project options.

# SUMMARY – POINTS ABOUT THE NEW TERMINOLOGY

## The farmer

- To survive in the future pig producers have to become more business-orientated.

- Thus they need better measurement terminology to help them take the right financial decisions.

- *The new terminology* takes cost and income into account when measuring performance, thus is profit-oriented, not just performance-oriented as at present.

- Capital resources are finite, can be difficult to acquire, and costly. The New Terminology helps rank them in their most cost-effective manner.

- If the salesman is going to start using these new terms, the farmer must ensure he understands the logic, so that he can critically assess what he is being told. Salespeople tend to stress the good points of their product and avoid mentioning the less good/unproven aspects.

- It is then easier for the farmer to assess the *value of what he is told*, and *compare it to* the blandishments and claims of other salesmen interviewed.

## The salesman

- Those customers who survive will be bigger and better businessmen to deal with and so the sales interview will increasingly be *financially*-based rather than *performance*-based.

- Salesmen must understand the *new terminology* so as to keep pace – and even think ahead of – their customers, because the new terminology involves *cost-effective* measurements, which help persuade the customer to put his money their way.

- Good quality products are not cheap, and can never be the cheapest on the market.

- The *new terminology* makes it easier to sell good quality in a highly competitive sales situation.

- The *new terminology* makes selling more interesting as it opens up new/novel sales approaches.

- Understanding *the new terminology* and when to use it effectively to make a sale will impress customers that salespersons have their eventual profit rather than the physical performance of their pigs as the centrepiece of the sales interview.

## The commercial company

- Has the advantage of highly professional sales negotiators fully conversant with business thinking and up-to-date terminology. Better-trained salespersons mean more sales for the same costs thus more profit to the company.

- Other commercial companies will be immediately placed at a disadvantage if they are unaware of the new terminology and how it can be effectively used both to make a sale and retain repeat business. This itself will have a debilitating effect on the other company's morale in feeling 'left behind', especially if the new terminology is quoted at them by potential *and existing* businessmen farmer customers.

  This means you/your Company, not they, will get the potential customer's orders and you/your Company will take their existing customers off them more easily.

  In the farmer's case, he will be able to use his limited capital more cost-effectively than other farmers not using the concept, and so make more money, or in difficult times – survive!

## The scientist

- In the end the farmer pays his salary. He therefore needs to keep abreast of the changing *commercial* aspects of pig production. So he needs to familiarise himself with the New Terminology as well as retaining, correctly, his use of precise physical performance measurements. In contrast to the busy farmer, he can measure and use performance accurately to help the farmer achieve the basic data so essential to his proper use of these new econometric-based terms.

Also, very often, by taking on board the New Terminology I suggest, the scientist is in a better position to prioritise his own applied research from the farmer's viewpoint. He can persuade his Departmental Head that his preferred line of research has an econometric benefit for the farmer's business and therefore is worth academic investment. Research bosses need just as much persuasion as a hesitant farmer or a recalcitrant sales manager!

# REFERENCES AND FURTHER READING

BPEX Yearbooks for 2009 and 2010.

Campbell, R.G. Personal communication, unpublished. (Effect of Bioplex Iron on a large industrial sow complex.)

Close, W.M. "The Effect of Bioplex Iron on Sow & Piglet Performance: Preliminary Results" (1998). Internal Res. Ref. 44.034.

Close, W.M. "Organic Minerals for Pigs : An Update" Procs 15th Annual Symposium in Biotechnology 1999, pps 51-60.

Gadd, J. Monograph: "The New Terminology : Guidelines for The Feed Trade &

Ancillary Industries Sales & Marketing Departments." Privately printed; available from the author.

Gadd, J. Monograph: "Pig Production Standards" (U.S. edition 1998, Metric Edition 1999.) Privately printed; available from the author.

Gadd, J. "Off With The Old, On With The New", Pig Farming **47**, 1 (Jan. 1999)

Gadd, J. "New Terms, Better Yardsticks", Pig Farming **47**, 2 (Feb. 1999)

Gadd, J. "Questions You Should Ask The Sales Rep", Pig Farming **47**, 11 (Nov. 1999)

Gadd, J. "Man of Straw . . .and Iron", Pig Farming **47** 7 (July 1999)

Janyk, Stanley W. Organic vs Inorganic Selenium Supplementation of Gestating Sows, Jan. 1998, Agricultural Research Council, Animal Improvement Institute, Irene, South Africa. (Presented at the World An. Prod. Conf. Seoul, Korea)

JSR Genetics (2009) "Meat per Tonne of Feed - Become a Convert". Press Release (example of an international company using the 'New Terminology').

---

To give the reader a clear insight on how the new terms REO and MTF work in practice, the following page numbers refer to some examples from the many research and farm trials cited in the text. While all the REOs are of course positive, the MTFs can be either positive or negative dependent on the trial result.

| **REO** | | **MTF** | |
|---|---|---|---|
| *Subject covered* | *Page* | *Subject covered* | *Page* |
| Purchased creep feeds | 7 | Alleviating stress | 121, 123 |
| Attended farrowings | 33 | Better growth rate | 342 |
| From the 'Imprinting' concept | 25 | Better birthweights | 50, 53, 54 |
| Veterinarians input on litter size | 30 | Postweaning check to growth | 58 |
| Matching diet to immune status | 101 | Matching diet to immune status | 101 |
| Too high a sow replacement rate | 107 | From correct batching & matching | 123 |
| From a good long induction period | 107 | Effect of various stressors | 121,123 |
| Sow longevity savings | 108 | Mixing before shipping | 124 |
| Proper cleaning down | 183 | Value of staff training | 285 |
| Complete sanitation protocol | 193 | Effect of better growth | 342 |
| Steam cleaning bulk bins | 211 | Improved performances in 2010 | 382 |
| Complete mycotoxin prevention | 220, 221 | Incorrect ventilation | 386 |
| Value of staff training | 284 | Vets input, weaning to slaughter | 387 |
| From on-farm mixing | 363 | Incorrect feeder throat adjustment | 392 |
| Feed raw material analyses | 363 | Having extra feeders at weaning | 399 |
| Veterinarians input on FCR | 385 | CWf v. same feed as dry pellets | 437 |
| Destocking by 15% | 431 | Effect of overstocking | 425 |

'PPTE' Price Per Tonne Equivalent

PPTE goes on to relate the advantages or disadvantages of the above new terms into what they mean in a lower or higher cost per tonne figures. Examples of these further calculations are on pps. 324, 342, 371. A PPTE figure really drives the savings or costs home in a form which means so much to the producer - **what they mean on a cost per tonne basis.**

# 11

# MANAGING PEOPLE - WHAT THE EXPERTS DO

## ERRORS COMMON TO MOST PIG PRODUCERS

I've been fortunate to visit about 4000 pig farms in my life and talk to their owners and managers. It is presumptuous and maybe impertinent of me to list their faults as I see them as in the list below, but perhaps it is worth you running your eye down them all the same. Pig farmers are surprisingly similar in any country.

On the positive side, pig producers are dedicated, hard-working, courageous, resilient, good-humoured, they do want to care for their animals within the bounds of convenience and cost, and many outside influences, such as bureaucracy and imposed legislation. Pig producers are tolerant, often to a fault.

Now for the bad news!

## ARE YOU HERE … ? SOME COMMON FAILINGS AMONG PIG PRODUCERS

✓ Ignorance of what is possible/how lamentably we fail to achieve current genetic possibilities.

✓ Not measuring things well enough on paper or online (recording) or in the piggery (monitoring devices/controls).

✓ Thinking you know best. 'Experience' often holds you back!

✓ Not being observant enough.

✓ 'Tailchasing' by too much time spent on daily chores.

✓ Doing too much hard work themselves thus …

✓ Not investing in automation to remove hassle, drudgery and thus jeopardising livestock care.

✓    Not understanding ventilation/the way air moves.

✓    Overstocking; a global failing, and its effect on performance and disease.

✓    Not using cheap, temporary – even 'throwaway' – housing to 'defuse' production bulges and isolate sick pigs.

✓    Wastage, especially food, and not realising the many, hidden ways of wasting it.

✓    Underestimating the importance of the post-weaning phase to finished pig profits.

✓    Not training labour to modern demands.

✓    Not being present at farrowing.

✓    Pushing gilts into production too impatiently.

✓    Not using the vet properly.

✓    Not using AI sufficiently; getting careless due to over-familiarity.

✓    Slow to espouse business partnerships/linkages/collaboration.

✓    Not supporting producer discussion groups.

✓    Falling behind in biosecurity requirements (the poultry industry is ahead of us in cleaning and disinfection, for example).

✓    Not treating pig production sufficiently as a business.

✓    Not realising how mycotoxins are 'hidden thieves'.

✓    Not spotting where or when to invest, *i.e.* poor prioritisation.

✓    Tending to delay spending completely rather than spending an affordable amount in the right place, and so building on that extra income to reinvest elsewhere.

Quite a long list. Even so, I find many people are in 50% of it. Please go back through this list again and think hard where you may be adrift.

## On the other hand …

During these 4000 or so farm visits I have also been privileged to sit at the feet of, marvel at and learn from, a couple of hundred top-class owners and managers. I wish I was as good as they!

Maybe this is what they have in common – and it is not so much an opposite list to the foregoing as you may think.

# PORTRAIT OF A PROFESSIONAL PIG PRODUCTION MANAGER – A CHECKLIST

Their one object is to *maximise profit*. Not necessarily physical performance or even income. Their methods can be divided up into short term and long term goals as follows:–

**Short term**
- ✓ Set production targets to achieve projected income.
- ✓ Maintain sufficient replacement stock.
- ✓ Breed to a pre-calculated production target.
- ✓ Reduce mortality and stillbirths.
- ✓ Wean a quality pig.
- ✓ Wean a sow suitable for prompt rebreeding.
- ✓ Maximise rebreeding effect (i.e. 500kg weaning capacity per SPL).
- ✓ Minimise rebreeding time (5 days or less).
- ✓ Minimise disease/maximise health.
- ✓ Reduce costs/identify waste & inefficiency.
- ✓ Improve animal and staff welfare.
- ✓ Motivate staff.

**Long term**
- ✓ Talk to various market outlets all the time.
- ✓ Select the correct genotype of stock for the market outlet.
- ✓ Maintain a recording system to **maximise pig flow**.
- ✓ Maintain a recording system to **identify problems**, especially to forewarn of **potential fall-off against targets**.
- ✓ Select, monitor and train staff.
- ✓ Pay staff adequately.
- ✓ Purchase feed correctly (dialogue regularly with a pig nutritionist).
- ✓ Sell pigs effectively.
- ✓ Plan maintenance and repairs at minimal disturbance to the manhours available.
- ✓ Make cost-effective alterations.
- ✓ Train himself.

*These are all key tasks*. Most top managers employ them.

# WHAT MAKES A GOOD PIG MANAGER?

Not long ago I visited, or interviewed at conferences, six of the most highly-regarded pig farm managers in the world. I summarise their priorities which were . . .

✓    Good planning - of pig flow and work flow

✓    Monitoring progress - records and measuring equipment

✓    Buying and selling - regular dialogues essential

✓    Motivating staff - very interesting, which I will expand on below

✓    Self training - keeping within the loop is vital

✓    Knowing the local and national market

✓    Passionate about pigs? - Not necessarily - more interested in business management

## Now for something nothing to do with pigs !

What top high street managers tell me . . .

In addition to this most valuable expert and highly practical pig-based experience, outlined above, I have visited the CEOs of four high street companies which have given us as a family superb service over the years, in order to ask them, as a very satisfied customer of their obviously very successful businesses, how they themselves manage their working day.

This is the gist of what they told me, but not in priority order. All these points I think are very applicable to a good pig manger.

✓    **Keeping a tight eye on sales** (at least half the day). Whether we as pig managers need to allocate so much time to this as the retail trade does is doubtful, but the subject is important and I discuss it below.

✓    **Having a good Personal Assistant** (quote: "A first-class PA provides the information I *might* need ready and waiting before I need it"). The reason why I mention this one is that a good accounts/records compiler called in to do just this work is a similar key assistant for a farm manager to a high street CEO's PA.

✓    **Good negotiating skills** (both inside - staff, and outside the company - suppliers).

✓    **Keep suppliers in the loop** (quote: "An essential part of successful negotiation and it costs nothing but a little time and trouble").

✓    **Staff training, training, training** (quote: "People make or break a business").

✓    **Taking the trouble to remember people's names** (quote: "I ask after their families and welfare).

✓    **Always have an open door** and suggestions/complains box. (quote: "An essential finger on the pulse enabling me to nip trouble - people trouble or product trouble - in the bud").

✓    **Make your staff always go more than halfway** when dealing with customers.

## Time and 'tail-chasing' in our pig industry

So there in a nutshell we have invaluable advice from successful managers in two very different areas - but pretty similar in context. Look at the attention to **people** for example.

Two pig managers who had over 100 staff to motivate and supervise mentioned how important it was to **make time** to handle people, which I find is lacking on the smaller, say 300 to 500 sow farm. This is because the manger has - or feels he has - too much to do, some of it involving manual work. Sure, when I was managing a pig farm in my youth I too found it a relaxation from brain and paperwork, not to mention the telephone (today the computer). It was a relief from the stress and responsibility of managing to go out for a while and get my hands dirty. But I see so many managers thinking they have to assist/get involved in a disproportionate amount of physical work otherwise the work-flow gets behind. 'Tail-chasing' is the bugbear of the pig production industry at both stockperson and manager level.

## So how much 'work', how much 'thinking' on the average medium-sized farm?

I put forward these suggestions for you to consider . .

I suggest that a manager of a 400-500 sow farrow-to-finish unit should spend at least 35 hours/week on non-manual management tasks, and more if possible. He needs to . . .

**Look at every pen of pigs** once a day - **two hours/day**, plus 30 minutes/day to check on equipment and measuring instruments like thermometers and controllers.

**Brief all staff** formally each day - **15-20 minutes**. Not only does this help to ensure that the right things at the sharp end are being done in good time, but it is also a useful finger on the unit's pulse, both where pigs and people are concerned.

**Plan for, and check on, pig flow against target - 2 hours/week**. This entails:-

**Monitoring performance. 1**. From the records, using the computer's predictive 'what-if' facility - grossly under-used by managers in this computer age. Every farm over 300 sows should have a part-time person dedicated to keeping what records the manager requires to meet performance and income targets. Managers do too much of this number-crunching themselves which soon gets them stale - as it did me. They do this because they don't trust anyone else to do these important collating tasks and the stock people don't like it anyway. Delegate this vital and time-consuming task to a hired part-timer who does nothing else. The top pig managers said that they become their 'right arms'. Remember how vitally those retail managers regarded their PAs to provide them with essential facts on which they could make the right decisions? A good pig manager needs to spend **2 hours/week** on interpreting the numbers - not having to add them up! Then printing off weekly against-target graphs so as to motivate staff. **That** is good management.

**Monitoring performance. 2.** Of measuring equipment on the farm, as the top managers said. Most farms haven't enough monitoring devices available anyway. **30 minutes/day**.

**Buying and selling.** Regular telephone and personal contact (using the computer to confirm things, not so much to dialogue). **One hour/day** - at least.

**Seeking advice/information exchange**. i.e. with the vet, the nutritionist in charge of ration design, other managers, with salespersons (always give time to those 'nuisances' but strictly limit them to 10 minutes - no more, as they can be good sources of what is going on locally). **One hour/day**.

**In this day and age - the bugbear of form-filling and bureaucracy**. Very variable between industries but in my country it takes a minimum of **an hour a day** just to keep your nose clean with the authorities and the law. Dreadful!

Finally allow yourself an hour a day for emergencies, accidents and - to use the vernacular - 'cock-ups' which need your urgent attention/decisions.

This little list - which in practice is probably minimal - totals about **7 hours/day, 33 to 42 hours/week**, all on checking, thinking, analysing and leadership.

## And training yourself?

That workload doesn't leave much time for self-training, does it? In my case it had to be out-of-hours reading up about technology - developments in science and what ideas and products might help on the farm.

Just as important is the local pig discussion group, which when well-run and **open only to bonafide producers** (guest speakers are politely ushered out clutching a bottle of Scotch after their presentation!) is a huge source of helpful information - and a filter for misinformation - shared between fellow pig producers in the same boat as yourself!

## The wrong way round?

I find many managers spend longer than this at the job, bless them, but were doing too many manual tasks themselves. Getting out and wielding a shovel/mending something is a very seductive diversion from what a manager really needs to do as outlined above. Bless them again, they were getting things sorted sure, but they were damned tired at the end of it.

Remember, the majority of the top pig managers I interviewed admitted that they weren't *that* interested in pigs but were absorbed by the business aspects of pig production, while most pig managers and some owners I talk to these days tend to be the other way round.

Where are you in that dichotomy?

# EFFICIENT USE OF LABOUR – YOURSELF

## Examples from my own philosophy ...

Know yourself and your job. Mine is writing and giving correct advice, so in my case ...

| | |
|---|---|
| **Writing** | The idea is more important than the writing of it. |
| | An idea needs to be published about four to six times to make the research and fact-checking involved cost-effective. There are many ways of writing up the same information. |
| **Getting and storing information** | A third of my working hours is spent on this aspect. Obtaining, cross checking and cross-referencing information so that I am pretty confident it is supportable. |
| **Computers** | Absolutely essential; to get and store vital information, to dialogue with people who know mre than I do, to save time, to see what others are doing, to train myself. However, so many of my peers are hooked on the computer drug – I am sure it is actually restricting their creativity, tying them down. Beware! |
| **Advertising** | Advertise a farm? Why not? If you are producing something you are proud of, tell people, even (especially) if it is "only" *your processor.* |

**Travelling**   Is a wonderful source of ideas on what to do (and what not to do) in pig production. If you don't go-see, you are working with one hand tied/missing opportunities/getting out of date. Travelling improves your sense of judgement and self-knowledge, and realisation that you are not alone.

**Training yourself**   I've always spent 15% of my annual income and 5% of my time being trained by others. Even today, after 55 years' work in pigs, I am never too old to learn – neither are you! I learn something new about pigs every day; something important once a month; something of quite earth-shattering impact once a year! Pig production is rocketing ahead – we all need to keep up with it.

That is why, being well past retirement age, I am still actively engaged in the world pig industry. How interesting it all is!

## DOING A GOOD JOB – A CHECKLIST

Think like this about your own job. The approach to what you do won't be so very different from my own experience – even if your job is.

✓   Question everything you do.

✓   Can someone or something do (some of) it better or cheaper?

✓   Just because everyone does it one way, is it really right for you? Think laterally.

✓   Be a people-person. Talk to everyone. Never be afraid/too proud/too embarrassed to ask if you don't know. For each rebuttal/refusal you'll acquire a whole crop of useful advice ten times over – and it will be free.

✓   Silently question everything you are told; there is a lot of misinformation/half-truths about.

✓   Keep people in the picture. Your staff, your bank-manager, your vet, your nutritionist, the accountant, your doctor, your family.

✓   Write things down. Busy people cannot remember everything. Then review your notebook regularly.

✓   Keep quiet about good ideas – or write a book about them!

*Think more : stop chasing your tail. Then you'll work less but work more efficiently.*

# SOME HOME TRUTHS FOR PIG MANAGERS

**Selling his output effectively** : The manager has to keep a very sharp eye on selling his pigs to the outlet which can influence his income in the most positive way – which sometimes may not mean the buyer who is currently offering the best short-term price.

The good manager realises his job is not pig production, *but meat production*. His job is to *make it easy for the processor to buy his pigs* which means maximising the output of exactly the quality of meat which the buyer needs, on time, in level deliveries (which is what a contract means, otherwise why bother to contract?) at the minimal cost to the farm enterprise.

### Cost control

In exactly the same way that the most efficient pig farm manager is the one who manipulates productivity correctly so as to maximise profit, so it is not necessarily the farm which spends the most money which makes the most profit. Control of costs is a vital management area in any business, and with a modern pig farm likely to invest more money per head of staff working it than most industrial businesses in Europe (surprising but true; output per man on 'the larger' pig units in Britain is over £230,000/year), spending the capital wisely is a vital area of eventual profit. The ways in which money can be saved, or better, redirected into areas which give a more promising return can be counted in hundreds, not tens on any pig farm, big or small. But here are the 'big three' in my book as they affect *cost savings* … I attack, on my clients' farms, when spending on **things** is involved, waste of *food*, waste of *warmth* and waste of *space*. Attention to all these areas rarely fails to provide the client with 20% (sometimes 35%) savings in costs which is a huge benefit to cash flow – and pays my fees 20 times over!

As to spending **capital**, the big three here – in my opinion – are on *precision feeding*, reducing *disease* and accurate *environment control*. An investment of one monetary unit on facilities which improve each one of these will yield never less than a three-fold return on *each*, and that's a huge hike in income.

### Where managers go wrong

I know of very few pig farm managers who don't do some manual work – only on the very biggest farms, perhaps.

### Some notes for owners

Trouble is, most pig farm managers *don't allocate enough time to the key areas above*, and get caught up in an increasing spiral of manual work. Sometimes it is not directly their fault – the owner expects it and they may be strapped into a job specification which is out-of-date for the 21$^{st}$ century. Thus I hope as many owners will read these notes as managers!

Secondly, ***managers are still very much output-minded*** to the exclusion of all else. When I managed a pig farm I was offered (or asked for) a very low wage but a 10% share of the profits. That taught me very effectively how to manipulate output so as to maximise profit. If I got it right I was well-rewarded, but so was the owner – 9 times more! Thus he should have been much happier than I – but of course he wasn't, as owners are never satisfied, are they? We employees have to live with this. If you work for somebody, it usually goes with the job.

Third, too many pig farms I visit are understaffed. ***This means the unit is always chasing its tail.*** Repairs and maintenance tasks in particular are one long round of emergency action and the manager often has to assist to ensure the work-flow comes back somewhere near on-schedule again. This happens nearly every day!

This is often as bad on the larger 800 to 2000+ sow units than the smaller farms. Such units should have a full time maintenance-man – electrician, plumber, builder, welder, either contracting out the work, or if large enough as an in-house employee. Managers should sit down and work out the cost-effectiveness of such a policy. Pigmen should be trained as professionals in ***pigmanship*** – it is too important a job to leave to half-knowledge about peripheral tasks because he or she is busy at something else which also won't wait.

## Absolute nonsense!

I often hear it said with pride that the modern pigman does indeed have to be stockman, electrician, computer operator, plumber, carpenter, welder, midwife, nurse – and so on. **This is absolute nonsense!** It is out of date and muddle-headed thinking. No modern production industry could ever survive on such a disorientated set of half-skills. The good manager isolates blocks of these skills and employs the correct professional for each one. Only on the smallest, non-industrial unit does the manager have to be knowledgeable and practically adept about all these.

### Understanding what the computer can do

Those managers who are using a 'what-if' computer program on feed nutrition intake, for example, have already discovered that the way the feed compounder specifies and formulates feed for him may tend to be in their interest and not necessarily his. A few progressive feed firms have seen the light on this. But the present system (to make manufacturing easier and cheaper) is, as you know, to fix nutrient specifications first and hope that by advising daily feed allowances target *performance* is achieved (not necessarily profit). A better way is to assess likely target *profit first*, then input the farm's specific details on pigs, housing and management – and *last, specify nutrient requirements to meet the target profit*. Think about it. This is a totally different approach, which will mean (as farms get larger) that these farms will have their own feed formulation which may change in nutrient specs as the profit target shifts due to the effect of market forces and as their pig's immune status changes. A good manager will pick-up technicalities like this from using computers and understanding what they can do for him – but he will never do so if he is dung-dozing manure or weighing pigs or mending things.

**The most difficult job of all**

Is checking upon what is going on – which means what people are doing – without raising antagonism and being fair and equitable in pulling things back satisfactorily. Very few managers (in any business; not just pigs) find this easy. I certainly found it very hard. It helps if the manager has been professionally-trained in handling and motivating people, because the first thing a course of this nature does is to teach you how to handle yourself! Often I find where there is a poor working owner/manager relationship it is the *owner* who needs training, not the manager!

As a consultant, I find that is a tricky message to get across.

**One of the secrets of a good people manager is to be in the right place at the wrong time – and make it look accidental!**

# THE TEAM SPIRIT / 'ESPRIT DE CORPS'

Pig farmers in my own country (Britain) have something to learn here, compared to the larger German, North American and Japanese farms. British farmers would argue that an average individual stockmanship is better in Britain than in any of these countries, which is often true, but they miss the point of those nations who are naturally more group-orientated. This is that team-wise, they take a great collective pride in the unit. As a result the whole farm is neater, tidier, and better-organised; it is *good organisation* on the larger pig farm which is a vital prerequisite of *good stockmanship* which follows on from it. The larger pig farm in the future needs both. The first-class manager can improve both the organisation and the stockmanship, but of the two his main role lies in organisational skills. The key task where stockmanship is concerned is for the manager to recruit good section heads and get the labour load right, then with both in-house encouragement and off-farm training courses/group discussion sessions the general stockpeople will improve by leaps and bounds, as it is a self-fuelling process.

# MANHOURS PER SOW, NOT SOWS PER MAN!

Some while ago  I got bored with people asking me "How many men should look after 'x' number of sows?"

I replied, "It all depends on how well the breeding pig technicians do their job. *Time spent with pigs* is quality time and 'sows per man' doesn't easily bring out this essential information. What matters is how well the staff use their time, which of course has to be paid for. Man-hours per sow - or per finished pig in the case of the feeder - highlights this.

So I started to record man-hours per sow per year against weaners achieved in the same period on as many of the farms I visited as possible.  To cut a long story short (i.e. using data from 158 farms) there seemed to be a rough correlation between man-hours per sow per year and weaners per sow per year, irrespective of weaning age.

Producers spending only 16-18 hours/sow/year tended to achieve 17-19 weaners/ sow/year, while at the other end of a sigmoid-shaped graph, those who were at 24-25 hours could show 22-25 weaners/sow/year.

I then recalculated the figures on the more pragmatic *weaner weight per sow per year* figure (yet another new and better term) and to cut another and similar story short, it wasn't much different when looked at this way, rather to my surprise.

These results encouraged me to go on to record how the man-hours were split between tasks on another 50 or so clients visited over some 10 years (1994-2004).  Table 1 expresses them in the interesting form of man-hours per sow.  I've seen similar figures expressed as a percentage of total hours worked, but not on the – what I consider critical – man-hours per sow basis. Critical, because how the sow performs is the basis of eventual productivity of the unit right through to shipping.  "If you don't get 'em bred, you don't get 'em shipped," as one American laconically remarked to me last year.

The total figures for each farm were taken off worksheets where available and the proportional splits agreed between staff and the management in each case.

What these figures suggest…

- In farrow-to-finish labour cost terms, economy of scale reduced labour cost/ sow by 8% (not as much as I expected).

- On the smaller units the amount of labour spent feeding is too high at 31.5% of total labour costs compared to 17.6% on the larger farms (mostly due, in their case, to automatic dry or CWF pipeline feeding systems).

**Table 1. Workload expressed as man-hours per sow per year.**

| | 40 farms 120-350 sows | | 10 farms 875-2040 sows | |
|---|---|---|---|---|
| Breeding to weaning: | | | | |
| Feeding | 4.2 | | 2.1 | |
| Serving | 3.5 | | 3.1 | |
| Care and attention | 2.5 | | 1.8 | |
| Moving | 2.0 | | 1.9 | |
| Cleaning and disinfection | 1.8 | | 1.9 | |
| **Total breeding** | **14** | **50.7%** | **10.8** | **57.5%** |
| Finishing: | | | | |
| Feeding | 4.5 | | 1.2 | |
| Moving and weighing | 2.0 | | 2.1 | |
| Cleaning and disinfection | 1.5 | | 1.1 | |
| **Total finishing** | **8.0** | **30%** | **4.4** | **23.4%** |
| Other tasks: | | | | |
| Repairs and maintenance | 2.6 | | 2.1 | |
| Records | 1.1 | | 0.8 | |
| Other management | 1.0 | | 0.6 | |
| **Total other tasks** | **4.7** | **17%** | **3.5** | **18.6%** |
| Building construction: | 0.9 | 3% | 0.1 | 0.5% |
| Total man-hours/sow/year | 27.6 | 100% | 18.8 | 100% |
| Finishing pigs produced/sow/year | 19.8 | | 20.1 | |
| Liveweight produced/sow/year (kg) | 1784 | | 1850 | |
| Labour cost/sow/year | €203.73 | | €187.53 | |

- Conversely, cleaning and disinfection allowances of 14.5% and 16% of total labour input seem disproportionately low in these days of high disease risk. Is only one seventh of your labour effort spent on biosecurity enough? Spending six times more on labour for everything else these days when disease is our main threat to profit (after the fickle pig price), must be walking a tightrope, I guess?

- Note the typically high costs of moving pigs around (16.3% and 21.3%). This is one area where simplification, better housing design and automation will pay dividends in reducing labour demand. The situation is much worse on the larger units (as one would expect).

## WHO DECIDES WHAT ON THE PIG FARM?

Here are some past, present and future assessments of who takes the decisions on a modern pig farm (percentages in all cases). The latter two have had to be subjective, of course – of who takes the more important decisions on both the large (2000 sows plus) and smaller farms (mostly around 500 sows) which I have visited mainly in Europe over the past 15 years- about 300 in all, I see from my records.

**10 to 15 years ago** – mostly day-to-day decisions

| | *Larger (1000 sows plus)* | *Smaller (300 sows or less)* | |
|---|---|---|---|
| Veterinarian | 10 | 2 | Vet rarely used. |
| Stockperson | 45 | 23 | Took a lot of decisions. |
| Manager/ owner | 45 | 75 | Owner dominant. |

**Present experience** – mostly important decisions.

| | | | |
|---|---|---|---|
| Veterinarian | 25 | 30 | Growing presence on both |
| Stockperson/section heads | 25 | 30 | Section heads dominant |
| Manager | 45 | 40 | Managers often preoccupied/ too busy |

**In 10 years time?**

| | *Larger (5000 sows plus)* | *Smaller (around 500 sows?)* | |
|---|---|---|---|
| Veterinarian | 40 | 40 | Much greater input |
| Pig technicians/section heads | 20 | 10 | Trained to follow advice/new technology |
| Manager | 40 | 50 | More professional |

In the same way that an agricultural salesman should always try to speak to the person on the farm who authorises the sale/signs the cheque, so a farm advisor like myself should also try to talk to the person who takes the decisions if his advice is to be acted on. Just like the salesperson, I always make a note of who these people are after I have visited a farm as it is useful to know should I return.

My guess is that the pig specialist veterinarian will play an increasingly important role (as he is doing already) but will involve himself more in measuring things both inside and outside of the pig, forward planning (pig flow and work flow, as both markedly influence disease status) and management. Those of us at the sharp end of consultancy work realize how much getting these decisions right affects disease prevention. I therefore welcome the veterinarian taking a greater part in the management of the business, and future owners must make provision for his time and costs in this developing situation.

You may wonder why I forecast that decision-making by the highly-trained pig stockperson/technician is likely to fall away in future? This is because that very training he/she is increasingly undergoing from now on is designed to use their stockmanship skills to maximum effect, the major decisions being taken by others – the manager, the section head and the veterinarian.

Another trend I have noticed recently is that in my last 35 farm visits, mostly to larger units, the owner was only present in six of them and for two of these they took little part in the discussions. To a certain extent I actually welcome this trend – if it is one - as it shows they have confidence in their managers and are letting them get on with it.

# MOTIVATING PIG FARM WORKERS

" It is so difficult to get good stockpersons."

" Having found someone experienced, it can be a real problem to retain him/her".

" I worry too much about getting hold of and then keeping good people".

Do the pig textbooks - in fact any agricultural textbook - cover the motivation of pig technicians adequately? Apart from Dr English's excellent book on stockmanship – no!

I've done quite a bit of work on labour problems in some Eastern European countries and coincidentally in other parts of the world too. In east Europe they have plenty of labour – relatively cheap labour too – but lack experienced pig technicians, and motivation was poor on most farms, both large and small. The smaller farms were mostly family farms employing one or more outside workers, while the big boys – 1000 sows upwards - had a large labour force with many workers, many of them bussed in from a distance, and taken home again at night.

The smaller units needed rather different guidance on motivation from the 'industrial' farms, which is probably true globally, I guess. Table 2 summarises my own experience across the world over the past 10 years.

**Table 2. Pig technician's priorities – in order of preference.**

| Large Farms. | Medium/small Farms. |
|---|---|
| 1. Getting on well with co-workers. | 1. Working conditions. |
| 2. Working conditions. | 2. Pay. |
| 3. Efficient, decisive management. | 3. Getting on with the boss. |
| 4. Pay. | 4. Time off. |
| 5. Promotion prospects. | |
| 6. Sense of belonging. | |
| 7. Sociable working hours. | |
| 8. Training. | |

## Basic motivation strategy

In the last chapter my interviews with the leading and most successful high street bosses all remarked how vital it was to talk to their staff. How might this apply to a pig farm, especially the larger ones?

Why not ask your staff....    What their priorities are.
What they least like about the job.
How would they improve it?

This is best done by an **independent person** with their replies **guaranteed to be anonymous** and that they **will be reported back to them**.

Do not do this yourself otherwise you will only get the answers that they think you want to hear.

Then... hold a **discussion session** with your staff and see how many responses are appropriate to your conditions. I've organised several of these and it was both illuminating to all involved - and motivational too. From these sessions Table 3 could be compiled.

Table 3. **What motivates pig technicians?**

| | | |
|---|---|---|
| Health and safety. | Good pay. | Job security. |
| Relationship with boss. | Good farm equipment. | Efficient, decisive mgt. |
| Sense of belonging. | Teamwork. | Being kept informed. |
| Achievement from work | Encouragement. | Involved in decisions |
| | Sociable working hours. | |

(Not in order of preference as priorities vary greatly among workers).

By the way, notice that I am calling stockpeople 'Pig Technicians', not 'farm workers' or 'labourers' or 'pigmen'. Not even 'stockpersons' although that is much better. They are skilled technicians and giving them a proper title acknowledges this, is part of the motivation process, and helps our industry in the consumer's mind that the modern farm animal attendant is not just a country yokel - unfortunately too many do - but a trained and caring person. The introduction in several countries, being led by the British as I write, of a certificate in pig stockmanship, if needs be with several grades, is a great step forward in motivation.

## How does your unit rate?

Of the 13 priorities in Table 3 – how many would you, as a manager or owner, award yourself full marks? The most surprising result of this little exercise was the number

of instances when the boss admitted he didn't really know or, having thought about it, wasn't sure whether he was providing them, and to what degree.
So much for the positive angle – what about the negative?

Table 4 is to my mind, even more revealing and many owners/ managers were quite embarrassed when shown it!

**Table 4. What demotivates a  pig technician?**

A sample of the most common replies:

| | |
|---|---|
| Unapproachability of boss. | Boss poor communicator. |
| Boss never wrong. | Boss incapable. |
| " I'm considered a second class citizen." | Unfair treatment. |
| Being over-controlled. | "My needs not considered". |
| Low standard of 'idiot-work' | Too much tail-chasing due to lack of |
| Overcriticism from boss. | investment. |
| Too much repetitive work. | Not trusted or treated responsibly. |

('Boss' means the employer or the stockperson's immediate supervisor ).
 All these replies were gathered from staff who had recently left their jobs – the majority, but not all of them - from previous employment on pig farms. Note that the predominance of dissatisfaction – sufficient to cause them to leave – **were sociological reasons** and that pay was rarely mentioned, although as Tables 2 and 3 indicate, pay is nevertheless a motivating factor. I deduce from this that not only do quite a few employers in pig industries across the world need training in people-management, but so do their section heads. Agricultural Colleges please note?

## Four key areas

From the above discussion it seems the keys to the sociological side of staff motivation are…

✓    Pig technicians to be given a reasonable workload.

✓    For their contribution to be recognised,

✓    And valued in a way that they can appreciate,

✓    And so consider themselves as a useful team member.

## Some motivating ideas which work

Space is understandably limited on this vital subject (a whole book could be written on it as it affects livestock farmers) so I'll go on to provide a list of motivating ideas and essentials, and later deal with three undoubted minefields which the agricultural writers shy away from.  These are the vexed questions of pay, bonus incentives and the dreaded – on both sides of the desk - job appraisal!

The author with "two of the best pig stockmen I've ever known" Gordon Spenceley (left) and Paul Christopher at Deans Grove Farm's record year as long ago as 1982, when the pig team achieved 27 pigs/sold/sow/year (2.5 tonnes of liveweight/1.74 tonnes deadweight shipped per sow per year). Despite this excellent (for the time) physical performance, the saleable output is relatively modest by today's standards as our contract demanded a light finishing weight of 88-90 kg liveweight.

# SUGGESTIONS FOR MOTIVATING PIG FARM WORKERS

✓   Always express performance results in graphical form. Very few of us piggy people like keeping records and even fewer are good with figures – which includes myself, by the way! Remember the old Chinese proverb about 'seeing and understanding'.  In my country we have a saying 'every picture tells a story'. I have done several modest tests with both farmers and college students where they were asked questions with the answers expressed in columns of figures and exactly the same data in graphical form. Correct answers were between 18% to 40% higher in the pictorial presentation (See Table 2 in the Records Section on page 300).

✓   When performance figures are linked to agreed targets in graphical form it can be amazing how this gets workers interested. I have known section heads get quite hooked on record-keeping - even staying behind after hours when using graphs. Remember that a computer can churn out data in coloured pictures just as fast and easily as it can print those boring old columns of figures.

✓ Next, when pictorial weekly achievements against target are pinned up every week on the rest-room wall, after a time curiosity gets the better of even the lowliest stockperson and they get involved too in following performance matters. An example is on page 297 in the Records section

✓ Owners must provide a decent rest-room. The need to provide relaxation in reasonable comfort and cleanliness is vital. Delegate a rota to keep it tidy – pig workers can be as messy as their pigs.

✓ The same with the shower room. Clean and kept clean with a good laundry service.

✓ Provide overalls with the farm logo on the back. A sense of belonging is a motivating factor and worth the little extra it costs.

✓ Have a planned training schedule. A good pig vet who knows the farm and its staff as well as its pigs is a marvellous trainer, either in a group session or on a one-to-one basis as he tours the farm. Managers should be prepared to cost this training into the farm labour costs.

✓ Have a recognised promotional ladder. Owners and managers should take advice on this as well as on the training aspect. A cleverly-designed bonus 'carrot and stick' format works well, tied in to a monetary reward. For example, targeted performance is based on a 'cliff and plateau' concept - if these are spaced intelligently in an apparently achievable manner, then as performance nears a wages increase 'cliff' which is tied to targeted performance, there is an incentive for the employee or team to climb the 'cliff' and reap the resultant financial reward. Then having climbed it, the threat of falling back down again to the previous wage level encourages the workers by their own performance efforts to distance themselves safely away from it – whereupon the next cliff looks tantalisingly close. And so on. It is a sort of both positive and negative bonus – clever! Done well it does work. The farm's part-time record keeper is the right person to administer the system as he/she is considered impartial The management skill lies in estimating the height of the cliffs and the length of the plateau between them.

✓ Consider insurance for paid sick or injury leave.

✓ Have job descriptions and even an 'employee handbook' to cover aspects of farm routine, health and safety, emergency drill and general instruction and guidance outside the more personal aspects of a job description list. The handbook should be written in a friendly 'look at it this way' style and not in a bureaucratic manner.

✓ Finally – I also tell owners **to use their imagination.** Such as giving or lending a sow to a pig technician either as a bonus reward or as a straight gift in order to retain a good worker. The worker keeps the profit from the litter.

# AN EMPLOYEES' CHECKLIST

Taken from a Gallup poll of employees – not confined to our industry - is nevertheless a useful run-through of what employees think about their employment

Owners and managers should use it as a matrix to find out what the employee really thinks of his job under your supervision.

It can be an interesting exercise for both of you during a job appraisal session, see page 281. Make out a composite form for all personnel and fill it in for each employee - *after* not *during* - the interview with tick, cross or question-mark. This is the information from 'the horses mouth' you need, in brief, on which to act before the next job appraisal.

## DOES THE EMPLOYEE AGREE OR DISAGREE WITH THE FOLLOWING...

1. I know what is expected from me at work
2. I have the equipment I need to do my work well
3. At work, I have the opportunity to do what I want to do best, every day.
4. In the last week I have received recognition or praise.
5. My superior seems to care about me as a person.
6. There is someone at work who encourages my personal development.
7. My opinions count at work.
8. My employers line of business makes me feel my job is important.
9. My colleagues are committed to doing high quality work.
10. I have a best friend at work.
11. In the past six months someone has talked to me about my progress.
12. This past year I have had opportunities at work to learn and grow.

Source: Extrapolated from Hooper (2010)

# THE DIFFICULT QUESTION OF PAY

How much should a pig technician be paid? Owners must examine their labour costs in relation to total production costs. In the developed pig industries of Western Europe, Scandinavia and N. America, labour costs in my experience lie between 12% -14% of total production costs. For Eastern Europe, the Pacific rim (excluding Australia/

New Zealand, Japan and Korea) about a quarter to one-fifth lower. If the owner is remunerating his pig technicians below these levels they could be underpaying and demotivating them, with the risk of damaging the performance of the pigs in their care.

I have evidence from many years of persuading owners about these problems – yes, usually they are hesitant but this reluctance is overcome by establishing a bonus system - to move their total wage costs up to these national benchmarks . In such cases the gross margin often rose by 15% within a year due to better-motivated staff.

I tell managers to "look at what this extra financial outlay will mean to production costs– it will surprise you as it won`t be all that much".

Owners should not increase wages suddenly, but increase them steadily over a period and try to link it to a sensible well-planned bonus plan.

## So should you pay a bonus?

A bonus is a payment incentive scheme. They always involve complex issues and are sometimes effective and sometimes counterproductive. Bonus schemes are now a common part of employment in the retail and industrial sectors (and overdone in the financial field, it is true!) where it is easier to measure performance than with livestock production. The agricultural world wide is lagging behind in this 'new' area and we must move with the times.

I was once paid mainly by a profit–sharing scheme for four years, perhaps the ultimate in any payment incentive scheme, and have been an advisor and designer of several bonus schemes for clients - most successful, some not.

This is what I have learnt…. The ones which work best are when individuals are rewarded….

- On the **team** reaching targeted levels
- From which individuals get a **proportionate** share
- Which has been **agreed beforehand.**
- **A bonus must never compensate** for a less than reasonable salary or wage

Other considerations are…

- Records must be seen to be adequate, and administered by a trusted and proficient farm recorder.

- Co-responsibility. Managers must make their subordinates believe that any problems - and successes - lie at the manager's door as well as their own. The manager must be the team leader, in other words, not just he who must be obeyed.

- Time. A year to assess the bonus is too long, causing disinterest; quarterly maybe too short for administrative purposes, so 6-monthly payments have seemed to be best.

## Setting targets

This is the task most producers find the most difficult and get wrong the most often.

Pre-set fiscal goals can be affected by market forces outside the farm's control, so it is still preferable to base targets on physical performance – but using the profit- oriented terms describe at the start of this Business Section. Nevertheless the size of the bonus can be based on what improvements in retained margin can be made from predicted achievement over target - as long as the bonus only swallows 10% of it - no more. Only in the most prolonged profit troughs due to circumstances outside the farm's control will this safety margin be breached to inflame cashflow difficulties severely. In such extremely critical times the staff will understand rather than the alternative of some of them having to be laid off.

Targets are based on the owner's decision but are tied to the best likely profit-earning sectors on any farm, such as conception rate; numbers born alive; weight of weaners per farrowing, and meat per tonne of food used from nursery to finishing.

## The 'One-off' incentive

There is scope for individual targets based on areas needing improvement at any one time. These 'one-offs' are a good idea but managers can be too impatient in setting incentive goals, when the incentive `cliffs` are too high to scale or are too short in the time given to do so.

Such specific incentive proposals should always have a set of 'fact sheets' given to the staff involved so as to provide them with a personal aide-memoire to remind them of the critical stepping stones they need to check out so as to be sure of their bonus. **Many of the checklists in this book satisfy this need.**

## The scaled bonus

A target financial increase from, say, pigs born alive per litter or numbers weaned/ weight of piglets weaned/meat per tonne of feed used in the case of the growing/

finishing staff. etc are easy to calculate and some managers design a sliding scale where the full bonus is based on 100% achievement and there is a progressive reduction as the achievement lessens, usually by  drops of 25%. across the fixed bonus period. This is a sort of graduated cliff to climb and can stimulate motivation.

## The vexed subject of the job appraisal

Everyone hates it, both managers and staff. Done badly it can be very demotivating. One background reason why I left commerce was a succession of I thought dreary and unhelpful annual appraisals, which in the end contributed to the feeling that maybe I was better off working for myself!

Done properly they...

- Keep employees informed. How they have done in the employer's eyes, what chance is there of advancement, what changes are likely in their workplace (or not!) and how they will be affected. Some of these findings may be tough for them to  accept but it is better to know than not know – I've been there, and it helped me with my future career change which has worked out so well.

- Help the manager assess and record employees' behaviour and attitude. See the checklist on the next page.

- Identify performance problems and forestall others.

- Provide a forum for expressing and resolving concerns – on both sides.

- Guide the appraiser on what instruction/ training is needed before the next appraisal is due.

- Enable job descriptions to be revised and updated if needs be, and reinforced in the employee's eyes. Job descriptions are the roadmap to success in the job.

-  No job description? Oh dear! One of the most vital documents on the farm. The appraisal is the time to ask the employee to agree with you what it form it should take. But the description should fit the job – not the employee.

Since my own unfortunate experiences on the receiving end of my own job appraisals 35 years ago I have been fortunate to sit in on several professionally-conducted sessions held by  experienced personnel officers. (One from one of the retail firms I visited). I sat and listened. This is what I have learned then and subsequently, when farmer clients have asked me to take part as an intermediary –'holding the ring' as they say - in the job-appraisal sessions they had to carry out.

# MANAGER'S CHECKLIST FOR A GOOD JOB APPRAISAL SESSION

✓  Educate new employees about what a job appraisal is beforehand. This takes some of the apprehension out of it.

✓  Choose a neutral venue. Not behind a desk - better out of the office altogether to avoid interruptions. Be friendly – smile and establish eye contact. Don`t fold your arms or pace about as I've seen happen!

✓  Why not start by asking the employee for a self-appraisal – this is a great ice-breaker. Then follow with the sort of checklist-cum-quiz featured on page 278.

✓  Concentrate on tasks, job related issues and attitudes to the workload required.

✓  Start by covering the employees successes before dealing with any under-achievements.

✓  Do not write notes during the interview. While these are essential for future reference, make time between interviews to do this. What is much better is to have a standard form for all employees (jocularly called 'the Death Warrant' by employees) to be completed by agreement between you, signed and dated. This provides an accurate record for later consideration, especially if disputes should occur.

✓  If the employee strongly disagrees with you, ask him to write out in his own time later what his views are, and promise another session to go into them.

✓  Set goals by mutual agreement, both short term, 3 months, and long term, 12 months. Use these goals on the 'Death Warrant' to review at the next appraisal, and if not met, discuss why *and how you can help.*

✓  **Two 'Donts.' 1.** Never ever hold 'peer appraisal' sessions (where an employee is asked to sit in and be the 'prisoner's friend' - the quickest way to lose good staff that I know. A modernist idea which is as daft as it is dangerous! Appraisals should be confidential, never the property of work colleagues – even the employee's section head.

  **2.** Do not use the appraisal to fix wage rises and promotions, this must be done at another time.

✓  Most annual job appraisals should last about an hour. Sounds long? Good sessions with both sides benefitting do last that long. Make time for it; they'll respect you and yes, the occasional chap will be bored stiff - in which case you will have learnt something about him/ her.

# TRAINING

A programme of planned training is a strong motivational factor and helps lift the animal attendant into the 'Pig Technician' status in society.

In addition, training in pig stockmanship and caring for pigs – especially the younger animals where the mortality levels are highest - improves productivity and profit substantially. The British, Danes and Dutch have long trained their pigpersons well and this is forging ahead with both in-house and off-farm courses funded partly from central government, the industry itself and a contribution from the stockperson's employer. When completed satisfactorily, this degree of formal training provides the stockperson with certificates of competence which not only helps them in a future career but also rewards the employer with improved performance and profit. (See Table 1)

The Americans, with a large number of Latin-American workers on their large pig units, have also done an excellent job of harnessing this cheap but lowly-educated source of labour, the majority of whom only speak their native language.
 Most interestingly - because this workforce knows so little about pigs - they are a fertile field for the `performing to rote,` idea and, given carefully written own-language notes describing what daily tasks to follow in sequence, do perform surprisingly well - possibly due to their having no preconceived opinions. But dedicated and caring section heads are absolutely essential for success in such cases.

## Plenty of pig training guidance and advice available

A whole book can be written on training pigpersons from the most junior employee to the manager and I do not attempt to go further into the excellent pig training booklets, the late Dr Peter English's first-class book on pigmanship and the various internet CDs which are available to all in the world who can speak English.
What did seem to be lacking in the past was some firm evidence of the benefits of training and I could find nothing at all on the likely payback in monetary terms from the cost of training courses to make pig technicians out of ordinary stockpeople.

## An important survey

After some searching I managed to find four progressive British farrow-to-finish producers who had already embarked on a programme of training their staff using recognised training bodies and their local agricultural college. A patient wait of 2 to 3 years to accumulate post-training performance evidence was then compared to their records of the two years prior to the start of training.

Correction factors were built-in to the comparisons to allow for normal improvements in performance which could have occurred over the passage of the five years studied, mostly from improved technology.

I was worried that varying disease incidence would distort the performance figures and make it all a waste of time, but apart from a bad but thankfully short flare-up of pneumonia on one of the farms, the disease pattern was broadly similar before and after. My statistician consultant was therefore happy enough.

The results are summarised in Table 5.

**Table 5. Average results from 4 breeding farms (average 370 sows) where 2 out of 3 workers (2 out of 4 in one case) attended day or block-release courses over 2 years.**

| | *Results from 2 years prior to programmed staff training* | *Results from 3 years during and after training* |
|---|---|---|
| *Performance* | | |
| Av. Sow nos. | 326 | 390 |
| Born alives/litter | 10.8 | 11.2* |
| Weaned per sow and | 22.3 | 24.1* |
| serving gilt/year | | (24.8 actual) |
| Farrowing rate | 86% | 89%* |
| Av condition-score at weaning | 2.25 | 2.53 |
| Mortality to weaning | 11.2% | 8.7%* |
| Av weaning weight (kg) | 5.21 | 6.27* |
| Live weaners produced per | | |
| tonne of all food (kg) | 89.2 | 116.2* |
| *Financial:* | | |
| Value of extra weaner weight | £103.47 | £134.79** |
| (at £1.16/kg) per tonne of | | + £31.32 ( + 30%) |
| all food. | | |
| Av. cost of training/tonne feed | - | £6.87 |
| (including extra called-in help) | | |
| Av. cost of additional wages (17%) per tonne of feed | | £4.07 |
| after training period  includes new bonus rewards) | | |

**REO 2.86:1**

* Correction factor of -3% allowed-for to accommodate natural performance progression across 3 years, due to better genetics health, nutrition etc.
** Additional correction factor for the 2.7 extra days weaned.
 Time scale (2003-2008).

This was certainly the most interesting amateur survey I have undertaken out of the several surveys into other subjects I have published in my career. This was because I got to know the pig technicians (as most of them certainly became once the courses were finished) intimately, with their fears and motivations, skills and workloads, some of which information helped form my views in this chapter.

## TRAINING CONCLUSIONS

From these painstakingly recorded figures, even after all training costs have been accounted for, the value of extra meat sold per tonne of food fed on these 4 units gave a 2.9 to 1 payback from improved performance after key staff had undergone off-farm formal training, with the local vet assisting on one farm.

The main benefit of the training was in the farrowing and post-weaning stages, less in the rebreeding areas.

An interesting spin-off from the exercise was that on two of the farms the owners felt that as they had invested such a considerable sum in the extra cost of training and subsequent increased pay awards, more time should be allowed the staff to complete the work to the standard set at the training courses. Interestingly again this did not increase the labour load but one of the benefits of the training was to enable the staff to work more efficiently, which itself is motivational.

Subsequently bonus systems were introduced on three of the units and an extra man employed on the fourth.

All staff appreciated the courses and seemed motivated by the experience.

## REFERENCE

Hooper, R.(2010) 'The Future Staffing of the Pig Industry' BPEX (UK) KT Event, Peterborough, England

# 12

# SHARING THE LOAD - BUSINESS PARTNERSHIPS

Way back in my career I was involved in business partnerships, so a good deal of my own experiences are featured in the Checklist I have compiled below.

All pig producers are being exhorted to become 'businessmen-farmers' these days. To keep abreast of this movement I now have quite a few books on business management on my shelves, and have put myself on one or two specialised business courses recently.

This doesn't make me an expert in the subject – rather the opposite, as it happens, as they revealed my own shortcomings only too clearly – but I was surprised that no-one seems to have covered the subject of choosing a business partner either on the courses I attended or in the textbooks I purchased.

## CO-OPERATION

Important? Yes, very, and not an odd choice of subject at all, because co-operation is now the buzz-word in many pig industries. Either co-operation up the marketing pyramid or alternatively (if you are are wary of the big battalions taking control) then forming localised mini-businesses where small groups of producers co-operate both to produce pigs and then market their own brand of pig products. This can also be very successful – and, more satisfying!

## THE FAMILY FARM HAS A FUTURE

All over the world I encounter family farmers apprehensive – even defeatist – about their chances of survival. OK, take that attitude if you must, but doing so tips you even further over the cliff. I've seen it happen with past clients of mine, now no longer in pigs or even in agriculture. Very sad.

In uplifting contrast, some clients have taken one look at that cliff edge and said "*No way* am I going to fall over it." Faced with the impossibility of making sufficient profit from, say, 150-300 sows on their own, they have gone one of three ways.

1.    Mixed farmers have tended to close the pig side and concentrated on other sectors of the farm, and/ or engaged in diversification off-farm (*see item 3*)

2.    Specialist pig producers have formed mini-businesses, usually 800-1500 sows or more, with other producers, sharing the profits. Increasingly, while no-one actually likes profit-sharing, the ***improved performance from business restructuring has more than made up for the need to distribute what profits there are inter alia.***

      (This is where co-operation/partnerships come in and why I am writing this section).

3.    Both specialist and mixed farms have diversified. On-farm if they can, but more often off-farm, even into becoming franchisors, i.e. running a different business with other's capital.

## DISAGREE

I totally disagree that 'there is no future in the family farm'. With the right attitude of mind and good guidance, family farms can have a future. There is no shortage of conventional or venture capital, interest rates are affordable; but you must create your own local market, or supply into a national or global market who can sell what you produce. So what you produce must be seen to be wholesome, tasty, safe, attractively packaged (do your own) welfare-friendly and available at a perceived value-for-money price. Price is still crucial. **People will pay more for something they perceive is good value.**

The latter, vital factor of securing a competitive price comes from co-operation and partnership. So let's talk now about partnerships.

## EXPERIENCE

I've been involved in two personal business partnerships, one was in pigs and the other wasn't. Both were successful, but are now very amicably dissolved as my partners wanted to move onwards and upwards, which they have.

Lately I've become involved as a background consultant to other multisite production ventures started by pig farmers. So – from their and my experience here is a list of what I think are basic principles for a successful partnership – something the textbooks never cover!

## A PARTNERSHIP CHECKLIST

✓ Both parties should be prepared to go more than halfway towards the other.

✓ Are you a 'sharer' and your partner likewise? If not, go no further. Sharing ideas, goals, workload and borrowings.

✓ Like and respect your partner. If not, go no further.

✓ Ensure the partner's family are supportive. If not, go no further.

✓ Clearly define, *in writing*, each partner's horizon of responsibility and agreed commitment.

✓ Do not interfere in your partner's agreed responsibilities. Discuss the partner's decisions, but be positive not negative.

✓ Talk daily; meet weekly.

✓ Do not have an open-ended agreement. Agree and define a time-scale, with a 'what-if' closure/ withdrawal clause built in. Each partner should have a 'silent' alternative partner or strategy in mind for their part of the business. This need not be talked about, but prepare yourself on what to do if a partner wants out. It can happen suddenly as opportunities occur.

✓ Mutually agree performance standards for each production section/unit and agree a 'what-if' action plan *supported by all parties* if one section is in trouble. One advantage of co-operation is that you can help each other, providing *each knows in advance* what this might entail.

✓ Accounts must be independently compiled and audited, and not involve any of the partner's family in their construction – so as to save money, for example.

✓ Do not have too many partners. Group businesses over 6 tend to fail in my experience, and 10 or more always have.

✓ One of the partners should be a butcher. There are plenty of these leaving their trade (sadly) so getting a good experienced man should not be difficult.

✓ Agree on a team leader, with (only if needed) a 3 year rotation.

✓ Use the big battalions, but keep them at arm's length. Never enrol them as a partner. If you do, which is becoming more commonplace, it is a good idea to have an informed third party comment on their proposals (which always favour themselves), so that you are forewarned of what might happen.

✓    Consider venture capital, but present a good case with an enthusiastic, united front. The venture capitalist will be a 'townie', so think your case into his shoes – *i.e.* his likes and dislikes about raising meat/food – from an urbanite's viewpoint.

✓    Plan well ahead on where the market may lead. Do your research; seek out the information – it is out there.

✓    If you can produce a good product, sell it to any outlet who (a) will listen (b) who clears the precautionary business hurdles (part of the research process) (c) never restricts your options or give exclusives (d) seek out concerns that need to steal a march on their competition.

# 13

# MODERN AND FUTURE RECORDS

It may seem odd to need to define record-keeping on the pig farm, but perhaps the most arresting definitions are, for once, negative ones.

(1)  Without records the control of productivity is guesswork.
(2)  Without control of productivity, profit can never be maximised.

## PROBLEMS WITH RECORD KEEPING

The first problem is that too many pig farmers are not good record-keepers. You still hear people saying about a stockman or owner – "he's good with the pigs but hopeless at records", although this criticism is not so common as it was. This is because pig production units are becoming specialised capital-intensive operations requiring full-time management. Record-keeping itself has moved much closer to being user-friendly with the arrival of the microchip, the desk-top computer and the internet to provide information. Records are less of a drudge than they used to be – *if* you use the right record system.

I go on to many farms where record keeping is disliked by the staff. In most cases the records are too cumbersome, the staff are probably keeping too many records and they have insufficient time left to do them well. In many cases they are not using them properly – not getting the most out of them. A lot of hard work for a minimal result, which is wasteful. If the right record system is used properly then keeping the records becomes – well, not enjoyable perhaps, because compiling amounts of numbers and events is always going to be a repetitive task – but at least tolerable. With a good system the stockman/manager actually looks forward to his weekly or monthly summary. Man is basically a hunter-gatherer and records satisfy the 'gathering' (or in the 21$^{st}$ century – collecting) instinct in us – it is satisfying to be able to compare current performance with previous achievements, and to see if past efforts have indeed borne fruit.

# CAN DO BETTER

Whilst not denying the great strides forward made by computerised record-keeping systems over the past 20 years, with my on-farm problem-solving work – which encounters a variety of such record systems – such experience suggests that the pig records industry has 'plateau-ed off' and much more can be achieved by them. Sorry – but I'm sure computer recording needs to move up a gear!

Table 1 lists the deficiencies I have run up against in consultancy work.

**Table 1. What is wrong with record-keeping today?**

| Problem | Result | What needs to be done |
|---|---|---|
| Inputting is too onerous | Risk of error | Re-examination to make it as simple, for example, as the original 'Pig Tales' method (page 297) |
| Too cumbersome | Offputting | Re-examination to include only essential information |
| Not graphical | (1) Slows up recognition of a problem<br>(2) Non motivational | Most columns of figures can be 'pictorialised' and the pictures up-dated frequently. Stockmen are better motivated/ trained by monthly/weekly graphs in the rest-room than by columns of figures. |
| No statistical overlay | Wrong assumptions made | Simple statistical flags appended to changes to signify :<br>"*No problem – green*" "*Problem may be ahead – amber alert*" "*Action required – red alert*" (Traffic Light System) |
| Over-dependence on physical performance | Wrong conclusions may be reached | Include additional data on econometric as distinct from physical performance (*e.g.* The "New Terminology") |
| Underuse of "*what-if*" facility | Opportunities lost | Most recording systems are only retro-active in this important area, depending on the farmer or his adviser to call it up. Better is to insert **auto-periodic** "*what-if*" exercises when productivity falls away, thus stimulating thought, consultation and, if needs be, earlier corrective action. |
| Weak in financial aspects | Overdependence on physical performance in determining strategies | To maximise returns, performance records **must be linked automatically** to financial records. Another opportunity here for more '*what-if*' promptings, at present rarely attempted, this time based on economic matrices. |

# WHAT RECORDS ARE NEEDED

I have read many worthy but complex flow charts on how record keeping systems should progress. All this theory is fine, but what I hope you, dear reader, will use as a take home message in this section is much simpler:–

1.  Correct and regular inputting is vital. (Cross-checks to pick up as many human errors as possible are needed.)
2.  **The software needs to collate only the essential information to be measured against the farm's preset targets.** (Most systems provide a plethora of largely unused data, I suspect from trying to 'outsell' each other.)
3.  The correct interpretation of what the figures show is critical. (Any computerized matrix can assist here.)

It can be seen from the above that both the stockperson at the start of the recording process and the manager (or section head) at the end of it are absolutely vital cogs in the productivity machine, and moreover have an equal responsibility in achieving profit!

A good record system makes the manager's respective jobs of collecting and interpreting the data *easier*, not more difficult, as some systems do.

## Some sound sense about targeting

A fourth area is equally important – that of targeting. You will find plenty of targets propounded, and indeed you will find them in this book, but at the end of the day it is *the business's own targets which are paramount*.

These targets are farm-specific and are tied to the farm's own capabilities and circumstances – no-one else's. Don't listen too hard to other people's targets; they are a useful sounding-board, but that is all. Bench marking, the latest buzz subject, is in my view, over-hyped.

Your targets will depend on what you / or your financial adviser / he who lends you capital / your immediate customer demand of the business.

Your targets will depend on the circumstances which affect you – your own required standard of living, your pig-housing status, your labour pool, your financial strengths or weaknesses, the stipulations of your immediate customers, the cost of money, credit facilities, disease status and so on. Targets have to be fixed for most of these – some flexible in a crisis, others immutably fixed whatever happens.

Having established your own targets, especially the output targets, a primary role of farm records is *to inform you of how you are progressing against target.*

Records will also allow you to re-adjust your targets if the initial decision on that particular aim / forecast was wrong. *Every* record should have a target attainment figure alongside it, because by studying these individually, two things appear.

1.    The area of under or over achievement is pinpointed. (Yes, *over*-achievement is important too – you may not have enough future housing space to accommodate a surge in productivity, for example.)

2.    A general under-achievement of a final target can be quickly traced to the sector of the business where the deficiency lies.

# SO … WHY KEEP RECORDS ?
# AN EXPANDED CHECKLIST

✓    *Daily activities* – Reminding and forewarning us of essential tasks. The most important reason is to identify basics which need to be completed on the farm (service dates, due return dates, farrowing dates, weaning dates) so as to maintain an efficient pig flow and work flow.

   Secondly – to record primary events so that your awareness of how things are progressing is accurate. Right decisions ride on accurate information.

✓    *Animal location* :  It is important to know where every animal is on the premises. This is vital where breeding stock is concerned but the nursery/growout operator may only need to identify a sample of growing/finishing pigs to obtain growth and carcase quality efficiency ratings.  On the other hand some producers carefully trace back market information to individual boars/AI and female lines, and recording the progress of individual growing pigs is again necessary in order to be able to do this.

✓    *Traceback:* Records are becoming increasingly important to the retail sector in terms of quality control right back to the farm and, from the farmer's point of view, in ensuring he gets his just financial rewards for investing in the means to provide high quality meat to the processor.

✓    *Monitoring Efficiency* : Records are used to identify both efficiencies – performance and economic – which influence possible profit. This is a huge subject dealt with later in this section.

✓    *Modelling* : Records provide data for the at present under-used practice of modelling. Modelling uses a computer program to establish a predictive "what-if" situation where different actions, target weights, production systems, market outlets, carcase grading goals, improvements to insulation or ventilation, feed scales or diets of different nutrient density, etc. are tested out against current performance. Modelling is a form of instantaneous forward trialling on paper and is becoming increasingly accurate in its predictions. Records therefore provide us with guidelines for future production targets.

✓ *To help those who can help you*. Two or more brains are better than one. The vet, the environmentalist, the feed rep/nutritionist, the accountant/financial adviser and the management consultant, like myself. Every business needs advice. Good advice rests on what good records reveal.

✓ *Diagnosing Trouble* : While modelling looks forward to predict the results of proposed action, good and accurate records are invaluable when seeking the past cause of poor profit performance. At present the potential is barely explored and I can see a much greater part played by the computer in future. At present much of my job as a farm consultant is taken up with studying figures provided by the producer who is seeking help with a problem. **My experience is that about half of the likely solutions can be discovered lying hidden in the records** – pointing to areas where attention is needed. Sometimes, by analysing records sent in advance of a farm visit, a solution to the problem can be suggested so that the visit to confirm the diagnosis is unnecessary, which saves me much time and the farmer some money. In my case this has happened in about two in five of my 'call-outs'; the rest do need a farm visit. But this pre-studying of the records I find essential so that what I see on the farm tour confirms or denies my preconceptual study of the 'on-paper' situation. Frankly I am surprised more consultants and veterinarians don't work this way. Records first – *then* a look at the pigs, not the other way round, which is commonplace.

## A future role for the computer

There is no reason why, when programmed in detail, a computer should not follow the same diagnostic processes used by the consultant when examining records. I can see the computer monitoring not only the producers' management and stockmanship skills as it does now, but signalling warnings and *suggesting remedial action* automatically and at no extra cost.

## Using the computer properly

The logic of this procedure is straightforward:–

• The pre-programmed computer will flag up below-target achievements in a wide range of performance subjects.

- **Traffic light system:** As mentioned earlier a statistical overlay can be included so as to 'grade' the performance shortfall as, for example, something to watch out for or something to act on (Warning Level / Action Level - I call it my 'Traffic Light System').

- If action is needed, guidelines are pre-programmed in to the matrix in the form of checklists similar to the examples in this book. Thus the computerized recording system does much more of the diagnosis work than is achieved by present systems. *It can even go on to suggest what to do*, and/or suggest further checks.

  The producer is alerted to the problem, advised of how serious or important it is, and pointed towards areas/actions he should consider as likely remedies.

  Some computer recording firms are failing us in not providing these built-in benefits, and I hope they read this section!

✓ *Monitoring Health :* This leads on to a very similar situation with regards to disease. Records aid the veterinarian to identify causes of disease and production problems and already computers are assisting the pig specialist vet, albeit in a limited way, to isolate the most likely cause of disease. At present the veterinarian works to a schedule of :

  ✓ monitoring the level of disease

  ✓ where and when it occurred

  ✓ what medicines and their dosage were used

  ✓ how the diseases that occurred might relate to the management, etc.

  Computers will markedly assist and maybe even revolutionise this approach in the future. We need a computer-programmed vet!

✓ *Motivating Staff :* Records can be a great stimulus to improving the morale and efficiency of the stockman. This may seem to be a paradox, as I have already said that producers and stock people usually dislike compiling and processing records, but their attitude can be completely changed if:–

  (a)　Compiling on-site information is made as simple as possible (Figure 1).

Continuing on with the staff motivational benefits .

  (b)　The computing aspect is done for them

  (c)　The information is presented in an attractive and easily-recognised form, which in the majority of cases means 3-colour graphics. (Figure 2)

  (d)　This form of presentation is updated weekly with performance against target demonstrated in an on-going form.

## Example of daily diary entry from
# The Farrowing House
## Sent to database on same day each week

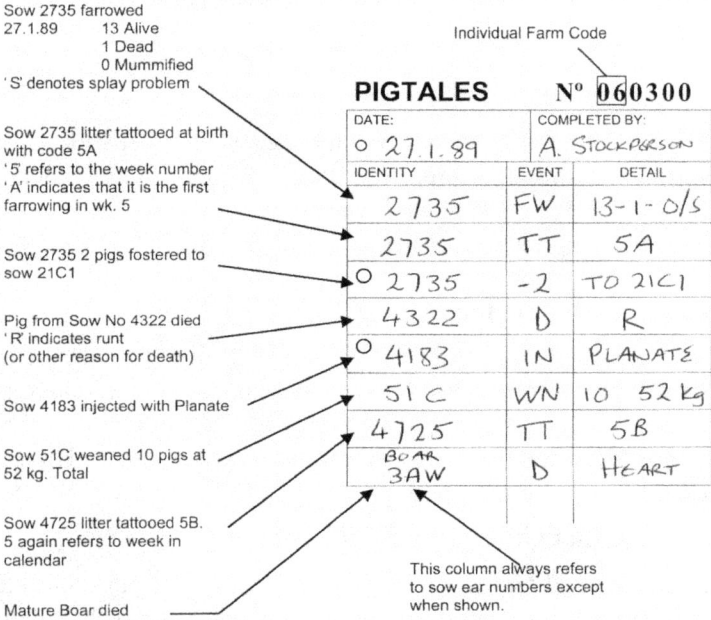

Sow 2735 farrowed
27.1.89     13 Alive
            1 Dead
            0 Mummified
'S' denotes splay problem

Sow 2735 litter tattooed at birth
with code 5A
'5' refers to the week number
'A' indicates that it is the first
farrowing in wk. 5

Sow 2735 2 pigs fostered to
sow 21C1

Pig from Sow No 4322 died
'R' indicates runt
(or other reason for death)

Sow 4183 injected with Planate

Sow 51C weaned 10 pigs at
52 kg. Total

Sow 4725 litter tattooed 5B.
5 again refers to week in
calendar

Mature Boar died

Individual Farm Code

**PIGTALES     N° 060300**

| DATE: | COMPLETED BY: |
|-------|---------------|
| O 27.1.89 | A. STOCKPERSON |

| IDENTITY | EVENT | DETAIL |
|----------|-------|--------|
| 2735 | FW | 13-1-O/S |
| 2735 | TT | 5A |
| O 2735 | -2 | TO 21C1 |
| 4322 | D | R |
| O 4183 | IN | PLANATE |
| 51C | WN | 10  52 kg |
| 4725 | TT | 5B |
| BOAR 3AW | D | HEART |

This column always refers
to sow ear numbers except
when shown.

*System devised by Pig Tales (copyright)*

**Figure 1. An example of a classic (and unsurpassed) inputting system from PigTales – 30 years** ahead of its time in the 1980s. Just a series of simple cards sent in to the computer, using if required, the stockman's own abbreviations/code.

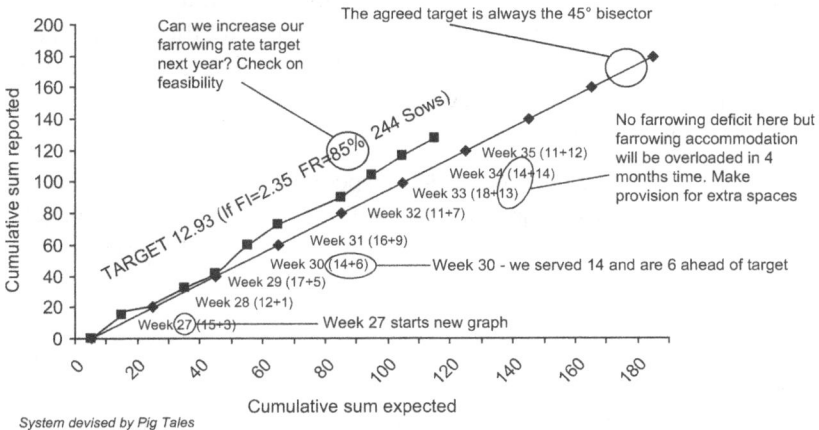

**Figure 2.** Breeder – Services Cusum. Another highly-motivational weekly diagram.

# ACHIEVING TARGETS – STOCKPEOPLE

Stockpersons don't like figures. Anyway targets make stockpeople nervous – an understandable reaction for most of us! I'm often told "The boss is always on about targets. He comes back from some meeting or other and says that we seem to be falling behind, and we – or indeed myself - need to do better."

## Four factors

On investigating further with management I suggest there are four main reasons which contribute to this 'consumer resistance' attitude by the stockperson. It is these areas which I suggest have not been given enough space in the several excellent books on pig farm stockmanship I have read.

1.   **Targets are often set too high**.

I don't necessarily question that the target itself is too high, but that the means of achieving it is too hurried. Attitudinal research among typical stockpersons shows that while they are naturally practical and forthright in opinion, they are also likely to be conservative, reserved and sensitive. As a result, target-setting for them should be imaginative and encouraging.

2.   **Not enough thought is given to presentation**.

Stockpeople need targets which interest them - not put them off or bore them. Figure 2 gives an example of a target Cusum (Cumulative Sum) graph for any important performance-indicating item (the target is always the 45° bisector and the week's achievement is plotted either above, below, or on, the line). For managers and owners, on a computerized system, any change can be given a degree of statistical significance, if required, to assist management as to the degree of importance that the deviation from target indicates which might affect the final profitability of the enterprise. Fortunately, for the stockpersons and section heads a simpler hand-updated Cusum graphs illustrating the weekly progress in the departments in which they work is all that is required.

This is a grossly under-used idea, especially for the stockpeople. Sure, pinned up in the rest room each week, the updated Cusum graph is largely ignored at first. But after a while curiosity sets in and the graphs are looked at, especially if a performance bonus is in the offing. A few more weeks and they become a centre for discussion - even controversy, with remarks, usually good-natured, flying about like "Well that's not our fault – it is the questionable quality of weaners we have just got in from you nursery guys" etc. An atmosphere of competition sets in which is all to the good.

3.   **Weekly goal-discussion sessions are essential.**

These are not so much target-setting sessions but 'where have we got to this week's meeting where the latest weekly results on the graph are examined by all the farm staff in a spirit of communal appraisal by everyone. I have seen this working so successfully on a few large units, as it does seem to nip problems in the bud and settle disagreements before they fester. I am amazed that every single pig unit, not just large ones, employing staff doesn't do this. – in my experience only one in 40 do so.

The low uptake could be because either the owner or manager hasn't realized the huge advantage that the use of the Cusum idea affords. It is unfrightening and easy to compile and follow. It does not rely on boring columns of figures -which accountants love and everyone else hates!

4.   **Bonuses for hitting targets are badly thought out.**

Is a bonus plan a good idea? With well-motivated staff such as described above – definitely so. With poorly motivated stockpersons it usually makes little difference and can upset and demotivate other workers. The important thing about staff bonuses at any level is that they should be…

1.   Based on a **team basis** where individuals are rewarded on the **team** reaching targeted levels.

2.   Designed so that individual workers get a **proportionate share**.

3.   Which has been **agreed beforehand.**

4.   Everything is **easily verifiable**; nothing is hidden.

These four criteria are essential for a trouble-free bonus system.

The textbooks seem terrified of the contentious subject of bonuses, which is covered in more detail on page 279. I speak from experience, as for several years I was paid almost exclusively from acheivement bonuses – a gamble which in my case worked, thank goodness.

The value of graphics in getting information understood is not in question (Table 2).

## Improvements needed

There is room for improvement in all three cases, but especially the first and last. There must be a hundred computer recording systems world-wide at the time of writing if those in-house variants used in veterinary practices or in the massive industrial units are included, yet I only know of two which meet all three criteria in every respect. The majority are pretty unimaginative, I guess.

**Table 2. Farmers understand pictures / graphics**

Farmers were presented with a farm problem hidden in:
- a)        columns of figures
- b)        exactly the same data in graphical form

They were asked to provide a numerical answer to each question.  Number of correct answers were:-

|  | (n) | Columns of Figures presentation | Graphical presentation |
|---|---|---|---|
| Class 1 | 30 | 42% | 78% |
| Class 2 | 34 | 57% | 84% |
| Class 3 | 37 | 36% | 67% |
| Agricultural Students | 20 | 80% | 90% |

Source: Gadd (1994)

# 'TRAFFIC-LIGHT' RECORDS

Of course the remark in the above paragraph will annoy many worthy people, far more numerate than I, who design and run such schemes – so let me give one example of what I mean about 'lack of imagination.

I have to consult many of the client's records both before visiting a farm problem (far preferable as it gives me more time to try to find out what may be the problem from the figures) or in the farm office (when I am often rather rushed before the owner understandably wants me get out and look at the pigs).

Having done the on-farm problem-solving job for 40 years or more, I have found that many clues to the situation lie hidden in the records and occasionally the likely solution can be found there too. No need to tour the pigs, except to obtain proof of the prognosis! The eye confirming – or not, which is equally important - what the records have suggested.

Weekly, monthly or quarterly changes to performance can be quite obvious – even alarming - but not influential, while others seeming quite minor nevertheless could be critical.

## A suggestion:

What I would like to see in a computerized recording system today are two improvements.

1. **A statistical overlay**

   This would group into three categories any changes, up or down, to all the main performance parameters, each performance measurement is given colour-coded `stars` as follows.......

   **Green** – The change is not significant in affecting the predetermined target. No need for any action.

   **Amber** – The change is also not significant – but if it occurs again next time to the same degree or worse, then this is an advance warning of an impending target shortfall -or alternatively overproduction which may impinge on pig place adequacy, for example.

   **Red** - Action needed. The change is statistically significant and will affect target. unless action is taken

   But what action?

2. **Action print-outs.**

   With every Red alert, a pre-programmed list of suggested factors to check and actions to take is provided for that particular performance category.

   If thought worthwhile a brief list can be provided for an Amber alert as well, along the lines of "You might check on the following things which may be about to slip."

   For years I have provided these suggested action lists (but not the statistical overlays) for clients to reinforce and confirm the advice given during the farm visit by means of visit reports – and specialist pig veterinarians have been doing this for decades too.

   **Feasibility.**

   There are statisticians, nearly all university-based, who could provide advice on the numerate side of things – something far beyond my own simple mind! Many pig academics could do this work, too.

   As to the 'Action lists', a whole host of advisers, veterinary, environmental, nutritional and managemental are available within and outside the government / university extension/ local advisory services.

Some college students will be reading these observations of mine – **there is a remarkably rewarding career field for you in this respect.**

I have come across no computer programmer pigmen yet!

The problem is that very few of the people who have the great numerate skills to be able to design the software, have ever worked on a pig farm. They therefore do not seem to fully understand the situations in which their presentation of the data is going to be used (where operators are busy, hassled and probably tired) or the mental capacity of the ordinary stockman to analyse or interpret numerate facts, especially when presented frequently and in bulk. Even experienced farm advisers and veterinarians welcome the presentation of information in clear pictorial form rather than in the columns and pages of printout with which we are presented!

**Please, computer programmers, go graphic in everything you do!**

# IDENTIFYING THE PIG

The first step to establishing a record-keeping system is animal identification through ear tagging, tattooing, notching or, more recently by electronic implants – and way into the future even by a DNA 'barcode'. It is not well done, for example in Europe 10 to 15% of all pigs are wrongly identified due to lost tags or illegible tattoos. A good identification system has the following characteristics:–

## AN IDENTIFICATION CHECKLIST

- ✓ It is unique to each animal. No two pigs have the same identification.
- ✓ It is readable from a distance.
- ✓ It is permanent over the animal's life.
- ✓ It is easily replaced if required.
- ✓ It is easily applied.
- ✓ It has a wide range of numbers and codes.
- ✓ It is humane.
- ✓ It is economical.
- ✓ It is tamper-proof.

## TATTOOING

Tattooing, usually of the ear, but occasionally of the neck behind the ear, is the most popular method at the time of writing. Pigs may also be slap-marked before slaughter. (In some countries it is expected that ear *notching* will be/is now banned on welfare grounds.) Because so many producers world-wide still use tattooing, I cover this rather outmoded technique here. (Tattooing can be used for growing pigs).

# A TATTOOING CHECKLIST

Size of needle letters/numbers:

Sucklers 3 day to 10 kg                     8 mm (0.31 in)

Pigs 10 to 100 kg                           10 mm (0.35 in)

Breeding stock and slap marking   16 mm (0.63 in)

*Paste :*        Use colour green for white-skinned pigs. A good brand should last 10 weeks or more. Black is usually used for slap marking before slaughter. Grate polish is sometime used. Dark skinned areas will need an ear tag or tags, or notching.

*Timing :*       3 to 5 days old if possible but ear size may be a limitation. Before 21 days is advisable.

*Positioning :*  Figure 3 shows where to place the tattoo for pig identification purposes.

Figure 4 shows how to record birth date if required.

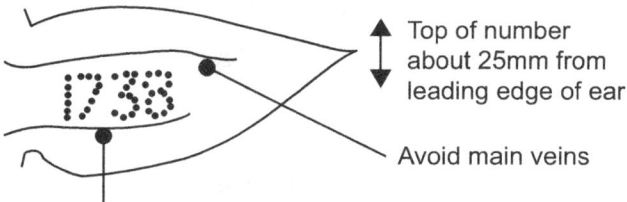

Top of number about 25mm from leading edge of ear

Avoid main veins

4 numbers only, otherwise convolution of inner ear is reached and numbers are difficult to read

**Figure 3.** Positioning of an ear tattoo

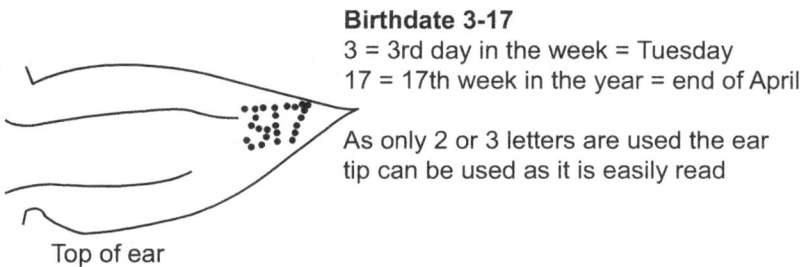

**Birthdate 3-17**
3 = 3rd day in the week = Tuesday
17 = 17th week in the year = end of April

As only 2 or 3 letters are used the ear tip can be used as it is easily read

Top of ear

**Figure 4.** How to record a birthdate

Make sure you have . . .

- ✓ Applicator with at least four sets of numbers 0-9
- ✓ Die applicator/toothbrush
- ✓ Surgical spirit
- ✓ Record book
- ✓ Disposable tissues
- ✓ Disposable gloves
- ✓ Die soaking / cleaning tray
- ✓ Die cleaning detergent
- ✓ Mild die / applicator disinfectant

Observations :

> Die pastes vary in persistence so it probably is worth tattooing both ears, one inside, one outside.
>
> Avoid ear veins as they confuse 5 and 8 and 7 and 1.
>
> Go carefully/methodically.  Mistakes cannot be rectified.
>
> Check you have 10 mm dies (0.35 in) if tattooing piglets under 6 kg.
>
> Tattooing is not too easy to do.  Ear notching is easier but is brutal/stressful. Tagging is easier still, but large tags can damage small ears; small tags are more difficult to read later in life, both ours and theirs!

---

# EAR NOTCHING

The author declines to write about this subject as he feels it infringes on his own personal welfare beliefs.  It cannot be humane.

If you are determined to ear notch – and ear-notch young pigs especially, you should consult your veterinarian so as to minimise shock, pain and the risk of infection.  In any case, ear notches can grow out, get misread after fighting and are tedious to read. Please don't do it.

# EAR TAGS

Plastic ear tags are a useful means of identifying breeding stock.  Gilts and boars should be tagged when selected as replacements.  A number of different tags are available, are relatively easy to read, but do have a tendency to pull out.  The large tags placed centrally through the ear seem to be the most effective. Tags placed around the edge of the ear tend to pull out more easily. Care must be taken to allow for growth of the ear.

Permanence of ear tags is largely dependent on the conditions under which the pigs are kept. It is worthwhile trying different tags to find one which is most suited to a specific situation. It is definitely advisable to put a tag in each ear, and if possible have them numbered top and bottom.

# CHECKLIST
# HOW TO CHOOSE A GOOD EAR TAG

At the time of writing there are about 17 makes in the world market, so some guidance on choice may be helpful.

✓ **Flexibility** : The composition of the plastic must be correct to stand up to chewing, abrasion and weather conditions, including ultra-violet light in hot countries. A nice flexible but durable texture is best.

✓ **Hygiene**   : Choose a tag where the ear piercing spear *is incorporated in the male half of the tag* and not in an ear-piercing portion of the applicator. Such an instrument has to pierce other ears and there is a distinct risk of infection, as it is impossible to swab-clean the ear surface well enough beforehand. The applicator should have a spring-back and surface-clearance action to avoid tearing the ear during penetration.

✓ **A sharp spear** : The brass spear point must be very sharp and have a gradual conical shape, *i.e.* a good piece of engineering turning. So when buying a brand of applicator, a magnifying glass helps as it does when choosing a drinker valve, which also needs good precision engineering.

✓ **A strong shoulder** : Some tags wear and snap off at the point where the flap meets the spear (the 'shoulder'). Choose a tag which is strengthened at this point.

✓ **Permanence** : While it is possible to get the tags pre-lettered or numbered by a heat-sealing process, some farmers prefer to make their own codes. A special pen is available for this purpose. Preferably, the ink should actually infuse into the polythene surface which is made porous to allow this, thus the mark is permanent and will not fade or rub off with time. Alternatively, tags can be obtained with a covering disc to slip over a own-written number on the plastic disc. These must be cleaned with methylated spirit first, as if the cap is put on greasy it will slip off.

✓ **Problems** : Are usually abscesses/infection. Did you use a clean applicator? Wipe the site first with surgical spirit? Are the lying areas dirty? Or too wet? Are the tags stored clean and dry? A few cases of allergy are reported – change the brand of tag?

✓ **Legibility** : If possible choose tags numbered top and bottom, especially in group-housed sows and gilts. As colour coding is useful in following parent lines and crosses, legibility depends on size and colour, bearing in mind the larger the tag the easier it is to lose. Research by Lecurier suggests the following colour contrasts as being the most legible from a distance in poor light, or under a light layer of dirt.

**Table 3. Order of preference of colour contrasts on ear tags**

| 1-12 Order of Preference | Decoration | Background |
|---|---|---|
| 1 | Black | Yellow |
| 2 | Green | White |
| 3 | Red | White |
| 4 | Black | White |
| 5 | White | Black |
| 6 | Yellow | Black |
| 7 | White | Red |
| 8 | White | Green |
| 9 | White | Black |
| 10 | Red | Yellow |
| 11 | Red | Green |
| 12 | Black | Blue |

Five colours in all. Do not use blue on ear tags unless a sixth background line colour is used. Blue is difficult to read in poor light.

## Losses

Welded iron rod partitions, sow stalls, scratching and rubbing (mange), fighting, outdoor fencing and huts and poor insertion are all common causes of loss. Use both ears and immediately replace when one tag goes missing. It is better not to use tags in growing/finishing pigs – use a tattoo. Generally other markers and sprays don't last, but a concentrated solution of crystal violet dye crystals used at 1 kg per 4.5 litres water and applied with a paintbrush can last 5 weeks on fattening pigs when aerosol sprays, though considerably more convenient, seem to last much less.

# SLAP MARKING

Producers selling pigs directly to processors may wish to use the packer's carcase classification information as part of their herd improvement program, by linking carcase weights, lean meat percentage estimations and backfat measurements to sires and dams, or as will be increasingly common, on different feeding strategies under farm test. This requires an individual tattoo for each carcase in addition to any registered brand. This number will be recorded against the weight and measurement and passed back to the producer. The individual number is applied using a rotary tattoo.

## A SLAP MARKING CHECKLIST

The slap mark should be applied to the shoulder of the pig so that the brand will be clearly readable on the carcase at the place of slaughter.

To achieve this:

✓ Use only approved branding inks, that is, carbon-based fluids or pastes. Non-approved substances such as aerosol stock markers, boot polish and even sump oil should not be used!

✓ Apply ink to the needles each time the pig is struck with the tattoo brander.

✓ Use a shallow tin with a sponge pad about 20 mm (0.70 in) thick to coat the tattoo needles with the fluid or paste.

✓ Make sure that all numbers and letters are placed correctly in the brander.

✓ Make sure that tattoo needles are clean, sharp and in good condition. Bent, broken or blunt needles will not work properly.

✓ Ensure that the slapper strikes the pig evenly, and when the pig is still.

✓ Ask the buyers of your pigs whether your tattoo brands are readable. Better still, regularly follow-up your carcases on the hook.

Make sure you have . . .

✓ Slapper and sufficient sets of numbers.

✓ Code/record book.

✓ Ink pad.

✓ Some restraint device is advisable as the pig should remain still to get a good mark.

# ELECTRONIC IDENTIFICATION (EID)

Safe and cheap electronic identification is replacing conventional methods of labelling all livestock in the near future, although some work has yet to be done on the concept, particularly on international standards and agreements.

Experts believe a more comprehensive system than conventional tattooing or tagging is needed, perhaps including an electronic identity chip carrying information which starts with the animal's birth date and the farm of origin.

Electronic tags are based on the passive or active transponder (more expensive) principle. At the stage of writing the experts seem to prefer the former because of cost. Required information is "read" on to the tag and then extracted by passing the detached tag over or through a reading machine or instrument. An active transponder transmits its own signal.

## What is a transponder?

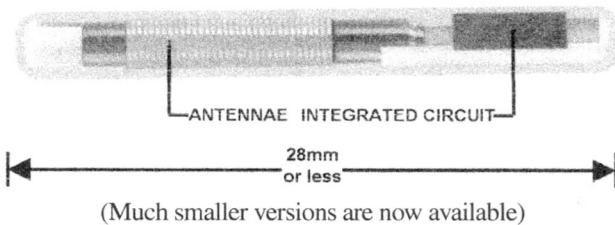

ANTENNAE  INTEGRATED CIRCUIT

28mm
or less

(Much smaller versions are now available)

An injectable transponder, also known as a subcutaneous semiconductor chip, consists of a small transmitter and antennae. When a readout unit is brought close to an animal, the transponder becomes charged with energy and transmits its identification number. The number appears on the display of the readout unit. The transponder is shaped in such a way that it can be easily injected using a syringe. It is important that the encapsulation protecting the transponder be accepted by the body and that it does not migrate within the body.

But existing ID tags are not perfect. The ideal system would allow relevant information to be read on to individual animal tags starting at birth on the farm, and whenever stock changed hands at store or the final slaughter stages.

# MICROCHIP IMPLANTS OR TAGS (BUTTONS)

Implants must not migrate through the body or break (early models were glass-cased!) and of course must be fully recoverable at slaughter. However, the carcase must still retain its identification through to final processing and dissection.

## Solving the processor's problem . .

The same transponder technology can be embedded in an ear tag or button which can be removed from the head/ear at slaughter and re-attached manually to the carcase with another pin, so that identity is registered both at the weighing and grading points. Thereafter individual primary cuts can be identified with the same number from similar bar coded pins.

## … But not the farmer's

The problem still remains that the relatively expensive transponder button/tag may be lost on-farm, as with any tag – especially in the hurly-burly of the finishing pen and that metal rails etc can disturb the read-out in the case of breeding stock.

It is now possible to reduce the transponder to a very small size (as dog/cat identification) but at the time of writing this device is for numeric ID only, and cannot yet be easily read by equipment at a convenient distance for a larger animal not under restraint. Again, its very small size may mean it might not be recovered at slaughter, or lost in the busy turmoil of the slaughter plant, and these small microchips are more expensive too, though mass-production will help with the cost angle, should every pig be so labelled in future.

As well as ongoing technical research to solve these headaches, therefore, governments need to co-ordinate their efforts to use the concept of an individual electronic animal passport in the form of a microchip so that traceability is made easier, especially as existing tags cannot be interfered with. The concept also makes much simpler and quicker the tracing of pig movements after an outbreak of Notifiable Disease.

The use of such microchips to incorporate weighing, electronic feeding, environmental controls and biological body monitoring as animal records seems relatively simple compared to reconciling the continued but basic headaches of cost, loss of the device on farm and recovery/registration from slaughter onwards.

## REFERENCE

Gadd, J. (1991) 'Getting the Message Across' Zen-Noh (Japan) Conf Proc on Pig Production, Tokyo.

# 14

# CONTRACTS - HOW TO MINIMISE BUSINESS RISK

A business management technique which anticipates possible difficulties and then plans to reduce their consequences.

At the time of writing the subject of Risk Management is poorly understood by pig producers world-wide despite an increasing number of training courses for all businessmen, very poorly attended by pig producers. Broadly it concerns two areas:

**Avoiding Risk**          Anticipating that a difficult situation could occur
                           and putting in place actions to avoid it.
**Minimising Risk**        If the difficult situation does arise, taking action
                           to reduce its impact.

Only the best ordered, most tightly-disciplined pig industries in the world are minimally affected by pig price and pigfeed price volatility – two of the most common risks affecting most pig producers.

The main risk areas are:–

- *Production risks*, including that of disease and a rise in the input costs, of feed and pig purchase price. These are the three main risks, affecting pig production almost on a daily basis.

- *Marketing risks* such as a fall in the pig sale price. This is the most influential risk of all and is likely to become even more volatile in future.

- *Financial risks* such as unfavourable cash-flow and inability to meet debt repayments.

- *Human risks* to your health or that of your employees.

- *Legal risks* through being caught out by failing to meet mandatory legislation/regulations.

The problem for all businesses – and especially the volatility-prone pig industry – is that these main areas overlap and can be difficult to disentangle to form a rational business plan.  (Figure 1)

How many squares are there
in this diagram?

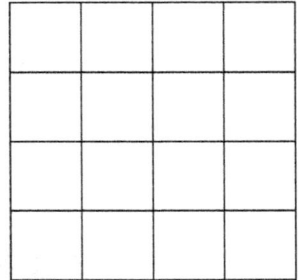

**Figure 1.** An example of how risks overlap (Martins, 2000)

At first sight it is easy to count 17.  Then on perusal, 4 more appear.  And so on.  After detailed study however, you will discover 30!

To help you disentangle the many risks you face, a Risk Management Plan is essential.

Here is a checklist of the factors which involve your business risk.

# RISK MANAGEMENT
# A BASIC CHECKLIST

To do a Risk Assessment Audit you should check:–

✓    Your information.

✓    How much you know about, and respond to, the market you are in.

✓    Your business flexibility (e.g. 'Tied' contracts).

✓    Records (performance monitoring).

✓    Records (economic monitoring).

✓    Your technical knowledge and willingness to adopt new technology.

✓    How much you talk to others (co-ordination), networking, attending lectures etc.

✓    Membership of key bodies – buying group, internet web , pig discussion group, NPA.

✓    How often you meet knowledgeable people (networking).

✓ How much of a gambler you are (see Table 1).

✓ How flexible your mind is. Thinking outside the box.

✓ How and how much you delegate responsibility.

✓ How much you have examined / thought about diversification.

✓ Your staffing / workload.

✓ Leasing not buying.

✓ Using industry credit.

✓ Use of forward contracts.

✓ And forward buying of feed.

✓ Use of fixed rate loans.

✓ Hedging/use of options. Careful! A minefield; some advice is questionable.

✓ Ensuring an adequate capital base.

✓ Keeping your bank informed.

✓ Your use of performance consultants.

✓ Your use of financial consultants – tax, etc.

✓ Your use of the vet as a <u>regular</u> adviser.

✓ Your use of the New Terminology.

✓ Compliance with regulations.

✓ Insurance (disease, key men, etc).

✓ Yourself as manager

✓ Carrying out a biosecurity assessment.

✓ Alarms and back-up systems.

✓ Key equipment maintenance / monitoring.

Do you see what I mean about 30 squares ? There are 32 here, and many interlock and overlap.

## COMPILING A RISK INVENTORY

But set out baldly in a list like this is not all that helpful. Let me now group these individual, disparate squares into larger multi-square groups, what is called a Risk Inventory, and put some brief action lists to them.

*You* **are the most important part of your business**. Your brain, however much others might deprecate it, is light years ahead of the most advanced computer yet constructed. So know yourself.

# RISK INVENTORY – YOURSELF, AND PRECAUTIONARY ACTIONS

| Risk factors | Possible actions |
| --- | --- |
| Health | Insurance back-up, delegation, training, medical check-ups. Know your staff's emotional issues : What happens if … audit on staff shortage, you fall |
| sick, have an accident. | |
| Are you a gambler ? | See Table 1 |
| Accidents | Insurance.  Key man insurance.  Compliance with COSHH, Health & Safety Regulations, *etc*. Unexpected shocks… Emergency Telephone numbers.  First Aid Training.  Fire/fire drill.  Flood / wind damage.  Pollution and its financial consequences.  Staff negligence.  (Do a "What happens if…" audit with the staff concerned.) Theft and theft protection, increasing these days on farms. |
| Deaths | Do a "What happens if…" audit |

## How much of a gambler are you ?

Risk-taking is, of course, a gamble.  So how much of a natural gambler are you? Table 1 (Harris, 2000) will give you an idea.  Fill it in and turn the page to get a psychologist's view of where you may be in this respect.

# A MORE DETAILED ACTION CHECKLIST

In compiling this series of Risk Inventories I give merely those items I've found useful in my own consultancy work with pig farmers.  There are bound to be others which I've not had need to consider or which will be specific to your business.  But among the variety of suggestions offered there may be one or two that you haven't yet thought about.

If you have a high gambling score you will need to review your protective actions very carefully – fate will ensure you'll need them!

**Table 1   ATTITUDES TO RISK   From Harris (2000)**

*Mark an 'X' in the box that provides the best indication of how you rate yourself on the scales below in relation to the statements given.*

|  | |  |
|---|---|---|
| I would not sign a fixed price contract because it could limit my income | ☐☐☐☐ | I would sign a fixed price contract if it provided me with a basic income. |
| I like to try out new ideas | ☐☐☐☐ | I prefer to stick to what I know |
| I will invest in new systems ASAP and develop them to suit my business | ☐☐☐☐ | I like to see how others get on first with new systems |
| I am happy to borrow to invest in my business | ☐☐☐☐ | I am not happy farming on borrowed money |
| I like to have lots of variety in what I do | ☐☐☐☐ | I tend to use tried and tested methods. |
| If thieves want something, there is not much you can do to stop them. | ☐☐☐☐ | I like to keep things locked and safe. |
| I think staff should be trained in more than just technical issues. | ☐☐☐☐ | As long as staff do what they are told, I am happy. |
| My ideas are not always practical. | ☐☐☐☐ | I am a practical-minded sort of person. |
| People see me as an 'ideas' person | ☐☐☐☐ | I like to deal with factual information. |
| I often lose things. | ☐☐☐☐ | I can be something of a perfectionist. |
| I take each day as it comes. | ☐☐☐☐ | I never leave things to the last minute. |
| I find it easy to put myself in someone else's position | ☐☐☐☐ | I don't back down in an argument. |
| I assess the likelihood of success of a project on individual merit | ☐☐☐☐ | Past events always influence the likelihood of success of a new project. |

Score – total of each column      ☐☐☐☐

## Answer to Table 1

The higher the total score in the two left-hand columns the more of a risk-taker you are.

The higher the total score in the two right-hand columns, the less of a risk-taker you are.

---

If towards the low (cautious) end, you will need to consider actions which may improve your chances of progressing your business without involving too much risk – in your case I'm sure it will involve, nevertheless, not spending too much money!

If you are of a cautious nature - consider:–

•   Getting better information on which to base decisions.

•   Joining a local web-site or buying group.

•   Keeping your bank better-informed. They are experts on risk so a good source of advice.

•   Examining what you can do to adopt as much as possible of a *full* biosecurity system (see Biosecurity section).

•   Leasing, not necessarily buying.

•   Spending more time on continually comparing sales contracts. The small print often makes them very different.

# RISK INVENTORY CHECKLIST - THE PHYSICAL PRODUCTIVITY OF YOUR BUSINESS

| Risk Factors | Possible Actions Checklist |
|---|---|
| ✓ Disease | Biosecurity. Correct use of the vet (induction etc). AIAO/Segregation. Batch production. Delivery and collection especially other people's contaminated transport vehicles. Vermin and birds. Drug records. Awareness of local disease situation. |
| ✓ Weather | Information. Frost and heat precautions. Weak structures re heavy rainfall and wind. |
| ✓ Services | (*e.g.*) electrical, water, carcase disposal: Back-up systems. |

| ✓ | Staffing/labour | Work flow. Training. Motivation. Sickness. Part-time help. Your *own* contribution *i.e.* time spent on management rather than helping out manually: measuring, checking, forecasting, quotations, record analysis, your own housing, updating yourself, talking to others. |
| --- | --- | --- |
| ✓ | Food quality | Home-mix / complete feeds? Wet *v.* dry? Know-ledge of and control of raw materials. Dialogue with nutritionist and vet (*e.g.* current immune levels). Wastage, spoilage (mycotoxins). Feed and raw material analysis. Farm trials. Availability and correct use of by-products. Water adequacy. |
| ✓ | Housing | Adequacy, internal monitoring (measurements), especially ventilation and power saving, use of consultant. Repairs and renovation. Fire risk. Provision of cheap overflow accommodation. Understanding basic technology – thermo-dynamics/air movement/pig behaviour. |
| ✓ | Records | Avoiding non-superfluous data. Action on data available (action lists). Graphics not numbers. Staff motivation. Outside inputting help *e.g.* accounts and tax help, and secretarial assistance. |
| ✓ | Contracts (termination or altered) | Good, up-to-date knowledge of alternatives available including altered transport costs. Use a 'profit-box' system to compare the basics of those on offer. Alternative options – like the US share-cropping idea. "Think beyond the pig price" (Thornton, 2000), *i.e.* market security could be as attractive. If others haven't considered it – propose your own contract/modify theirs/if I do this - can you do that? |
| ✓ | Carcase quality | Good feedback from processor, then close liaison with nutritionists. Close liaison with geneticist. Concentrate more on male genes? Keep your vet in the picture. 'Profit-box' recording system on each batch. Liaison *i.e.* (checking on) slaughterhouse/grader, especially if charges occur. Campaign for payment on lean. Stress audit. Check on transport competence/lairage time/cleanliness. |
| ✓ | Productivity | Diagnostic records, use of consultant in interpretation. Understanding and using the New Terminology. |

| | | |
|---|---|---|
| ✓ | Levies and impositions | Join your local organization to assist strength of industry voice. Help organise local agitation in cases of illogicality/unfairness, but get facts right first. Authorities do listen if there are enough of you - rarely if it is only you. |
| ✓ | Reducing exploitation by suppliers | Diversification, *i.e.* contract processing, farm shop (with other foods), local brands, exploit brunch/sandwich market, joint ventures, home mixing/wet feeding, more AI use, weaner gilts, commodity buying, internet buying (buying 'clubs'), Farmers' Markets (in future, Farmers' Supermarkets, *i.e.* open always) |

## RISK INVENTORY CHECKLIST - REGULATIONS

| **Risk factors** | | **Possible action** |
|---|---|---|
| ✓ | Non-compliance | Information on COSHH, HACCP, Drug Use, Health & Safety, Home Mixing, Fire, Stock Movement, *etc.* |
| | | Do an audit on each. Discuss with staff. Secure hazardous/restricted substances (insurance claim). |
| ✓ | Hidden shocks | Employ specialist consultant/share him with a group. Tap in to E.U. local Government internet service (advance warning of possible unfair/restrictive legislation). |
| ✓ | Consumer pressure | Seek information. Listen. Be quick to exploit fears if public opinion seems justified as the big battalions and governments are often reluctant, disinterested, uncomprehending, self-protective, thus slow to respond. Keep an open mind; some past apparent disadvantages have turned out to be benefits, farmed correctly *e.g.* outdoors, sows in groups, bedding, swing-away farrowing crates, liquid feeding, nurse sows etc. |

## RISK INVENTORY CHECKLIST - ASSETS

| **Risk factors** | | **Possible action** |
|---|---|---|
| ✓ | Fire, theft, unexplained losses | Do an audit on each. Farmwatch. Rapid-contact phone numbers. Insurance. Guard dogs/kennel & chain. Alarms/Farmwatch/Notices/Electronic farm road gates. |

| ✓ | Livestock | Perimeter fence; pigs away from public access. Leasing/'share-cropping'. |

---

# A RISK MANAGEMENT TOOLBOX

Here are some tools to deal with likely risk. While directed at the British pig producer, and reflecting his weaknesses at the start of the 21st century, there are lessons here for any pig farmer across the world.

## Your own attitude to risk - what is crowding in on you

* The number one threat to your business is the increasing amount of price risk in pig production.

* Price volatility is worsening, doubling every 5 years.

* Not only are costs (input prices) rising, but returns (output prices) are more volatile and trending downwards, or at best only keeping pace with inflation.

* Production constraints (welfare, drugs, pollution, the effects of BSE and salmonella control - campylobacter in the future) are increasingly affecting intensive livestock production in NW Europe.

* It is essential that you keep abreast of what is happening not just locally or even nationally, but for your long-term planning purposes (5 years plus) in critical sectors of the global pig industry like Europe, Russia and Eastern Europe, North America, mid- and mid-west USA, mid & Western Canada, Brazil, Mexico and China. This is especially true of longer-term planning (10 years+) as it could well be your best option to take your production expertise overseas from now on and/or link up with overseas processing marketing opportunities which could well be supplying a share of your home market by then.

## So the first tool in your toolbox is:

### *Information*

You must put by more time in future to collection information on:

*Markets* : local, national, European, global.

*Herd size and growth* : British pig census data here has tended to slip over the past 20 years and reinvestment and reorganization is under way.

*World market information:* is much better and, at the time of writing, various data sources are available, such as the excellent PigSite internet source and the UK B.P.E.X.

and N.A.D.I.S. bulletins. The Danes do this superbly for their own producers. Copy them!

***Output price trends*** : There is always a reason for a pig-price movement and you need to keep abreast of what causes them. Informed retrospective comment is good in parts of the UK agricultural press and such weekly or monthly newspaper's pig sections *e.g. Farmers Weekly* and *Pig World* should be read regularly, eventually to build up a good market knowledge and a 'feel' for making the right strategic decisions.

***Competitors' behaviour*** : One competitor, well recognized but often overlooked by pig producers nevertheless, is poultry meat. Keep an eye on what they do and the way they think, from production to marketing. You will learn much and could be forewarned of any new impact on pig production. The poultry boys are still ahead of us - just!

***Input price trends – retrospective monitoring*** : Feed contracts and forward buying are valuable cushions against unpleasant price rises at little cost to the producer. Also, a record of feed price movements, while less cyclical than 30 years ago, is still a valuable graph to have on your office wall as it can suggest reasonable predictive success up to 6 months ahead. *I see almost no such evidence in the pig farm offices I visit*; when I was buying I monitored raw material price graphs assiduously, and it was well worth it.

***Pro-active buying*** : One or two clients have become so experienced in this technique that they have entered the feed raw material market themselves/ on behalf of others/ their buying group, normally on 25 tonne minimum parcels. This has been worth, after admin costs are deducted between 3% to 6% reduction in final feed costs, and while this could be below a wholesale or brokerage discount, it is nevertheless a reduction in purchasing risk if managed well. The main risk remaining is to be landed with a substandard parcel, hence membership of a buying group or farm partnership is valuable. Experienced operators know the right questions to ask of the right brokers – on origin; length of the market; declared critical analyses; dry matter (if applicable, *i.e.* on by-products); storage/warehousing conditions and reason-for-sale. Again, taking time and patience to acquire information before action is taken is critical, and itself is an insurance against risk.

***Action*** :    Work much harder in knowing ***your*** market, your immediate purchaser's market, ***his*** buyer's market and the world market. ***You are much less likely to be exploited***, even if many events are outside your control. You narrow the exploitation-gap, or see how it might be narrowed.

**The second tool in the risk management toolbox is:**

*Co-operation/co-ordination/integration*

This needs a degree of selflessness which, it must be admitted, is not a strong feature of the global pig producer and especially so in Britain, but we are getting better at sharing risk. Are you a sharer? Of information, of risks, of profit? You need to be. **Isolation, euphemistically called 'independence' is a non-starter these days.** So many proud 'independents' I knew have gone bust.

Several parts of the world are well advanced in co-operation where pigs are concerned, such as Denmark and the USA, with Holland, Spain, Canada and even Mexico showing progress in the intention and then determination by farmers to band together to share risks and secure less volatile markets, or at least have a calming influence on future local instability.

Modern pig production needs a co-operative mind-set. Choose **sharers** as partners. A sharer is prepared to go more than halfway in a partnership, those sort of partners markedly lessen risk. Again, I speak from experience!

**The third tool in the toolbox is:**

*Relationships, not transactions*

Following on from this, relationships are more important than transactions. Stop and think about this. An auction is a transaction. While it can be a satisfactory method of buying or selling it is still a temporal, ephemeral act, not one which is necessarily committed forward to a long term arrangement. **To investigate business relationships you need to talk.** Talk and propose. Talking has always been a prerequisite of negotiation.

Negotiation in its turn has always been a preliminary to co-operation. Proposition – making suggestions – is also a positive part of co-operation, too. So think positively and anticipate in advance the likely partnership problems which are always there, his and yours.

Pig farmers are not good at co-operation because they don't talk enough – to each other or to their primary customers, the processors/retailers. Too often any talking there is just revolves around problems. If the possibility of problems are too prominent in the discussions the issues become soured and defensive attitudes occur.

**Other people have problems too.**  Understanding them and providing positive suggestions will get to a solution more quickly than highlighting (justified) grumbles or erecting barriers.  Get round to them later but only after the advantages have been well-aired and understood.

# NEGOTIATING SKILLS

Any owner or manager must be a good negotiator.

We've all been negotiating since we were babies, trying to wheedle our parents, grandparents, brothers and sisters to give us what we wanted, or not having imposed on us what we didn't like!

Since then we have negotiated our way through life with varying degrees of success.

Years ago I put myself on a negotiating course – one of the best things I've done, because I not only learned a lot, but had to *unlearn* even more!

## Curb those negotiating instincts

- Don't talk too much. The more you talk the more you are likely to give away in the end. Encourage the other fellow to talk, he may reveal something about his position. The key to communicating is listening. This sounds odd, but try it and you will see how true it is.

- In the same vein, do not give away your position at the start of the negotiation. Come to it gradually through discussion and questioning.

- Farmers are too forthright, impatient, anxious to get to the object of the negotiation.

- Instead, start assessing the other side's expectations as soon as you can. This will give you an idea of what concessions you may need to make. And sometimes this will tell you that there is no need for any at all!

## What to give away

- With regard to possible concessions you can afford to make, prepare in advance and have clear in your mind which concessions will cost you a little, but could be worth a lot to the other side. Knowing the market you are both in is always homework worth doing, especially in the other side`s field of expertise.

- If you are like to be asked for a price reduction on what you hope to secure, have a list of questions to ask them if to find out if their real concern is something else. These might be, from my own experience when interviewing salespersons.........

- Poor sales; cash flow; advance contracts; uneven orders/deliveries; unexpected expenses; bulk/ combined deliveries; labour costs; transport costs; workload; ordering on-line; method of payment (e.g. direct debit/standing order).

- Do not give up anything without getting something in return

- Negotiating with employees is rather different from negotiating business-to-business. Staff negotiations tend to be more formal – they have to be, as unlike a business deal you as employer are in a dominant position. In such cases they involve the outcome you *both* want, as the relationships are on-going and not concluded at the end of a commercial deal when the business negotiator disappears at the end of a confirmatory email.

- Leave haggling to the concluding moments, not early on. Haggling is for the end-game when you have to bridge a final gap. Do it too soon and you squander the fruits of skilled negotiation. Haggling is the opposite of negotiation. Farmers love to haggle.

- Similarly, don't go for the full amount too soon. Lead up to it. If you need to buy 100 units, ask the price of 50 first, which puts you in a position to negotiate a discount on 100.

- Finally – agree things *in writing* after the handshake.

## 'If' and 'Then'

One final observation. Good negotiators tell me – and I have had it done to me – that there are two key words which usually help construct a satisfactory deal. `If` and `Then.` " *If* I do this, *then* can you do that?" That little subtle approach when stalemate looms can make all the difference.

## CONTRACTS

Contracts for pigs minimise risk. They are, if you like, more of a transaction in farmers' minds, but a good relationship helps materially, like this :
Develop a *relationship* with the buyer by ...

- Listening to his needs/problems. Ask what they are/may be – this often starts you both off in a positive mood.

- Selling a product based on his requirements (*He* is the customer, not you).

- Setting up a linkage which helps him defray costs and for you to obtain a guaranteed outlet/obtain a premium.

- In this way both his and your risks are reduced.

- Suggesting opportunities to jointly develop products so that his customers are pleased.

- Thus all parties benefit – you, him, his customers.

So **think beyond the price!** Sure, price is of great importance, but demonstrating a positive attitude to a relationship as well is crucial. UK pig farmers complain to me that the processors won't talk to them. This is partly because they suspect that pig farmers, beset as they have been over the recent years with economic impositions, will be negative and 'difficult'. Remember a good partnership tries to go the extra mile and a positive approach to a negotiation is one way of doing so. Positive talking does not cost very much, and usually nothing at all.

## BEING POSITIVE

Here are some examples from some of my clients across the world.

1. Written evidence from a local butcher that a certain small number of your carcases are taken each week because his customers pay a premium (especially over supermarket prices) for taste and appearance.

2. Evidence from your farm's AI/breeding policy that carcase quality is uppermost in your mind and the essential genes to achieve it have already been invested in your business – and how much/kg this has cost you. 'Marbling' (interstitial fat) is genetically-influenced and improves eating quality (succulence). If you have it - promote it!

3. Evidence that you try to produce meat *'cleanly'*. A demonstrable biosecurity protocol can convince the processor of this in your case.

4. Use organic selenium in your food as it can cut the retailer's drip-loss by up to 10%, worth 2p/kg to the supermarket in more pork *sold* in relation to that *purchased* by them. The research and econometrics have been published – so use it as another positive contribution you can make with your pigs to *his* business, *and* how much it costs you, about 0.5p/kg or whatever, **on his behalf**.

5. Evidence of how your pig flow is planned to allow for evenness of supply both in number and weight. Evenness is crucial to factory processing. Several of my clients did this very well, but never mentioned it to the buyer!

## "JUST SIGN THE CONTRACT!"

Producers often claim that similar approaches to those given above are all very well but all the buyer is interested in is your getting on and signing the contract to help fill his procurement targets. This may be so in some cases, but more and more contractors are responding to the longer term relationship aspect of contracting – they don't want their procurement base constantly changing, which is inefficient, stressful, adds to their costs and puts them into a risk situation themselves. Remember, talk to these people and provide positive suggestions. You are the seller, they are the customers and you need to be pro-active in the selling sense. Being pro-active in any negotiation puts you in an advantageous position anyway.

## A PIGMEAT CONTRACT CHECKLIST

| | | |
|---|---|---|
| *Deliveries* | ✓ | Number per week. |
| | ✓ | Delivery time window. |
| | ✓ | Variability in numbers? |
| | ✓ | Confirmation of time and date needed? |
| | ✓ | Late delivery penalty? |
| | ✓ | Transportation requirements. |
| | ✓ | Dirty pig charge? |
| | ✓ | Growth promoter ban? |
| | ✓ | Antibiotic residues? |
| | ✓ | Haulage costs. |
| | ✓ | Biosecurity clauses? |
| | ✓ | Estimated time in lairage. |
| | ✓ | Estimated time of weighing-in lairage. |
| | ✓ | Situation re liability for strikes, government restrictions, flood, fire, etc. Situation in holiday weeks? |
| *Price* | ✓ | Base price? How is it calculated/fixed? |
| | ✓ | Location of base price in relation to weight band penalties and % outside agreed weight band. |
| | ✓ | Valuation of underweights? |
| | ✓ | Valuation of overweights/uncleans. |
| | ✓ | Non-contract price. |
| | ✓ | Deductions  – Processor's classification |
| | | – Standard levy and/or |
| | | – Promotional levy (called 'check-offs' elsewhere) |
| | | – Meat Inspection Charges |
| | | – Residue Testing Charge |
| | | – Waste Offal disposal charge |
| | ✓ | Feed price rise/fall allowances/deductions (+ VAT situation, called 'sales tax' elsewhere) Pricing alteration situation. |

*Other*            ✓   Length of contract.
                   ✓   Length of notice of alteration.
                   ✓   Length of notice of termination.
                   ✓   Length of payment delay.
                   ✓   Insurance situation.

# THINGS TO GET STRAIGHT IN YOUR MIND ABOUT A CONTRACT

## Comparing contracts

Pig farmers are too haphazard in comparing contracts, which must be done continuously with all the outlets within a feasible haulage range, not just when approaching contract termination.

A manual – or much better a computerised – spreadsheet should be compiled for all these possible outlets, on to which the farm's current actual and the past year's average performance can be overlain. A nett cash flow benefit is then obtained based on the actual performance of the pigs sent in.

The differences between current UK contracts can be substantial. An exercise I did recently on a contract involving 130 pigs/week compared to three others showed differences of between –£1.47p/pig to +£1.93/pig nett income over each 6 month sample across a 110 mile transport range compared to the farmer's chosen contract. In gross return terms for readers outside the UK this was –2.2% to +2.9% – a difference of 5.1% between the best and worst contract – over 33% of nett profit on the farm concerned. About 87% of these differences come from the positioning of the base price and the implemented or decreased price/kg in relation to backfat measurements. Only about 13% of the price differentials obtained came from other deductions/costs.

However when differences in transport distance were taken into account, and building-in accepted shrinkage losses, the nett profit difference overall in theory might have narrowed to 20% overall and to +9% in the most favourable contract. Even so, the exercise revealed surprises in the financial differences between the contracts and the importance of carefully comparing all costs and benefits together with transport distances when examining contracts.

## Negotiation

On presenting our calculations to the existing contract supplier, price adjustments were made by them to bring it in line with the most favourable competitors, in mid-term too. On another occasion the contractor looked at the farm's track-record and also increased the premium allowed. Experiences like this confirm that talking, negotiation and presenting positive information is important, as evidence, even within the very competitive industry of pigmeat procurement/processing, can change things in your favour. So work at it.

# CONTRACTS – A SUMMARY

- Be positive, look forward, not back. "Everyone has a horror story over a contract." (Strak, 2000).

- In your business, what areas of your production costs *could* be contracted? Pigmeat, feed, veterinarian, cleaning?

- Study various contract price formulae. Work at it. Become knowledgeable about other people's contracts.

## The Profit Box Concept

The Profit Box (see Fig. 2) is an excellent device in three important ways.

1.  It enables you to compare how your existing contract compares with any others available – especially as your terminal date approaches.

    The profit boxes from other contract possibilities can be 'laid over' your typical achievements to see whether more or less of your performance history is likely to fit into their boxes – i.e.. the **number** of finished pigs falling inside the edges of the competitive box and **where** they appear. As many pigs within the top right-hand corner area as possible will give you the greatest income from any contract. Give yourself three months advance grace to work this out because there are other questions to ask the alternative outlet(s) before you make a final decision.

    It may or may not pay you to change. However, comparing profit boxes will give you 80% of the critical information you need to compare contracts (Fig. 3).

2.  The profit box, filled in from your processors monthly returns, keeps you informed of how well you are doing – how many finishers from each batch shipped are inside your contract box and where they are – how close to the top right corner. Being a visual presentation the situation is very apparent and any changes in your performance quickly evident.

3.  But it has another valuable use. If the breeding lines (especially the progeny of boars/ AI semen) are tagged on the display each month, then the success or otherwise of the genes you are using emerge, as the better genetic lines show a larger number of finishers towards the top right hand corner. Here the profit

box is not used to rank contracts – but different breeding stock. I have seen this applied to feed trials too.

**MOST PROFIT FROM THE TOP RIGHT CORNER!**
Try to record each batch of finished pigs on scattergrams like this . . .

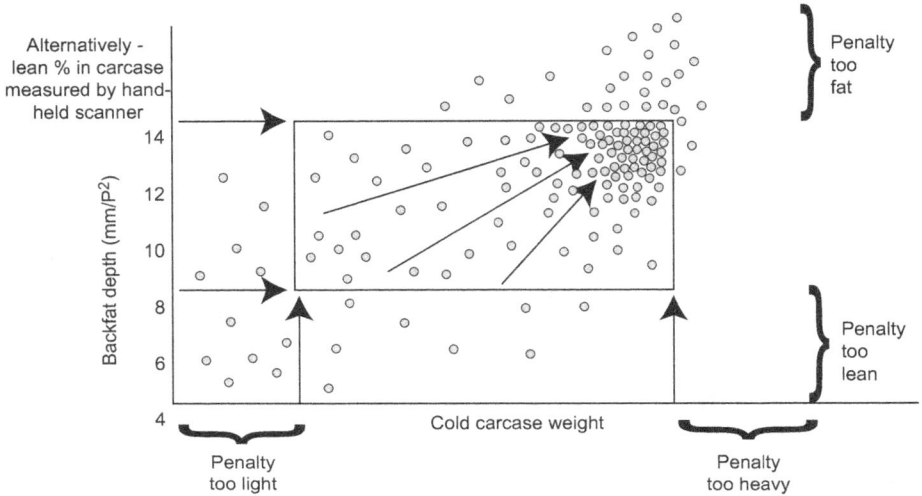

**Figure 2.** The Profit Box

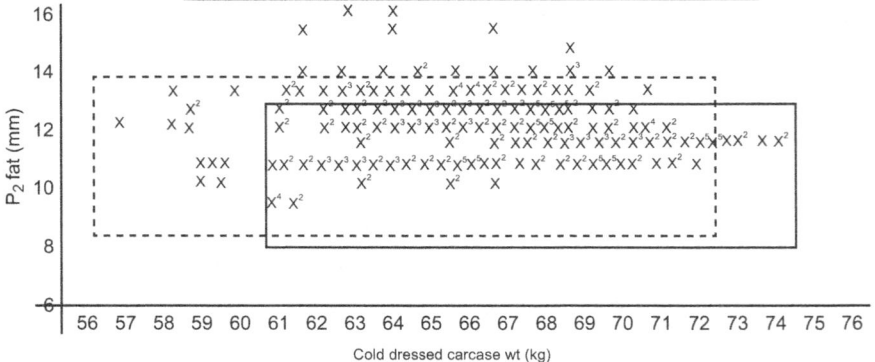

**Figure 3.** A comparison of $P_2$ backfat thickness and cold carcase weight.

## Not being used

Frankly, it disturbs me to discover that so few producers use the idea. Of the last 82 finishing farms I have toured or contacted, mainly across Europe since 2001, only nine were using the concept! Latterly I started to ask the others - once I had shown them the diagrams on my laptop – why they weren't taking advantage of such a good idea?

Most of those who had heard of the system said "Too laborious. Entering every pig from the processor's feedback as to weight and carcase quality takes up too much time".

I have to disagree – this is a quite simple job which a farm secretary in the office can do perfectly well - for 500 pigs shipped/ month on average, it never took more than three hours/month entered on to a graticule spreadsheet, so the converted said.

Using a profit box is one tool the efficient manager cannot ignore.

*   Using a Profit Box layout, start a contract comparison spreadsheet. Update it at least monthly.

*   Consider production costs (are yours realistic?) and long-term margin shares. Don't be afraid to talk about margins, his and yours.

*   A contract is give and take, both sides.  Be wary of contracts where the other side is cagey – they could let you down.  Redouble your efforts in probing, checking, getting confirmation in writing.

*   Look beyond just the price.

**Action** :   Talk, talk, talk.  Work at it.  Spend more time researching and negotiating, thus start early.  Once you've contracted      for a period, start looking at other options to use as future ammunition or reasons to change/re-negotiation next period.

## LIFTING THE LOAD - CONTRACTING OUT

You should examine employing others to grow out post-nursery stock.  Under these agreements the person finishing your pigs provides labour, power, buildings, straw, water and insurance.  The owner provides pigs, feed, vet and med. and management input if needed.  This allows you to dispose of perhaps 25-30% of your labour costs and/or concentrate more fully on the – let's face it – more labour-intensive and skilled aspects of breeding and nursery work as distinct from the not so onerous and less risky skills of the grow-out process.  Some opportunities also occur for contracting out the nursery rearing aspect alone, but only if the contractee has the correct housing and labour expertise and time available for looking at every weaner in his care at least twice daily.  Nursery rearers have taken pigs from 7-35 kg at a £3.00 - £3.50/ head fee, while finishing fees seem to be around £5.00/head.  Overall savings of £5 to £6/head (7 to 8.5p kg/dwt but more typically 5p/kg) have been made on fixed costs and you still have the pigs to sell at the end of the period (these are UK 2008 costs).

The system does have all the disadvantages of a transport move at a critical stage of the pig's growth curve, but this is usually more than recouped by the advantages of segregation for the finishers, and the benefits of batch production to yourself as the breeder.

## CONTRACTING OUT CHECKLIST

✓ Insist on a **proven** track record. Ask for records, especially sale dockets if available, and evidence of performance.

✓ By judicious questioning, get an idea of how much time the contractee is able to spend on stockmanship. This is absolutely vital in the case of nursery contracting-out. For example what other tasks/jobs is he/she likely to be engaged in?

✓ Check the buildings to be used very carefully, especially ventilation adequacy.

✓ For example, has he his own pigs? This in my experience is unwise (risk of disease).

✓ Tour his premises with your vet and listen to his advice.

✓ Stipulate a proven biosecurity protocol; you to supply the products used.

✓ **Draw-up a written, signed contract**.

✓ Sale dockets go to both of you, but pig cheques go to you as the owner, with agreed payment dates **signed in advance by both of you**.

✓ Ensure insurance cover is in the contract and who pays it.

✓ Keep your market outlet in the picture.

✓ Visit regularly. (Unannounced visits however can cause friction.) Have clear *written* guidelines on the agreed management protocols.

✓ Check on weighing and loading-out facilities and how smoothly it is likely to be done.

✓ Never contract with those who have other pigs on the premises or adjacent to them. Consult your vet if in doubt.

✓ Renew the contract annually, not longer, by including a satisfaction or target-clause in the contract.

✓ To ameliorate this requirement, you can make provision for a modest performance bonus. Incentives are often worthwhile in a contracting agreement, and could be essential if you have found a conscientious partner and need to retain him.

* Partnerships succeed when the terms are clearly written down and signed by both sides. They can always be modified next time.

* Partnerships prosper when both parties go 51% towards each other.

# FUTURES (FORWARD PRICING)

After introduction in 1984 and compared to vibrant pig futures markets in Chicago and Amsterdam, this has not been a success in the UK, the scheme barely lasting a decade. The demise was due to insufficient trading seemingly caused by disinterest in risk-spreading from both processors and retailers who, in the UK market, have largely been able to control their own input price structure so as to minimise their own risk in this area.

However, increasing globalization of pig trading could involve large UK pig producers in an EU Futures market, and a watching brief on developments in this direction is advisable if a UK Pigmeat Futures Market based on a better pricing structure than AAPP can be devised. There are good sources of advice and information on the situation from various sources *e.g.* BPEX and Euro PA at the time of writing.

While International Futures markets may seem remote, there are benefits in learning how these markets operate.

Input price risk spreading on feed ingredients (especially, for pig producers, grains and soya) presents a happier picture. The UK feed trade already has forward buying contracts on complete feeds and straights, some of them even providing capped forward options where, for a form of insurance premium, pig producers could protect themselves against upward price movements of grain and soya and yet could still benefit from a price drop. In 2000, only 5% of all farmers used this facility.

Strak (2000) reports as an example: "It would only cost a fraction of a penny/kg pigmeat not to be caught out by the price of soya doubling." In such cases there is surely no need for farmers to engage too deeply in futures or options for feed ingredients when the UK feed trade can be approached for the same risk reduction at what seems to be a reasonable cost.

# SUGGESTED ACTION

• Larger producers should subscribe to the relevant information services/updates available.

• All producers should explore the costs and attractiveness of feed ingredient forward option schemes within the EU feed trade rather than attempting to beat the market by speculating on their own. At the same time a watching brief on spot price movements on raw materials not covered by such schemes can be rewarding in making your own forward buying decisions.

# INSURANCE

Insurance cover has always been a popular way of reducing risk, and some companies have specific risk insurance policies. However, disaster-recovery policies don't cover price risk.

Farmers with fields close to public roads are not natural insurers – for example, I found few outdoor breeders have Public Liability Insurance and many field vehicles used very occasionally on roads are not covered. Recent floods and gale damage caught many people out.
All the time reputable insurance companies are designing new policies and it is worth keeping an eye on them by means of a once-yearly visit from their salesperson. You don't have to bite, as the cost may be still too high, but a counter-proposition may yet be accepted by them. Test the market. Explore. Propose. Talk to them.

# REDUCING RISK BY MANAGING YOUR BUSINESS BETTER

## MINIMISING CAPITAL RISK CHECKLIST

✓    Work out a ***Debt Management Plan*** before you borrow money. Prepare a range of likelihoods where pigmeat prices, production costs and interest/ exchange rates[1] could move.

✓    Prepare contingency plans for each scenario.

✓    Review your borrowing intentions on each of, say, three situations. Optimistic, Forecasted Reality, Pessimistic.

✓    In the light of this preparatory work, use your accountant or financial advisor to work out ***Gearing Ratios*** for your business, *viz.* Debt-to-Equity, Debt-to-Convertible Assets; Debt-to-Disposable Income. Monitor these ratios frequently and discuss significant changes with your financial advisor.

✓    Assess your ***True Asset Position***. Identify liquid assets which can be converted to cash quickly.

[1]    Exchange rates have a direct impact on prices. For example ± 10% in the Euro/sterling rate produced ± 3% change in the UK pig price from 1993-2000 and 9% between 2000-2010.

✓ **Debt Management Capability** : Establish what funds you have available to meet operational costs on equipment, housing, breeding stock and food, all of which will determine the payback period. Review loan arrangements in relation to expected sales and cash-flow.

✓ **Research Borrowing Options**. Your present deal may not be the best available. Discuss with your financial advisor how they compare. For example : **Venture Capital** – used particularly for major expansion or diversification, this can provide an alternative to conventional borrowing or complement it. Explore. There is quite a bit of this money about if pigs are profitable!

## MANAGING INCOME CHECKLIST

✓ Keep accurate and up-to-date[2] **financial records** *viz.* :

- **Timing** : Projected and actual cash flows.

- **Net Equity Position** : Balance sheet of assets and liabilities.

- **Net Profit or Loss Position** : Profit and Loss statement (earnings and expenses)

✓ Be careful to separate **living costs** from **business costs**. (The author has two separate bank accounts.)

✓ Be careful to separate **diversification income** and off-farm income from the on-farm income.

✓ Consult your accountant to have **tax management and planning** strategies in place. Taxation is a significant risk in cash flow management, and risk can be mitigated by tax-smoothing arrangements. If necessary seek advice from a taxation specialist.

## MANAGING YOUR ASSETS CHECKLIST

✓ Review **insurance cover,** not only on disaster protection, but also health and accident to all personnel, including yourself/your family. Also review areas of possible loss not covered or poorly covered by government/EU compensation.

✓ Draw up a list of **improbable/rare, possible and likely losses** and review the need for cover and its cost.

---

[2]    On the past 100 farms the author visited and questioned globally, these were on average 9.7 months in arrears.

✓   Set up an annual *Risk Assessment and Insurance Review.* Both the farm, political and the insurance industry circumstances can alter radically in 12 months. Any good insurance company will guide you on the current position.

# MANAGING LEGAL RISKS CHECKLIST

✓   Discuss your *business description* with your financial or legal advisor, *e.g.* sole owner/trader, partner, limited company, trustee.

✓   Check your *taxation, accounting and financial reporting*, as well as any auditing obligations and succession arrangements. Also what mandatory government returns (annual census, *etc.*) are required.

✓   Check that *COSHH, HACCP, IPPC, Animal an Staff Welfare and Farm Safety requirements* are in place and being followed.

✓   Check that your buyers' *Codes of Practice/Quality Assurance Rules* are being followed.

✓   Check all contractual arrangements. Make sure you have the *original documents safe and quickly accessible*.

✓   If applicable (*e.g.* farm shop) check that all *trading practices*, including fair trading and product claims, are in order.

✓   Footpaths, outdoor sows, farm shops *etc.* Have you *Public Liability Insurance ?*

# DIVERSIFICATION

'Having another string to your bow' is an attractive option for pig producers who have suffered the pain of reduced cash flow due to price volatility. This is especially true if the alternative source of income can be linked to the farm itself (on-farm income) such as bed & breakfast, gites/chalets, excursion centre, horse-riding, B&B plus stabling, farm shop, organising producer marketing, e-business / IT centre for other pig and livestock producers, AI service, vermin control, golf driving and shooting ranges. Areas like these are not far from the expertise or facilities a pig farmer already has.

Outside the farm (off-farm income) can seem to be limited in choice and also daunting due to the unpredictability of the novelty involved, apart from merely offering oneself as contract labour or doing another job. But there maybe more ambitious opportunities in the wider world of business where franchising something completely different is a route through the minefield of a stand-alone on or off-farm business.

Generally speaking farmers are not good at marketing. This is just what most accredited franchisors with a track record are very good at, so a well-chosen franchisee will have had the marketing of a product/service done for him, with the likelihood of a trade name already established. Nevertheless the franchisee has to have operating organizational skills and, particularly if a pig producer has been a successful breeder he will already have had the natural ability in organizing and forward planning to contribute to the success of the franchise.

## REFERENCES

Martins, A. (2000) Risk Management Initiative Course, ADAS Wolverhampton. National Pig Association.

Thornton, K. (2000) Personal communication.

Strak, J. (2000) *The UK Pig Industry - Can we Compete*? JSR Healthbred Technical Conference

# 15

# GROWTH RATE

Growth is the progressive increase in size of a living thing. Growth rate on the other hand is the rate of increase in body weight for a unit in time, e.g. grammes per day (g/day).

It is better to measure growth rate metrically (g/day) rather than in oz/day or lbs/week as the Imperial system is too imprecise, especially for small pigs.

In practice Average Daily Gain (ADG) or Daily Liveweight Gain (DLWG) are the preferred terms but in nutritional and genetic research papers you will also encounter Lean Tissue Growth Rate (LTGR)

## *TARGETS*

Table 1 gives an idea of the growth rates possible on a farm where good lean-gain genetics are used (especially in the male traits). In this case, the AIAO housing is well-designed and operated (with a move at 10 weeks) and the disease levels low. The pigs are fed ad-lib on a multiphase system.

Table 1 may surprise producers, but they are achievable targets for the next 5 years.

If Table 1 gives the top end of the growth spectrum. Table 3 illustrates the bottom end. If a producer approaches this level of DLWG an investigation is needed.

## SOME THOUGHTS ABOUT ADG

• An important measurement, as fast growth to slaughter generally needs less food. A slaughter pig finishing a week faster saves 7 days food, at maximum food intake, from its total feed requirement.

**Table 1. Target growth rates/day, top stock and conditions**

| Age | | Liveweight (kg) | Pigs growing at (g/day) | Weight put on per week (kg) | Growth rate from 21 days (g/day) |
|---|---|---|---|---|---|
| Days | Weeks | | | | |
| 21 | 3 | 6* | – | – | – |
| 28 | 4 | 7.20 | 171 | 1.2 | 171 |
| 35 | 5 | 9.80 | 357 | 2.6 | 271 |
| 42 | 6 | 12.85 | 435 | 4.25 | 326 |
| 49 | 7 | 16.50 | 521 | 3.65 | 375 |
| 56 | 8 | 21.25 | 679 | 4.75 | 436 |
| 63 | 9 | 26.10 | 710 | 4.97 | 479 |
| 70 | 10 | 31.35 | 750 | 5.25 | 517 |
| 77 | 11 | 36.75 | 771 | 5.40 | 549 |
| 84 | 12 | 42.49 | 820 | 5.74 | 579 |
| 91 | 13 | 48.65 | 880 | 6.16 | 609 |
| 98 | 14 | 55.16 | 930 | 6.51 | 638 |
| 105 | 15 | 62.09 | 990 | 6.93 | 668 |
| 112 | 16 | 69.16 | 1010 | 7.07 | 694 |
| 119 | 17 | 76.65 | 1070 | 7.49 | 721 |
| 126 | 18 | 84.86 | 1115 | 7.81 | 751 |
| 133 | 19 | 93.09 | 1175 | 8.23 | 784 |
| 140 | 20 | 101.49 | 1200 | 8.40 | 828 |
| 142 | 21 | 110.59 | 1300 | 9.10 | |
| 154 | 22 | 120.34 | 1400 | 9.80 | |

Notice how (as with FCE and MTF) it is important to have start and finish reference points when examining DLWG, and especially when comparing them with other quoted performances.

*Many weaners are at 7 kg or more at 21 days. These can suffer from less post-weaning check to growth and arrive in the 105 to 120 kg liveweight band some 4 to 7 days sooner.

- Generally speaking you cannot grow a pig too fast to 30-35 kg because of the young pig's superior food converting ability. Beyond that growth rate has to be balanced with food conversion and grading (sufficient but not excessive fat deposition) to maximise income and keep costs to a minimum.

- It is advisable at least to keep a *weekly* check on growth rate of groups of pigs in each separate environment. In my experience fewer than one in five producers do so. But a proper representational weighing system of both food eaten and growth rate is much more important to profit than most producers are as yet prepared to accept. The reasons will be discussed later.

- When balancing the advantages of reduced days to slaughter, food conversion and grading to achieve maximum return, many trials rightly emphasise the importance of the reduced food effect on improved FCR but omit mention of overheads saved. These can be between 35 to 50% of any food savings from faster growth (Table 2) which is a significant improvement to profit. **Don't forget overheads!**

**Table 2. Overheads are often overlooked when examining faster growth**

Assumptions:. Pigs 35-105kg Dressing per cent 75% Av. food cost/ tonne £160

| | | Overhead costs include capital depreciation | | | |
|---|---|---|---|---|---|
| Days to Slaughter | Food eaten (kg) | Overall FCR | Overall ADG (g) | Food cost /pig | Overheads cost/pig |
| 97 | 203 | 2.9 | 725 | £32.48 | £23.28 |
| 88 | 189 | 2.7 | 800 | £30.24 | £21.12 |
| 80 | 161 | 2.3 | 875 | £25.76 | £19.20 |

| | Savings per pig | | |
|---|---|---|---|
| | Food | Overheads | Overheads as a % of total savings |
| 97 | - | - | - |
| 88 | £2.24 | £1.16 | 52% |
| 80 | £6.72 | £ 4.08 | 68% |

**Comment.** Overheads these days are a substantial, and rising, portion of cost of production and presently stand at around 42% of total production costs (range world-wide 38%-47%) The quicker pigs can be shipped due to better growth rate the lower the overhead demands per pig will be.

## WHAT CAUSES POOR GROWTH RATE?

The causes of these poor action level growth rates in Table 3 will be found to be:

1. A larger than acceptable *post weaning check* (see the *Post-Weaning* check section).
2. Lack-lustre growth once the post weaning check is surmounted. This is usually due to *disease*, with *respiratory infection* the most common cause in my experience. As well as veterinary consultation, an audit of the ventilation system especially in winter is advisable.
3. *The 12-16 week check.* The reasons for this phenomenon are unclear but the following checks are advised. Those producers recording weekly growth rates do see it on their graphs while it may go unnoticed otherwise.
4. After 16 weeks to slaughter the pig's immune status, peck-order, appetite and thermoregulatory system should be well-established. The areas to check here are disease, again respiratory disease, ileitis/colitis and overcrowding. Too stuffy air rather than a too cold environment is usually the culprit, both in summer

**Table 3. Action level growth rates/day**
These can be considered as borderline performance levels

| Days | Age Weeks | Liveweight (kg) | Pigs growing at (g/day) | Weight put on per week (kg) |
|------|-----------|-----------------|-------------------------|------------------------------|
| 21 | 3 | 5.5* | – | – |
| 28 | 4 | 6.6* | 157 | 1.1 |
| 35 | 5 | 7.8 | 171 | 1.2 |
| 42 | 6 | 9.5 | 243 | 1.7 |
| 49 | 7 | 11.5 | 286 | 2.0 |
| 56 | 8 | 14.5 | 429 | 3.0 |
| 63 | 9 | 18.0 | 500 | 3.5 |
| 70 | 10 | 21.75 | 536 | 3.75 |
| 77 | 11 | 25.75 | 571 | 4.0 |
| 84 | 12 | 30.25 | 643 | 4.5 |
| 91 | 13 | 35.0 | 679 | 4.75 |
| 98 | 14 | 39.75 | 750 | 5.25 |
| 105 | 15 | 45.25 | 786 | 5.5 |
| 112 | 16 | 51.0 | 821 | 5.75 |
| 119 | 17 | 57.0 | 857 | 6.0 |
| 126 | 18 | 63.0 | 865 | 6.0 |
| 133 | 19 | 69.0 | 871 | 6.1 |
| 140 | 20 | 75.2 | 886 | 6.2 |
| 147 | 21 | 81.4 | 893 | 6.25 |
| 154 | 22 | 87.65 | 893 | 6.25 |
| 161 | 23 | 93.9 | 893 | 6.25 |
| 168 | 24 | 100.15** | 893 | 6.25 |
| 175 | 25 | 106.4** | 893 | 6.25 |

Notice how in this fairly typical example the pigs are slow in getting away after weaning and again how later growing pigs sometimes get 'stuck' at around 16 weeks of age when their potential growth should still continue to escalate by 5% to 6% per week.

*These are critically light weaning weights and can well account for the slower subsequent growth. Today target weaning weights would be 7 kg at 3 weeks and 8.5 kg at 28 days on the best farms. These alone would go a long way towards achieving much of the better growth rate in Table 1.
**Such slow-growing pigs are often sold off lighter to free up replacement pen space. This is called 'Topping' (See page 351)

and winter. Occasionally nutrition is at fault, in my experience poor amino acid balance can be responsible, but also overfeeding protein in the last month before market weight is rather too common. This protein would have been better fed in the first third of the pig's life.

5.    Temperature is often at fault (See *Food Conversion* and *Hot Weather* Sections)

## POOR GROWTH CHECKLIST

### *SLOW-UP IN GROWTH OR 'STOP-START' GROWTH BETWEEN 12-16 WEEKS OF AGE  (USUALLY BETWEEN 30-45 kg)*

Consult the relevant sections in this book for remedial advice.

✓    Check stocking density.

✓    Assess ease of access to food and particularly water in warm weather.

✓    Check for wrong-mucking / pen-fouling / tailbiting.

✓    Check for unevenness. If distinctly obvious, then pen-splitting rather than re-mixing of individuals is wise.

✓    Straw-based pigs can develop mange at this time which in its early stage can cause low-level but continuous stress and affect food utilisation. Greasy pig can occur at this time but this is more noticeable early on.

✓    Food change? Where food is concerned the specs may be satisfactory but *palatability and texture* may cause reluctance to eat and you should also keep freshness at the back of your mind.

✓    Following on from this, and possibly because the pig's desire to grow is accelerating around this time, *moulds/mycotoxins* could be having a direct or indirect effect (*i.e.* palatability). Producers able to measure feed intake per pen on a daily basis with a C.W.F. pipeline layout have picked this up promptly, a distinct advantage of the concept, not sufficiently emphasised by the manufacturers?

✓    Housing change? Again, direct and indirect effects have been noticed. *Direct* – *i.e.* the new house's environment is not up to standard because the pigs are cold, being placed in *too much spatial volume in relation to their group body weight.* Temporary lids and hovers help here. *Indirect* because the pigs are slow to adjust to new conditions – feed hoppers, dry to wet feed, fewer water points, less dunging area, a change in their socialization pattern. This may take longer-term planning to get better.

35 years ago we did some work which showed *any* change of housing cost about 3 days' growth, but with care this could be halved.

# HOW GROWTH RATE AFFECTS MEAT PRODUCED PER TONNE OF FEED (MTF)

Table 4 shows the enormous reduction in output – saleable lean meat in relation to the food fed to obtain it – between the target excellent growth rates in Table 2 and the poor growth shown in Table 3.

**Table 4. Reduction in MTF between action level and excellent growth rates**

|  |  | *Excellent* | *Action level* |
|---|---|---|---|
| Days 7kg -105kg | ( i.e.. from weaning to slaughter) |  | Days 6kg -105 kg |
| Overall FCR |  | 2.5.1 | 2.9.1 |
| Food eaten |  | 245 kg | 287 kg |
| Pigs / tonne of food |  | 4.08 | 3.48 |
|  | (Killing out % for both groups 74%) |  |  |
| MTF therefore |  | 317kg | 270kg |

*Effects of food*
Cost of poorer growth of
47 kg less MTF/tonne of feed                                            £ 56.40 PPTE*
   @£1.20 kg deadweight

*Effects of overheads*
But this does not include the extra overhead costs
   at £0.24/pig/day                                                    £33.60 PPTE*
Total PPTE from slower growth.................                         £90.00 PPTE*

*= PPTE Price Per Tonne Equivakent – see Business Section page 243
(PPTE is a way of converting several performance shortfalls to a cost per tonne basis)

**Comment**
While these are both ends of the spectrum, I still come across these differences today. By expressing the penalty of slower growth in these terms, it is immediately apparent that in the example above, the penalty is equivalent to paying half as much again (£90/tonne) for an average cost of a food fed from weaning to slaughter of £185/tonne. Absolutely staggering!

# KEEPING A CHECK ON DAILY GAIN

Keeping an on-going record of liveweight gain in the growing pig is important, yet only perhaps 15 to 20% of producers keep a ***constant*** finger on this pulse. Most people are content to assess average daily gain (ADG) – also expressed as daily liveweight gain (DLWG) – at the end of each month, or even three months. By then factors which led to any worsening are historical and may be difficult to identify and remedy.

Again this is most important as we know daily feed intake is affected by many variables :- temperature, stocking density, disease, stress, trough cleanliness and several others. The differences can be as much as 20%

Genetics also comes into it, as the appetite potential of different lines can vary by 15% even within the lines from the same seedstock house. One recently disclosed the two popular strains of theirs were 2.75kg/day and 2.35kg. This is quite normal, the strains having other attributes.

**Inform the nutritionist regularly of your growers' daily feed intake.** The modern pig nutritionist designs a feed in nutrients per day to meet the genetic potential of the strain to be fed. He therefore needs regular information from the farmer on how the conditions on his farm affect and thus change that basic information in terms of daily intake.

With this information the nutritionist then sets to work to revise the formula to provide the correct level of nutrients to supply the daily levels the pigs need.

If he has to make an informed guess because – **as is usually the case**, this information is not provided – then he could well oversupply the nutrients making the food more expensive than required , or alternatively undersupply them and affect performance to the same degree. Both ways the producer loses out by 15%, or whatever is the error. It is not the nutritionist`s fault.

## So how often?

I suggest 3 times a year which should take care of seasonality. This data built up over a couple of years (but much shorter see below) will enable the nutritionist to design a more cost-effective diet, either a cheaper one or a better-performing one – usually both, for each individual farmer.

The information is easier to collect if growing pigs are batched on an All-in/All-out basis. Better still, if a CWF system is used then the computer will not only record the daily feed intakes but can automatically send them direct to the nutritionist,which must be a benefit of the concept if used like this  But  quite a few are not used in this way.

## Is the cost of 'hand collection' worthwhile?

If you haven't got a CWF system, that is.  Of course it is. Not to do so risks losing 10-15% performance and/or raises the cost of the feed cost to slaughter by a similar amount. Too big to ignore.

# NOT SPOTTING THE CAUSE OF SLOW GROWTH PROMPTLY IS COSTLY

Failure to detect and act on a 10% reduction in daily gain in pigs of 60 kg across a 4 week period results in 15 kg less MTF *i.e.* 15 kg less saleable meat for every tonne fed then and thereafter. This is equivalent to a £18/tonne increase in the cost of grower's food, or about a 12% price rise. Add typical overhead costs to this and the 12% becomes 18% - on some farms 22%.

**So slow growth is dreadfully costly when looked at in this way**

Even a 5 % shortfall in growth (say around 40g/day or a quarter of a kg/week) on what is feasible today is barely noticed by the producer - but it is equivalent to having to pay £4.50/tonne more for his growing finishing food. Many producers I know would argue half the night to obtain that amount of discount from his feed salesperson!

# SIMPLE FORMULAE FOR ESTIMATING GROWTH RATE

A variety of formulae exist based on a rolling average inventory usually across three months *i.e.*
*Average output weight minus average input weight divided by days in the system* for each of the past 3 months ÷ 3.

Table 5 gives an example:–

**Table 5. Typical simple method of recording daily gain**

| A | Jan: | 96 kg | − 31 kg | (65,000 g) | ÷ | 77 days | = | 844 g/day |
|---|------|-------|---------|------------|---|---------|---|-----------|
| B | Feb: | 94.5 kg | − 29 kg | (65,500 g) | ÷ | 79 days | = | 829 g/day |
| C | Mar: | 92 kg | − 30 kg | (62,000 g) | ÷ | 78 days | = | 795 g/day |

*3 month average (A-C)  823 g/day*

| D | Apr: | 95 kg | − 30 kg | (65,000 g) | ÷ | 74 days | = | 878 g/day |
|---|------|-------|---------|------------|---|---------|---|-----------|

*3 month rolling average (B-D)  834 g/day*

Table 5 immediately reveals the disadvantage of this approach (even if a computer is used).

1.   A month elapses before figures are available.
2.   Casualties and mortality can distort the figures.
3.   As can factors such as underweight marketing.

# CHECKLIST - 12 FACTORS AFFECTING WEIGHT AND/OR PROBE VARIATIONS AT SLAUGHTER

Not in order of importance as incidence varies from farm to farm.

1. **Birthweights.** 1g heavier at birth = 2.34 g heavier at 21 day weaning = 20 to 30g heavier at slaughter? i.e. 100g heavier at birth translates into 2 kg heavier by 106 kg. In my experience these figures at slaughter can be minimal.

2. **Birth to slaughter.** 50% of runt pigs can reach 7 kg at weaning if fed and managed selectively.

3. **Weaning.** Aim for 4 kg difference in pen weight, lightest to heaviest, in 21-28 day weaning, 4.5 kg in matched pens averaging 6 kg, and 5 kg difference in pens of 7 kg+. These differences enable the pecking order to be established more rapidly so that the submissives get away sooner into full growth potential.

4. **Adequate trough space\*.** Vital, especially post-weaning.

5. **Feeder gap space\*.** Check every day. In comparison, only about once weekly – weight variation at slaughter widened by 20%.

6. **Overstocking\*.** By 15% also widened variation at slaughter by 20%.

7. **Genetics.** Preponderance of dam lines influences variability in both growth and grading. Mixing male lines may increase variation similarly.

8. **Environment.** Too hot, too cold, incorrect ventilation affect variation.

9. **Feed intake\*.** Can vary by 20% within a pen and by a similar amount between pens. It is essential to provide your nutritionist with estimated daily intakes from each farm (or even from each house if a CWF pipeline is installed) each quarter so the nutrient intake from the diet can be revised. This helps take care of seasonality (and to a certain extent health changes).

10. **Water provision.** Accessibility can be as important as adequacy.

11. **Seasonal effects.** Probe and carcase weights are lower in the summer months.

12. **Health.** Good health reduces the spread of weights within pens.

\*  Profit Box figures (see page 328) are available for these four failings, both before and after rectification  For example, attending to feeder gap space daily improved the number of pigs in the top right-hand profit box quartile by 17%; giving 50% more trough space for 10 days post-weaning by 19%; and destocking to the advised maxima, by a massive 28%.

Many of these influences are self-evident - even so, despite ample guidelines being available (such as for factors 5, 6, 8, 9 and 10) my farm visits reveal about 33% are not being followed.

# REPRESENTATIONAL WEIGHING

We need a better system, and while weighing is a highly unpopular task, there is no substitute for representational weighing of pigs. This is because we need to identify as quickly as possible when our test/sample pigs fall below the target daily gain graph.

Because environment can have such an effect on growth rate, it is important to select pigs from a typical piggery.

Ideally, each house containing a different environmental system should have 3 sets of two pens weighed to the schedule below. One pen near the coldest end (or in the tropics the warmest side) and another pen, in cold localities, in the middle of the house.

## Suggested representational weighing schedule

*First* : Feed intake is measured over 12 days in two pens at *each* of these weight ranges: 25 to 35 kg; 55 to 65 kg; and 85 to 95 kg, *i.e.* six pens in all. A pen would normally contain a minimum of 10 pigs. In N. America these ranges could be, in their terms, 50-75 lb, 120 to 150 lb and 180 lb to slaughter weight.

*Second* : Record body weight at the start and finish of each 12 day period.

*Third* : At the start of each 12 day period, record how old the pigs are in days.

*Fourth* : Keep an environmental temperature record as a back-up. You can then plot feed intake/pig in kg/day against observed live bodyweight. From these data you can plot a representational feed intake curve in kg/day and a pig growth curve of bodyweight against days in the piggery.

**From this extremely useful farm-specific data the nutritionist can design your diets far more accurately than just supplying formulae off fixed price lists.**

He now has at least a guide to ... The genotype you use – the pig's feed intake – their lean tissue growth rate needs – an idea of their current immune status in your circumstances – and what your environmental conditions are from season to season.

Producers should take a decision on how closely to attain this ideal with the workload involved. If this proves difficult right through the growth period, representational weighing is most valuable between 11-13 weeks of age (30-50 kg).

I have experience of five producers who do this full programme and their experiences can be summarized as:–

- The extra work involved resulted in an (average) increase of labour load of 12.5% (2 men weighing 30 pigs/week in two houses). On average this extra work increased the cost of producing one finished pig by 0.625% (Range 0.28% to 0.9%).

- While at first it was difficult to convince the staff that this extra chore was worthwhile, within 2 full batches of pig all 5 farms agreed that the information it provided gave a new dimension to their work. However, *time was put by to do the job* and it was not forced upon already over-worked stockmen.

*Users' comments :*
Those farmers who weighed pigs and food representatively on a weekly basis have written to me …

> *"How fickle even an experienced eye can be in estimating growth rate. While 'faster' or 'slower' growth is noticeable on a pen basis, we were often wrong in quantifying it. Before measuring it, we **under**-estimated a slow-up in growth by 50% rather too often!"*

**Comment:** Remember what as little as a 5% error could cost! (page 344)

> *"Changes in pen growth on a weekly basis seems largely linked to feed intake" (see below)*

> *"Plotting weekly growth rates of one pen suggested, from the final graph, that pigs grow in gentle but perceptible waves, often about 14 days in 'wavelength'. If you look **down** a completed growth rate curve from one side and at an angle you often see it."*

> *"Where we had a slight (not an acute) problem, the drop-off below target was immediately noticeable in the test pens. However, we did pick it up before the weekly weighing in about 50% of the cases, but in the other 50% it made us go back to that pen and look again. This often seemed to occur when the pigs were too warm in summer/too airless on a cold night. Hot pigs can look "puffier" and this fools the eye – but not the weighing machine."*

# WAS IT WORTH IT ?

The labour cost of weekly representational weighing is equivalent to the cost of food for 1 kg of liveweight gain *in one pig* – taken as £0.38 in this case.  Quite modest.

## Did it pay?

During the two years that the 5 farms have been engaged in the process, the feed cost/ kg gain has fallen by an average of 1.8p/kg/ (corrected for feed price movements) on 4 of the farms *for each kg gained*.  In other words a 90 kg finisher has saved £1.62/ pig, while the weighing only cost 37.5p, an REO (payback) of 4.3:1.  It is not known how much of this improvement can be put down to alert stockmanship, but these figures from this exercise helps illuminate any cost-benefit position.

# DIFFICULT!

I have found it difficult to persuade farmers to test-weigh, especially weekly.  In fact the 5 farms concerned represent only a 5% success rate of those approached!

By far the most common objection has been the difficulty in squeezing in this extra job.  The next, persuading people that it is worthwhile anyway.  If the cost is around 0.6% more on the production cost of a finished pig, and the return is 2.8% more income, it does look to be a good bet.  An REO (*See Business Section*) of over 4.6:1.

Speed the day when one pen in the piggery will have a weighing *platform* so that a group of pigs can be driven there and weighed en masse; this is only available for weaners at present.

Calculations have suggested that the device will not save on the 0.625% extra labour cost, as the capital investment will erode much of that, but it will make the concept of representational weighing more practicable, and the possibility of that 2.8% more income more likely.

Alternatively the advent of more wet feeding systems provide evidence that growth rate is tied closely to feed intake.  Computerized wet feeding systems can record feed intake very precisely on a daily basis, or even part of a day.  If there is a link between growth rate and feed intake then fallaways in daily gain can be picked up within hours, not after one week or more.  The basic pen-by-pen system is here with us today in some CWF systems.

## The future

In future, *each* pig could be electronically tagged and weighed as he stands at the dispenser, so both feed intake and weight gain could be measured and recorded for each pig on a comparative basis. Moreover pen ambient temperature can be measured and controlled by the same device, and temperatures altered as well as nutrient allowance to compensate accordingly.

I have a feeling that by the time the current edition of this book is published, we will be starting to see this new technology in use on some of the larger pig units and breeding company test farms.

With over 40 direct or indirect causes and effects a growth rate problem is indeed multi-factorial!

# THE REAL COST OF WEIGHT VARIATION AT SLAUGHTER

The textbooks don't cover this well at all. There seems to be too little research on a situation I see every day on my farm visits – the problem of under-occupied finishing pens at slaughter, holding up the re-occupation of the finishing building. The difference between clients who minimize their 'close-out' period compared to those who didn't seems to be a rise in production cost per finisher of 4% - nearly all due to the higher cost per m2 of the unutilized pen space.

Mixing the laggards together elsewhere to free up this empty space did not improve matters, as the slow-up in performance due to the pigs fighting cost about the same in lost income.

So in every case we had to tackle the reason why there was too much variation in reaching the minimum contact weight threshold.

## Do the experts stress the wrong thing?

Possibly. Much is made in the advisory literature and stockpersons' training courses on the importance of skilled 'batching and matching' on entry to the nursery accommodation, and to a lesser extent on leaving it. Sure, this is important, but however well this is carried out - including the fascinating skill of assessing the 'do-ability' of different batches of weaners (which weigh the same but don't look the same – what Prof. Colin Whittemore calls 'thinnies') I guess the following have far more of an influence on slaughter weight variation.

So far I have attempted to measure what these other areas cost.

1.    Stocking density.

You do not have to be an Einstein to realize that unrestricted access to food all through the growers life is a major influence on even growth within a group. Overcrowded pigs are not only more stressed, which impinges on their ability to convert food efficiently, but the submissives are likely to eat less food than the dominants and so start to lag behind in weight-for-age terms.

I've hammered on about incorrect stocking density for decades  because I still see it on a third of the pig units I tour, even on the best farms, where it has been at least 15% too much. The average econometric benefit I measured on three farms from adjusting it back to the well-published levels overrode the cost of the extra space needed by 6:1. This is enough of a payback to make us all think.

So does overstocking also increase slaughter weight variation? I'm sure it does, and having failed so far to find clear evidence in the literature of what it might be, I would welcome any information readers might have and be good enough to send on to me.  In the meantime I'm trying to collect some farm figures myself, but this will take time.

2.    Adequate trough access.

There is now enough evidence of the penalties of insufficient trough space, and detailed allowances have been well- published . Inadequate trough space has cost, in my experience, about €3/pig.

3.    Correct feeder settings (i.e.. feeder throat gaps).

Various N.American researchers have done good work on this, such as Patience, Gonyou, Dritz, Tokach, and Dean Boyd (himself a progressive large-unit farm manager). So advised feeder settings and the amount of feed visible at any one time on the feeder plate/in the ad-lib feed trough are also available and I cite this work on page 391. The workers have also suggested what the performance penalties could be from not getting it right but which has not yet got into the textbooks. Neither has the effect this may have on weight variation by slaughter. From the N. American figures, under European 2010 costings, I calculate this cost us €1.80/pig, which accords well with the 4% production cost increase mentioned earlier.

## Evidence from a 'real farm' as distinct from research conditions

My own experience tells me that farm close-out variation is considerable - and costly. To this end I was called last autumn to a large farm with 4 separate nurseries looked after by different section heads. There was a distinct difference in the performance between two of them and the main difference seemed to be that the one with the greatest variation at slaughter was in charge of a section head who only occasionally checked his throat settings – this was all to obvious from looking at the feeder trough contents, some were overfull and some quite bare. I compared his close-out performance with another section head who obviously paid great attention to altering each throat gap, if necessary, several times a week. The poorly adjusted throat nursery had 87% of the pigs remaining once the first 10% were ready for shipping while the 'good' nursery only had 36% laggards once 10% of his pigs were ready. A big difference.

## TOO MUCH VARIATION IN SLAUGHTER WEIGHTS - DOES 'TOPPING' HELP?

We all get this problem to a greater or lesser extent. Is 'topping' the answer?

'Topping' is the term given to removing pigs to a spare pen or even on to slaughter which are 5 to 10% ahead of the rest of their companions in a pen. This is sometimes done about a week to 10 days before slaughter, when the finishing pens start to look distinctly overcrowded. As it seems to be a growing trend, I did some measurements in 2010 with co-operating farmers as I was unsure whether some forms of topping are a good idea. The results encourage me to say….

Topping or 'Selective Removal' is done for a variety of reasons:-

(a)  Removal to avoid overweights. Of course this is essential, but those chosen must be loaded straight on to the transporter and not kept mixed in one pen and thus handy for the morning collection.

(b)  Deliberately moving 7 to 10 days before estimated shipping date some of the faster growing pigs to another small pen - if such pens are available. Two producers constructed straw bale pens for this purpose but were careful not to be tempted to mix growers from different pens together to avoid skin damage from scrapping and antagonistic stress negating any potential improved growth rate. Topping adherents claim that the growth rate of the moved pigs improves by 20g/day and of those left behind, due to more space and less competition for food, by up to 80g/day. This we found to be mostly correct.

(c)  Conventional Topping. Sending some of the faster-growing pigs to slaughter about a week sooner than normal. These pigs are shipped within the contract weight range but well towards its lower end. We tried this, and dependent on the terms of each farms contract, some 4.5 kg of potential deadweight return was sacrificed costing 6.30 euros/pig, say 13 euros for two of these bigger pigs sent on early from each somewhat overcrowded pen of 15. The remaining 13 benefited from more room and better access to food - even ad-lib food, but they still only saved half a day occupancy from their faster growth plus 2 euros- worth of food for the whole 13.

Yes, to add to this, a weeks food was saved from those shipped early, but this only recouped one third of the 13 euros lost income per pen from the meat they would have put on during that week.

**Conclusion:** The heavier you can ship every pig up to, but under, the contract overweight cut-off point the better. On all of these farms, topping carried out sensibly and well, did have its physical benefits as claimed by its protagonists, but it did not seem to pay. A classic example in that while the published physical performance benefits may well be as claimed, when the econometrics are added in these favourable impressions need to be questioned.

* It seemed to work better when pigs were overstocked – but they shouldn't be in the first place! It is cheaper not to overstock than to be forced into topping – see the overstocking cost table on page 421.

*Topping also works better when there is spare accommodation available – which ideally should be filled anyway!

So until I am persuaded otherwise, topping is not for me.

# WHAT AFFECTS GROWTH RATE ?
# A SUMMARY CHECKLIST

We need a more planned approach to a slow growth situation. The following checklist will help

| | | |
|---|---|---|
| ✓ | *The pig* | Genetics. |
| | | Age. |
| | | Sex. |
| | | Docility. |
| ✓ | *Its food* | Nutritive specifications and balance. |
| | | Raw Material Quality / Availability to the pig. |
| | | Daily intake. |
| | | Palatability. |
| | | Water. |
| | | Growth enhancers. |

Wet/dry & fully wet feeding.
Adequate trough space.
Feed texture.
Access to hoppers troughs.

✓ *Its surroundings* Temperature over the skin.
Air speed (draughts).
Air positioning.
Humidity.
Floor surface.
Bedding.
Insulation of all surfaces, especially the floor.
Gases (not necessarily toxic).
Airborne dust.

✓ *Its companions* Pen shape.
Position of furniture.
Stocking density.
Group size.
Weight variability.
Docile genes.

✓ *Disease* Biosecurity.
Diseases present, including moulds/mycotoxins.
The degree of immune protection needed.
Precautionary measures in place.
Curative measures in place.
Veterinary supervision.

✓ *Management* All-in / All-out (AIAO).
Batch production.
Continuous production.
House changes.
Batching and matching skills.
Weaning weight/size.
Representational weighing/ records.
Computerisation (measuring, monitoring, highlighting action needed).
Monitoring and maintenance of environment equipment.
Weekly (daily?) progress chasing/actioning.

✓ *Stockmanship* Quality of person.
Time to do the job.
Ongoing education / training / demonstration by peers.
Daily briefing.
Observation and recording.

## PLANNING AND PRIORITISATION –
## A CONSULTANT'S GUIDELINE

The most effective way to solve a slow-growth problem is, in my experience, carried out like this. While certain sectors below may seem obvious, *it is easy to assume things which may not be true or necessarily evident*. Everything should be checked out so we have a firm base on which the best remedies can be decided.

Knowing how a good adviser wants to work will help both you and him solve the problem.
Approaching the problem

1.   Have we a problem?
2.   What is the evidence for it?
3.   When was the problem noticed, and how did it come to light?
4.   Have the assertions you are told about been checked out? It is wise for the consultant to check them for himself too.

Now we know this . . .

5.   How serious is the problem ?
6.   What is this costing? (Useful in deciding which action is most cost-effective and which to attend to first. Use MTF and PPTE guidelines as in Table 4, page 342).
7.   Is there evidence of 'stop-start' growth?
8.   So-called compensatory growth?
9.   How 'even' do the pigs look?
10.  Are any recent veterinary reports/observations available?

Table 6 provides some evidence of which areas in the preceding checklist are most likely to be involved.

## PRIMARY AREAS TO LOOK FOR IN
## POOR GROWTH RATE

1.   Inadequate or unbalanced nutrient intake on a daily basis.
2.   Pigs too hot or too cold.
3.   Inadequate air movement and poor positioning of air.
4.   Disease, stress and likely immune demand.
5.   Any changes – of environment, of feed, of comparisons, of management, (possibly) of stockpeople, of outside weather conditions, pen soiling.
6.   Unawareness – *i.e.* of weight, temperature, internal climatic fluctuations, feeding times, water supply, and poor batching & matching at weaning or after the nursery stage.

**Table 6.** An analysis of 137 cases of inadequate growth rate investigated over some 25 years in 14 temperate-zone or colder countries. 90% of the cases were followed up within 6-9 months (Figures in %)

| Area thought to be involved | | | Resolved or mostly resolved over a period of time | Unsolved |
|---|---|---|---|---|
| 1. | Nutrition | a) – dietary imbalance | 8 | 5 |
| | | b) – feeding system/ allowance | 9 | 1 |
| | | c) – palatability | 2 | - |
| 2. | Temperature | | 12 | 4 |
| 3. | Ventilation | | 14 | 8 |
| 4. | Disease | – pathogenic | 4 (*referred to veterinarian*) | 5 (*referred to veterinarian*) |
| | | – poor hygiene | 6 | - |
| 5. | Multifactorial, *i.e.* several causes suggested. | | 7 | 3 |
| 6. | Other, *i.e.* outside N°s. 1-4. | | 8 | - |
| 7. | Minor or no problem on investigation | | 2 | - |
| 8. | Client ignored advice | | - | 2 |
| | | | 72% | 28% |

**Note**: Items 1b and 2; 2 and 3; 3 and 4 were often inter-related

**'Other'** Includes stocking density, water adequacy, stockmanship errors, too many changes – all thought to be primary factors affecting growth rate on the 8 farms involved.

**Comment**: Despite the author being employed as an animal feed firm troubleshooter for 60% of the period, notice that the feed or the way it was fed was only implicated in under 1 in 5 of the slow growth complaints resolved. Notice also the high proportion (28%) of environmental errors responsible. A very different causal list is seen in tropical countries, where in my experience genetics/appetite are primary causes.

# SOME OBSERVATIONS ON CHECKING UP ON THESE PRIMARY DISTURBANCES TO GROWTH RATE

## *Nutrition*

Compound (ready-made) feeds

***Errors in nutritional specifications*** are much less common than 20 years ago due to computerization in the formulation office and in the feed mill. Mistakes do occur, largely due to errors in assumptions of nutritive value of the raw materials used. However feed manufacturers now analyse raw materials for primary nutrients to a much greater degree than they used to, but energy achievement seems to be a possible weak point, even today.

*To a certain extent sub-quality samples* have to be used-off in manufacture and it is hoped that this is done *very* gradually especially in the case of weather-damaged items or spoilage. If not, a reduction in growth rate is quickly noticed. However short-cuts are (occasionally) taken and farmers should be aware of this possibility. If this is proved – or maybe even suspected, an immediate change of supplier to one more trustworthy is wise.

Be extremely careful when 'using-off'. Seek professional advice.

# OVER-USE OF CERTAIN RAW MATERIALS

The writer is old-fashioned enough to believe that a good mix of raw materials to make-up the dietary specifications provides the best growth rate. Conversely however the Americans get excellent results with almost universal corn/soy diets (with a min./vit. supplement) for growing pigs. Such diets are dry and palatable and as a result mycotoxin levels could be lower than in some countries.

Feed compounders do have maximum raw material constraints and purchasers should ask what they are. Over-use of wheat, when price and availability is favourable; wheatfeed, tapioca (manioc), biscuit-meal and rice bran is known. Maximum levels of DDGS ingredients and canola meal are now established. Bakery and confectionary by-products are widely used now by the feed trade. We must trust them to use these blends wisely.

# FARM-SPECIFIC DIETS IMPROVE GROWTH RATE

The feed trade is still moving only slowly in the important area of matching nutrient specifications to the immune status of the pigs. All nutritionists are well aware of how this affects performance when it is badly adrift (up to 40% in growth rate, and possibly over 50% in protein deposition), but in the writer's experience some Sales & Finance departments seem to be holding things up on cost grounds. Therefore there is a reluctance to change to a radically new way of selling pig food so that immune status can be assessed and the diets supplied to cater for it.

Producers with growing finishing pigs can help themselves by talking directly with their supplier's pig nutritionist, and by providing him with details of the current health and biosecurity status of their grow-out facilities. He can then get specifications closer to being right with a custom-mix. However to utilise the possibilities to the full, such a producer will eventually have to espouse a CWF system (Computerized Wet Feeding) and co-operate with the feed supplier on a "Challenge" or "Test" Feeding programme so that his pigs' lean gain growth curve can be measured and the diet adjusted accordingly. At the time of writing lean growth itself could be a pointer to the immune demand.

## PALATABILITY

Sometimes growth rate can be affected by lack of palatability. Most nutritionists have palatability tables and build in the necessary constraints when including likely unpalatable ingredients. However what is less well recognized is the *cumulative* effect of the milder unpalatable ingredients in the diet, especially if other aggravating factors like over-coarse or too-fine grinding, soft pellets, mould residues, added fat levels and some unpalatable chemical additives (*e.g.* nitrofurans), high mineral levels (limestone) are also present. General lack of freshness is also an appetite-reducing factor in small pigs.

In most of the above cases the inclusion of an aromatic taste-enhancer may not work, and most firms selling palatants do not necessarily claim that they do.

Ingredients known to be unpalatable are sorghum (milo), rapeseed, olive pulp/cake, wheatfeed in excess (>40%), any food containing moulds or mould residues (mycotoxins), too much DDGS, very hard particles of wheat or maize, most minerals in excess (limestone flour is sometimes over-used), food ingredients which have been oxidised, especially rancid oils/fats, and all ingredients with pungent odours – some confectionery waste can contain such smells. They are not unpleasant to us humans but the pigs definitely object to them.

Suggested maximum inclusion levels are given in Table 7.

If palatability is a suspected cause of slow growth, do not resort simply to flavour enhancers, but immediately change the diet to one known to contain fresh, and a wider variety, of ingredients.

## BY-PRODUCTS OF BIOFUEL MANUFACTURE (ETHANOL)

These are DDG/S ( Distillers Dried Grains with Solubles), glycerol, corn gluten etc. DDG/S is by far the largest source of by-product material issuing from the biodiesel industry, the majority of it to date coming from corn (maize).

**Table 7. Suggested maximum levels for common feed raw materials. Above these levels growth could be affected (% as fed and of good quality)**

| | *Sows* | *Weaners* | *Grower* |
|---|---|---|---|
| **Ingredient** | | | |
| Barley | *No limit, but watch dust at high levels* | | |
| Wheat | 50 | 33 | 40 |
| Maize | 40 | 40 | 25 |
| Oats | 40 | 10 | 25 |
| Tapioca | 25 | 10 | 20 |
| Wheatfeed | 35 | 15 | 30 |
| Soya | 25 | 30 | 30 |
| Full fat soya | 15 | 20 | 20 |
| Meat meal | *(not advised due to risk or fear of BSE in certain countries)* | | |
| *Otherwise* | 7.5 | 2.5 | 7.5 |
| Fishmeal | 7.5 | 10 | 5 |
| Bread waste (63% DM) | 40 | 30 | 40 |
| Biscuit waste (86% DM) | 40 | 40 | 30 |
| Cake waste (85% DM) | 40 | 30 | 40 |
| Confectionery waste (98% DM) | 20 | 7.5 | 15 |
| Wheat starch syrup (DE12 MJ/kg) | 25 | 15 | 20 |
| Potato waste (Steamed) (11% DM) | 25 | 15 | 20 |
| Fodder beet (17.5% DM) | 20 | 5 | 10 |
| Maize gluten meal (DE 13 MJ/kg) | 20 | 5 | 10 |
| Rapeseed (DE 12 MJ/kg) | 12 | 5 | 15 |
| Skim milk (DM 9%, lysine 0.23%) | 30 | 30 | 30 |
| Whole milk (DM 13%, lysine 0.27%) | 60 | 80 | 80 |
| Yoghurt waste (DM 14%-20%) | 25 | 15 | 20 |
| Whey (DM 5.5% DE 0.85 MJ/kg) | 25 | 10 | 25 |
| Brewers yeast (18% DM) | 25 | 10 | 10 |
| Extr. Rice Bran (88% DM 16% fibre) | 15 | 5 | 10 |
| Linseed cake (33% F.P.) | 10 | 2.5 | 5 |
| Maize Germ Meal (DE 13 MJ/kg) | 20 | 15 | 10 |
| Distillers Draff (23% DM) | 5 | 0 | 5 *(after 50 kg)* |
| Distillers Grains (89% DM) | 5 | 2.5 | 5 |
| DDGS** | 30 | 10 | 30 |
| Lupins (88% DM, DE 17 MJ/kg) | 10 | 5 | 10 |
| Sugar beet pulp (dried) (88% DM)* | 10-20* | | |
| Peas | 15 | 10 | 20 |
| Sorghum | 20 | 0 | 10 |

* Care needed due to swelling when moistened.
**See below

DDG/S is produced by a dry milling process, 100 kg corn producing approximately 31 kg distillers grains and 42 litres fuel ethanol. The ethanol is added to petrol at around 10%, or if in the form of biodiesel at under 6%. This saves on coventional sources of crude oil.

## How ethanol is made

The production of ethanol is fairly straightforward. Whole corn is ground, water is added and the mixture cooked. Enzymes are then added to turn the starch into glucose Lastly yeast is added, fermenting the glucose into ethanol.

The residue of protein, fibre, and minerals, etc., remain as solubles and are dried into DDG/S, usually abbreviated to DDGS.

The DDGS fraction is used as animal feed and competes with the established corn by-product trade (corn gluten feed and corn gluten meal) with their lower yield of 100 kg corn producing approximately 19 kg of corn gluten feed and 4.5 kg of corn gluten meal.

Glycerol is another future by-product of biodiesel manufacture with a good energy content but is high in salt. Used up to 10% in sodium-controlled diets, it can lower the cost of cereal energy and can also improve drip loss.

## Problems with DDG/S

There are two current problems with DDG/S. Its variable quality issuing from the many manufacturing plants, as well as high mycotoxin levels and high fibre levels. So consultation with a pig nutritionist is essential before inclusion in pig feeds, especially creep and link feeds.(Link=immediate post-weaning feed) where the level and quality of fibre needs to be under control.

Most of the advice to date on using DDG/S, together with advised maximum inclusion levels, are based on US research data and therefore relate to their relatively simple corn/soya formulae. The addition of certain enzymes can affect performance positively depending on the ingredients in the 'host' feed – another reason why a nutritionist should be involved.

Overuse due to the tempting price savings when cereal costs are high have led to poor carcase quality, so maximum levels depend on the formula make-up of the host diet. Yet again expert advice is needed and seemingly authoritative published maximal data you might read about should not be adopted 'out of the blue' but be referred to a pig nutritionist first.

**Moral: DDG/S –Sure, a useful ingredient – but be careful!**

Some water sources (*e.g.* in central Canada and Mexico) can contain excessive levels of trace elements which may affect appetite.

# WHAT TO DO IF YOU SUSPECT THE FOOD

If you suspect your delivered feed is the cause of your slow growth …

• Immediately take a 2 kg (5 lb) representative sample of the suspect feed.

• Keep at least a 1 kg (2 lb) sample of the food refrigerated, for future use if needed.

• Contact the feed supplier and speak to the company nutritionist. *Use the local representative or sales department staff only as intermediaries*. Generally speaking samples should be taken by them *in a properly representational manner* , half given to you, and a telephoned, then written report by E-mail, fax or letter sent to you. This should take hours after the arrival of the sample, not days.

• Check that your bulk bins and feed hoppers do not contain obvious mould or mould residues on the 'unpolished' surfaces, *i.e.* how clean they are.

• Assemble and provide proof, if needed, of the batch number of the delivery involved, with all paperwork attached.

• Provide evidence of the performance loss, with dates.

• If requested allow the feed manufacturer full access to your premises and equipment.

• Employ an independent pig consultant or pig nutritionist for a second opinion if the feed supplier is dilatory or seems evasive. Allow him also to tour the premises.

*__Remember__* : Do not necessarily blame the food.  Look again at Table 6.  In this survey only one in ten of poor growth problems probably or certainly involved feed quality.  This is in contrast to two out of three cases where the food was immediately blamed by the producer for slow growth or poor FCR.

# THE ON-FARM MIXER

The four major problems seen with home mixers which affect performance, including growth rate, are:–

1.  Failure to provide their nutritionist with details of the expected (or actual) analyses of the home-grown or bought-in raw materials used, especially cereals.

2.  Opportunist buying of what appear to be good value raw materials without assurance from the vendor of a nutritive declaration on at least the more important nutrients – "buying blind".

3.  Failure to keep the mixing area clean enough, so that the risk of the damaging effects on performance of residual mycotoxins are high.

4.  Inaccurate measuring of inclusion rates.

## RAW MATERIAL ANALYSIS

The majority of on-farm mixers use their own grain, or grain from an adjacent farm. Not surprisingly these parcels can vary enormously in nutrient quality and quantity (Table 8).

**Table 8. Typical raw material variations**

|  | Average Protein (%) | Range Min (%) | Max (%) |
|---|---|---|---|
| *Off-farm* | | | |
| Barley | 11.2 | 7.8 | 13.9 |
| Wheat | 11.9 | 8.3 | 16.4 |
| *Bought-in* | | | |
| Wheatfeed | 16.0 | 13.7 | 19.7 |
| Soya 44/47 | 41.1 | 34.0 | 47.0 |
| Full Fat Soya | 36.1 | 33.6 | 45.0 |
| Best fish meal | 70.3 | 64.0 | 73.5 |

Source : UK feed trade (various sources)

While the nutritionist can make fairly close assumptions on bought-in materials, producers should *always* provide him with the declared analysis of purchased goods so that he can narrow the variation still further.

Not to do so is negligent, or at best lazy, as the information is often provided free by the vendor.

## THE COST OF GETTING IT WRONG

Many producers think that "taking the rough with the smooth", the quality evens itself out over the lifetime of the growing pig. This is definitely not so. Dealing with feed complaints on protein/ vit./min mixes and where careful analysis revealed the shortfall in formulation, I found the improvement due to subsequent correction was never less than £2/pig, and one supplier published a figure of £3.

Even taking the lower figure of £2/pig, at 5 pigs to the tonne of feed, this is a £10/tonne leeway. Assuming the average home-mixer produces 5,000 finishers/year thus consuming 1,000 tonnes feed/year, the shortfall could be £10,000/year!

This allows considerable scope for analysing at least the parcels of grain used, if nothing else. So what might this cost?

# THE COST OF GETTING IT RIGHT

Table 9 suggests current analysis costs . . .

Table 9.  Current lab analysis costs

|  | £ |
| --- | --- |
| Dry Matter | 4 |
| Protein | 7 |
| Oil | 7 |
| Fibre | 8 |
| Lysine | 30-35 |
| Mycotoxins | 25-40 |

A few years ago I contacted all the major UK feed compounders and asked them how much they spent per tonne of feed sold on *raw material analyses alone*. The average was £1.40 on feed costing about £165/tonne or 0.8%. In today's prices this would be at least £2.20/tonne on a £175/t diet (1.43%).

Figure 1 is interesting as it charts the performance shortfall of grow-out home mixers over grow-out complete feed users across a ten year period in terms of FCR – a similar deficiency occurs in growth rate.

How much of this leeway is due to the formulator/nutritionist (the same person in some feed companies) having different conceptions of the raw material analysis he has on hand?

If we assume from Figure 1 that the average shortfall is now £9.00/tonne (at 5 pigs/tonne, £1.80 pig) what can we afford to spend on grain analysis to help the nutritionist to give us a more accurate specification using the bulk raw materials we possess in storage or are buying each month? Producing 5000 finishers/year, even taking this cost at £1/pig this is £5,000, but I find an adequate cost is nearer to only a fifth of that.

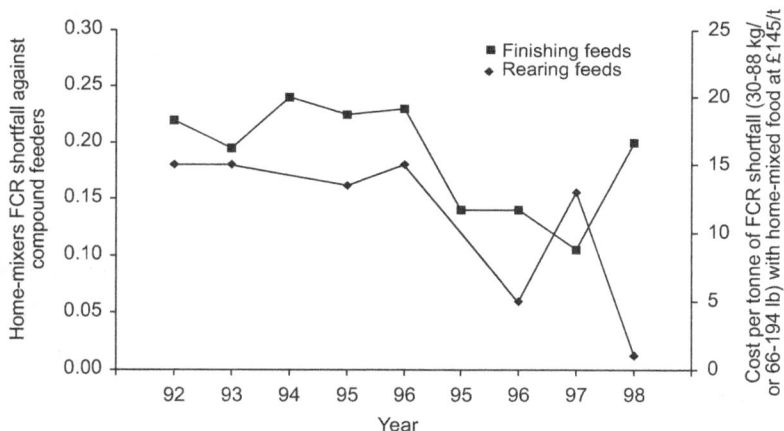

**Figure 1.** Records show a consistent shortfall in FCR for homemixers v. compounds

**Source**: Extrapolated from various recording schemes and compound feed trade data

**Comment**: This was 12 years ago. While things are getting better, there still seems to be a £9/tonne (6%) difference in poorer performance of home-mixed grow-out feeds to those provided by the feed trade. The fact that the same nutritionists are often involved in the dietary specifications suggests that the home-mixer is adrift - by not knowing the analysis of his on-farm ingredients.

**Table 10. How much to spend on raw material analysis?**

|  |  |  |  |  |
|---|---|---|---|---|
|  | 12 lysines | at | £38 | £420 |
|  | 15 D.M.sat | £4 | £60 |  |
|  | 15 oils | at | £7 | £105 |
|  | 15 fibres | at | £8 | £120 |
| *Optional ...* | 6 mycotoxins | at | £40 | £240 |
| Contingency at nutritionist's request ... |  |  |  | £100 |
|  |  |  |  | £1045 or 21p/pig |

Spending 21p/pig to recapture what the evidence suggests is a discrepancy of somewhere between £2 and £1.80 pig is an excellent REO – some 8 to 10:1. Alright, some laboratories charge more, but even at double the cost, the payback is over 4:1. And if your nutritionist suggests it, you can even afford to pay for more comprehensive raw material analysis!

# BUYING BY-PRODUCTS : A CHECKLIST

Many by-products – just because they look cheaper on a feeding cost/day basis (as indeed all are) – are purchased 'blind' or 'on trust'. There can be more variations in critical nutrients than with grain or bought-in straights, especially when the liquid fraction is high, *i.e.* skim and whey.  Some by-products are produced to a nutrient specification, which helps greatly in calculating dietary daily intake, but most are not.  The following checklist is helpful in getting value for money.

**Some questions to ask before buying**

A.    What is its source (a by-product of what?)

B.    If originally designated for the human market why was it rejected?  Is there a toxin factor? i.e. Mycotoxins?

C.    If past its sell-by date, how stale is it?  (Has an anti-oxidant been used?)

D.    Is it raw or is it cooked?

E.    Is it of a single natural source or is it a mixture?  If a mixture, what are its constituents?

F.    Is it palatable and digestible?

G.    What is its analysis and will it be consistent?  Always get a salt level on whey and whey concentrates and a DM declaration on skim and again especially whey.  (*See Table 12*).

H.    Is there continuity of supply or is it a one-off?

I.    Will it be delivered fresh and what is the delivery cost, the cost per tonne of dry matter, the cost per unit of energy and protein?

J.    Who else is interested?  Who's turned it down?

K.    What is it really worth?  Do the sums; they matter.

L.    **Finally, always consult an independent nutrition specialist.**

# VARIATION IN LIQUID BY-PRODUCTS – HOW IT AFFECTS GROWTH RATE

Both skim and whey are variable products owing to the quantity of water (and in whey's case salt) allowed to enter the collection tanks at the milk and cheese factories.

## HOW TO CHECK OUT SKIM MILK

With skim one expects a 9% dry matter content, with a specific gravity of 1.033.  Specific gravity (sp.gr.) *is the weight of a substance compared with the weight of an equal amount of some other substance taken as a standard*. For liquids the standard taken is usually water. Thus a sp. gr. for a sample of skim milk of 1.033 means that one kg of skim milk weighs 1.033 times more than water, and can be quickly measured by a hygrometer – a simple and inexpensive instrument.

Table 11 gives a simple on-farm guide to the value of skim milk relative to its sp. gr. Low sp. gr. suggests dilution with washing water at the factory, and the balancer meal/supplement will need nutritive compensation if growth rate is to be maintained, which if these low sp. grs. are consistent, should be drawn to the supplier's attention and appropriate financial compensation claimed for the extra balancer meal required. Talk to your supplier beforehand about these sorts of QC matters. He could just want to dispose of it while you are buying it as a nutrient substitute.

**Table 11. Relationship of specific gravity, dry matter and composition of skim (as fed basis).**

| Specific gravity | Measured dry matter % | DE (MJ/kg) | DCP % |
|---|---|---|---|
| 1.036 | 9.5 | 1.58 | 3.5 |
| 1.033 | 9.0 | 1.50 | 3.3 |
| 1.031 | 8.5 | 1.42 | 3.1 |
| 1.030 | 8.0 | 1.33 | 2.9 |
| 1.028 | 7.5 | 1.25 | 2.8 |
| 1.027 | 7.0 | 1.17 | 2.3 |
| 1.025 | 6.5 | 1.08 | 2.2 |

# HOW TO CHECK OUT WHEY

Because whey is a high-energy food, and provides a reasonable amount of good quality protein as well, variation in sp.gr. is a very important factor in purchased deliveries. Again a hygrometer is essential to keep abreast of delivery variance and Tables 12a & b show how only small differences in sp.gr. (of only 0.002 sp.gr. or 1% dry matter) per batch can affect nutrient intake, and how a typical whey balancer meal designed by a nutritionist on the *average* expectancy of 1.022 sp.gr. (5% DM) needs to be increased or decreased per litre of whey fed to rebalance the nutrient capability of the pig. Tables 12 c&d illustrate this difference on a per pig basis and what failing to do this over quite small movements in sp. gr. from batch to batch costs in lost growth rate and FCR.

**Table 12a.  Relationship of specific gravity, dry matter and composition of whey (as fed basis).**

| Specific gravity | Dry matter of whey | DE MJ/kg | CP g/kg | Lysine g/kg |
|---|---|---|---|---|
| 1.027 | 6.5 | 1.07 | 10.4 | 0.65 |
| 1.025 | 6.0 | 0.98 | 9.6 | 0.60 |
| 1.023 | 5.5 | 0.90 | 8.8 | 0.55 |
| 1.022 | 5.0 | 0.82 | 8.0 | 0.50 |
| 1.021 | 4.5 | 0.74 | 7.2 | 0.45 |
| 1.020 | 4.0 | 0.66 | 6.4 | 0.40 |
| 1.019 | 3.5 | 0.57 | 5.6 | 0.35 |
| 1.018 | 3.0 | 0.49 | 4.8 | 0.30 |

**Table 12b.　Change in Balancer Allowance required for equal nutrient intake with varying Whey compositions.**

| Specific gravity | Change in balancer allowance |
|---|---|
| 1.022 | No change |
| 1.021 | + 5 kg/1000 litres or 50g/10 litres |
| 1.020 | + 10 kg/1000 litres or 100g/10 litres |
| 1.019 | + 15 kg/1000 litres or 150g/10 litres |
| 1.018 | + 20 kg/1000 litres or 200g/10 litres |
| 1.023 | − 5 kg/1000 litres or − 50g/10 litres |
| 1.025 | − 10 kg/1000 litres or − 100g/10 litres |
| 1.027 | − 15 kg/1000 litres or − 150g/10 litres |

**Table 12c.  Example (20 pigs in a pen).**

| | S.G. of Whey | | 1.022 | | 1.018 | |
| | Whey litres/day | | Balancer kg/day | | Balancer kg/day | |
| Liveweight kg | Per pig | Per pen | Per pig | Per pen | Per pig | Per pen |
|---|---|---|---|---|---|---|
| 25 | 3.0 | 60 | 1.0 | 20 | 1.05 | 21 |
| 40 | 4.5 | 90 | 1.4 | 28 | 1.50 | 30 |
| 60 | 7.0 | 140 | 1.6 | 32 | 1.75 | 35 |
| 80 | 9.0 | 180 | 1.8 | 36 | 2.00 | 40 |

**Table 12d.** Measurements of growth performance from two sections of one farm where a hygrometer was or was not used to readjust whey balancer meal according to shortfalls in specific gravity readings.

| | ADG 20-88 kg (g/day) | Average variance detected below norm* DM % | Total lysine g/kg |
|---|---|---|---|
| Hygrometer used and meal allowance increased to Table 14b levels. | 759 | 1.3% | 0.15 |
| Hygrometer not used | 721 | – | – |

\* *On 35 days out of the 90 day period.*

**Comment** : By adjusting the meal allowance when necessary on a wet-feeding circuit cost 95p/pig more but the quicker growth therefrom saved £1.25/pig food at slaughter. Technically, the producer using the hygrometer, by recording his readings and sending them to the supplier, could claim from the supplier of the whey the 95p/pig spent on extra meal to compensate for the shortfall of 1.3% dry matter below the agreed standard of 5% DM, thus benefitting by the full £1.25/pig. Without a hygrometer no such claim can be substantiated. Always use a hygrometer, and tell your supplier why. At least it may improve the consistency of your deliveries!

# NUTRITION AND GROWTH RATE

The correct balance of the primary nutrients in the diet can affect growth rate. Of these lysine:energy ratios are important. While supporting amino acid balance is also critical this has to be left to the nutritionist to get right.

I am not a nutritionist and wisely leave the intricacies of ration design to those who know what they are talking about! I quote from Mick Hazzledine, one of the world's leading nutritionists especially in the field of the growing/finishing pig:-

*"Most of the major EU pig-producing countries use net energy systems and have done so for some years. Many use different NE (net energy) values for sows and growing pigs to allow for the higher fibre digestibility in older animals.*

*In the UK 10 years ago a typical finisher feed fed from 60 kg wouould have had a DE of 3.27 – 3.32 Mcal/kg; today it would be be 3.15- 3.30 Mcal/kg and there will be inncreasing pressure to reduce this as DDG/S volumes increase. ."*

Hazzledine: Banff Pig Conference (2010).

The application of nutritional research like this is these days based on commercial application and I find varies considerably from country to country. My suggestions overleaf are therefore ballpark figures which although somewhat dated are still serving me well on farms across the world. They direct the producer's attention to checking from the information that his country has that his animals are getting what the his experts recommend - and often I find that from going through Table 12's guidelines with him that they are not. It may be daily intake – which in my job as a management advisor I am able to assist, or it can be in dietary design which is the nutritionists job – not mine. Having alerted him to any apparent leeway outside management which could be due to diet composition influencing daily intake, I then recommend that they consult a nutritionist.

Which is as far, as a non-nutritionist, I am prepared to go. My job is rather like to being at an airport indicating to the client which could be the best flight for him to take to get him safely to his destination at a reasonable, or least, cost according to his means. To point him in the right direction. My job is not to fly the aeroplane – that needs an expert, and this is the nutritionist.

## LYSINE: ENERGY RATIOS CHECKLIST

These are often seen as straight ratios averaged out for all three 'sexes' – entires, castrates and gilts. (For example, Table 13).

**Table 13. Recommended overall lysine : energy ratios and daily intakes for all growing/finishing pigs – past advice (early 2000's).**

| | | | | | *Genetically improved stock* | | | | |
|---|---|---|---|---|---|---|---|---|---|
| Body weight (kg) | 7 | 12 | 16 | 20 | 40 | 60 | 80 | 100 | 115 |
| Target growth rate/day (g) | 250 | 400 | 500 | 650 | 800 | 900 | 950 | 1000 | 1250-1300 |
| Lysine needs/day (g) | 4.6 | 7.6 | 9.6 | 14.0 | 20.0 | 24.4 | 27.0 | 28.5 | 30.2** |
| Approx* energy needs/day (MJ DE) | 4.5 | 7.3 | 9.9 | 15.5 | 24.5 | 30.0 | 34.0 | 36.5 | 38.0** |
| Lysine: DE ratio (g/MJ DE) | 1.02 | 1.04 | 0.96 | 0.90 | 0.82 | 0.81 | 0.79 | 0.78 | 0.82** |

* *Dietary DE can change dependent on appetite, especially of nursery pigs. Note: DE, not NE. NE requires local application.*
**For very high lean deposition genotypes over 106 kg liveweight.*

While Table 13 gives a guide to correct daily nutrient intakes of total lysine and digestible energy for all pigs, further research suggests the true picture is much more complicated. This can be due to:–

✓ Differences between genotypes, often influenced by appetite capability. (Table 14).

✓ Differences within the genotype. (Figure 2).

✓ Difference between the sexes (Table 15).

✓ The effect of immune demand. (see Immunity section*).*

**Table 14. Appetite and lean gain per tonne of feed of 4 major European breeds (pigs 55 – 90 kg)**

|         | *Appetite Difference to Breed A* | | *Lean Produced/Tonne Feed (kg) Difference to Breed A* | |
|---------|-----------|------|-------|--------|
|         | *(kg/day)* | *%* | *(kg)* | *%* |
| Breed A | 2.84      | –    | 301   | –      |
| Breed B | 2.78      | - 2.11 | 268 | - 1.96 |
| Breed C | 2.79      | - 17.76 | 274 | -8.97 |
| Breed D | 2.51      | - 11.62 | 253 | - 15.95 |

*\* Same standard feed fed, but adequate diets fed* ad lib *throughout.*

Econometrics: When the diet of Breed D was adjusted in nutrient density based on a 12% lower appetite potential to that of Breed A, lean per tonne exceeded that of Breed A by 2.6%, and while Breed D's diet cost 8% more, the extra yield of lean recouped 80% of this extra dietary cost.
Source: RHM (unpublished)

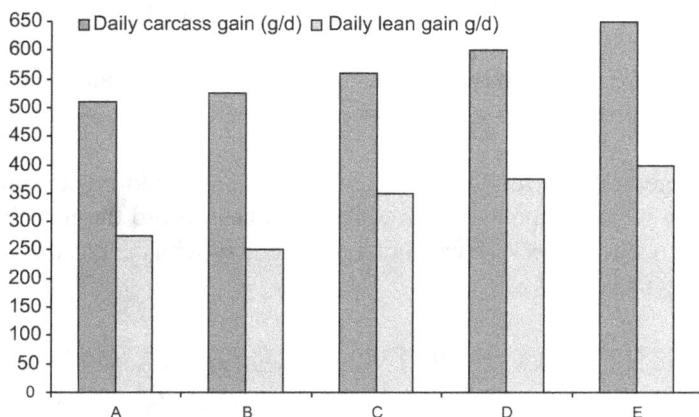

**Figure 2** Variation in growth characteristics within a single genotype
(Source: Owers, 1994)

**Table 15.   By having to cope with a high disease challenge, genetically improved[†] pigs 6.3-27.2 Kg (14-60 lb) eat less, grow slower and have a poorer quality carcase**

|  | *Immune Stimulus Required* | | |
|---|---|---|---|
|  | *Low* | *High* | *Difference\** |
| VFI (kg/day) | 0.97 | 0.86 | 12.8% more |
| ADG (g) | 677 | 477 | 42% more |
| FCR | 1.44 | 1.81 | 25% better |
| Protein gain (g/day) | 105 | 65[†] | 62% more |
| Fat gain (g/day) | 68 | 63 | 8% more |

Source: Stahly *et al* (1995)
\* In favour of low immunity needs
[†] The leaner the genotype the more the protein gain is damaged
Note: Both sets of pigs could be considered "healthy". A high disease challenge is described typically as a 'pig-sick' building and a low disease challenge environment as 'all-in/all-out/multisite' scenario, properly disinfected.

## And is there a tropics factor ?

I also suggest that certain genotypes ('breeds') may need different lysine:energy ratios when kept under very hot and humid (tropical or near-tropical) conditions, but at present this is surmise on my part but gathered from experience in such conditions. It could be that some high-lean/'low-appetite' genotypes are unsuited to these near-tropical conditions when fed diets formulated to "European" specifications – and this area would repay research. We may even need a 'tropical' genotype of pig with its own dietary specification, and probably do. I come across too many cases of very slow growth in the tropics among genetically-improved, very healthy pigs on clean farms, which nevertheless do well in cooler climates.

Figure 3 suggests that a fourth source of information should be consulted – the seedstock (or semen) supplier. This dialogue is best left to the nutritionist; the producer just needs to check on whether he has done so with the breeding company on the producer's behalf.

# THE EFFECT OF IMMUNE DEMAND ON GROWTH RATE AND LYSINE NEEDS

Groundbreaking work in America (1993-97) showed us how much of certain nutrients are 'stolen' away from growth rate where the pig is faced with disease challenge and needs to reinforce its immune shield in order to remain healthy (Table 16).

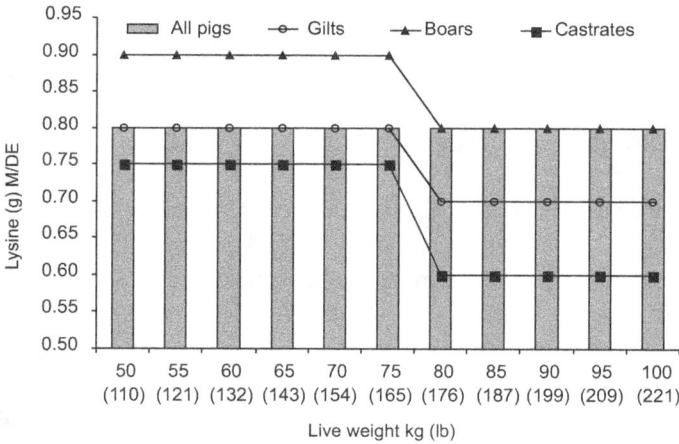

**Figure 3**  Lysine energy ratios from 50-100 kg established by JSR Genetics Research for their own genotype.  (Penny 2000)

While the Iowa State work in Table 14 could be said to have measured the extremes in the early growth stage, the economics from subsequent trials to slaughter-weight have maybe shown a very serious potential shortfall in income where the growing pig's diet was not altered, particularly in the lysine levels on which supporting amino acids are based, to more closely match the immune demand, or lack of it (Table 15).

**Table 16.  The monetary cost of poorly matching diets to immune status (Results averaged from various trials)**

|  | | *Extra overheads, feed and turnaround costs* | |
| --- | --- | --- | --- |
| 1. | Extra days to 100 kg | 18 days more | £ 6.89/pig |
| 2. | [M.T.F. Meat per tonne/food (kg) | 18 kg less | £21.60/tonne PPTE[2]] |
| | *Reduced income pig* | | |
| 3. | From food not utilized | | £ 6.79 [3] |
| 4. | From needlessly higher-priced food | | £ 2.21 [4] |
| | Total reduction in margin/pig (1+3+4) | **£ 8.14** (-35%) | |

This suggests that just feeding one diet to healthy, high performance gene pigs without taking immune status into account could cost the producer at least £8/pig

Notes:   1.   At 24p/day overhead-costs/pig.
2.   PPTE = Price per Tonne Equivalent.  This figure reveals that by getting it wrong the reduced performance is equivalent to a rise in feed costs of £21.60/tonne, or about 12%.
3.   At 3.18 pigs/tonne
4.   At £6.60/tonne more expensive.
UK Costings at March 2010

# HOW CAN THE FARMER SOLVE THIS
# SEEMINGLY INSURMOUNTABLE PROBLEM?

## Portable scanning is now feasible

Let me explain. After all, how can the farm staff tell how high or how low is the current immune shield of the pigs they are looking at? Apart from intelligent guesswork that is - and a wrong guess could make the performance situation worse. Challenge Feeding goes a reasonable way towards solving this uncertainty, but the cost up until now of using muscle-scanning equipment to give a satisfactory answer in the form of a lean growth curve measured from a sample of the pigs present to send to the nutritionist (which the American work suggests is a good indicator of immune status) has been a real problem    This is because the height and shape of the lean curve being laid down during its growth reflects the pigs use of nutrients which it has to divert automatically from forming lean meat into the process of building up its immune shield sufficiently robustly to protect itself from the intensity of disease challenge it is experiencing.

However one equipment firm now have on the market a portable, hand-held scanner which can do this lean assessment task sufficiently well for the nutritionist to assess a likely lean accretion curve, and  then design a farm-specific diet to fit it.  Thus the nutrients supplied to the pig per day can now follow the immune status of the pigs quite closely - in any case to match it more closely to what has been done up to now, and we saw from Table 15 that by ignoring immune status this could be costing one-third of the potential gross margin in 2011, as this book goes to press.

This future development fits in well to the changes in the way pig feed is marketed, because many feed companies are already working on farm-specific diets and have shifted away from selling branded products on a price list to contract manufacture of the farmer's own feeds . Also, sudden farm-specific dietary changes which the nutritionist may advise are no problem, now that so many of their feeder-producer customers have CWF pipeline feeding equipment allowing any changes to be effected at the farm mixer within seconds, not days.
Unlike the sophisticated body-scanning equipment needed previously and developed for human use in hospitals,  this smaller device can be available to the farmer or the former feed salesperson to collect the data regularly at a fraction of the cost and email it to the nutritionist, so to bring challenge feeding back into the realms of practicability.

Toplis has shown how the genotype's effect on growth rate is magnified when the immune demand is high or low (Table 17).

**Table 17.** **The effect of genotype on growth in response to immune  activation**

| Lean Potential | Daily gain, g, to slaughter Disease Challenge | |
| | Low | High |
| --- | --- | --- |
| Low | 680 | **599** |
| High | **826** | 625 |

Source : Toplis (1999)
British genetically-improved pigs tend to be in the bold sectors, thus we have most to gain.

The very considerable differences in lysine needs throughout the growth period, influenced by high or low immune status, is again illustrated in Figure 4.

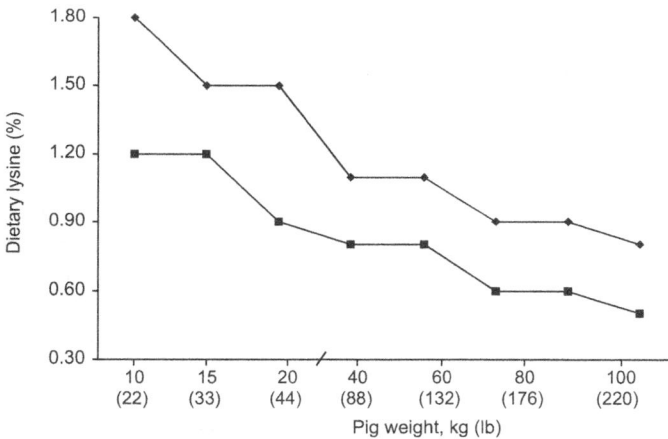

**Figure 4** Dietary lysine concentrations to fit immune status.  Dietary lysine concentrations to optimise efficiency of feed utilisation in pigs with a low or high level immune system activation. Data derived from castrates with a moderate genetic capacity. (Adapted from Williams *et al.*, 1997)

# HOW CAN THE PRODUCER ESTABLISH HIS IMMUNE STATUS ?

Having recognized that tailoring the diet to suit the immune demand of growing pigs is likely to increase profitability substantially, if you can't manage what you can't measure, then how does the farmer measure immune status? Do his pigs need a high protective wall or a low one? If a low one, how much of a better diet do they need? If a high one, how much can lysine be reduced so as not to waste it by using energy to excrete the surplus lysine (de-amination) and so go on to unbalance the correct lysine:energy ratio and lower growth rate still further?

There are three ways of solving the problem. I hasten to add that we are at the start of 'something big' here, and that any suggestions at this early stage are tentative. But the problem is so important to profit that we ***must*** address it. We need to try the suggestions out and see how they work.

1. **Serology**. This method uses the vet, doing routine blood tests, to determine the level of disease present. Three snags. First, even state-of-the-art serology only covers certain diseases, and so may not include the one doing most of the challenging. Second it is expensive. Third, it could be time-consuming.

   By all means use the vet to disease-profile your herd, especially the breeding herd, to establish correct protocols to prevent disease, but for determining the degree of disease challenge – this is probably not yet an option.

2. **Challenge or test feeding**. Whatever the size of the herd, take 50 pigs and feed them a 'non-dietary-limiting' feed, designed by your nutritionist. Record growth and take periodic lean-scan measurements from the end of the nursery period to slaughter and record carcase measurements after slaughter. This will measure their level of lean accretion and enable the nutritionist to design a diet based on the assumed performance of the whole herd (or section of the farm where the groups are housed) thus automatically taking into account both the disease status as well as the genetic type used.

   If the environment is radically different within the herd, select 50 pigs from each of the different environments.

   The snag in the past lay in the expensive deep-scanning equipment needed to measure lean deposition, so it was then really the province of a feed manufacturer who sells his food on a herd-specific basis – a marketing strategy for the future, though a few firms are already trying it in different ways.

   Recently, a much cheaper if simplified version looks as if it will do the job for the farmer himself. At the time of writing the jury is still out on this, but it needs progressing.

3. **Measure growth rate**. Recording daily gain accurately, while in theory less accurate than Challenge Feeding, seems from initial reports to be worth trying as there may be a simple correlation (linkage) between immune status and growth rate. Of course, the snag here is that other things besides immune demand (environment, stress, appetite, wet-feeding, the design of the diet itself, *etc.*) can alter speed of growth, but this possibility is so simple and farmer-friendly that is must be explored.

# ACHIEVING A LOW IMMUNE STATUS
# A CHECKLIST

A low immune status - in the growing/finishing pig - is a <u>good</u>, not a bad thing. A low immune threshold suggests that the pig is not being overly challenged by disease or stress, so it can divert more nutrients into growth, not fortifying it's immune shield.

| | *Action* | *Further information* |
|---|---|---|
| ✓ | **Minimize Buying in of Stock.** | *Weaner gilts* |
| ✓ | Keep visitors away. | |
| ✓ | **Adopt AIAO and batch rearing.** | *See relevant section* |
| ✓ | Establish strict reliable vehicle sanitation and driver discipline. | *Biosecurity section* |
| ✓ | Establish farm delivery and collection areas well away from pigs. | |
| ✓ | *Never* allow a knacker/casualty disposal firm to cross the farm boundary. | |
| ✓ | Incinerate all casualties. | |
| ✓ | **Put in place a complete biosecurity system.** | *Biosecurity section* |
| ✓ | Appoint vermin control/bird control stockpeople with responsibility for these areas. | |
| ✓ | Realise the value of routine 'fogging' to keep respiratory disease as low as possible. | *Biosecurity section* |
| ✓ | **Induce an adequately-long and well-planned induction protocol for any new breeding stock.** | *Empty Day section* |
| ✓ | Avoid a 'needle-happy' vet. | |
| ✓ | Use a veterinarian experienced in the techniques of naturally-acquired immune protection. | *Using a vet effectively section* |
| ✓ | Have isolation facilities, preferably off-farm, for all new stock. | *Choosing a gilt section* |
| ✓ | **Do frequent stress audits.** Stress lowers resistance | *Stress section* |

| **Action** | **Further information** |
|---|---|
| ✓ Realise how important it is to wash away looseness and scour immediately it is noticed. | *Pen Soiling section* |
| ✓ Check all diets are adequately provided with zinc, especially organic zinc. | |
| ✓ Get your ventilation checked over and then establish a monitoring procedure to ensure fans, etc. are kept up to scratch. | *Ventilation Section* (See my previous 'Pig Production Problems' textbook, pps 483-553) |
| ✓ Does your slurry drainage flow *out of* the unit to the nearest exit, or through the unit? | |
| ✓ Farrowing house and nursery slurry pits also need sanitation when AIAO is practised. | |
| ✓ Site dung heaps off-perimeter and cover them. | |
| ✓ Have a clean and tidy rest room. | |
| ✓ Never allow farm staff to consume pork/meat products on-farm. | |
| ✓ Replenish foot dips frequently and use the correct disinfectant. | *Biosecurity Section* (See the plan on page 196) |
| ✓ Have a laundry system for clothes and washing kit. | |
| ✓ On outdoor units, the match is a vital disinfectant. | |
| ✓ On outdoor units, beware overuse of land, *e.g.* duplicating hut runs for weaners. | |
| ✓ If you have more than one farm unit, do not exchange tools/vehicles etc., and change *your* clothes when visiting. | |
| ✓ Monitor space allowances frequently to avoid over-crowding creeping up on you. | *Stocking density* |
| ✓ Keep domestic and farm animals out especially sheep and chickens. | |
| ✓ Cats and dogs should be kept inside the unit and not encouraged to roam adjacent fields. | |

# THE POST WEANING CHECK TO GROWTH

If we 'abrupt wean' between 16 days and 28 days all the weaners will slow up to a certain extent in growth rate. Post-weaning nutrition is primarily involved. Disease, poor management and incorrect housing are also concerned but the **primary cause is nutritional**.

This is covered in the section on 'Problems at Weaning – Nutrition' where several checklists will be found.

# COMPENSATORY GROWTH

One eminent pig researcher has stated …

*"There is no such thing as compensatory growth. It is the last refuge of producers who have not managed pigs properly. Experiments have shown that there is no regaining of lost time and no improvement in growth efficiency in piglets whose growth has been hampered."*

My field experience leads me to suggest modifying these forthright opinions! I don't disagree with the 'last refuge' statement but I have found on several farms that if the postweaning check is not too severe, say around 5 days of measurable growth reduction compared to other weaners, these pigs can finish within a day of the less affected pigs at slaughter weight. I know because I weighed them all on the same day!

This can only mean that in this (and other cases reported to me by clients) the pigs *did* seem to catch up in liveweight terms. Certainly it does not seem to happen often. How often I know not.

However, when the producer and I examined the killing-out percentages of the respective 'checked' and 'non-checked' pigs at weaning, the processor's returns showed, without fail, a reduction in the KO% of the checked pigs compared to the unchecked (or less-checked) pigs. The KO% was down between 0.2% and 1.06% suggesting that in *liveweight terms* the (moderately) checked pigs may indeed catch up, but in deadweight terms (which is the measurement which matters in profit terms) they do not do so? *Lean gain* doesn't seem to catch up or compensate - which is the result which mattered to us, and of course, matters just as much to you as a producer.

Whether there was any difference in liveweight or deadweight feed efficiency I don't know, as it was not possible to measure it under farm conditions, but we were meticulous over our weighing in the week after weaning and at slaughter, and the elapsed days were identical between checked and non-checked pigs.

The fact that the checked pigs had less dressed-out meat suggests to me that their FCR would be worse, as food into lean in weight terms is a good bargain, meat containing as it does so much water.

All this may be semantics as the end result in profit or income terms is the same, but the subject needs airing, I guess. Surely the correct statement is that "*compensatory lean gain is a myth*" not "*compensatory growth is a myth*".

# 16

# A NEW LOOK AT FOOD
# CONVERSION RATIO

## TERMS

Food Conversion Ratio (FCR) Food Conversion Efficiency (FCE)

Liveweight Food Conversion (LFC) *is the number of kilograms of food required to produce one kg of liveweight, e.g. kg feed per kg liveweight gain (or lb feed per lb liveweight gain, USA).*

Deadweight Food Conversion (DFC) *is the number of kilograms of food required to produce one kg of dressed carcase weight i.e. whole carcase less head, selected internal organs and digesta, and in the case of entires, the sexual organs.*

Lean Tissue Food Conversion (LTFC) *applies the same principle to* lean tissue *formation as distinct from deadweight, and is used principally in research.*

## SUGGESTED TARGETS FOR
## GROWING/FINISHING (TO 105kg)

Note: Live FCR.
Targets for ad lib fed genetically-improved pigs kept within
their LCT/ECT temperature zones

| Age (weeks) | Target FCR in that week | Overall, from 3 week weaning (6.5 kg) | Action level |
|---|---|---|---|
| 10 (30 kg) | 1.70 | 1.30 | In the present |
| 11 | 1.80 | 1.40 | economic climate a |
| 12 | 1.90 | 1.50 | 10% worsening of |
| 13 | 2.00 | 1.60 | the overall FCR |
| 14* | 2.10 | 1.70 | (30-100 kg) |
| 15* | 2.25 | 1.80 | figure opposite could |
| 16 | 2.30 | 1.90 | be considered an |
| 17 | 2.35 | 2.00 | action level *i.e.* |
| 18 | 2.40 | 2.10 | more than 2.5:1 |
| 19 | 2.45 | 2.20 | (6-100 kg) |
| 20 | 2.50 | 2.25 | or over 2.2:1 |
| 21 (100 kg) | 2.55 | 2.30 | (6-65 kg) |
| 22 (105 kg) | 2.60 | 2.35 | |

Overall FCR 2.35:1 (30-105kg)

* Assumes a change of house or pen-size in these weeks

# SOME THOUGHTS ABOUT
# FOOD CONVERSION EFFICIENCY

• Food conversion worsens as the pig ages – gets heavier – due to older pigs needing more food for body maintenance (Table 1).

**Table 1. Change in feed conversion with liveweight**

|  | Feed consumed/day | Used for maintenance | |
|---|---|---|---|
|  | (kg) | (kg) | % |
| 60 kg pig | 1.75 | 0.7 | 40% |
| 90 kg pig | 2.50 | 1.10 | 44% |
| 120 kg pig | 2.9 | 1.35 | 47% |

• The effect of higher lean tissue growth deposition due to generations of genetic improvement is changing the traditional profile of the FCR graph from more of a 'V' to more of an 'L' shape. This means that the producer has more scope for *ad-lib* feeding to higher liveweights before grading suffers.

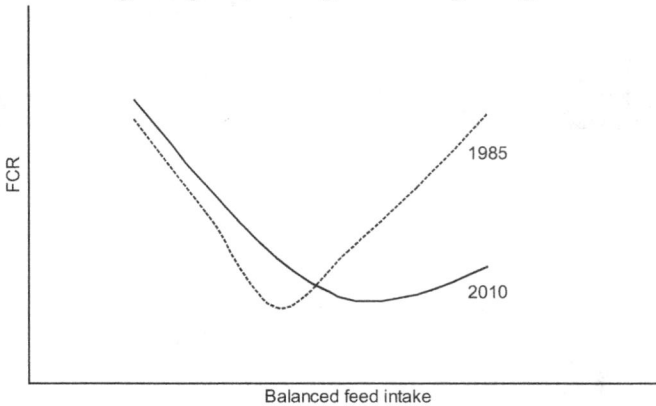

**Figure 1.** How FCR is changing on-farm

• In Figure 2 the longer, flatter shape to the FCR line also means a wider range of feed given is possible (A) over which little change in FCR is apparent compared to less improved strains (B).

• While FCR is a very important measurement, it must be taken into account along with daily gain and carcase grading – all three affect profitability.

Nutritionists now have predictive computer models, which *if given accurate information* can put these three important factors into econometric perspective.

This means that a worsening FCR must always be referred to a competent pig nutritionist.

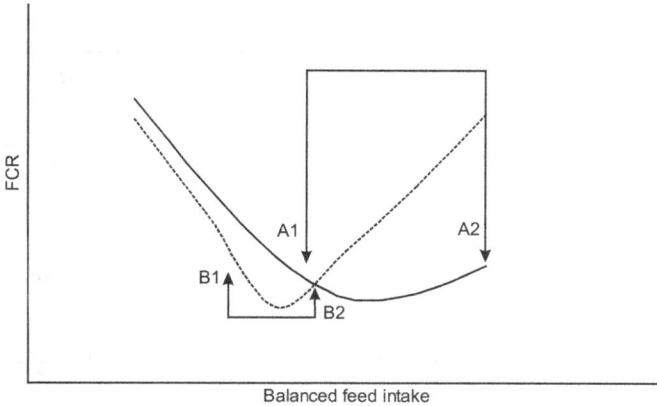

**Figure 2.** Perceptible changes in FCR allied to feed intake. Note that under modern genetic conditions, the difference in FCR variation is over a much wider spread of balanced feed intake (A1-A2) than over the same FCR variation in the past (B1-B2)

- One problem with FCR is that it is difficult to measure accurately enough on a busy farm (Table 2). While this table is over 25 years old now, recent analysis of FCR complaints concerning feed suggest that any error is only a little less, if at all. 7 recent investigations (1999-2008) revealed 0.2 LFCR differences above or below what was claimed. This is equivalent to a 18% rise or fall in the current price of feed.

- **If FCR is poorly estimated under real conditions, the MTF measure (Saleable Meat Produced Per Tonne of Food Fed) is a better yardstick – it is easier to measure and thus likely to be more accurate.**

This is fully described in the Chapter on New Terminology. Use both measurements if you wish, but you will soon find yourself preferring MTF if you do, leaving FCR to the research worker who can measure it accurately.

**Table 2. Farmers calculated and actual FCR's taken from 5 complaints over poor performance (1975 - 1977)**

| Farm | Weight range (kg) | Farmer's estimated FCR | Actual FCR based on careful measurements on-farm | Likely reason for error |
|------|-------------------|------------------------|--------------------------------------------------|-------------------------|
| 1 | 6-28 | 2.9 | 2.71 | Input food not weighed |
| 2 | 20-91 | 3.2 | 2.86 | Poor recording |
| 3 | 30-90 | 2.9 | 2.81 | Mistake over input batches |
| 4 | 25-86 | 2.6 | 2.92 | Guesswork (!) |
| 5 | 30-64 | 2.6 | 2.45 | Poor recording. |

Average error of 0.22 FCR over 48 kg (106 lb) LWG, an 8% error
Source : RHM Agriculture (Unpublished) 1977

**Table 3. Example of feeder pig performances, UK 2010 (from AgroSoft 2010)**

|  | Weight put on (kg) | FCR | Food eaten per pig (kg) | Pigs/ tonne | KO% | Saleable meat per pig (kg) | MTF* (kg) |
|---|---|---|---|---|---|---|---|
| Top third | 97.3 | 2.56 | 240 | 4.17 | 77. 0 | 78.0 | 325 |
| Average | 91.3 | 2.43 | 222 | 4.51 | 76.2 | 75.1 | 396 |
| Bottom third | 87.3 | 2.40 | 210 | 4.76 | 72.7 | 68.9 | 328 |

Performance rankings...... Top third 2   Average 1.   Bottom third 3

Extrapolated from the above figures....
Econometrics (Prices/tonne feed:Top £168.71; Average £194 83;Bottom £212.56).

|  | Cost of food per pig (kg) | Expressed per per tonne fed (A) | Income/tonne/food from MTF* (B) | Earnings per tonne fed (B-A) |
|---|---|---|---|---|
| Top third | £34.32 | £143.11 | £ 390 | £247 |
| Average | £37.07 | £167.19 | £407 | £240 |
| Bottom third | £41.16 | £195.92 | £394 | £198 |

Economic rankings.........Top third 1   Average 2      Bottom third 3.
Pig income based on a pig price of £1.20 deadweight (i.e..saleable meat).

* MTF (saleable) Meat per Tonne of food fed (av. 7.5 -98.2kg) from weaning to slaughter.
Comment:

1.  Notice how using FCR as a yardstick can give a different impression (as well as different proportional differences) to the use of MTF, a figure which is more profit-based than FCR and in my experience is less subject to on-farm collection/ recording errors.

2.  The differences per tonne of feed may not look dramatic, but for a producer shipping 5000 pigs a year, the differences are considerable.

3.  These figures by an acknowledged recording system are based on daily gains (g/day) of Top Third 669, Average 637 and Bottom third 598, between typically 7.5 to  98 kg liveweight.

    (These ADGs will look low to many producers across the world due, in the author's opinion, to the  reluctance/inability of many British producers to update their growing finishing accommodation after a succession of punishing economic crises and the reluctance of many banks to lend).

4.  The above figures do not include differences in overheads which are largely dependent on daily gains, with the bottom third producers pigs racking up 6 days more overheads/ pig than the top third. Again with 5000 pigs to ship per year this is another 30,000 pig/ days to pay for.  Taking overheads as low as £0.24/pig/day, then for one year's trading this is a £7200 extra cost burden, equivalent to foregoing any income at all on 205 finished pigs (33.000 ÷146 days at 598g/day) for a bottom third producer compared to one in the top third.

# FACTORS WHICH AFFECT
# FOOD CONVERSION RATIO (FCR)

**IMPROVE IT**

- Younger pigs
- Genetically superior strains especially terminal boars /AI
- Nutrient-dense food
- Correct amino-acid/energy balance
- Adequate clean water
- Pigs within their LCT/ECT comfort zone
- Appetite – as food intake rises in fast-growing pigs less food (in % terms) is used for maintenance
- Freedom from stress of many forms
- Sex – entires have better FCR
- Wet feeding in correct environmental conditions
- Well made pellets over meal
- Low immune demand
- Segregation and batch rearing
- More available trace elements in feed
- Good biosecurity protocol
- Growth enhancers

**WORSEN IT**

- Age
- As fat deposition increases
- Overfeeding energy
- Overfeeding protein
- Too few or unbalanced amino-acids
- Castration
- Overstocking & aggression
- *Ad lib* feeding too long in later growth period when appetite is high
- Pigs of low genetic merit (lean tissue deposition potential poor)
- Cold (seasonal changes a clue). Also cold draughts
- Too dry or too humid air
- Ill health
- High immune demand
- Feed wastage
- Too dilute wet feed/by-product feeds
- Particle size
- Incorrect feed
- Floor feeding
- Dusty feed
- High gas levels
- Continuous occupation of pens
- Poor cleaning & disinfection
- Poor control of self-feeder and wet/dry feeder settings
- Mycotoxins
- Worms/parasites

# DEALING WITH AN APPARENT FCR PROBLEM

## A PRELIMINARY CHECKLIST

To save the embarrassment of the adviser finding your figures/assumptions are incorrect, double-check the FCR situation. Common errors discovered on-farm have been…

✓   Input food dockets not matching with output pig numbers or weight.

Note: Using MTF in place of FCR removes the need to weigh any pigs or food.

✓   Bulk bin residues not taken into account.

✓   Surges or dips in pig-flow not taken account of in monthly output averages.

✓   Double-counting of pigs and food deliveries.

✓   Inaccurate recording by staff – cross-check yourself just to be sure. This includes weights of pigs and foods.

✓   If feeding by volume *e.g.* when pipeline-feeding, failing to take account of feed bulk density – see the relevant paragraphs in this chapter.

✓   Computer inputting errors.

✓   Failure to record mortality removals from the tally.

## ARE YOU FEEDING WHAT YOU THINK YOU ARE?

Investigations into complaints of poor FCR have revealed that one in six producers are simply using the wrong – or a less beneficial – diet! At the time of writing most producers are feeding on the '3-step' principle – Post Weaner (or Starter) / Grower / Finisher – which in future will eventually change to a 5 Phase or even Multiphase system under tighter control *e.g.* automatic dry feeding or fully wet feeding, with pigs raised in batches *i.e.* in one particular environment.

Until that situation occurs it is vital to check that the correct diet is chosen for the class of stock kept, and their age or weight at transfer from one environment to another.

With these new high-lean gain genes, especially beyond 100 kg live, it is most important to choose the correct ileal lysine to Net Energy ratios and daily intakes of all other nutrients.

In an ideal world there are four advisors who should be consulted, two of whom should have on-the-spot knowledge of the farm.

<u>*The Geneticist*</u> : Your breeding stock supplier should be able to give you target nutrient requirements for his genetic lines. This must be used as a basis for dietary design and daily intake (appetite is a critical factor these days when *ad lib* feeding).

But this basic data needs further qualification from:–

<u>*The Nutritionist*</u> : Pig producers do not consult the person designing their feeds anything like frequently enough. Four times a year is minimal. Only the nutritionist knows the analysis of the raw materials he adds to his feeds and these change. If on-farm mixing or custom/contract mixing by the feed manufacturer is practised he needs to know, from you, what the critical analyses of the on-farm ingredients are, so at least 2-3% of feed costs must be invested by the farmer in getting the less-established ingredients (especially cereals) analysed.

**Most important too, he needs to know your typical <u>daily feed intake</u>/pig.**

The nutritionist doesn't need to visit the farm but needs far more information than he is currently being given. I estimate that the average nutritionist, because he is working on assumptions – and this includes overages to protect performance – costs his clients 0.15 on their FCR (6-100 kg) in so doing. This is not his fault – *it happens because he hasn't sufficient dietary analysis information to practise the precision nutrition of which he is capable and uses safety margins to ensure his advice achieves results*.

<u>*The Veterinarian*</u> : The effect of disease challenge on FCR is underestimated. In my experience it can be as much as 20-30% and this is confirmed by workers like Stahly (1997) (25%). The concept of the drain on nutrients to build a protective immune wall in healthy but challenged pigs is discussed in the checklist section on Immunity. So the presence of an experienced pig veterinarian reporting back to the feed design team at least twice a year on the likely immune status of the herd is vital, as it rises and falls across time. Where this has been done effectively gross margin has risen by £8/pig for an extra veterinary cost of £1.30/pig, including the tests needed – an REO of 6:1.

<u>*The Environmentalist*</u> : He is the second expert who should visit any pig farm once a year.

*First* : To analyse any environmental deficiencies, tell you what they are costing you and to suggest cost-effective improvements so you can prioritize your capital expenditure needs.

Ventilation is the most critical area needing advice like this.

**Second :** To revisit periodically and review progress, suggesting short term adjustments if needs be.

The effect of substandard environment on the FCR of the growing/finishing pig is considerable, and like the problem of matching nutrient intake to immune challenge, is generally underestimated. This is covered in the Growth Rate section of this book, but Tables 4 and 5 give two examples of how poor environment affects FCR.

Table 4. **The cost of inadequate temperature**

*TOO COLD : Pigs kept 1°C (2°F) below LCT (100 sows' progeny)*

|  | *Extra feed/day* | *Extra feed/year* | *Extra days to slaughter* |
|---|---|---|---|
| Weaners 6 to 20 kg | 8 g | 680 kg | 2 |
| Growers 20 to 100 kg | 25 g | 6000 kg | 3.5 |
| Total penalty (if not increased) | Food conversion worsens by 0.03:1 | 6.68 tonnes for the herd | 5.5 days overheads for each pig |

Many pigs are kept 3 to 4°C below LCT, especially at night, costing at least 0.1 higher FCR           (6-100 kg)
Poor control of temperature costs typically 8 kg of lean meat per tonne of feed (8 kg MTF)          (6-100 kg)

Table 5. **The cost of poor ventilation on pigs from 6-100 kg**

| *Depression re:* | *Not achieving UCT or LCT* | *Appetite* | *Stress** | *Health* | *Total* |
|---|---|---|---|---|---|
| Food conversion | 0.1 | 0.1 | 0.05? | 0.20 | 0.4 worse |
| Daily gain, g/day | 30 | 20 | 10? | 50 | 110 g/day slower |

\* *Difficult to measure, minimal estimate only. Poor ventilation alone can cost 30 kg MTF (6-100 kg) i.e. lean meat per tonne feed.*

# REGULAR VISITS ESSENTIAL

Like a veterinarian who can improve MTF in the growout pigs by around 10 kg by minimizing the drag on FCR through better health and advice on more closely matching dietary design to immune needs, the environmentalist can set in train improvements which can provide *another* 15 kg MTF.

Thus at current feed and pig prices, an extra 25 kg of MTF is equivalent to a reduction in all nursery and grower-finisher feed price of 20%. Most of these savings are in lower FCRs.

## No, it doesn't cost too much!

Many of my clients protest that to call in these experts costs a fortune. Maybe it looks like that, but 'Think-on' as they say in Yorkshire. – a hard-headed lot of pig farmers are Yorkshiremen!

My records from before-and-after annual visits suggest that regular veterinary attention throughout the weaning to finishing phase can provide 10 kg more MTF. At the time of writing and a pig price of £1.40/kg dead this is worth £14.0 tonne of feed.

At 4.5 pigs/tonne this is £ 3.11/pig - a producer selling 1000 finishers/ year would have an increased income of £3110.

Plenty left over to pay for extra veterinary attention in the grow-out herd.

The ventilation engineer's payback is even better. Here the income increase is nearer £6.39/pig - £6390.

Now I'm not necessarily saying that the ventilationist, with all his measuring equipment and computer models, is twice as useful than the vet in the grow-out house. But thermodynamics are very <u>measurable,</u> and resultant adjustment can give a rapid response to improvements in FCR/MTF compared to the more difficult and lengthy task the veterinarian has in getting disease costs down in the same circumstances.

While a regular supervisory veterinary visit is requested by at least 25% of typical producers, the figure for an environmentalist (to monitor thermodynamics and air movement) world-wide is less than 1%. **In my view a serious oversight.**

## A PLAN OF ACTION

After double-checking the degree of the problem and the facts from which it is calculated, ***contact the qualified nutritionist who designed your feed***. Beware of itinerant feed salespeople and 'specialists'; use them solely as communication links. If you have FCR problems, you need specific scientific advice based on fact and calculations, not 'opinion' or 'sales experience'.

# A FOOD CONVERSION CHECKLIST

**The 10 Essentials : Provide the nutritionist with the following :**

✓    *Type of diet fed plus expected analyses.*
He may – or should – ask for check-analyses.

✓    *Type and method of feeding system used.*

✓    *Proof of amount fed.* The nutritionist works on **expected daily nutrient intake details**, so your MUST provide him with this. Also the bulk density of your feed could be critical, especially if wet feeding. See the section on *bulk density* in this chapter.

✓    *Weight, numbers and ages of the pigs fed* and if segregated by sex.

✓    *Your current daily dietary changeover points* and whether associated with other changes, like housing.

✓    *The genotype(s) used.* This information must come from your breeder/supplier and/or AI source.  Only 20% contact their breeder when an FCR problem occurs.

✓    *The health status of the pigs*.  This information must come from your veterinarian.

✓    *The environmental details.*  An environmentalist would supply these data, but in his absence, the nutritionist needs to have an idea from you of:–

- Floor type; bedded, concrete, slats, etc.
- Wetness of floor or bedding.
- Stocking density allied to weight/age.
- Roof and wall insulation.
    - Ventilation adequacy *i.e.* air changes/hr, air speed over the pig's back at the hottest/coldest part of the 24 hrs, air flow directional plan (cross section), typical temperature range. You cannot see air movement. Use smoke generators or tubes to do this.
    - In summer, cooling facilities available.
    - In winter, likelihood of draughts, especially nocturnal. (Be honest about this; draughts affect overall FCR considerably).
- Likelihood of dust.

✓    *The degree of biosecurity.*  This means the method of cleaning, disinfection, mycotoxin and vermin control.  Again, assessment can be made from expert biosecurity firms like DuPont and a veterinarian is a help here too.

✓    *Type of feed receptacles, positioning and use*.

✓    *Estimate of wasted food.*

# THE FEED'S TO BLAME !

90% of FCR problems will be either wholly or partly in these 'non-feed' areas. The other 10%? Regrettably it does have to be poor quality feed ingredients or spoilation in manufacture. Many farmers, when complaining about poor performance, suspect it to be the other way round. It is not; the 90:10 ratio is about right, in my experience, both inside and outside the feed trade.

Some other factors which can impinge on FCR are:

## Deamination

Deamination occurs when the pig has eaten an excess of amino-acids in relation to energy intake and so cannot use all of them. They therefore have to be reprocessed by the pig to be excreted as nitrogen and this takes up food energy. The resultant alteration to the amino acid:energy ratio worsens food conversion efficiency to a surprising extent in pigs which otherwise look healthy enough.

Thus it is essential to get the proportion of essential available amino acids to net or digestible energy level correct according to the pig's age and weight, otherwise FCR suffers.

This is a nutritionist's job. Your job is to ensure you *are* feeding what he thinks you are/tells you to!

## Bulk density of feed

As this can vary substantially in both dry and especially wet feeding, FCR can be affected.

Table 6. **Does feed vary in bulk density ?**

| Product | Mean density (g per l) | Range between deliveries |
|---|---|---|
| Pig breeding and finishing pellets | 646 | 616 to 700 |
| Pig breeding and finishing meals | 479 | 456 to 497 |
| Cattle nuts | 611 | 532 to 675 |
| Broiler pellets | 645 | 560 to 696 |
| Layer's meals | 561 | 481 to 728 |

The overriding influences in the case of pressings are the grinding and pelleting conditions: in the case of meals, the choice of raw materials and grinding conditions. (Figures supplied by one feed compounder)

More producers are adopting fully wet-feeding and fail to allow for changes in bulk density. Two vital checks are needed on an ongoing basis, both for liquid and dry (pellet) feeding.

# 1. CHECK THE DRY MATTER CONTENT OF THE LIQUID DILUENT (IF NOT PLAIN WATER)

Both skim milk, but especially whey, can vary considerably from delivery to delivery, with other factory carbohydrate by-products (such as yoghurt) being less of a problem. It is essential to do a simple hygrometer test on each delivery after obtaining from the supplier an expected range of dry matter density, thus securing an average. If this varies by more than 10% batch-on-batch, then errors in predicted or target FCR are likely. In any case, if 10% below expected average, an equivalent price adjustment can be requested for that delivery.

Specific gravity is the weight of a substance compared with the weight of an equal amount of another substance – in this case water. If the specific gravity of water is 1, and if a sample of whey shows an SG of 1.022 this means that whey is 1.022 times heavier than water, the extra weight in the case of whey is primarily the nutrients it contains. **The use of a hygrometer is very simple, rapid and inexpensive.**

Table 7 shows typical specific gravity differences in whey deliveries across a year and around a SG 1.022 norm.

Table 7. **Relationship of specific gravity dry matter and composition of liquid whey (as fed basis)**

| Specific Gravity | Dry Matter of Whey (%) | DE (MJ/kg) | Lysine (g/kg) |
|---|---|---|---|
| 1.027 | 6.5 | 1.07 | 0.65 |
| 1.022 | 5.0 | 0.82 | 0.50 |
| 1.018 | 3.0 | 0.49 | 0.30 |

Liquid whey needs a special balancer solid diet. If the SG of whey drops from 1.022 to 1.018 and a 60 kg pig is fed 7 litres of whey/day, then a typical balancer meal allowance must be increased from 1.6 to 1.75 kg/day otherwise the overall FCR will increase by 0.05.

# 2. BULK DENSITY OF DRY FEED, ESPECIALLY PELLETS

The bulk density of feed ingredients varies considerably – we have often seen this when an empty 15 tonne bulk bin barely holds a 12-tonne bulk delivery!

Table 6 shows by how much individual deliveries of bulk meal (wet feeding) and pellets can vary. 5% variation is common and if the food is fed volumetrically either 5% nutrients are wasted, thus affecting F.C.R. by 5%, or 5% are under-supplied, leading to a 5% reduction in daily gain and perhaps a 4% reduction in FCR.

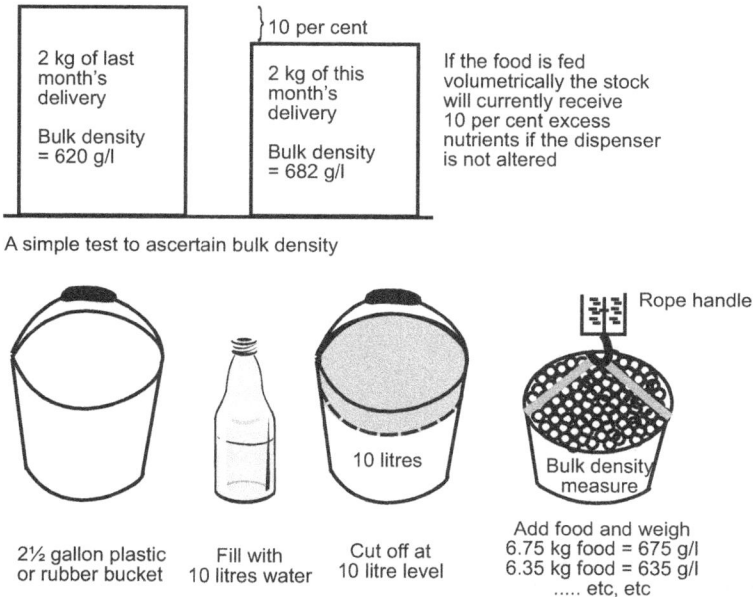

A simple test to ascertain bulk density

**Figure 3** How to check for bulk density of dry feed

## FREE-ACCESS FEEDER OPERATION

Evidence continues to accumulate on the advantages of correctly-operated wet/dry or single-space feeders over free-access dry feeders. From 18 trials surveyed in the late 90's the benefits lay between a 0.15 and 0.28 FCR improvement, mostly due, it is thought, to less waste.

However, when the single-space feeder is not frequently adjusted to meet individual pens of pigs' feeding style, or differences in feed texture, then the drawbacks can be considerable.

## FEEDER SETTINGS

Research done in 1994 (Table 8) suggests that the difference between 'adequate' and 'too little' flow-plate adjustment can be as much as 18 kg MTF (30-100 kg), equivalent to a 14% rise in feed price at the time. The difference in FCR was 0.23, or 6.6% worse, showing that careless or lazy attention to single space feeder settings can completely negate the economic advantage of the wet/dry feeder over the simpler and cheaper dry feeder.

The correct setting depends on the design, so consult the manufacturer for guidance. In general each pen of pigs should be made to 'work' for food, and this extra action does not seem to carry a performance penalty at all. If they don't 'work' for it, the food can be wasted.

**Table 8.  The effect of feed settings on pig performance**

|  | *Low* | *Medium* | *High* |
|---|---|---|---|
| Food intake, kg/day | 1.97 | 2.14) | 2.21 |
| Liveweight gain, g/day | 727 | 797 | 845 |
| Food conversion efficiency of carcass gain | 3.70 | 3.58 | 3.47 |
| Backfat thickness at $P_2$, mm | 10.6 | 11.1 | 12.1 |

From Walker & Morrow (1994)

The effect of poor feeder setting on pig behaviour and pig performance is given in Table 9 also from Walker and Morrow's pioneering work and Table 10, extrapolated from the same research.

**Table 9.  The effect of feeder settings on pig behavior**

|  | *Low* | *Medium* | *High* |
|---|---|---|---|
| N°. of feeder entries/pigs/24 hr | 51.5 | 45.6 | 42.2 |
| Feeding time/pig/24 hr (minutes) | 110 | 78 | 87 |
| Queuing incidents/pig/24 hr | 70 | 45 | 26 |

**Table 10.  The penalty of incorrect feeder adjustment  (pigs 30 – 100 kg)**

| Feeder gap | *Too restricted* | *Ideal* | *Too generous* |
|---|---|---|---|
| Days to 100 kg | 95.3 | 87.88 | 81.3 |
| Food eaten/pig (kg) | 190 | 188 | 207 |
| MTF (kg) | 270 | 279 | 248 |
| Av. backfat at $P_2$ (mm) | 10.4 | 11.0 | 13.2 |

Comment:    I have kept to Walker and Murrow's pioneering work of over 15 years ago as several similar trials on wastage and feed settings in the intervening years have produced  similar results.

All these properly conducted trials confirm how extremely important it is **to get these throat settings right** – whichever make and design of feeder is used. I find it disturbing to have to report that fully half of the grow-out units I am asked to tour have feed hopper management way below the dedication and standard needed so as not to waste considerable amounts of food.

Failing to check feeder settings daily is a global weakness among pig producers and matches their errors over stocking density (see next chapter)

## How much loss of income?

Expressed in MTF terms – as it should be - the correct setting against an overgenerous gap gave 31kg more saleable lean per tonne of feed fed.

At a deadweight price of £1.20/kg this is £37.20 not secured from hoppers which oversupplied feed For the pigs not getting enough, this was 9 kg less lean sold/ tonne of food fed - £10.80 per tonne of food

Why express this way? Because one feed dispenser should suffice, say, 3 batches of 15 pigs each eating about 220 kg feed, or about 10 tonnes of feed put through that feeder each year.

If the trouble is taken to get the throat settings exactly right could be worth as much as a weighted average of £30/tonne reduced income from less meat sold ('weighted,' as my own observations on dozens of farms since 2005 suggest that the error is two-thirds towards oversupply rather than the throats being too restricted) and if each feeder costs £ 150, then the payback would be about 6 months (£150 cost of the feeder divided by the amount saved per year, 10 tonnes at £30/tonne equivalent)

A couple of batches of pigs feeding from properly adjusted hoppers should pay for the feeder itself - an excellent return, with an AIV – see Business Section – of 3 to 11 **Evenness (i.e. little or no weight variation at slaughter) is also affected by poor throat adjustment.**

### An interesting experiment

On a large Eastern bloc farm in 2009 which had 4 big nurseries, I noticed one of them was very well managed and the another, under a different section head fairly new to the game, much less so.

It was obvious that there was a difference in evenness between the two groups of pigs at 30 kg.

Now the owner is a remarkable man. Instead of rushing off to bawl out the errant stockman, he agreed much to my surprise to leave things as they were for just one more batch and try to measure, in the interests of science, any difference in the variation at slaughter between the good and not-so-good nurseries.

This enabled me to put some econometrics to the age-old problem of too wide a close-out window and consequent inefficient use of expensive finishing accommodation, not to mention being obliged to ship-off underweights.

Or being forced to put the laggards into a spare pen elsewhere which inevitably results in fighting and condemnations! He could use this evidence as an object lesson for all of his (132!) workers. Table 11 suggests that the inexperienced nurseryman's casual attention to feeder management was costing my client 11% in underused housing costs alone.

**Table 11.   Effect of careful feeder throat\* supervision on liveweight evenness at shipping.**

|  | *Pigs remaining\*\* at less than shipping weight at shipping (%)* | *Median variation in liveweight* |
|---|---|---|
| Poor adjustment – Less than once weekly? | 87% | 12 kg |
| Daily supervision and     adjustment if required | 36% | 5 kg |

\* Plate feeders
\*\* Once 10% of the batch were estimated to have reached contract weight.
<u>Source</u> :Client information 2009

## So how often should the hoppers be checked?

Each hopper once a day. This is rarely done and I've seen it done in too much of a hurry when it is.

## And what throat settings?

This will depend on <u>trough throat design</u> and the <u>physical properties of the pellets</u>. Special settings are needed after weaning. Depending on the type of feeder (consult the manufacturer for their advice) the following settings are advised…….
<u>Weaning to 7 days after</u>

Allow two thirds of the trough to be covered after a feeding session.
Note my advice (Tables 10-12) on having plenty of feeding space available especially at this critical time.

It is easy to allow too much feed to be on offer at any one time as in this case. Research suggests that this was costing the producer the equivalent of a 17% rise in his growing feed bill.

Small pens are still the norm on many intensive farms, but they need to be wider than they are deep so as not to stint on available space per pig and still achieve the 20% trough unoccupancy level which produces the best performance.
**Stocking density involves adequate eating space/pig as well as floor area/pig.**

Weaning from 7 days to 8 weeks after.

Reduce to one third trough area covered

To slaughter

Never more than one quarter covered, less if possible.

It is here that the 6 to 12% wastage occurs and is where daily attention to the amount of food available pays off handsomely. So too will careful observation of feeding behaviour of each pen at least once a day, especially if some of the less dominant pigs seem to be gathering around the trough, when the throat should be opened a little until the food remaining starts to exceed half the trough surface. This is good stockmanship and once again I repeat - needs TIME to do it well.

## Feeder access

Wherever possible the feed station should be as near as possible to the access passage - if it is easy to get at, it can be attended to frequently.

This is not always possible and I have seen some ingenious long distance 'throat-altering rods' welded by the staff on to pen divisions, operated from the access passage, which get over inaccessibility difficulties. They do seem to get jammed due to congealed feed or dust but attention about once a week sorts it out.

It is a pity that hopper manufacturers don't devote more time to ideas like this, offering it as an 'extra'. When I have raised this matter with them at Expos. they all bring up the question of extra cost, and did not seem to have taken on board the researcher's findings in Table 9 where attention to correct daily settings paid off so handsomely. When the likely economic benefits were shown to them they fell back on the rather lame excuse that "The hoppers should be placed next to the access passage anyway", which is best of course, but in many cases not possible.

## Feeder management

When the feeders are being introduced for the first time the gap should be raised to allow the pellets to cover well over half the plate or trough base.
Come round again in the afternoon and check two things - that the pellets are flowing freely or that they are not flowing too easily so as to encourage wastage. Pellets will tend to bridge in feeders especially if high in fat or slightly dusty, so have a rubber truncheon handy, preferably with a right-angle piece of metal inserted into the other end to facilitate reaching up into the throat gap to clear lower congealment, the rubber portion being used to thump the side of the feeder to dislodge bridging. Again, surely the manufacturers are slow in not having such a useful tool on the market. I used one I made myself out of thick rubber pipe - day in, day out - for 6 years.

## Enough trough space

Fortunately this much argued-about subject seems to have been resolved (Tables 12 and 13) Looking back through the evidence on insufficient feeder space allowance per pig, the range of penalties incurred seem to be about 9 to 23 kg MTF foregone from not allowing the pigs of any age to eat comfortably without stress. This was mostly derived from food conversion rises of between 0.15 to 2.0:1 (30-103 kg) when feeding space was below the advice given below.

We need to allow pigs enough time and space to feed without undue stress. I have used the widths in Tables 12 and 13 for several years and can vouch for them.

Table 12. Recommended feeder space widths

| Pig weight (kg) | Shoulder width (mm) | Feeding space allowance (mm) |
|---|---|---|
| 5 | 110 | 120 |
| 30 | 200 | 220 |
| 50 | 230 | 255 |
| 60 | 250 | 276 |
| 70 | 270 | 300 |
| 80 | 280 | 310 |
| 90 | 290 | 320 |
| 100 | 300 | 330 |
| 110 | 320 | 350 |

**Comment**: So as to accommodate different genetic phenotypes a variation of +/- 4% might be made for pigs with slimmer or heavier forequarters. This table also needs to be viewed in relation to whether the pigs are rationed,i.e.. limit-fed, or ad-lib fed. If rationed with distinct timed feeding episodes an overage is needed. Table 13 below addresses this situation.

Table 13. Trough space according to feeding method

| Pig weight (kg) | Trough length per pig(mm) | |
|---|---|---|
| | Limit fed | Ad-lib. |
| 10 | 130 | 35 |
| 20 | 160 | 40 |
| 50 | 215 | 60 |
| 90 | 260 | 70 |
| 110 | 275 | 75 |

Figures based on those suiting the biggest pig in the pen, not the average.
Source: Young (2006)

## Eating time

Many stockpersons who are interested in their job are also interested in how long it takes pigs to eat their fill, as this affects docility and stress and through that performance, including FCR.

Dr. Gonyou who has researched matters in this area when in Canada, suggests that 80% feeder occupancy during daylight hours probably supports maximum growth rate, thus stockpersons should be ensuring 20% of feeder space is free at any one time.

Eating time is important to avoid overoccupancy at the trough . If a normal eating period during the day was 18 hours and the average eating time for pig in one pen was 60 minutes across the day, then to allow for 20% inoccupancy at a feeder as

Gonyou advises, then the feeder should support 14 pigs at most. Longer eating times of 70 and 80 minutes per day would reduce this to 11 and 9 pigs.

But most stockpeople are much too busy trying to measure eating times, so what this work suggests **is never to have too little feeder space**.

Incidentally I wonder how much of the penalty from overstocking (see Table 2 page 421) is due to insufficient feed intake at the feed trough/hopper? I suspect quite a bit.

Pigs generally will not catch-up by feeding at night unless they are extremely hungry, and we are not talking about that - what happens is that the submissives further down the peck order will just follow the satiated bigger companions off to bed. They will not necessarily be hungry as such but will not have eaten what they should have eaten to use their genetic capability to lay down meat.

## Leave the lights on longer?

If feeder space is below the advised levels above, this doesn't seem to make much difference. Other work shows that the growing pig needs a period of dark to sleep just as we humans do and that leaving lights on all night worsens performance.

## What affects eating time?

Age. Nursery pigs take longer than older pigs to eat a set amount – about 10% slower.

Feed type. Pellets are eaten faster than meal  Hard/brittle pellets and the opposite, dusty feed of any sort, are eaten at least 33% slower and sometimes not at all. Fats unless carefully chosen and kept from going stale/rancid can improve appetite when fresh, but very much the opposite when not.

Taste and smell. For example, pigs are good at detecting mycotoxin presence and rancid fat and will 'mess-about' with suspect feed. Some drugs are unpalatable.

Wet/partly fermented feed will be eaten, when fresh, very much quicker than ad-lib or limit-fed pellets. CWF adherents have to be careful that less dominant pigs get their share, which has stimulated such users to explore special formulae for ad-lib wet feeding. We had a lot of trouble with this when I was with Taymix, the wet feeding pioneers. We found that pigs needed more room at the ad-lib trough as they will approach the trough from different directions. The Danes have since quantified this in providing not less than 35 cm per division compared to the accepted wisdom of allowing 30 cm of width between divisions at a single space feeder. These are usually provided at one feeder to 15 pigs. See-through divisions encourage uptake and stop wasteful lateral nosing and competitive 'I'll have yours' behaviour.

## So....never stint on feeders

Table 14 shows a farm trial on 14 kg pigs I did recently where there was certainly little evidence of Dr Gonyou's 20% inoccupancy in the nursery!

**Table 14. Performance of two batches of 150 pigs from 10 to 25 kg and then to slaughter where an extra nursery feeder was put in when the existing feeder was seen to be periodically crowded**

|  | *Feeder retained* | *Extra feeder put in at 14 kg* |
|---|---|---|
| Daily gain (g/day 10-25 kg.) | 410 | 432 |
| Performance to slaughter (25-106 kg) | | |
| FCR | 2.58 | 2.42 |
| Daily gain (g/day) | 726 | 803 |
| MTF (kg) | 348 | 375 |

Comment: The substantial 27 kg more lean meat sold per tonne of food used by the 147 and 148 pigs/group remaining when sold at slaughter weight would have paid for the whole nursery to be fitted out with spare feeders!

Do **not** stint on feeders!

## 'Competition' Feeders.

These with wet or dry dispensation are particularly useful for younger pigs - indeed due to the success of one of them, this type of feeder is often called a 'transition feeder.

Their success has been due to the inclination for pigs to feed as a group and engage in a degree of competition, which speeds up eating time and encourages intake and thus growth rate, which in the young growing pig is all to the good – it is almost impossible for them to overeat if the diet is correctly designed to meet their lean meat accretion curve and predicted appetite fixed by genetics.

The earlier models (shown in the illustration) made from galvanized or stainless steel, are now much smaller and made largely from metal-reinforced polypropylene. They suit both small pen sizes up to the increasingly popular 'Big Pen' concept, where 200 or more growers are kept in one group. As so many variant feeders of this type are on the market, the manufacturers should be consulted as to number of pigs per feeder and their position in the pen. But most big pen layouts accommodate about 60 pigs per feeder of this design.

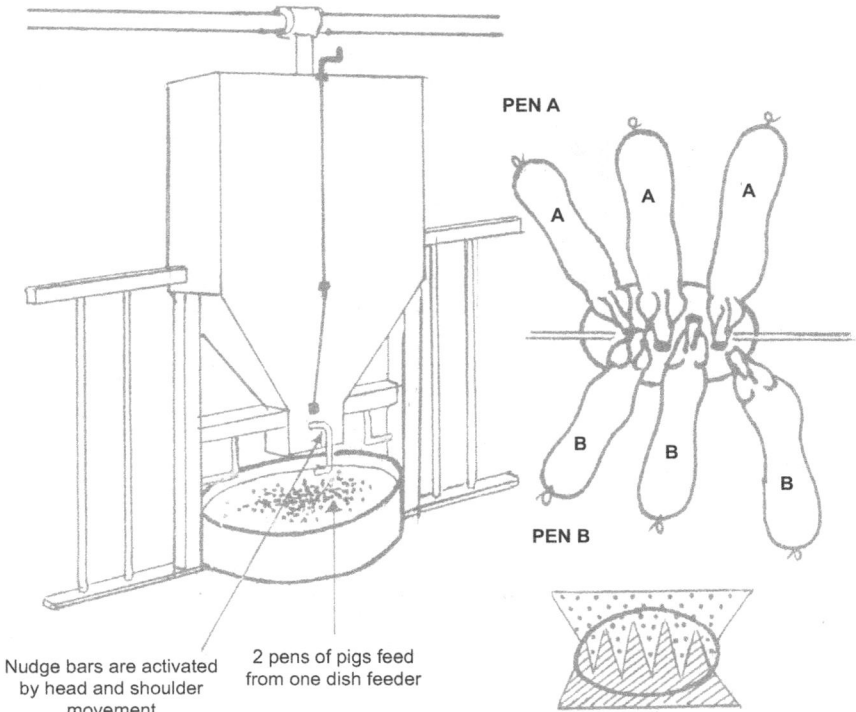

Nudge bars are activated by head and shoulder movement

2 pens of pigs feed from one dish feeder

PEN A

PEN B

Studies of feeding behaviour suggest that even long after weaning, pigs in groups of 25 or more prefer to eat in groups of 4-6.

These groups tend to compete with other groups in the adjacent pen and from each other's uptake area. This stimulates and increases feed intake, shortens feeding time and lowers trough wastage.

The principle of "competition feeding"

### Are competition or plate feeders better?

I have studied over 20 results – farm and manufacturer's - and the average improvement over the conventional 'long' hopper of old was 17% better daily gain to slaughter and one whole FCR point improvement.

Doing a quick MTF exercise against cost (2010) then by replacing all the conventional hoppers would have paid back in 13 months for small pens and 10 months for big pen layouts, including installation costs. Yes, better indeed.

### Dry semi-wet/dampened/dry pellet Competition feeding v. Fully wet(CWF).

Being a confirmed pipe/fully wet feeding man myself, I have compared these results to what I have on record for conversions to wet feeding over the same recent time period. While the performance improvements are not so good as changing from dry food hoppers to fully wet feeding by computer (CWF), the conversion costs of installing both types of competition feeder (dry, or dampened by the pig) are much lower than CWF. I discuss what these might be in the chapter on CWF.

However, the long term advantage of CWF's novel possibilities for both today's circumstances - but especially for the future - make it advisable to study both options carefully when contemplating expanding/updating/ renovating a pig farm. Should these CWF developments I describe in the chapter on CWF materialise (surely it is only a matter of time) then conventional competition feeding - except for the 'transitional'' phase of early post-weaning growth - will never be able to compete.

You must judge for yourself.

Figure 4, from video-recorded data, shows where pigs tend to rest or congregate in a pen with a single space feeder set laterally. Compare this activity pattern with the diagrams in the stocking density section.

## FURTHER READING

For information on other factors which affect food conversion, check out the sections on:–

*   Stocking density.
*   Temperature, air movement and draughts (Growth Rate and Ventilation sections).
*   Dust and gases.
*   Biosecurity (cleaning, disinfection and mycotoxins).

- Stress and stressors.
- The post-weaning check.

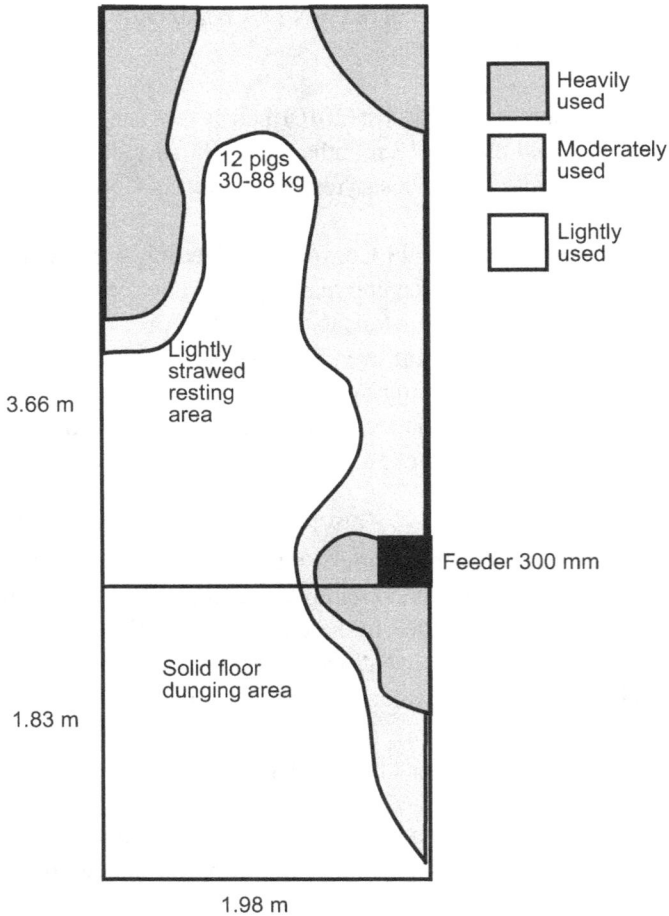

12 pigs
30-88 kg

Heavily
used

Moderately
used

Lightly
used

Lightly
strawed
resting
area

3.66 m

Feeder 300 mm

Solid floor
dunging area

1.83 m

1.98 m

Video'd activity as:

|  | Mean % | Range % |
|---|---|---|
| Asleep/dozing | 4.95 | 20-80 |
| Feeding/drinking | 9.9 | 4 - 18.4 |
| Trying to feed | 2.6 | 0 - 15 |
| Social activity | 30.2 | 6 - 48.5 |
| Fighting/playing | 7.8 | 0 - 25 |

**Figure 4.** Good single space feeder pen

<u>Note:</u> Personally, I would much prefer a 'squarer' shape - but retaining the same propotional pattern.

# REFERENCES

Hall, S. (2010) Personal communication.

Thoday, M. (2010) BPEX (UK) Personal communcation.

Walker, N. and Morrow, A. (1994) Some observations on single space hopper feeders for finishing pigs.  Hillsborough Research Station Report. Quoted in *NAC Pig Unit Review*.

# 17

# MIXING PIGS WITHOUT TEARS

Pigs may have to be mixed so as to facilitate pig flow, reduce stocking density or make maximum economic use of pig space.

## TARGETS - GROWING PIGS

In growing/finishing pigs it is far preferable never to have to mix separate pens together, in view of the antagonism this causes. Howard Hill, one of the world's greatest pig managers, once famously said:

*"Don't mix those damned pigs!"*

## MIXING GROWING PIGS

However this has to be done when batching and matching weaners into or out of the nursery or when a production bulge necessitates some thinning down, and also when laggards need penning together so as to free-up pens in the final few days before the slower growers become heavy enough to be shipped.

The older the pigs, the more disparate their weight, the greater the difference in the new environment, the more fighting is likely to occur, especially when there is insufficient movement space and where feeding and watering is radically different. Good stockmanship and forward planning minimises the need for mixing, which can hold up growth surprisingly. (Tables 1 and 2)

**Table 1.** Enforced mixing of pigs from 10 days before shipping can slow growth rate considerably

*Pigs varying in weight by 10.8 kg (av. wt. 82.1 kg) growing at an average of 760 g/day were mixed from 4 pens into one pen of 15 pigs until average shipping weight of 92.2 kg (contract 90 kg), and were compared to pigs of similar weight which remained in pen groups until shipping.*

|                            | Mixed pen | Unmixed pens |            |
|----------------------------|-----------|--------------|------------|
| Av. daily gain, g          | 696       | 805          | (+13.5 %)  |
| Av. feed intake/day, kg    | 2.05      | 2.21         | (+ 7.2 %)  |
| Av. FCR (25.1 to 92.1 kg)  | 2.94:1    | 2.73:1 (– 7.7 %) |        |

Source : Clients' records

**Comment**: The marked reduction in growth rate of the mixed pigs seems due to the lower-end pigs in the new peck-order eating less, thus pulling the pen average growth rate down badly, probably combined with increased anxiety stress reducing the FCR of the food they did eat.

Alternative options are to leave the laggards in the 4 pens (ensuring they were warm enough). At today's capital costs amortized over 10 years, locking up the growing space thereby was a roughly similar cost at 1.2p/kg deadweight compared to the extra costs of food and overheads from mixing the 4 groups – at 1.1p/kg in this trial. On the other hand, selling the laggards along with the others at 82 kg instead of 92 kg reduced income in this trial by as much as 12p/kg dead and usually would not be an option, as every producer knows.

**Table 2.** Mixing weaners immediately penalises growth ADG 20 days postweaning, g

| Litters kept together, not mixed with other litters | 350 |
|-----------------------------------------------------|-----|
| Litters mixed                                       | 240 |

Varley(2001)

# WHAT DOES MIXING COST?

Pope (1996) reports, compared to unmixed pigs (controls), average daily gain over 2 weeks was 19% less when the pigs were mixed at 84 kg just before sale. Mixed pigs spent 17% less time eating than the controls on the first day and 24% less during the second.

19% less daily gain from 84 kg to sale weight at 96 kg reduces the saleable lean produced/tonne feed (MTF) by 5.45 kg, equivalent to a rise in feed cost over that final period of 4%.

Mixing young pigs – as distinct from the postweaning check to growth where the gut surface is damaged – in the author's experience, does not usually penalise days to slaughter very much, and sometimes not at all, as the mixed pens if remaining healthy can display compensatory *liveweight* growth by slaughter despite what the experts say. However, careful measurements of saleable lean from the processors' returns I have seen where this had occurred (to the producer's relief) suggest, however, that the compensatory growth after the *mixing* check to growth suffers in respect of *lean dressing percent* ranging from 0.5% to 0.72%, average 0.61%. At 5 pigs/tonne feed and a typical 74% dressing percent this is a reduction in MTF of 3.05 kg, equivalent to a 2.2% price rise in all grower finisher food fed from nursery to slaughter.

As Howard Hill says, 'Try not to mix pigs' so . . .

- *Minimise the need to mix pigs in your pig flow plan.*   So can you put them elsewhere – in temporary but satisfactory accommodation? Can you secure a satisfactory market for your lighter pigs? This need not then cause so much of an income loss as expected. I have records of several case-histories where the loss of income from selling lighter pigs averaged only 4p/kg deadweight, but mixing them with pens of heavier pigs and selling them at the normal sale weight cost 5p/kg. This was due to a check to growth in both the newcomers and the indigenous pigs, plus aggression damage (scarring and tailbiting) causing condemnations which accounted for a substantial part of the reduced income.

# CHECKLIST
# IF YOU ARE FORCED TO MIX PIGS

✓    The big error made by most stockpeople is to *fail to allow sufficient space* for submissives to avoid dominants, and not allowing the dominants in each group to get the challenges over with minimal damage and stress to either party.

✓    The answer to this is *not* to assume that if the to-be mixed groups have existed satisfactorily at an acceptable and approved stocking density, that they will be able to so so on being mixed. You must allow them more space, and at least +20% is suggested. This is why smaller groups settle down better than larger groups in relatively constricted pens, such as in a typical flat-decked nursery and why much larger groups can often be mixed with no discernible trouble in straw-based yards where there is much more getaway space, of if needs be 'sleepaway' space for the first night or two. Another example of the BIG PEN concept.

✓    'Sufficient space' certainly involves voiding/exercise and resting areas *but also feeding space* and is certainly one reason why two drinkers (or more) in a pen is always preferable to just one. The majority of aggressive incidents occur at or near feeders and this is why an extra feeder or temporary trough/hopper seems to help.

✓   If practicable, reduce the feed allowance slightly for the to-be-mixed and recipient pigs, say by one-third, from the morning of mixing day.   Do not overdo this as it will increase aggression at mixing.  Just get all the pigs slightly hungry, but not overly so.

✓   If withholding some food prior to mixing, extra trough space is vital as two-thirds of aggressive acts are directed at pigs trying to approach the feeder. Feeder seeking is so important that subordinate pigs can attack dominants. Put in a temporary feeder/trough so the chased-away pigs can gain access to it.

✓   Introduce the pigs just before dusk, certainly by late-afternoon.

✓   Move pigs quickly but *gently*.   Stressed pigs aggress more especially in new surroundings.

✓   Allow ample food once the pigs are moved.

✓   Immediately spray all the pigs with lavatory freshener aerosol or detergent. Sump oil is much too messy. Don't use it unless you want a divorce!

✓   If using bedding, re-bed amply just before mixing.

✓   Place some of the to-be-mixed pigs faeces in the voiding area.

✓   Check drinker flow and adequacy.  Submissives will be last to eat – they tend to tank-up on water as they wait and this itself may induce tailbiting.

✓   Do not batch *too* evenly, especially young pigs out of the nursery. A 4 kg range is better than identical weights (Lean, 1985) for smaller pigs and maybe up to a 6 kg difference for older pigs.  Why so much of a difference? This allows dominance to get established more quickly and the pen settled down sooner. At 21-28 days, a 1 kg range may be good within each batch of, say, 20.

✓   Check you aren't exceeding stocking density – a common error.  In fact, try to put in say, one less pig per pen if practicable, *i.e. destock* by 15%.  Remember mixed pigs need fleeing space.

✓   In typical intensive nurseries small groups (<20) mix better than large ones, and 12 is better still.

✓   In larger groups, *i.e.* 'big pens', now gaining favour, once group size reaches 50-100 animals, the drop off in weight gains after mixing begins to disappear.

✓   Recheck ventilation.  Getting the two groups to 'rub together' in the first night is important, so an ample warm and dry sleeping area helps this natural socialization.  Submissives tend to 'sleep out', get cold and stressed and may tailbite next day.  A temporary cover or lid across the pen has been known to help, but don't place it flat – raise one end by a few centimetres to circulate the air over the sleeping pigs.

✓ Observe, observe, observe. Mixing can awaken a latent persistent tailbiter – often, I find, a small female. Watch for this pig, which must be removed and penned with others displaying a similar trend where, in large units, a sedative can be given. Such pigs rarely convert well, and whether to keep such individuals is debatable.

## MIXING TARGETS - SOWS

Sooner or later, even on the best-run batch-served breeding unit, sows have to be regrouped. No producer needs to be told how much damage such strong and heavy animals can do to each other and every attempt must be made to minimise the need to mix groups together or introduce individuals into established groups. The writer has evidence that on some farms new to grouping skills when the gestation stalls had to be replaced with yarded systems, extra culling rates of from 5% to 11% were directly attributable to fighting and/or vulva biting.

## MIXING SOWS

Because Swedish and British breeders were in the van of moving from individual stalls to sow groups in yards under mandatory welfare legislation imposed in their countries, much valuable information on grouping sows has been accomplished by research farms in both nations, in particular Cambac, MLC (Stotfold), MAFF, ADAS (Terrington) and others in the UK, and the Swedish Pig Research Centre (Svalov). Recently the Danes and the French are contributing to our findings, and now - at last - the North Americans.

My own experience, like many breeders, has been largely by trial-and-error. My check-list below doubtless will be improved upon as the years progress and we all share our experiences, but it will serve to illustrate the current advice available at the time this section was written.

## MIXING SOWS – A GENERAL CHECKLIST

✓ Cater for the least dominant sow in the group. If you get the conditions right for her, not only will she benefit materially but the whole group will settle down more quickly. "Farm for the most timid sow" is as sound advice as Peter English's "Farm for the smallest pig in the litter."

✓   Allow a minimum of 3.5m² per sow.

✓   Adequate fleeing space seems vital. MAFF reports that 75% of aggressive sows will not chase another sow beyond 2.5 metres (range 0-20 metres). While the jury is still out on ideal pen shape, it could be that a larger, narrow pen could allow more fleeing space than the same stocking density in a square or near-square shape. While the latter is generally favoured, if you give them enough space I guess it doesn't matter.

✓   A specially designed mixing pen (Figure 1) may be a useful idea, especially where a gilt has to be introduced into an established group. The pen should be straw-bedded and floors provide adequate grip *i.e.* not a soft, spongy under-surface.

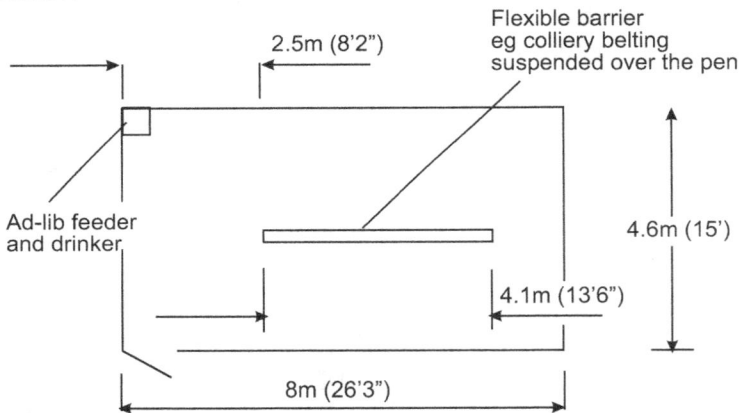

Note: All measurements internal

**Figure 1.** A specially designed mixing pen

✓   The mixing pen can be used for about 24 hours. Aggression is worst in the first 4 hours from introduction, so periodic suspension is wise during this time, say once an hour. (Extrapolated from Kay, 1999.)

✓   Pen size should be matched to age and weight of the animals housed.

✓   A mixing pen adds 6% to the breeding units capital cost; about 0.5% to total production cost/ year (500 sows).

✓   Ad-lib feeding reduces aggression. (Peet, 1993 and ADAS/SAC n.d.)

✓   Competition for food is a significant trigger. If you can, provide sugar beet pulp from single space feeders. (Figure 2).

✓   Both wet-feeding and to a lesser extent, trickle feeding (Hunter, 1998) minimise aggression (Table 3).

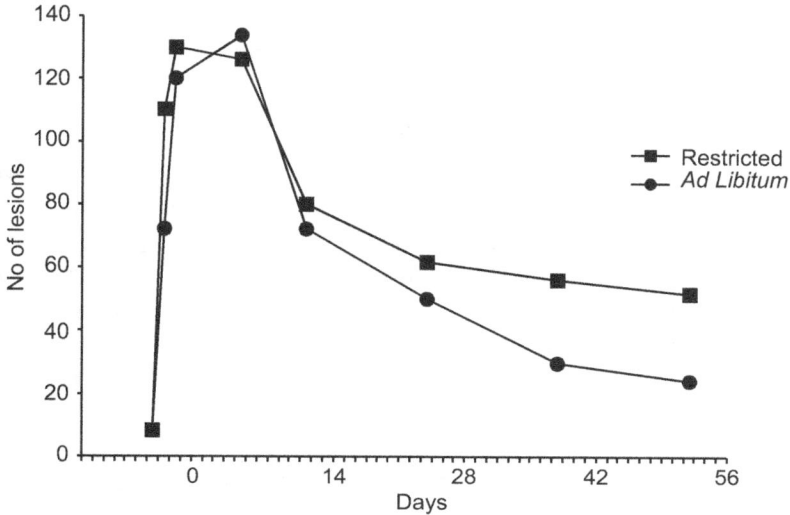

**Figure 2.** Number of skin lesions on gilts fed ad libitum or fed restricted
from an ESF feeder

**Table 3. Percentage sows settled one hour post feed-drop**

| Study number | | | Percentage settled (one hour hour post feed-drop) |
|---|---|---|---|
| 1 | Wet fed (conversion) | 10 | 56% |
| 2 | Wet fed (purpose-built) | 10 | 75% |
| 3 | Trickle (converted) | 6 | 56% |
| 4 | Trickle (purpose-built) | 5 | 52% |
| 5 | Dump fed (converted) | 55 | 22% |
| 6 | Dump fed (purpose-built) | 20 | 21% |
| 7 | Dump fed (service house) | 12 | 52% |
| 8 | Drop (converted) | 6 | 15% |
| 9 | Floor fed | 72 | 15% |

Source: Hunter (Cambac Group) (1998)

✓ Despite careful measurement of subsequent performance in one trial (Burfoot & Kay n.d.) when sows were mixed from 1 week to 6 weeks after service (to cover the implantation period in depth) there were no significant differences in total born, born-alives or total litter weight born.

✓ *However*, it is still current advice *not* to mix sows during the implantation period, and the writer can quote several cases where returns to service and poor litter size responded quickly when it was suggested that it was a good idea to move batch introduction of new gilts into the herd's farrowing profile, rather than their introduction into what is called dynamic breeding groups (*see below*). This could be valuable in outdoor breeding paddocks even though, paradoxically, there is much more room for escape.

✓    Mixing in the evening may be beneficial.

✓    Adding fresh straw  at mixing tends to delay the settling of peck-order as the more dominant sow's attention is distracted for a while. *However* adding fresh straw at evening or late-afternoon may be beneficial in aiding socialization – it depends in my experience on whether the sow genotype is basically placid, the space available and the frequent presence of a familiar stockperson.  If some or all are borderline, fresh straw sure helps!

✓    Avoid re-mixing sows whenever possible, if so, a mixing pen as described is very worthwhile.

## IF YOU HAVE TO FARM IN DYNAMIC GROUPS ...

Dynamic groups are large groups of 20 to 30 sows upwards, where sows are added and removed on a regular basis, usually weekly, on the demands of farrowing, weaning or service.  Batch-farrowing, growing in popularity reduces mixing problems.

## A DYNAMIC SOW GROUP CHECKLIST

✓    Introduce **more than** 3 animals in the sub-group, whatever is the size of the main group.

✓    It is vital to have adequate feeding, sleeping and watering space for all the sows.  Providing extra *ad lib* feeders is now considered essential.  Remember, "farm for the most timid sow/gilt."

✓    Introduce after implantation is complete, technically 28 to 35 days after service, but after 28 days in my experience seems satisfactory.

✓    Take time to introduce gilts to a dynamic group, and preferably grow them towards 130 kg before this is attempted.  Ensure the gilts are not bullied off their feed by partially bucket feeding them and standing by for a few minutes; this helps in flushing anyway.

✓    Individuals in the gilt sub-group should have been introduced to each other in electronic feeder training pens, a specially-designed mixing pen, or a holding pen near the main sow group. Gilts are excitable pre-oestrus and need familiarization with each other so as to minimise stress rather than excitement.

✓ The use of sedatives (*e.g.* amperozide) seems to me to merely delay aggression, though this maybe needs further exploration.

✓ The presence of a boar in a dynamic group does not seem to reduce inter-female aggression.

✓ Sows may be able to remember each other's place in the social hierarchy for 4 weeks, so within this time-scale "individuals may be returned to groups of sows with which they have previously been housed without any in-fighting" (Arey, 1998).

✓ If you have trouble in introduction, try penning off an area and keeping the new entrants there for a few days. Pen that area off away from access by main group sows for a couple of days before the sub-group is introduced. (This is a very useful idea.)

✓ Breaking up the lying area with divisions is sometimes advised. I recommend dividers are essential with electronic sow feeders on hard standings as seen in the continent of Europe, but my experience suggests that this is unwise in deep straw-bedded yards (see Figure 3), as it tends to disturb the natural territorial occupancy of the chosen resting area by the various sub-groups.

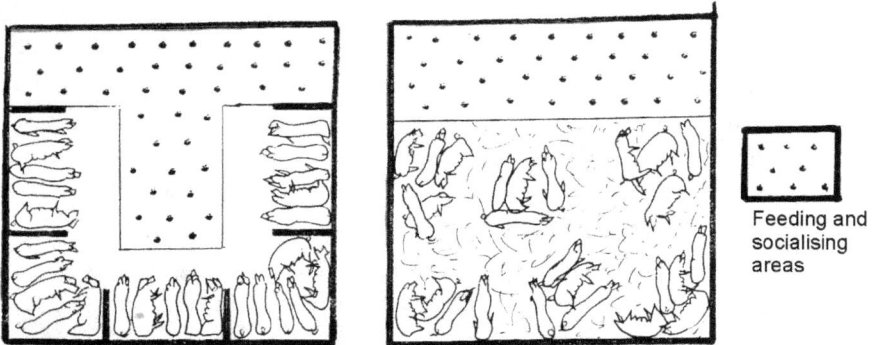

Feeding and socialising areas

**HARD FLOORS**

Divisions are essential. Sows will lie facing the feeding area many in the 'Dog-Crouch', semi-sternum position

**DEEP STRAW RESTING AREA**

Sows lie in family groups, the most dominant group(s) nearest to the feed area. Divisions will tend to hinder this. Notice many more sows are in the fully recumbent position.

**Figure 3.** Two yards . but two totally different layouts

✓ Be careful not to have protruding fixtures (feeders, drinkers) or sharp corners, especially on mixing pens where there is much sudden movement.

The author recognises that much about sow group behaviour is far from being fully understood and that the advice given in these pages may be modified as experience accumulates.

# SOME TIPS FROM A LIFETIME'S EXPERIENCE

I first kept sows in yards in 1949. We found that . . .

* A docile breed was a godsend! Saddlebacks and Berkshires, then Accredicross (Seghers) were easy to manage. (They still are!)
* We soon discovered the benefit of ample space.
* And of the straw yard.
* But both were expensive, so we used old converted (but warm) buildings and I suggest using moveable partitions to alter divisions. Docile sows and good stockmanship need nothing fancy, so save your money!
* There was always the timid sow or gilt. We had a couple of feeding stalls for these, in which they could be shut in daily. It took more time and trouble, but was always worth it – up to 4 more pigs per year for these ladies.
* After a while we used extra *ad lib* feeders as routine, as is recommended now some 60 years later! Things go full circle, don't they!
* Sows in yards need much better stockmanship than sows in stalls – and more time.
* With electronic feeders I much prefer the double-yard system as once a day you get all your sows up and moving past you. (Figure 4) Before they settle down again you have time to examine things like udders and vulvas and mark any which need an eye kept on them subsequently. Also those which are slow to rise can be noticed and catered for.
* Enough time, enough space and careful planning are the three routes to trouble-free grouping of sows, whichever breed you choose or in whatever climate and housing system you raise them.

# REFERENCES

Arey, D. and Turner, S. (1998) *Housing pigs in large groups.* SAC Report.

Burfoot, A. and Kay. R. (1995) *Agression between sows mixed in small stable groups.* ADAS Report.

Hill, M. (1997) Allen D Leman Swine Conference. University of Minnesota.

Hunter, L. (1998) Behaviour and welfare of sows in group housing. CAMBAC/JMA Research Report.

Kay, R. (1999) Sow agression under spotlight. *Farmers Weekly*, 8th October.

Lean, I. (1995) Matching for size increases fighting. *Pig Farming* (April). pp 68.

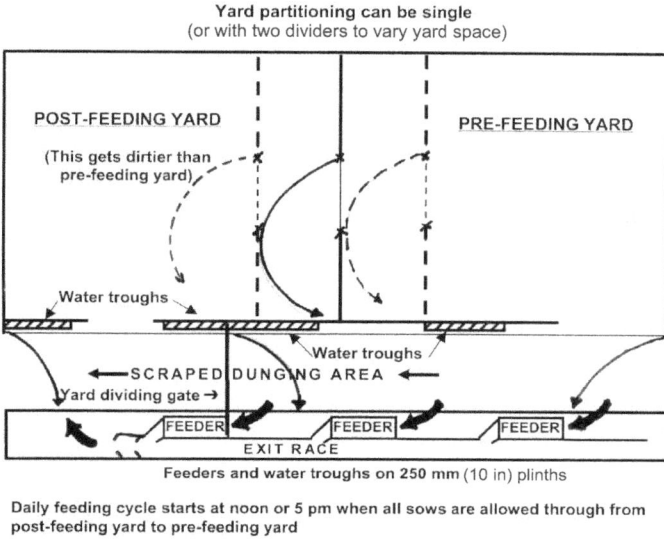

**Yard partitioning can be single**
(or with two dividers to vary yard space)

POST-FEEDING YARD

(This gets dirtier than
pre-feeding yard)

PRE-FEEDING YARD

Water troughs

Water troughs

←——SCRAPED DUNGING AREA ←——

Yard dividing gate →

FEEDER    FEEDER    FEEDER

EXIT RACE

Feeders and water troughs on 250 mm (10 in) plinths

Daily feeding cycle starts at noon or 5 pm when all sows are allowed through from
post-feeding yard to pre-feeding yard

**Figure 4.** Sketch of rotary station feeding layout for 120-150 sows

**Figure 5.** One of the reasons by the 'BIG PEN' concept is gaining ground is that
feeder space, unoccupancy level and eating time constraints are removed at a stroke,
which the author has measured as sufficient to pay for any extra housing or bedding
cost of this semi-extensive system, often raised as an objection.

MAFF (1993) *Pig welfare advisory group booklet series*. UK Ministry of Agriculture (now DEFRA). London.

Peet, B. (1993) Sow pen design agression and feed. *Farm Building Progress*, October, pp 3.

Pope, G. (1996) The cost of mixing pigs. *Pig Industry News* (Australia).

Varley, M. (2001) More space boosts piglet performance. *Farmers Weekly*, 22nd June.

Whittaker, X. and Spoolder, M. (1990) *The effect of ad-libitum feeding on a high fibre (sow) diet*. ADAS Research Report.

# 18

# STOCKING DENSITY - A WORLDWIDE FAILING

The minimum amount of space required by pigs so as not to restrain performance or aggravate aggression, and maintain welfare.

## A COMPLEX SUBJECT

The space requirements of pigs can be divided into various groups :

| | | |
|---|---|---|
| *Body-occupation space* | - | a pig lying down takes more space than one standing up, a supine pig takes up more space than one in the semi-sternum position. |
| *Body-activity space* | - | the space required by body posture changes, like getting up or down, lying supine, turning round/grooming itself. |
| *Social space* | - | the space required for socialisation with other pigs or access by stockpersons. An important part of this – often undervalued – is 'fleeing space' to mitigate aggression. |
| *System space* | - | the space required by different management systems *e.g.* straw v. slats, gestation stalls, groups in yards, wet feeding v. dry hoppers, etc. |
| *'Dead' space* | - | the space required for partitions, passages, corners and pen furniture. |

Thus the ideal commercially viable space requirement of any pig is complex and variable. Most of the categories above are two-dimensional, but three-dimensional space – volumetric space – should be borne in mind as low ceilings often predispose towards respiratory disease and may merit fewer pigs housed per square metre of floor space (or the use of improved air flow) to combat the build-up of infection.

# *TARGETS*

## SPACE REQUIREMENTS FOR GROWING FINISHING PIGS
*(for creep-housed piglets see page 20; for breeding stock see pages 430 and 431, at the end of this section.)*

One has to start somewhere, and Table 1 is based on advice from several countries.

Table 1. **Recommended slatted (left) and straw-based (right) space allowances for nursery-to-finish pigs. Compiled from various sources worldwide. Per pig housed**

| | Slatted & partially slatted/dung scraped | | Straw-based* | |
| kg | $m_2$ | $ft_2$ | $m_2$ | $ft_2$ |
|---|---|---|---|---|
| 5 | 0.1 | 1.1 | 0.25 | 2.7 |
| 10 | 0.15 | 1.6 | 0.4 | 4.3 |
| 15 | 0.175 | 1.9 | 0.45 | 4.8 |
| 20 | 0.2 | 2.2 | 0.5 | 5.4 |
| 25 | 0.25 | 2.7 | 0.75 | 8.1 |
| 30 | 0.3 | 3.3 | 0.8 | 8.6 |
| 35 | 0.325 | 3.5 | 0.95 | 10.2 |
| 40 | 0.35 | 3.8 | 1.0 | 10.8 |
| 45 | 0.375 | 4.0 | 1.1 | 11.8 |
| 50 | 0.4 | 4.3 | 1.2 | 12.9 |
| 55 | 0.425 | 4.6 | 1.25 | 13.5 |
| 60 | 0.45 | 4.8 | 1.3 | 14.0 |
| 65 | 0.475 | 5.1 | 1.4 | 15.1 |
| 70 | 0.5 | 5.4 | 1.5 | 16.1 |
| 75 | 0.525 | 5.7 | 1.55 | 16.7 |
| 80 | 0.55 | 5.9 | 1.6 | 17.2 |
| 85 | 0.575 | 6.2 | 1.65 | 17.8 |
| 90 | 0.6 | 6.5 | 1.7 | 18.3 |
| 95 | 0.65 | 7.0 | 1.85 | 20.0 |
| 100 | 0.7 | 7.5 | 2.0 | 21.5 |
| 105 | 0.72 | 7.75 | 2.1 | 22.6 |
| 110** | 0.74 | 8.0 | 2.2 | 23.7 |
| 115 | 0.75 | 8.1 | 2.25 | 24.2 |
| 120 | 0.76 | 8.2 | 2.3 | 24.8 |

*Notes*

- *Assumes pens are no longer than 2½ x 2, trough space not less than 100 mm (4 in)/pig and not more than 20 pigs per single-space feeder.*
- *Many producers on solid floors with no or minimal bedding would allow 10% more space to accommodate ease of dung removal (except for Straw Flow designs).*

- *Some research suggests no improvement in performance was obtained from the same space allowances on straw and on slats. However, in the author's experience this is not tenable in many or even most farm circumstances, and deep-bedded straw-based pigs in groups should have at least twice to up to 3 times the advised spatial levels for those on slatted floors.*

\* Minimum depth 10 cm \*\* UK Welfare Regulations require 1 m$^2$ per pig over 110 kg.

# WELFARE MINIMUM STANDARDS

The guidelines in *Table 1* can be considered just that – guidelines. They match or exceed the Minimum Welfare Standards which are already in place in several countries like the U.K., Sweden, Australia, Denmark and Canada. For example the U.K. Welfare of Livestock Regulations (1994) implement the E.C. Directive 91/630 which lays down minimum standards for the protection of pigs are as follows:–

**Annex Schedule 3 of the U.K. Welfare & Livestock Regulations (1994)**

This states …

The unobstructed floor area available to each weaner or rearing pig reared in a group must be at least:

- 0.15 square metres for each pig where the average weight of the pigs in the group is 10 kg or less.

- 0.20 square metres for each pig where the average weight of the pigs in the group is more than 10 kg but less than or equal to 20 kg.

- 0.30 square metres for each pig where the average weight of the pigs in the group is more than 20 kg but less than or equal to 30 kg.

- 0.40 square metres for each pig where the average weight of the pigs in the group is more than 30 kg but less than or equal to 50 kg.

- 0.55 square metres for each pig where the average weight of the pigs in the group is more than 50 kg but less than or equal to 85 kg.

- 0.65 square metres for each pig where the average weight of the pigs in the group is more than 85 kg but less than or equal to 110 kg.

- 1.00 square metres for each pig where the average weight of the pigs in the group is more than 110 kg.

This method of expression in bands or 'jumps' has encountered criticism in being too rigid and "Is at odds with the fact that pigs grow continuously and not in steps.

It is illogical that a pig weighing 20 kg needs 0.2 m$^2$/pig while a pig weighing 21 kg suddenly needs 0.3 m$^2$." (Morgan, 1997)

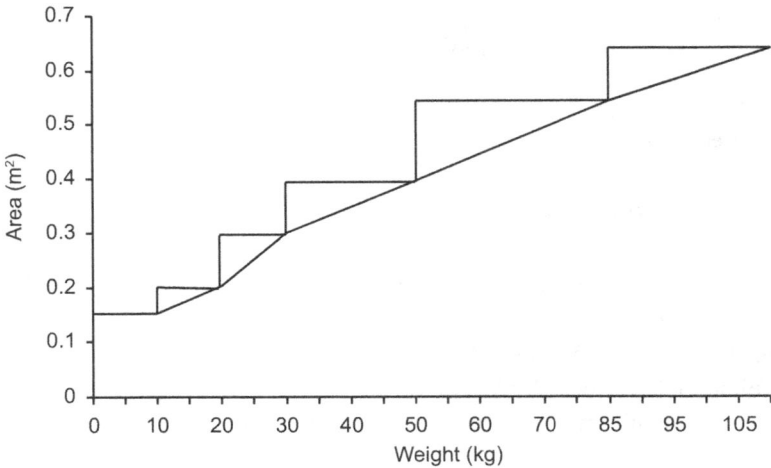

**Figure 1**. The minimum permitted space allowances under the EC Directive compared to the average expected weight of the pigs in a group.
This is depicted as a rising line.

There are also economic drawbacks to this method – a stepped system means that in order to comply with the Regulations, producers may have to move pigs regardless of whether their routines and pen sizes allow them to do so.

*For example* : Under the Welfare Regulations, if pigs stay in a pen to any weight over 50 kg they must then be given the same space as 85 kg pigs. If pigs are normally taken out of a pen at 55 kg (a little over the 50 kg mandate of 0.4 m$^2$/pig *i.e.*, say 0.42 m$^2$/pig) then this 0.42 m$^2$/pig allowance is adequate. But the Directive says each such pig must be given 0.55 m$^2$/pig – a difference of 0.13 m$^2$/pig. This would be an extra capital cost of £30/pig if new pens are needed to satisfy this destocking level below the needs from a graphical curve rather than a stepped system. On some units operating an all-in/all-out policy the management system could be disrupted and group sizes altered. In addition, such unnecessary destocking could increase the need for heat input in cold weather or at night which, if not done, could then compromise the pig's welfare or health!

## HOW MUCH DOES OVERSTOCKING COST?

Over a period of 10 years I recorded the pen measurements and stocking densities of all the grower/finisher houses I entered. In general some 38% were overstocked by 15% or more, which is putting 14 pigs into a pen designed to hold only 12 – obviously quite an easy thing to do.

We then carried out a carefully measured trial on three farms where we deliberately destocked half the pens on each farm to the correct densities as listed in Table 1. One farm had spare accommodation to take the surplus, and it was summer in the case of the other two, so these surplus pigs could go into yards or outside kennels.

Table 2 shows a typical result.

**Table 2. Likely costs incurred by overstocking a nursery and a finishing house by 15 per cent**

| | *Pigs 6-35 kg* | | *Pigs 36-100 kg* | |
| | *Correct density* | *+15 per cent* | *Correct density* | *+15 per cent* |
|---|---|---|---|---|
| Daily gain, g | 518 | 480 | 844 | 848 |
| Days in pen | 56 | 60 | 77 | 77 |
| Overhead costs @ 24p/day (£) | 13.44 | 14.40 | 18.46 | 18.46* |
| FCR | 2.02 | 2.12 | 2.42 | 2.63 |
| Total food eaten in period (kg) | 58.6 | 61.5 | 157.3 | 171.0 |
| Total food cost (£) | 11.13 | 11.69 | 27.53 | 29.93 |
| *Extra costs/pig (£)* | *1.85 plus* | **2.40** | **Total 4.20** | |

Savings in 15% less housing cost per pig (at £8.20/pig)**Savings £1.23**
**Costs  £4.20**

The average REO for deliberately destocking to guideline levels on all three farms was 3.5:1.

## Factors affecting stocking density in growing pigs

Check the following …

*Pen shape :* Long and narrow versus square. A ratio of 2 to 1½ is satisfactory, and 3:1 is not good if the shorter length comprises the feeding or dunging space. This is often seen in wet-fed pens where tailbiting can arise.

*Temperature :* Allow for 15% more space in hot weather (24°C or 75°F) dependent on airflow and cooling devices. An example is that Boon (NIAE) showed that given the option, the lying space occupied by growing pigs increases by some 15% as air temperatures rises by 6°C from their LCT.

*Draughts :* A pen with adequate space allowances but which causes pigs to huddle to keep warm can cause the same drag on performance as overcrowding.

*Feeder space :*   Penny (JSR Genetics, 2000) reports … "Offering pigs additional feed area can override the negative effects of reducing floor space. Looking at the performance of 396 grower pigs from 20-40 kg over a 28 day trial period, reducing pen space from 0.4 m² per pig to 0.3m², both with 50 mm per pig feeding space allowance, can affect performance." . . . "A decrease in floor area results in poorer liveweight gains and lower feed intake, but this can be overcome by offering more feeding opportunities. Raising feed areas to 100 mm per pig can override many negative responses from increasing stocking density."

(See Feeder space allowances, page 396).

15% overstocking is very common, as seen here. Just removing two pigs from this pen recouped 3.5 times more income at slaughter than the increased housing cost of 15% (see Table 2 for the economics)

## Solid floor dunging : sleeping space ratio

If the pigs' sleeping area leaves room only for the dunging passage space/slatted area, the pigs are technically overstocked. In the author's opinion the ratio of sleeping, plus socializing/feeding to dunging space should not be less than 3:1, or 25% more than the resting area alone. This is borne out by Edwards who writes:– "An experiment (was effected) where the size of pen was changed each week in relation to the weight of the pigs. At the lowest space allowance the pigs were given only as much space as was necessary for them all to lie down. This was compared with three progressively more generous space allowances, giving 12, 25 or 42% additional area". The results are shown in Table 3.

Even with constant group size, good temperature control, *ad libitum* feeding with plentiful hopper space and ready access to drinkers, individual pig performance improved as space allowance increased.

**Table 3. Effect of different space allowances on pig performances**

|  | *Lying space only* | *Other space allowance* | | |
|---|---|---|---|---|
|  |  | *+ 12%* | *+25%* | *+42%* |
| Daily gain, g | 844 | 862 | 882 | 897 |
| Feed conversion | 2.70 | 2.56 | 2.60 | 2.59 |

(from Edwards, Armsby & Spechter,1987)

When the economics were examined, it was found that the higher feed cost per kg of carcase produced more than offset any saving in housing cost at the lowest space allowance. Since this trial was carried out in such good conditions it is likely that even greater effects would be seen in day-to-day practice on commercial farms.

Edwards recommends +25% total space as an advised level over lying space.

## Pen shape

Many dung passages are not broad or deep enough as this saves construction and cleaning costs. This is especially true of some wet feed pen designs where pens are much too long and narrow. The 5 to 7:1 length/breadth ratios allow a large number of pigs to feed at one time but pens remain wet from misplaced voidings and wet-feed spillage, so the pigs lie uncomfortably/lose more body heat to the floor.

This is why wet-fed pens to the Suffolk design (trough to a broad front) tend to give better results than the traditional long/narrow design with the wet feed trough to one or both sides. However Suffolk pens tend to be 18-24% more expensive in capital and running costs, so such long-term housing costs money.

***Uneven pigs***. Differences in bodyweight of more than 3 to 5% per group may need destocking. However pigs are very adaptable; but providing greater spatial allowance during such disparities will help minimize check to growth. In nurseries the author finds a 15% easement beneficial. In older pigs the provision of screener boards may help but only makes matters worse in overstocked pens where fleeing space/hiding ability is constrained as a result.

## Pen furniture

In pen groups which follow the minimum stocking densities outlined in Table 1, careful positioning of pen furniture will help reduce aggression and improve performance. Figure 2 shows two pen layouts using a single feeding point.

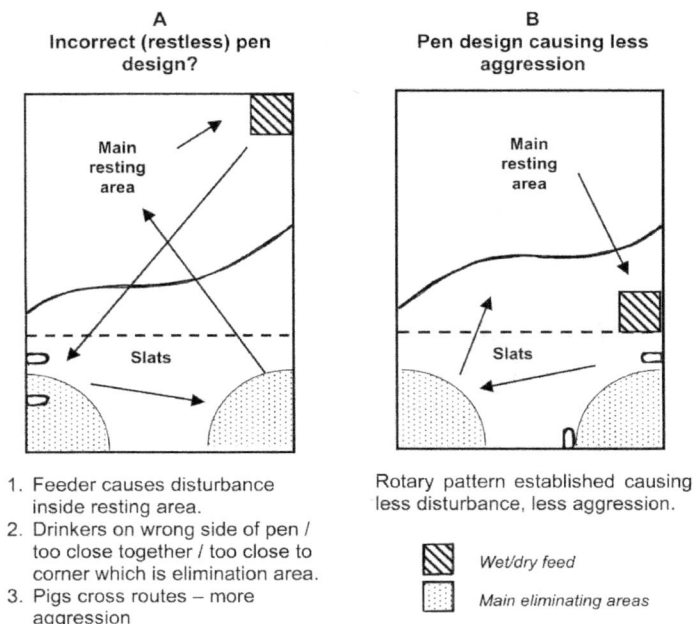

A
**Incorrect (restless) pen design?**

B
**Pen design causing less aggression**

1. Feeder causes disturbance inside resting area.
2. Drinkers on wrong side of pen / too close together / too close to corner which is elimination area.
3. Pigs cross routes – more aggression

Rotary pattern established causing less disturbance, less aggression.

*Wet/dry feed*

*Main eliminating areas*

**Access passage at top of diagrams. Arrows refer to pig movement sequences**
**Figure 2.** Even if stocking density and pen shape are correct - the positioning of pen furniture can calm or inflame aggression.

Table 4 illustrates what happened when the remainder of a new finishing house was built to the left hand in place of the preferred right hand design (to make hand-filling the feeders easier).

**Table 4. Performance comparison of pens A and B (12 per side)**

| (*All-in/all-out*) | (A) | (B) |
|---|---|---|
| Daily Gain, g (35-90 kg) | 567 | 608 |
| LFCR (35-90 kg) | 3.0 | 2.86 |
| MTF (kg) | 394 | 416* |
| Stockman's estimate of pig's time spent resting during daylight hours | *c.*60%-70% | *c.*80%-85% |

*The extra 22 kg meat sold/tonne feed was equivalent to a 14% reduction in feed cost at the time.

## Misconceptions about stocking density

*"Pigs aggress more when overstocked."*

Not necessarily. Pigs can adapt to overstocking and recover performance levels given time.

Research suggests that up to a certain degree of crowding all measures move in the expected direction and then a reverse trend occurs. The animals seem to adapt to over-crowding. Even so, there is always performance loss (or worse, *i.e.* tail-biting) until they do.

*"Giving pigs more room improves performance."*

On the whole it does but it of course depends on how crowded the pigs were in the first place. Since Powell and Brown's pioneer work of the mid-1990's there have been similar trials which are mentioned in this Stocking Density chapter, but as the work is often cited for it's beneficial effect on FCR, I extrapolate the physical findings into some interesting modern econometric (measurement of cost-effectiveness) figures, taking UK 2010 costs and returns as a base. These especially involve housing use – as these days good quality housing from the technical point of view is now much more expensive and becoming a larger proportion of total costs.

**Table 5. Effect of different average space allowances per pig across the growing period, both in relation to meat produced and use of buildings.**

<u>Assumptions:</u> House holds 1000 pigs from 7 - 102 kg. 4 days for batch clean-down and shipping-lag. 75% dressing per cent. Av. food cost £0.18/kg. Av. overhead costs £0.24/day per pig. Pig price received (deadweight , dcw) £1.20/kg 15-year amortised housing cost £36m² Basic housing space available 500m²

*(A) Physical performance.*

|  | Daily gain (g) | FCR | Batches put through per year | dcwt. meat sold per m²/year (kg) | MTF (7-102kg) (kg) |
|---|---|---|---|---|---|
| 0.4m² | 662 | 2.67 | 2.47 | 378 | 301 |
| 0.5m² | 731 | 2.55 | 2.72 | 388 | 316 |
| 0.6m² | 681 | 2.75 | 2.53 | 356 | 293 |
| 0.8m² | 640 | 2.86 | 2.40 | 342 | 281 |

*(B) Economic performance.*

|  | Total costs /pig | Income /pig | Margin / pig | Margin m² | Earnings per house per year |
|---|---|---|---|---|---|
| 0.4m² | £80.28 | £90.00 | £9.72 | £430 | £215.000 |
| 0.5m² | £76.68 | £90.00 | £13.32 | £466 | £233.000 |
| 0.6m² | £80.58 | £90.00 | £9.42 | £427 | £213,500 |
| 0.8m² | £84.48 | £90.00 | £5.52 | £410 | £205,000 |

<u>Comment:</u> Allowing the pigs more, i.e.. too much, space reduced performance significantly beyond the generally-accepted average of 0.5 m²/pig. Crowding them in also had a worsening effect on FCR. From these figures it can be seen that with the cost of modern grow-finish housing being so high, that daily gain – in order to facilitate throughput of pigs and increase batches put through the premises, is becoming as important as FCR is in achieving a high MTF.

## Daily gain v. FCR?

However with increasingly high housing costs, daily gain is also becoming more important than FCR in ensuring as many pigs are sold per expensive square metre as possible. Overheads today are climbing past one third of total costs of production very quickly and will soon be at 40% or more in all but the very low-cost units.

This state of affairs also highlights the importance of <u>even</u> growth as well as fast growth. The 'shipping lag' (waiting for laggards to catch up and achieve minimum contract weight) can drastically reduce the house throughput and give the same sort of margin penalties as keeping the pigs too crowded or not occupying the floor space efficiently. Producers must have contingency plans for reducing the impact of uneven growth on expensively long house turn-round times.

One of the commonest jobs I do on touring a grow-out farm is to check pen stocking densities – sadly it is wrong on so many farms.

Pigs which are over or understocked tend to 'fall-away' from their siblings, especially if overstocked - I guess due to being denied full trough occupancy rather than stress, an interesting area for the behavioural scientist

Now that the 'Big Pen' concept (see pages 408 and 415) is gaining favour, I notice reduced shipping lag-times in those of my clients who have adopted big pens (150 - 250 grower/finishers together) because of lower housing cost/pig. It is regarded as a spin-off from the concept and an important one at that.

*"Stress, as measured by plasma cortisol, increases with overcrowding."*

It often doesn't. Again, the pig is very adaptable. Stress is still very difficult to measure scientifically.

*"Growth promoters work better when pigs are overcrowded."*

Research suggests the effects are similar.

*"Overstocking is a behavioural problem."*

Sure, stress must play its part. Much more likely is that overcrowded pigs eat less because they cannot get enough time at the trough, and are indeed more stressed which may or may not show behaviourally thus the transmission of disease is increased, possibly by a reduction in immune status. The combination of less feed intake with a greater and altered demand for nutrients due to the need to raise the immune defences in overstocked pigs combine to give the performance reduction. Also, once the immune system readjusts/adapts, the performance may recover even though the stocking density remains high.

## Correct stocking density and understocking.

Some people suggest the deliberate understocking (within reason), while increasing the cost per m², improves performance to such an extent that income is actually increased. In my experience this happens only rarely.

Look again at the work in Table 5. The extra liveweight produced on the 'accepted' space allowance of 0.49m²/pig which in fact still gives a 22.5% overage from the 0.4m²/pig minimum welfare standard on an average herd weight of 50 kg (5-100 kg) – is dramatic. Compared to 0.81m²/pig but using the following reasonable assumptions and basing the performance comparisons on an econometric basis, Table 5, this time expressed in costs per m², reveals a much less dramatic picture, even though the conclusion is the same – it does not pay, in general, to understock. In fact any more than it does to overstock (*see* Table 2).

Table 5 shows how much it pays, under good management to adhere to the correct target stocking density minimum quoted in Table 1.

## Variable geometry – the crusher board

The crusher board for smaller growers is cheap, flexible, is simple to use, easy to clean and the need for adjustment makes the stockperson look at the comfort and condition of their pigs more frequently.

It is used as a simple divider board in a passageway creep area (to discourage the newborn piglets from wrong-mucking in a heated creep) right up to pigs as heavy as 65 kg (to provide a rising scale of cost-effective area per square metre of space available).

Moreover, it cuts down the workload; pens don't get fouled so frequently.
In technical terms:

• By maintaining the correct social space, it keeps young pigs warmer (many pigs are still kept below their LCT in winter or at night).

• So they grow faster.

• It keeps pens cleaner (as young pigs with too much room may dung at the back and too-hot pigs will dung/urinate around themselves to provide an indoor wallow).

- So they are less stressed, less prone to aggressive behaviour and thus convert better.

## Some suggestions:

The most effective housing in which to use the crusher board idea is the kennel or bungalow, especially if the design is long and narrow (like the monopitches) or wide and shallow (like the conventional Suffolk design – not the 'zig-zag' version – and especially the wet-feeding variety with a long outside trough). Conventionally one stocks these simple pens at 30 pigs and thins down to 20's or 15/16's. Even so, for 40 per cent of the time the pigs are either understocked or overstocked to a greater or lesser degree – hard on the pigs at some times, or on the pocket at others.

- Even so, crusher boards do not allow you to put more pigs into what space you have available but they help you manage the temperature better and more evenly. Thus for reasonably low-lidded pens a laterally-moved crusher board should reach up to the lid itself so as to cut down on the air volume. This allows for a reduction in the air volume especially in cold conditions to be circulated as well as reducing the floor space to be occupied by the young pigs. This assumes correct air movement within the air space provided.

- If the roof line is high (and for young pigs it shouldn't be, allowing for ease of inspection) then a rick-sheet batten-frame slotted in to the upper part of the crusher board cuts down on weight and cost markedly – as long as it is out of the pig's reach, of course.

- Always locate the board to the pen division, but use hooks or pegs only on the *board*, and round eyes or recessed lugs on the *pen wall*, in order to avoid damage to the pigs' skin as the board moves back up the pen leaving the fixtures exposed.

- Do not use the board to such an extent that late-to-bed pigs are forced to lie right in a kennel doorway and thus are exposed to a night-time draught. Move it one notch backwards to accommodate them and/or use flexiflaps in the pophole itself.

- Try not to angle the board diagonally across the pen – pigs may dung in any acute-angle deadspace you create. A right angle allied to a correctly measured space allowance is best.

Crusher boards can be templated to fixed equipment, like troughs and hopper lengths, *provided the inside of the feeder or trough is also shuttered off.* This is not onerous or expensive; for example a wet feed trough can have a moulded moveable concrete plug (with handle) placed at the junction with the crusher board. Stale feed negates the whole exercise, so seal off the unused portion of any feed receptacle.

Finally, a plea to the housing manufacturers. Could you design, as an optional-extra, a crusher board device into each pen/kennel division? Only a very few in the world seem to do so. And also give more thought to variable-geometry pen divisions? We haven't explored this cost-saving idea nearly enough. With intensive housing costs now rising past 15% of the cost of producing a finished pig, we must attempt to make better use of spatial investment.

*Exercise* : 25 x 15 kg second stage weaners to be thinned down at 25 kg to 12's, then grown on to 70 kg if required. Pens are 3m x 2m.

Using Table 1, 25 x 15 kg growers need 25 x 0.175 = 4.375m² on entry and will occupy approx 75% of the space available, so place the crusher board at 1.5m from the side wall (Figure 3) and reduce progressively until all the 6m² space is occupied, in this case at 25 kg.

Then (Figure 4) thin down to 12's or 15's, the former allowing occupation until 70 kg, the latter to 50 kg.

In the case of 12's the crusher board, after cleaning both it and the pen thoroughly, is positioned at 3 metres from the side wall, and in the case of pens of 15, at 2 m (as approximately only 2/3rds of the initial space is taken up) and again moved back progressively until all the 6m² space is taken up at 50 kg liveweight – when the group is rehoused to the grow-out facility.

## Value of crusher boards

Table 6 gives an indication of the value of a simple crusher board used in nursery kennels.

## How to use crusher boards

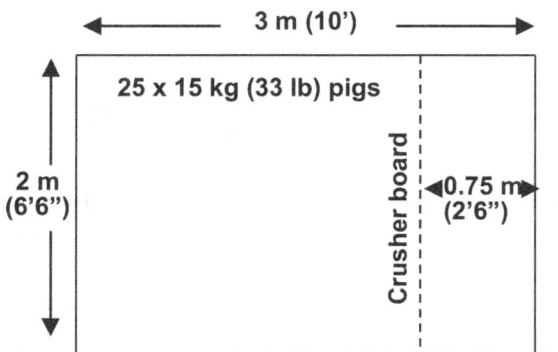

**Figure 3.** Position of crusher board on entry of weaner pigs.

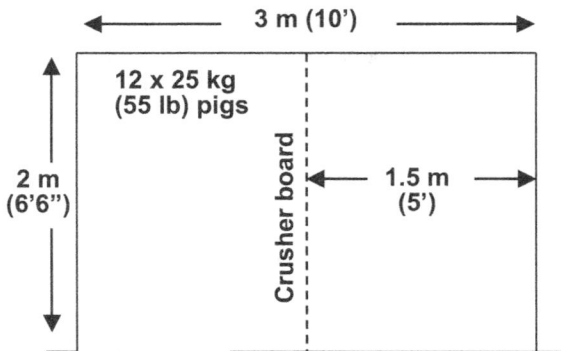

**Figure 4.** Crusher boards re-introduced on thinning down. In squarer shaped pens crusher boards can run either way - up or down/side-to-side.
Conversion: 1 metre = 3.281ft

**Table 6. Performance and econometrics to slaughter from using crusher boards (12-88 kg)**

|  | *Before* | *After* |
|---|---|---|
| LFCR to slaughter | 2.94 | 2.91 |
| ADG to slaughter, g | 613 | 631 |
| MTF, kg (lb) | 291 | 293 |
| Value of extra meat and saved overheads (2 days) (per pig) | – | £0.89 |

Pigs put through crusher board regime/year  820
Extra value from use of crusher boards  820 x £0.89 = £730 ($1175)
Cost of 10 crusher boards (home-made)  £400 ($644)
Payback therefore 0.54 years – under 7 months.

# STOCKING DENSITY FOR BREEDING STOCK - GILTS

Most gilts are selected or bought-in from 85 to 100 kg and placed into a wide variety of housing varying from kennelled yards on concrete to bedded straw in groups.

For examples of pen design (plans) see Brent, "*Housing the Pig*" (1986), a textbook which has stood the test of time and experience excellently, although I understand a re-write is planned.

Numbers vary, usually from 4 in a pen to 20 in a yard; I prefer a maximum of 6 housed together.

There are a number of basic precepts to follow so as to allow the animals to settle-in rapidly, avoid injuring each other and ensure good signs of oestrus.

## CHECKLIST ON GILT ACCOMMODATION SPACE REQUIREMENTS

✓   No gilt pen should have a side less than 2m long.

✓   Lying area alone for new entrants at 85-100 kg should be not less 0.6m²/gilt.

✓   If kept in groups until first farrowing allow 1m²/gilt lying area.

✓   Many tractor-scraped solid dunging areas allow a 1:1 ratio with the resting area: this itself is generous but very satisfactory.

✓   If the pen includes a slatted dunging area, this must be 25% of the lying area given above, and this slatted area have not less than a 1m side. However, from experience, the author finds a minimal total area of 2.8m²/gilt (as for sows) to be preferable.

✓   Providing individual feeder spaces is desirable. If so their width will depend on the body shape, especially the shoulder width, of the genotype chosen, but generally 450 mm to 540 mm is adequate. Gilts at first service are already larger (135kg), as longer induction times catch on.

✓   Gilts may have to be moved to the boar. If so, doorways must be a minimum of 900 mm and have rubber or plastic corner-protectors if made from brick or block walls. Passageways should be 1.2m.

✓   Wherever possible do not overcrowd gilts – remember the 2.8m²/animal advised. Allow adequate 'fleeing-space', this helps reduce aggression and makes first service easier and more effective.

## SOWS

As with gilts the range of satisfactory accommodation for sows is enormous – and this includes gestation stalls and the mistakenly maligned farrowing crate!

Table 7 may help breeders who require some minimal standards where considering building alterations and design. Those marked * are taken from the *UK Pig Animal Welfare Group* publications listed below.

## TARGETS - SOWS

**Table 7.  Some basic total space allowances – sows**

|  |  | *m²/animal* |
|---|---|---|
|  | Sows on slats, loose-housed | 2.8 |
|  | Stalled sow | 1.5 |
| PAWG N°.4 | *Cubicle & free-access stalls | 2.3 to 2.9 |
| PAWG N°.5 | Yards and Individual Feeders | 3.26 to 3.73 |
|  | Farrowing crate & pen | 4.6 to 4.8 |
| PAWG N°.6 | *Trickle fed or wet fed, yards or kennels | 2.3 to 2.79 |
| PAWG N°.7 | *Floor fed yards or kennels | 2.33 to 3.73 |
| PAWG N°.9 | Single Yard Electronic Sow Feeders | 2.66 to 3.18 |
| PAWG N°.9 | Twin Yard Electronic Sow Feeders | 2.7 |
|  | Group mixing pen | 3.5 |
|  | Outdoor Sows | 15 to 20 sows per hectare |

For further information see the Mixing Pigs section.

# BOARS & SERVICE ACCOMMODATION SPACE REQUIREMENTS

Many boars are badly housed in pens which are too small, badly positioned, too cold and uncomfortable.

**Table 8. Space allowance targets - boars**

| | |
|---|---|
| ✓ | Minimal area per boar  7.5m². |
| ✓ | Preferred area per boar  9.0m². |
| ✓ | Sides should preferably be 3m long; with a minimal height of 1.5m. |
| ✓ | Service pen (and boar accommodation if combined with a service pen, not advisable due to slipperiness) to allow adequate movement and minimize abrasion – 10.56m², i.e. 3.25m x 3.25m. |
| ✓ | Allow for 350mm gussetted corners, *i.e.* blanked-off corners, not a 90° right angle. |
| ✓ | Boar pens should preferably be fitted with a stockperson's escape pole, which duals as a boar rubbing post.  This should be close enough to the corner not to allow the boar to get jammed in. |

# BABY PIGLETS

For creep areas and their design, see the Creep Feeding chapter.

# REFERENCES

Brent, G. (1986) *Housing the Pig*. Farming Press (UK).

Edwards, S., Armsby, A. and Spechter, M. (1988) Effects of floor area on performance of growing pigs kept on fully-slatted floors. *Animal Production*. **46**, 553-459.

Morgan, D. (1997) Beware, the Devil is in the Detail. . **Pig Farming**, 16-17.

Penny, P. (1999) JSR Genetics Technical Conference. Press Reports.

# 19

# WET FEEDING
# - GATHERING PACE

Liquid feeding of pigs uses a liquid vehicle – primarily water, skim milk or whey, but any suitable liquid by-product will do – to carry solid nutrients (usually in meal form) in suspension, or alternatively some co-products in solution, to the point of consumption, which is usually a trough.

Liquid feeding, also known as pipeline feeding but more recently as Computerised Wet Feeding (CWF), should not be confused with wet/dry feeding (sometimes called, erroneously, 'single-space feeding'), where water is available from a displacement or nose-press valve operated by the pig itself to dampen or liquify meal, and sometimes pellets, into a shallow pan to a consistency of its choice.

Both methods of feeding are excellent systems when operated properly, with wet/dry feeding particularly popular at present for the younger grower, and fully liquid feeding for all classes of pigs, including breeding stock.

This said, the author believes that the liquid/pipeline system has advantages over the wet/dry system as detailed below.

This experience leads me to forecast that liquid feeding will eventually become the predominant method of feeding all swine across the world.

## CWF (COMPUTERISED WET FEEDING) PAST, PRESENT AND FUTURE

"I must declare an interest" as every politician talking about a commercial interest should say! So I confess that I have been a lifelong convert to the wet (pipeline) feeding of pigs.

For myself it all started on the Taymix farm in Dorset, England, when with some 12,000 growing pigs to feed, the ram pump of our wet feeding system broke down and the replacement park took 3 days to arrive.

All hands, including myself as technical director and my poor typist (who nevertheless donned overalls and enjoyed the change enormously) were mustered to carry bags of feed to each of the 800 pens. Twice a day.

I finished at two in the morning and had to be back at it again at six am. For three days until the spares arrived! That brought home to me what a labour-saving device CWF was!

## Learning curve

And what a learning curve it was, too! Not only to have a spare or two for every working part you could think of but, just as important, to have a back-up plan ready should more than a 12 hours breakdown occur. Over the years we had to face and overcome, largely by trial and error - problems such as flocculation (separation of particle sizes in the pipeline, causing blockage), cleanliness (at trough, mixer and delivery lines), overeating (the pigs just loved the liquid 'soup' and got too fat), how to balance the host of cheap by-products available (difficult, as everything was so variable in analysis), correct pen and trough shape (a pig just loved to fit snugly inside the generous troughs we provided for 50 pigs to all eat at once - providing a very effective dam - like as not with mouth open waiting for the next delivery of feed from the exit pipe!) causing a tidal wave of valuable food to exit all down the slats. We redesigned the trough to prevent this (the famous Taymix trough). Then there was concrete floor erosion (acid from the skimmed milk and whey available at an unmissable 'come and collect it' price), never to feed warm whey, the difficulty of pumping enough dry matter far enough (thickness of the mix) especially for younger growers, and how to feed farrowed sows without her piglets enjoying the sticky if thoroughly enjoyable bath in her feed trough - or drowning, or being eaten - that happened too!

## Early days - so many problems

Then there were the joys of dissuading farmers from doing obvious things like pumping downhill to save energy costs and so extending the pumping distance - but always to pump uphill to avoid blockage from settling-out. And from not doing essential things like putting an 'Oxford Union' at all right angle bends, so that rodding-out if ever it was needed (and it was) could be done in seconds rather than in a prolonged, swearing, frustrating hour of trying to rod round a bend 10 metres away. The food didn't block but a rat or bird could get in (and of course did). Never to seal the pipes crossing under concrete passageways but install them in a trench covered by a removable metal cover. And in winter to lag exposed pipes with a 20 cm collar of straw in a plastic sack - effective down to -12°C in a 30 kmph gale.

And the difficulty of convincing the producer that wet feeding was - surprise, surprise - an (atmospherically) wet process and that the ventilation must be uprated professionally to take account of this and not to blame the system or the change to wet food for fits of coughing.

The ventilation legacy still exists today, by the way. Recalibrated ventilation, especially to cope with winter conditions, must also be included in the conversion costs. Something some equipment suppliers tend to forget - I suggest you allow 8% more on the conversion cost so as to upgrade the ventilation. That was all of 30-40 years ago - now what about the present?

## CWF TODAY

While pipeline pig feeding, the old name for CWF, is commonplace now, considering the benefits are well-established and impressive (Tables 1 and 2 et seq.) and the limitations not insuperable (Table 3) you may well ask why the concept is not established on 90% of world farms rather than the 10% to 60% in the countries I visit.

**Table 1. What you might expect from computerised wet feeding**

|  | *CWF* |  | *Dry pellets* |
| --- | --- | --- | --- |
| *Growing/finishing 35-105kg* |  |  |  |
| FCR | 2.27:1 | (11.5% better) | 2.53:1 |
| Daily gain (g/day) | 796 | (+6.9% faster) | 745 |
| Days to finish | 88 | (6.4% quicker) | 94 |
| MTF (kg saleable meat/tonne feed) | 339 | (+12.8% more) | 300.6 |
| Average probe (mm) | 10.9 | - | 10.8 |
| *Lactating sows* |  |  |  |
| Born alive/litter | 12.44 | (+17.1%) | 10.62 |
| No. weaned/litter | 11.59 | (+19.6%) | 9.69 |
| Av. weaning weight (kg) | 8.75 | (+7.8%) | 8.10 |
| Av. litter weight (kg) | 101.2 | (+28.9%) | 78.5 |

From: MLC Workshop Report, Feb 2005)

Take the USA, for example, which to my mind shows an uncharaceristic refusal by these go-ahead people - both their farmers and academics - to have anything to do with CWF. North American farmers (but not so much Canadians especially in Ontario) still remain firmly wedded to their dry feed hoppers despite the appalling wastage which is so evident (12%, and even 6% on the best units I have been privileged to be shown). When I ask them why they are not interested in CWF, the replies have been "Don't really understand it"; "I'm happy with what we do"; "Don't like the cost of changing over"; "It will increase the slurry volume" (answer - not necessarily); "My feed supplier/vet/extensionist says no".

**Table 2. Performance and cost benefits from liquid feeding**
(UK prices converted to approximate 2009 equivalents in Euros). Figures for a 2000 place growout unit, pigs growing from 30-105kg. Av. dwt. 77kg. Figures based on Stotfold wet feeding Trial 1 (of four)

| Assumptions | Dry feeding | Wet feeding |
|---|---|---|
| *Physical performance* | | |
| ADG g/day | 754 | 796 |
| Days to 75 kg | 102 | 96 |
| No. of days per batch (inc. washdown) | 109 | 103 |
| No. of pigs produced/year | 6697 | 7087 (+390) |
| *Capital investment* | | |
| Complete feeding system (€) | 6390 | 77,304* |

*Includes € 50,760 new building cost. Hammer mill plus elevator installation cost € 10,200 and tanks, pipeline and feeders cost € 16,344. On this latter basis the completely new CWF plant and equipment cost approximately 4 times more than the existing dry feed system. This of course will vary from installation to installation.

| | | |
|---|---|---|
| *Cost of production (CoP)*** €/kg deadweight | 1.19 | 1.14 |

**Includes cost of weaners, feed, labour, power, water, bedding, mortality, waste management and capital investment. Buildings depreciated over 25 years at 6% interest. Even on this basis . . . of the specialised unit.

| *Liquid feeding compared with dry* | |
|---|---|
| CoP saving/kg deadweight | € 0.05 |
| CoP saving per pig | € 4.25 |
| CoP saving per year | € 30,120 |
| CoP saving per pig place/year | € 15.06 |
| Payback period for investment in CWF | 2.6 years |

Acknowledgement: Extrapolated from the comprehensive body of work published by MLC researcher Dr Pinder Gill and research farm facility manager Lisa Taylor (2002-2005) see References.

Knowing the Americans well after 30 of visiting their farms, attending their conferences and conventions and appreciating the friendliness and generous spirit of the American pig producer, one day the penny will drop and then an avalanche of enthusiasm will descend. Just as it is beginning to develop now over their keeping sows in groups rather than the growing public aversion to gestation stalls. Americans are like that; when they get convinced - do they go! There are some good youngsters coming on with more open, adventurous minds and they will lead the charge to CWF when it comes - if father and the bank will allow them!

I only hope they will benefit from our years of experience of both concepts (CWF and group housing of sows). Neither are 'as easy as falling of a log' and there are many down-to-earth practical skills in the 'do and don't do this' school of hard experience.

However, when all this is got right - as so many have done - Table 1 from a completely independent government-funded source who went into things in meticulous detail shows the typical benefits of CWF over dry pellet feeding.

**Table 3. Some of the drawbacks to wet feeding by pipeline (a summary - these are discussed more fully on pages 452-453)**

---

**Installation cost.** Considerable, largely depending on existing suitability of piggery conversion. Many reports show that payback has been within 30 months - some 36 months but no more. After that the benefits have average around € 4 to € 5 per finished pig. See Table 3.

**Quality, skilled and well-trained staff** are essential. This is particularly true when starting pigs as young as 6 weeks (2 kg) where strict cleanliness is essential and fresh, little and often feeding practiced.

**Care has to be taken** to monitor bushel weight of ingredients and to keep a check on co-product nutrient density and shelf life.

**Malfermentation** can occur with weaners and sows but is not a serious problem with grower/finishers. Care and experience will avoid most of it.

**Whey bloat.** Older pigs can bloat, especially on whey. Anti-bloat whey balancers can help but in the author's experience 1% more finishing pigs may be lost to this problem. Allow for this.

**Ventilation** needs to be professionally checked at installation wet feeding is a wet process.

**Frost** is rarely a problem if the installation is designed to accommodate frost and wind chill. Learn from the Canadians, Swedes and Finns.

**Blockage** can happen, but very rarely in a set-up installed by a professional who will design-out danger spots and install easy remedial action should a blockage occur.

**Overfat pigs** can occur after changing from dry feed. Initially this is quite common especially on ad lib short troughs and at first dismays the novice, but is rectified by referring matters to a pig nutritionist who will readjust nutrient density to match increased appetite.

---

Table 2 gives further results from a series of trials designed and supervised by a world-acknowledged researcher/nutritionist on a specially constructed new-build pig growers unit. This was done so as to exclude the understandable variables which can be found with manufacturers' trials on commercial farms. Being a research built from new, the CWF capital costs of over € 50,000 were very high - this is dealt with below in the discussion on costs.

So those are the sort of performance results which can be expected from pipeline pig feeding. However, the main drawback to the newcomer being converted to change to a CWF system has always been capital set up or conversion costs, so let me go into this in some detail.

## The problem of the cost of installing CWF

One of the barriers to converting to CWF has always been the cost - it still is. Sure, as a lump sum it is d*****d expensive. It was, but less so, in the past and capital costs are likely to remain a sobering barrier in the future. Compared to a modern dry feed system, a modern CWF set up from new can be between 4 to 8 times more costly in capital needs. Of course this is daunting, although a good deal of this can be in the buildings advised, including ground plan and cover over. In Table 1, for example, the cost of the bare minimum CWF equipment is nearly two-thirds less (€ 40,000) than the figure quoted by the two authoritative bodies, **who have gone into things thoroughly and who have included complete new building costs into the equation**. I therefore show Table 2 as an economic worst-case scenario, as even here the payback is a proven 2.5 years. All the CWF clients who have shown me their figures have not exceeded 3 years and several cite 15 months when using converted existing buildings. After that, cost of production savings are in the region of € 4-€ 5/ pig - and more if co-products are well bought.

I suppose I have been involved with over 100 wet feeding plants in my career and 90% of them have used/modified existing housing, not needing to build anew. Their costs based on 2010 prices in the UK varied between £25-£30 per pig place.

## Does CWF depend on the use of co-products?

Not at all, although this is a common question from the newcomer. Water is a perfectly satisfactory means of moving bulky feed, saving onerous and unpopular labour and reducing dust. However, if liquid co-products should be available, use them to reduce feed CoP still further.

Liquid or liquefiable by-products are simple to add, are a lot cheaper per unit of energy or lysine than conventional dry raw materials **if balanced carefully from declared nutrient specifications** provided hopefully by the seller. They can vary substantially in nutrient density and this could be another drawback.

Knowing as far as possible the nutrient make-up of the co-prouct is a worthwhile goal to reach in practice. An agreement paying on dry matter content (DM) and checking the DM content on delivery (a very simple and rapid test done in the office) is important until suppliers further improve the consistency of their products, which is happening now.

Really bad DM under-deliveries can then be referred for a price reduction on that batch and/or passed to a nutritionist to reformulate the balancer meal. Is such 'diet tweaking' worthwhile? It seems so, as where this has been done food conversion has improved by 3% and MTF (saleable Meat per Tonne of Feed) rose by 9 kg. That doesn't sound much but it is equivalent to an immediate 8% fall in the cost of a finisher feed, definitely not to be sniffed at. Again in practice, if the supplier's product varies that much then it is useful for the nutritionist to supply a range of standard reformulations to save inputting a new ration formulation every time - the CWF computer can do this in a couple of minutes and as often as you like. Just press buttons or let the nutritionist do it from a distance.

## FUTURE REASONS FOR USING CWF

Before I list the impressive number of benefits which are being secured from CWF, let me tempt you with a longer-term view. If the technical advances of the past 20 years are continued at the same pace, the future possibilities of CWF seem limitless - if only from what we know is possible today but which is not being put into practice due to a variety of practical hurdles which have not only put farmers and the feed companies off, but has also stunted investigatory research whose scientists feel there might be no market, at least for a while, so research funding is deflected elsewhere. In addition there are a number of future possibilities not yet explored and surely others yet to be discovered - who knows what lies around the corner.

If some of these suggestions below seem fanciful - think back to the present advances in pig nutrition, management and disease control which were unimaginable 50 years ago when I began my career in pigs.

## SOME IDEAS WHICH, WITH FURTHER RESEARCH, CWF COULD UNLOCK AND GO ON TO EXPLOIT

* The holy grail of feeding grass, brassicas, discarded vegetable haulms and even hedge-clippings to pigs.

* In the tropics, banana leaves and other exotic plants likewise, at present discarded.

* Synthetic amino acids not needing to be dried (and thus are 28% cheaper, I'm told) if available in liquid form added at the farm.

* Likewise existing protein sources with far less pre-processing needed and delivered by tanker. Vitamins too in liquid drum concentrations.

- Enzyme-rich ingredients like liquefied triticale which is predigested before it reaches the pig's stomach.

- If the Challenge Feeding concept eventually surfaces, potentially of huge significance in matching diet to immune challenge, then CWF is ready and waiting to accommodate the on-going dietary changes needed as the immune threshold changes.

- As it is for the Blend Feeding concept, where the manufacturer can mix 300 different diets from just two bins, maybe a small third one for additives. This brings the farm-specific diet into play (another holy grail) which removes the need for the feed mill having to stock an expensive product range and warehousing costs. The farm specific diet concept is already under way from progressive pig feed manufacturers and a CWF set up on customers' farms facilitates it.

- Phase Feeding seems to have stumbled recently as not all the research is positive and the reasons need more research. Especially into Multiphase Feeding where a protein accretion curve is followed daily and frequent changes made to the nutrient specification. This can be handled effortlessly by today's CWF computer which can change the nutrient specs daily if it is ever needed.

- We already are seeing ESF-type technology applied to automatic weighing (tomography), and sorting, heat detection and adjusting diets to environmental variation. CWF is even better-placed in the first and last as it is predominantly involved in the growing/finishing pig where these concepts have the greatest impact on production cost.

- The possibility of (advance) disease detection is an exciting new area - CWF could take this on perfectly well as it is present in every pen every day.

- Lastly, Menu Feeding for nursery pigs and Choice Feeding (see below) for all pigs are ideas which have blossomed and seem to have faded, mainly due to logistical difficulties which may have discouraged the researchers. If this was so, then CWF could re-address these snags. Time for another look?

# THE ADVANTAGES OF CWF AS THEY EXIST TODAY

Let us now look at some of the present possibilities in more detail. Those here now or on the cusp of commercial practicality and being tried on commercial farms as I write. In this respect CWF has long been ahead of its time, as it provides the on-farm technology and practical application to accommodate all the following benefits.

# AHEAD OF ITS TIME

## What are the advances in more detail?

*1.* *'Challenge' or 'Test' Feeding* *(See also Immunity Section)*

Solves the problem of wide differences in protein accretion curves *among pigs of the same genotype* on different farms caused particularly by differences in disease thresholds and also by variations in environment between individual farms. A small sample of growing/finishing pigs are tested regularly on a non-nutritionally limiting diet and carefully monitored. The results are computer-modelled and a Farm Specific Diet (FSD) is least-cost formulated for the whole herd based, for example on the lean accretion curve revealed by the test results.

<u>Value</u> : Between 20-40 kg more saleable meat/tonne feed is suggested. Feed cost/tonne rises by 6% to 8% but gross margin increases by 10%-13%; nett margin by as much as 20%

Wet feeding best suits this concept

*2.* *Blend Feeding*

The obvious problem created by Farm Specific Diets (FSD) is the multiplicity of diets demanded of the feed manufacturer. Initially some compromises can be made, for example by having a range of diets with nutrient densities most closely fitting the commonest protein growth curves, or supplying different basic diets for high/low disease status.

In future however full FSD will be available – one for each farm, reviewed regularly – and to avoid multiplicity of formulations, all can be made from only *two* diets delivered to the farm and placed in separate bins. One is of high nutrient density, one of low nutrient density, with a small premix hopper in between.

By blending varying amounts of the two primary feeds into a wet mixer, *every farm variant can be made on-farm* and the product range the feed compounder needs to carry is drastically reduced. Formulation and blending is all computer-controlled.

Wet feeding can accommodate this process in the most economical way.

*3.* *Multi-pen feeding*

In the short term future (5-10 years) many piggeries will continue to contain up to 9 different weight bands of growing/finishing pigs in one house. (Beyond this, multi-site production will adopt the poultry concept of batch rearing similar weights of pigs all placed in one house).

Only wet feeding by computer can easily, economically and accurately feed
9 weight categories of pigs with up to 14 different diets

## 4.   *Multiphase feeding*

At present we **Step-feed** (3 diets and only 3 nutrient ratios for a starter, grower, finisher) which is very inefficient.   **Phase feeding** (about 5 early steps, 3 later steps) is better, but not ideal.   **Multiphase Feeding**, where there are 30-50 nutritive-ratio changes across a pig's growing life, gives slightly improved performance over Phase Feeding but marked reductions in N+P pollution.

**Multiphase Feeding and FSD** (under research) is likely to give much improved performance as well as modest improvements in pollutants, but considerably less slurry **volume** as less water is needed to metabolise more efficient protein use.

Only pipeline feeding can cope with this degree of sophistication, again
effortlessly and accurately.

## 5.   *Choice feeding*

Up to now all these developments involve the **nutritionist** deciding when to change dietary allowances and nutrient values.   Choice Feeding allows the **growing pig** itself to make the dietary changes – quite accurately, it seems where the all-important protein intake is concerned.

While choice feeding for older pigs is still under research, a development of this – Menu Feeding – has worked well with nursery pigs.

## 6.   *Menu feeding*

Provides two feeds of slightly differing nutrient densities to be available at any one time all through the nursery period of 6 to 25 or 30 kg by which time a total of 6 diets have been on offer.   By a process of 'leap-frogging' one diet is changed every 7 to 9 days.   The diets also vary in flavours to further stimulate appetite as young pigs may quickly get bored with one added flavour throughout. Dramatic increases in FI (24%) and DLWG (23%) have been obtained in the nursery period.   However improvements in the all-important FCR are usually modest (1%-2%).

The real benefit occurs at the end of the finishing period even if the pigs are fed conventionally from 25/30 kg to slaughter.   Menu Feeding early on can give between 7 to 21 days quicker to slaughter.

And an increase of up to 20-40 kg saleable lean/tonne of food.

While the different diets can be added dry, by hand or auto-control, it is
easier and cheaper to do this by pipeline to avoid hassle and mistakes.

## 7.   Wet feeding nursery pigs (gruel feed)

Still under development, and on some farms now routine.  It is well established that the suckling pigs' intestinal surface, once weaned, is far less damaged if transferred to a thick wet gruel than to dry or even wet/dry feed.  (Figure 1).

So the weaners get away faster and reach slaughter sooner.

*A pipeline system is needed to achieve the degree of wetness needed.*

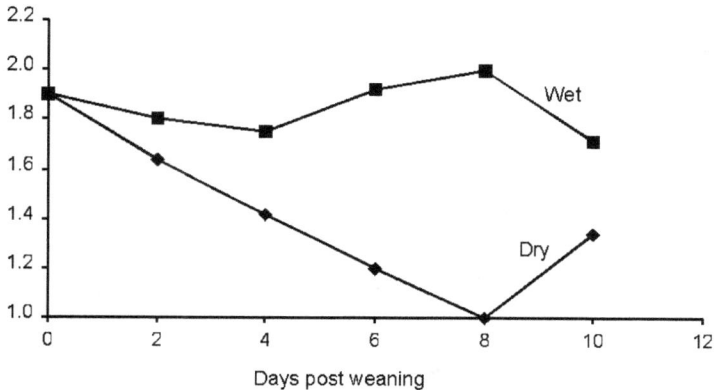

**Figure 1.**  Villus/crypt ratio in the distal jejunum of piglets fed a wet or a dry diet

## 8.   Inoculation by probiotic bacteria, with or without enzyme addition.

Inoculation of wet feed for weaners with fermentative bacteria markedly raises acid levels in the stomach thus reducing susceptible pathogens to harmless levels. The feed is held soaking with the inoculum for several hours before feeding, thus both physical softening and enzyme formation helps predigest the food in the relatively poorly-developed digestive tract of the weaner, itself under severe stress trying to cope with solid feed rather than milk. (Figure 2).

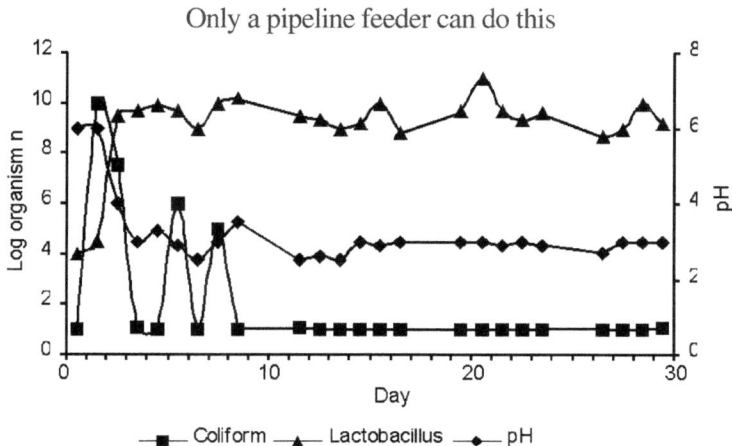

**Figure 2.**  Effect of *Lactobacillus* spp. growth on pH and *E. coli* numbers in liquid feed.
Source: Brooks (1997)

***What the graph shows:*** The satisfactory drop in pH (*i.e. rise in acidity*) of the digestive tract after 4 days. This makes it difficult for hostile coliforms to survive beyond day 10 while the acid conditions create favourable conditions for many beneficial bacteria *e.g.* lactobacilli.

Phytase is rendered more active and thus more phosphorus is released when it is included in wet feed (Figure 3).

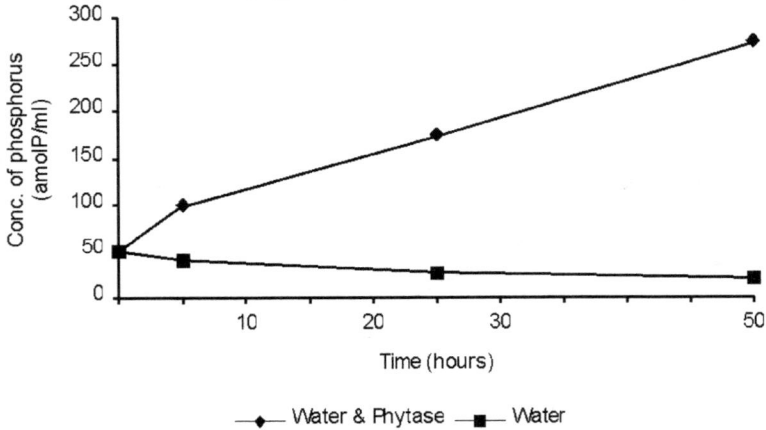

**Figure 3.** Effect of steeping soyabean in water or water-phytase. Source: Brooks (1997)

## 9.   *Nutritional Biotechnology*

We are seeing this area grow. It will grow more in future. Often, with nutrients like organic selenium (0.3 ppm) and organic chromium (200 ppb) only tiny quantities are involved – but they do a lot of good: far greater than their inclusion levels suggest.

Only pipeline mixers can cope with very small amounts accurately without recourse to unnecessarily bulked-up (thus more expensive) carriers.

## 10.   *Liquid feeds*

Crystalline supplementary amino-acids cost a lot to dry into a free-flowing powder-form. To add them wet, as made, is much cheaper.

Future enzyme technology will allow us to use 'wet' foods, presently too indigestible for pigs, like grass, grass silage, brassica tops, banana leaves, potato haulms and possibly even forest foliage.

Several factory by-products from the petro-chemical industry, the canning / sugar / confectionery industries can be used for pigs, to add to those from the dairy industry in use now.

Only a pipeline system can use all these materials and more

# PRESENT ADVANTAGES OF COMPUTERISED WET FEEDING

1.  **Pigs do it better.** A survey of the literature I did several years ago showed:

    | Effects of wet feeding | Growth rate | FCR | Carcase quality |
    |---|---|---|---|
    | Improvement | 37 | 32 | 8 |
    | Deterioration | 4 | 5 | 2 |
    | No significant difference | 12 | 16 | 16 |
    | No information | - | - | 27 |

    *To 2009.* Since then, 28 opinions of farmers who have made the change report a similar trend, with grading - subsequently remedied by a nutritionist - the only initial problem. A typical observation is "growth rate and MTF (Meat sold per Tonne of Feed) are noticeably better"

2.  **Less food wasted.** Wasted food can be direct (down the slats; trodden-in; lost in dust) or indirect (wrong nutritive ratios etc). Wasted food from pipeline troughs is not significantly different to that from wet/dry feed hoppers. Feed waste is literally money down the drain. Many dry feeders waste 6%, some as much as 15%.

    Savings when transferring to CWF. 5000 finished pigs/year (tonnes/year)

    | | With wet feed | Without wet feed | Savings to transfer to wet feed |
    |---|---|---|---|
    | Farm 1 | 770 | 838 | 68 tonnes |
    | Farm 2 | 803 | 900 | 97 tonnes |
    | Farm 3 | 984 | 942 | 42 tonnes |

    i.e. Dry pellets v. the same diet fed wet. Before and after figures taken from the purchase invoices of each farm.

    Comparison from one farm where wet/dry and fully wet systems were run in parallel for a while

    | | Fully wet n =20 pens | Wet/dry* n = 20 pens |
    |---|---|---|
    | Food collected as waste (dry matter basis) | 2.1% | 2.0% |

    i.e. wet-fed v. the same diet fed from wet/dry feeders. * Plate feeders to a proved design. Feed waste was collected from grilles under each feed station. Ad lib feeding in both treatments. Source: Gadd (2003)

3.  **Pigs eat more, convert better.** Appetite can be a limited factor these days. This is a problem in the modern hyperprolific sow and gilt, and in all pigs in hot conditions. For finishing pigs, FCR (30-105 kg) is often 0.1-0.15 better. For sows: up to 1kg/day in lactation more with piglet mortality to weaning down 1.7%, farrowing index up 6%, sows served by 5 days or less 23% higher, weaner weight/sow/year +17% (from 126 to 148 kg). These figures come from several clients' records (1997-2005).

4.  **Sow farrowing crate occupancy is better.** Weaner weight per crate +11%. Important as a square metreage of the modern farrowing crate is the most expensive space on the farm.

5.  **Sow condition is better.** More important than it sounds as a 'nose-dive' in fleshing and/or fat cover down through lactation puts a brake on rapid reservice and prolificacy/survival of the following litter. Table 4 is one of several comparisons I have filed over the years.

**Table 4. The effect of wet and dry feeding on sow condition and performance**

|  | Sows fed wet | Sows fed conventionally |
|---|---|---|
| Average condition score | 2.6 | 2.5 |
| Pigs per sow per year | 21.4 | 19.1 |
| Weaner weight per sow per year, kg | 147.7 +17% | 126.1 |
| Weaner weight per crate space per year, kg | 773 +11% | 696 |

Source : Clients' records

6.  **Wet feeding is particularly advantageous to sows in both hot/dry and humid tropical conditions.**

**Table 5. Sow trials on wet v. dry feeding of sows in the tropics**

| Farm 1 | Farm 2 | | (average of 2 trials) | |
|---|---|---|---|---|
|  | Wet | Dry | Wet | Dry |
| Farrowings | 136 | 130 | 161 | 85 |
| Av daytime temperature, °C | 28 | 28 | 30 | 30 |
| Feed eaten in lactation, kg/day | 6 | 5 | 6.2 | 5.5 |
| % sows served by day 5 after weaning | 64 | 46 | 60 | 50 |
| Weaner weight per sow per year, at 21 days, kg | 148 | 126 | 120 | 94 |
|  | +17% | | +28% | |

Source : Clients' records, Thailand, (1993)

7.  **Dust is markedly lower.**

Meal 14-79 mg/m$^3$. Pellets 5-23 mg/m$^3$. CWF 0.5-14 mg/m$^3$.
Source: Cermak (1978)

Carpenter reported that airborne micro-organisms are 3 times higher in the pens of dry-fed pigs and Robertson found that 45% of dust particles were in excess of 10 mg/m$^2$ - the exposure limit for the UK safety regulations during milling and mixing operations on-farm. With CWF the mixing is done in a tank thus removing up to half the dangerous dust problems affecting operator's health as well as lowering the risk of dust explosion.

8. **Healthier stockpeople.** Today on wet fed farms there must be fewer coughs, eye, nose and throat irritations, and thus less days off work. Table 6 shows the serious position of stockpersons' health incidence when pipeline feeding was only 6% at that time. Now it is some 6 to 7 times higher the incidence of the first 5 symptoms seems to have been halved (recent correspondence with 4 UK health authorities in rural areas).

Table 6. Pigmen reporting symptoms

| | | |
|---|---|---|
| 1. | Cough | 58% |
| 2. | Phlegm | 39% |
| 3. | Chest tightness | 26% |
| 4. | Throat irritation | 39% |
| 5. | Nose irritation | 39% |
| 6. | Eye irritation | 25% |
| 7. | Fatigue | 35% |
| 8. | Muscle pain | 22% |
| 9. | Joint pain | 23% |

N°s 1 - 6 are directly attributable to or aggravated by dust in piggeries.
Source : Watson (1978)

9. **Happier stockpeople.** One laborious task removed. <u>Manhours/week</u> spent preparing food and feeding 5000 pigs. Dry 20-30; Wet 5-6. <u>Staff turnover/year</u> 42% dry; after wet feeding 10%. <u>Easier to recruit</u> youngsters due to promise of computer use. Clients' information (2000-2009).

10. **Better use of labour.** World-wide, producers complain about the difficulty of getting and keeping good replacement labour. Moving food around has always been a heavy, onerous task (Table 7) and liquid feeding removes this completely. Hydraulics do the work!

Table 7. The amount of food handled by pig producers in a year for every 100 sows

| | *Feed handled per year (tonnes)** |
|---|---|
| Breeding herd | 142 |
| Weaner herd | 127 |
| Finishing herd | 271 |
| Total | 398 |

* *Farrow to finish – 100 sows, 20 gilts, 5 boars, 22 pigs/sow/yr sold*

**Table 8.   Effects of dry and wet feeding\* on labour issues.  (average of three farms corrected to 5,000 finishers at any one time)**

|  | *Manual delivery* | | *Wet fed by pipeline* |
| --- | --- | --- | --- |
|  | *Individual pens* | *Ad lib groups* |  |
| Man hours per week (feeding) | 50 | 20 | 5 |
| Labour cost per finished pig\*\* | £3.07 | £2.38 | £2.01 |
| Hours lost to sickness/non attendance (per year | 270 | 212 | 89 |
| Staff turnover across 5 years (%) | 64 | 58 | 10 |

\*  Farmer with 5,000 finishers to feed twice daily in five piggeries.
\*\* Including all other tasks
Source: Clients' records (1990s)

Wet feeding by pipeline reduced labour load on the finishing farms between 10 and 4 times.

11.   **Quick and accurate medication**. Literally within seconds even at very high dilution rates, affording 50% reduction in medication mixing costs (Taylor, 1976). Water is a far quicker and more diffuse substrate than meal, especially for minimal addition rates. These days additive manufacturers can advise as little as 250g (or 250 ml) per tonne of feed or liquid and the liquid fraction in CWF can deal with this low level.

12.   **Less stress.** Now a major 'hidden' problem in all pigs. Time spent dozing/ sleeping (pigs 20-50kg). Wet-fed 53%; Dry pellets 45%. After a wet feed 70% of grouped sows settled down to rest within 45 minutes compared to 80 minutes on dry pellets.  Several farms report a permanent disappearance of tail biting when changed from dry feed to a thick (3:1) mix).

13.   **Less mycotoxin damage.**  The other 'hidden' problem in all pigs. Because the mixing tank and pipelines can be/should be regularly sanitised and the pigs 'polish' a trough when feeding is finished, residual mycotoxin build up is lowered or even eliminated.  However, ad-lib troughs and sow troughs need careful attention and it is advisable to add a mycostat or mycoabsorbent binder to these wet feed mixes as a precaution.

14.   **Reduced slurry volume**. Logically one would have thought there was more slurry, not less from liquid feeding.  I find this is often not the case (Table 9).

**Table 9. The effects of wet and dry feeding on slurry volume. (Slurry removed from two dry sow houses in winter [Sept-May].)**

|  | *Wet-fed* | *Dry-fed* | *Wet-fed* | *Dry-fed* |
|---|---|---|---|---|
| Per sow per week, litres | 126 | 148 | 117 | 115 |
| Tanker loads/herd | 4 | 5 | 2 | 2 |

Source: Gadd (unpublished)

**15. Easier to recruit good labour.** Surveys of stockpeople have shown that a progressive farm is more likely to attract good employees (Table 10).

**Table 10. Ranking of reasons for accepting or quitting employment on pig farms in the UK. (Attitudinal surveys of skilled stockpeople.)**

| *Accepting* | | *Quitting* | |
|---|---|---|---|
| Modern attitude* | 11 | Work is hard, dirty, repetitive | 12 |
| Convenient location | 10 | Nobody listens to me | 10 |
| Automation* | 10 | No future/lack of time off | 8 |
| Wages/money | 10 | Not keen on pigs | 6 |
| Benefits | 8 | Don't like co-workers | 3 |
| Working hours* | 7 | Need a change anyway | 1 |
| Need the job | 3 | | |
| | 59 | | 40 |

Source: Staffing Agency (1988)     Source: Gadd, Survey (1990)
*Includes being able to use new technology, *e.g.* computerised wet feeding

**16. Less chance of salmonella.** From Denmark, the processors Steff-Houlberg and Danish Crown showed there was less risk of salmonella on wet feed (Table 11), which rather puts paid to the objection that wet feeding breeds pathogens. Of course it ***does*** if the facilities are filthy, but as the Danish work suggests, under decent conditions the risk (of salmonellosis) seems less. An interesting finding, backing up the view that with a one-hour soak before feeding, the acidity rises sufficiently to penalise salmonellae (and possibly other pathogens though only salmonella was measured).

Note that this research shows the risk of having over 33% samples positive for salmonella in meat was five times greater for units feeding dry feed than for units feeding wet feed.

**Table 11. The extent of salmonella in wet- and dry-fed herds**

|  | *Wet feed* | *Dry feed* |
|---|---|---|
| Over 33% positive | 4 (0.85%) | 92 (4.2%) |
| Under 33% positive | 466 | 2189 |
| Total | 470 | 2281 |

Source: Steff-Houlberg (1998)

17.   **Wet-fed pigs are more contented.** Everybody likes quiet, happy pigs. Bishop Burton Agricultural College in Britain did some interesting student work on growing pigs a few years ago as shown in Table 12.

**Table 12. Proportion of time pigs spend in various activities  20-50 kg during wet and dry feeding**

|  | Wet-fed (%) | Dry pellets (%) |
|---|---|---|
| Asleep/dozing | 53 | 45 |
| Feeding/drinking | 7 | 12 |
| Social activity | 35 | 32 |
| Fighting/playing | 5 | 11 |

Cambac Research drew this out most elegantly with group-housed sows, who settled down dramatically faster after being wet-fed (Figure 4)

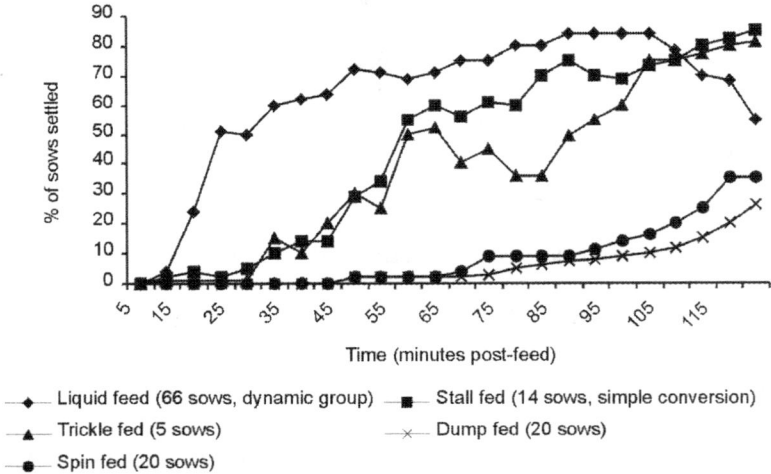

— Liquid feed (66 sows, dynamic group)   —■— Stall fed (14 sows, simple conversion)

—▲— Trickle fed (5 sows)                 —×— Dump fed (20 sows)

—●— Spin fed (20 sows)

**Figure 4.** The effect of differing sow feeding systems on post-feeding behaviour

# THE SNAGS AND DRAWBACKS TO PIPELINE FEEDING: HOW TO AVOID THE PITFALLS

About 20% of the problems encountered have been at the installation stage, and another 20% have been due to home-installed circuits where the producer himself, 'good' with plumbing, welding (both metal and plastic) and machinery, nevertheless got things wrong, sometimes invoking rectification which cost more than any savings he enjoyed from not employing an experienced installation team.

Table 13 summarises my own experiences from dealing with liquid feeding problems.

**Table 13. Analysis of 62 complaints/call-outs over the past 25 years involving the installation of liquid feed systems**

| | |
|---|---|
| Overfat pigs | 13 |
| Blockage | 16 |
| Food wastage | 7 |
| Dirty conditions/wrong machinery | 9 |
| Respiratory problems | 2 |
| Poorer performance | 8 |
| Computer origin | 6 |
| Trough fouling / lying in troughs | 3 |
| Young pigs / feeding messiness | 2 |
| Food delivery (valves) | 2 |

Some are duplicated *i.e.* more than one problem was encountered on the farm.

*Note*: The vast majority of these problems were sorted out, some very quickly. There were 4 complaints about frost which did not need a visit to rectify. Problems with FLF are not included as this concept is still under development.

Many problems have occurred with wet feeding because insufficient 'homework' was done before adopting the system. As with any radically new concept in pig production, ***do your research first***. Take time to visit, compare, enquire and argue/ negotiate. The pig industry has at least 40 years experience of basic liquid feeding of pigs, ***the answers are all there***.

# A START-UP CHECKLIST

✓ ***Choose a reliable manufacturer with a proven track record.*** If your research throws up hearsay problems, question him assiduously on a "what if" basis (mechanical and computer breakdown etc) and mull over his replies, especially valuable if he can direct you to customers who know about or have encountered the problem in particular.

✓ ***Be careful about new gimmicks.*** Ask for evidence that the new idea/ development/cost saving has been thoroughly tested under farm conditions and where, then go and see it in action. Ask yourself 'do I really need the modification/update?'. My maxim has always been... 'Simplest and most rugged is best'. Having said all this – I am impressed with the innovative track records of most of the firms involved in liquid feeding today.

✓ ***Make sure the installation team is qualified and experienced.*** The best equipment, poorly installed, will give trouble. If the manufacturer has his own installation division, this is excellent. If he subcontracts out – even under his own supervision, question him closely about the evidence of their experience and do a telephone check of customers who have used the subcontractors *recently*. Be especially vigilant if local electricians/plumbers are employed as sub-contractors.

✓    ***Ensure advice, spare parts and prompt service are available.*** You and your pigs will come to depend entirely on the system, so a prompt rectification/service is essential. *Author's note*: Especially over the long Christmas/New Year break! With others, I once had to feed 12,000 finishing pigs by hand for 3 days (and nights!) over Christmas because a spare couldn't be located. Never again!

✓    ***What facilities exist for initial training of your stockpeople?*** Will the supervisor stay on for 2 or 3 feeds to ensure you/they get it right? How good is the instruction manual? Does it have a 'what-if' section? Insist you get taken through it – make notes. Is your supervisor available on call/mobile subsequently? He knows your installation in detail, and can spot errors at once.

═══════════════════════════════════════════════════════════════

Having shown you an impressive list of the 'pros' of CWF it is equally important to recognise the drawbacks - the 'cons'.

# WHAT ARE THE SNAGS OF WET FEEDING?

These fall into two categories: True Snags and Perceived Snags. Perceived or apparent snags are those the newcomer worries about (justifiably) but can avoid with know-how and foresight. True snags have to be dealt with and absorbed.

## True snags

- *Capital cost.* I have discussed what it can cost today, but what has been the past experience? Installing a pipeline is expensive in capital needs, especially in converting an existing dry/wet-dry farm to it. Experience from 137 farms over 30 years suggests it raises housing costs, normally 9% to 11% per pig, to 11% to 13% across a ten-year amortization period.

  Also, updating the ventilation system (see below) will put between another 1.5% to 8% on top of this.

  For this you should get - average figures from many of those 137 farms:

  - A 0.1% improvement in FCR at least on every pig sold. Often more.
  - Or another 20 kg more saleable meat/tonne of finisher food.

This latter is a convincing sales point in liquid feeding's favour, and helps put the somewhat daunting extra capital cost in perspective, like this:-

You know how much income you should get per kg of saleable (*i.e.* dcw) meat. Next, a good pipeline installation should last 10-15 years (I've known early models, *e.g.* Taymix, last 27 years). Now calculate how many tonnes of feed the circuit will handle over 10 to 15 years and relate the value of 20 kg – or even 10 kg – more meat sales per tonne of feed handled, to the total capital cost. If this figure doesn't convince you, nothing will – so I'll leave you to do the sums!

Other economic benefits can be added to this:

- vet/med costs down by up to 33%

- a happier, healthier workforce staying longer and easier to recruit (Table 9)

- between 4% to 6% lower labour costs overall or much better *use* of labour (Table 6)

- 5% to 9% less annual housing maintenance costs (fans, structure)

- a minimal-cost entry into home-mixing, itself worth up to 18% cheaper feed-cost/tonne (25% with by-products).

From this equation you can see it easily pays for itself. Farm data collected since the 1990s show an REO of between 2:1 and 6:1 (average 4.1:1). The main difficulty is often finding the capital in the first place among other pressing needs.

- *Ventilation*: Wet feeding is a wet process! Ventilation will often need to be reviewed by an agricultural engineer to remove the excess humidity, especially in winter. Don't forget to check on this.

Sadly, the max/min ventilation rates advised in liquid feeding growing/finishing buildings, coupled with the need for accurate air placement, is often adrift. In my experience *all* the complaints of poor or disappointing performance can be laid at this particular door, and I find them operating years after the original conversion to liquid feeding. It is a ventilation problem NOT a wet feeding one.

- *Bloat*. Wet fed (older) pigs tend to bloat, especially with whey. Anti-bloat formulations can be obtained but it is still a problem. You will lose 1% more finishing pigs due to this, and I have never managed to get it much lower. You can ameliorate its effect, but rarely erase it entirely.

- *Young pigs – less than 25 kg.* Quite a few problems here – initial inappetance (odd, but it can happen), messy eating/wastage, lack of dry matter intake, oedema, and bed-wetting are all problems particular to young pigs. All these can be eventually overcome by trial and error alterations to management and improving nursery living conditions.

**GOLDEN RULE: The eating troughs of young wet-fed pigs MUST BE KEPT FRESH AND CLEAN!**

## Perceived or apparent snags

All the following snags are commonly encountered and made much of by the adherents of dry feeding. In fact they are rare in a well-installed pipeline system.

- *Frost* is also rarely a problem. Design the layout to accommodate frost and wind chill however acute. The degree of insulation the Swedes and some experienced Canadians use is surprising, but totally effective. If you are really worried, then drain the system after use *as long as the circuit does not have sagging pipelines*. If so, keep them charged full but insulate well.

- *Pipefouling* is rare if the equipment is used daily. Use water flushing and a dump tank to keep the circuit clean. Anyway sanitisation (pigs present or absent) is possible. To avoid mixing tank contamination of its upper, unscoured surfaces, use the whirlspray device periodically.

- *Blockage* can happen, but very rarely. Design the problem out at installation, so if a block does occur you can easily remove it. For example, a rat could get in.

- *Never pump downhill*. Pumping down is a logical idea so that gravity saves a little electrical energy or increases pressure at a distant pen. Trouble is, unless the circuit is flushed out and left with air (frost) or plain water, the feed particles flocculate down to the foot of a vertical or acute slope, leaving the supernatant liquid as a cushion protecting the solid plug below when the CWF mix hits it from above. But by pumping *up* a vertical, the turbulent liquid/solid mixture eats into the plug from below dispersing it into the liquid above and freeing any overnight blockage due to settling.

- *Do not have loops (sags) in horizontal runs*. A good installer will design this out.

- *Install an Oxford Union at each right angle bend*. Then all straight runs can be rodded out if needed.

- *Never seal a pipe underground*. Place in a channel covered with metal plates.

- *Overfat pigs* often occur on changeover as the diets aren't altered to allow for the overeating due to improved palatability, which can occur on *ad lib*. Consult your nutritionist. The same phenomenon occurs when changing from dry feed to wet/dry feed, and the solution is the same – adjust the diet. All newcomers encounter this problem of initially downgraded, fat pigs, and tend to be dismayed by it. A nutritionist can quickly address the problem. On ad lib CWF pigs just LOVE the 'soup' and the nutrient density needs readjusting a little.

- **Water deficiency.** Always have a drinker available. Liquid feeding is not a substitute for water, merely a physical method of moving bulky feed. Indeed, whey and skim 'carriers' are thirst-making in themselves, and whey concentrate is high in salt, as are some other edible industrial by-products. Supplementary water is absolutely essential.

- **Ironing out the snags pays hands down.** Finally, here are the consistent results across 3 years from 3 farms who had needed a good deal of attention and rectification in their early stages, having not got several of the perceived snags ironed-out before converting to liquid. They were on three different manufacturers' systems.

**Table 14. Feed consumption and FCRs on farms before and after wet feeding (av 30-88 kg) 1994-1998.**

| | Feed eaten kg kg | | | FCR | | Yield of extra saleable meat Per tonne feed kg |
|---|---|---|---|---|---|---|
| | *Before* | *After* | *Improvement** | *Before* | *After* | *kg* |
| Farm A | 168 | 154 | 8.3% | 2.89 | 2.61 | +29.3 |
| Farm B | 180 | 161 | 10.6% | 3.05 | 2.72 | +31.9 |
| Farm C | 197 | 188 | 4.6% | 2.81 | 2.69 | +10.9 |
| **Average** | | | | | **7.8%** | **21.0** |

\* The average of 7.8% wastage accords well with Dr Mike Baxter's work which suggested most producers waste 6% of their dry feed

# FUTURE REASONS FOR USING PIPELINE FEEDING

Up to now I have listed some of the evidence why large numbers of pig producers have already switched to wet feeding. But the future of pig nutrition is even more exciting and is changing quickly. See page 441.

Pipeline feeding is superbly poised to accommodate these developments because, as distinct from dry, or even wet/dry feeding . .

\*   It is remarkably flexible and adaptable.

\*   The computer technology needed is already here.

\*   The equipment is ready and waiting as are distributorships and spares/service facilities in many countries.

\*   The know-how / track-record of companies in this field is considerable, *e.g.* Big Dutchman have over 5000 installations world-wide.

# REFERENCES

Brooks P H, Geary T et al (1996); Procs. PVS (*Pig Journal* pps 43-67)

Carpenter, G A; *J Agric Eng Res* 1986 **33** 227-241

Cermack, J P (1978); *Farm Buildings Progress* **51** 11-15

Chesworth, K; Procs. Australian Pig Science Assoc. Conference, Adelaide Nov 2001

Gadd, J; 'Pipeline Feeding' in '*The Pig Pen*' Vol 4 N°4 Jan-March 1998.

Gill, P. (2004-2005) Finishing Pig Systems Research (Liquid Feeding) Reports 1 to 4. *Important fundamental and exhaustive research trials on liquid feeding, including econometrics.* UK British Pig Executive, Meat and Livestock Commission.

Lumb, S. (2002) Liquid feed research tackles co-product value. *Pig Progress* **18**, 7, 12-14.

McKeon, M. (2008) Cut your slurry costs. *Pig International*. Oct. 2008, 22-23

Robertson, J F; Dust in farm mill & mix plants (1991) *Farm Buildings Progress* **106** 14

# 20

# BATCH FARROWING

A deservedly popular strategy which has grown in popularity across the globe over the past 10 years. It consists of assembling the breeding herd from continuous production where animals are served, farrowed and weaned almost daily as they come round in the breeding cycle, into weekly batches where the primary tasks of serving, farrowing and weaning are all carried out sequentially in one period, usually a week for each of them. This has several advantages of which are a more efficient use of labour, improved herd health and lower piglet mortality.

To describe the procedure in detail would take up 20 pages in this book so I am going to describe the basic, salient points of the technique and pass on those which may help the reader new to the idea, better understand what the concept involves.

## How the idea works

Progesterone is a steroid sex hormone which initiates heat in the female. Its action can be blocked by a synthetic hormone which mimics the secretion of progesterone, thus delaying the cycle until the sows or gilts are at the same point in time, when treatment stops and the cycle can begin with all the females in the group at the same stage in the cycle.

It merely puts heat on hold and does not affect it once it is initiated. It can be used for indoor or outdoor breeding.

Added by a squirt-dose dispenser to the sow or gilts food daily, it ensures that the animals treated come on heat together when the treatment stops and a so group is formed which start their breeding cycle at the same time.

The product 'Regumate Porcine' (Janssen Animal Health) is the most popular in Europe, available from your veterinarian whose advice and supervision of the carefully-planned conversion schedule from a continuous to 3 , 4 or 5 week batch farrowing is essential. Conversion to a fully-operational batch production system takes about six months.

## Main advantages

Here is an impressive list I have collected from clients across the last 5 years. By converting to batch farrowing......

- Compared to traditional continuous production there is more efficient use of labour leading to more contented staff working more effectively.
- Despite this, man-hours per sow per year is not changed all that much. However the productivity from the same number of hours worked is greater (Table 1).

**Table 1. Results before and after converting to batch farrowing**

| | *2 years previously (5 parities)* | *18 months ( 3 parities) after the batch system was fully operational* | |
|---|---|---|---|
| Average litter size | 10.36 | 11.21 | |
| 24 day weaner wt/ sow/ year (kg) | 124 | 133.6 | (+ 7.7%) |
| Vet/med costs per kg weaner weight (GB£) | 0.090 | 0.074 | (- 16%) |
| Labour costs per kg weeaner weight (GB£) | 0.27.2 | 0 25.1 | (-18.0%) |

**Summary:** Pigs carried on to slaughter weight......
Total benefit to batch farrowing £0.12.4/kg sold.(12.4p/kg)
**Less** extra cost of farrowing accommodation needed  £ 0.012/kg (1.2p/kg) sold
**Net gain £0.11.2/kg = £ 9.67/pig (about +10%)**

**Source**: Gadd (2005)

Figures corrected for pig price and labour cost changes over the 4.6 years recorded.

- The benefits can also be carried on into the finishing stages (Table 2).

**Table 2. Comparison of growing/finishing pigs weaned weekly compared to weaning every third week**

| | *Weekly weaning* | *Weaning every third week* | *Improvement* |
|---|---|---|---|
| Daily Liveweight  Gain | 490 g | 547 g | 12% better |
| Feed Conversion Ratio | 2.36 | 2.26 | 4% better |
| Drug cost per pig | £3.07 | £1.83 | 48 % lower |
| Mortality weaning to slaughter | 11.5% | 6.6% | 41% lower |
| Financial advantage at the time | **£8.48/pig, or 8.7% more income** | | |

**Source:** Extrapolated, financially, from Kingston (2002)

- In addition Janssen reported (May 2010) that batch farrowing was providing one extra pig per (gilts) litter.

- The major tasks of insemination/ use of boars, farrowing and weaning always fall on fixed days.
- There is no overlap in these tasks as in continuous production, when some of all three usually have to be done in one day, thus staff work more efficiently and are more satisfied in their work.
- For example, taking a 5-week batch system and 3 week weaning – becoming popular – there are 2 busy weeks in the period, but in 2 weeks of the 5 there are no major activities such as service, farrowing or weaning, when maintenance and catch-up work can be done, and 8 of the 10 weekend days in the cycle are free of any major task. The two 'ease-off weeks' – if that is an allowable expression – is why this particular version of the concept seems to be growing in popularity. Table 3 illustrates the system.

**Table 3. Workplan for 5 week batch farrowing and 3 week weaning**

| *Day* | *Week 1 (Weaning)* | *Week 2 (Breeding)* | *Week 3 (Farrowimg)* | *Weeks 4 and 5 (No major activity)* |
|---|---|---|---|---|
| Mon. | Treat piglets | Insemination and scanning | - | - |
| Tues. | Move piglets | Insemination | Scanning | - |
| Wed. | Prepare gilts for synchronization | Insemination | - | - |
| Thurs. | Weaning | - | Farrowing | - |
| Fri. | Clean farrowing houses | Sows to farrowing houses | Farrowing | - |
| Sat. | Clean flatdecks | - | Farrowing | - |
| Sun. | - | - | - | - |

**Source**: Extrapolated from Janssen Animal Health (2010)

- Batch production allows the farrowing and post-weaning staff more time to attend to their two critical areas.
- There is greater control over fertility management. AI semen can be ordered in batches, cutting deliveries and reducing these costs, and which assists the AI process and QC procedures.
- More piglets are available for cross-fostering as more farrowed sows are present at one time which facilitates piglet swapping when needed.

- Pigs moving onwards in more uniform batches provide more even groups at slaughter which the processors want. It also speeds up housing turn-round time once the pens are empty.
- Improved turn-round time makes All-in/All-out easier to plan and manage.
- Batching groups suits the trend to raise large groups in yards/courts on bedding which is welfare-favoured by the legislators.
- Large groups make traceability easier as pigs are not mixed up when transferred to nursery and grow-out units.
- Batch production makes it possible to monitor feed and water consumption on a continuous basis so that changes in the pattern of consumption are noticeable. Such changes give advance warning of the onset of disease. This can be done using computerised equipment such as DICAM (Farmex UK) or load cells and water meters.

## Some hints from the experts

- Gilts must be well-grown (see Gilt chapter) and you must have enough of them - this is why a well-run gilt pool is important.
- They should be put on daily treatments of Regumate 18 days before the sows farrow in the group they are about to join.
- It is important that the gilt/sow eats its daily dose. Some people squirt the dose on to a slice of bread, which is more likely to ensure the pig eats it at once and avoids any other sow getting some of the dose in her food.
- You are going to operate larger groups of pigs so you need to give due advance warning to your supplier of gilts, similarly numbers in each batch of finishers being shipped to your processor.
- As new gilts will be coming in as groups, any harmful organisms new to the farm will be arriving in bulk too, so a high level of biosecurity is needed. Quarantining and induction (acclimatization) skills are important – see the Gilt Chapter.
- Batch production is a disciplined and somewhat rigid system, so when a building all has to be emptied by a specific date, it is helpful to have two outlets with different weight requirements, one with your primary market contract, the second being a 'safety-valve' outlet source, as flexible as can be negotiated.

## Overall view of the financial returns against continuous production:

The cost/sow/year of the treatment (Regumate) for a herd of 250-300 sows (2 men) should be under £5 (2010) and the benefit £20 sow/year on current prices - giving a comfortable REO of 4 to 1.

## What about the extra cost of buildings?

A difficult one, this, as farms vary a great deal on what new accommodation is available – a gilt pool for example. You will need a certain amount more of farrowing and weaner accommodation too, and how much this will be will depend on what is already available on the vital principles of **a complete group must fit into one or more of the houses** and that the houses **should not share the same air space,** so they do not want to be crowded together. Moreover each batch of pigs needs its **own separate pen accomodation** as it progresses through the unit.

My findings are that you could need 5%-10% more farrowing pens and 10%-15% more weaner space.

Building costs as a proportion of production cost also vary widely but if these are, say, 12% (amortised at the half-way stage of 20 years) on the quite expensive housing costs of the 21st century, then on the conversions I have studied in the past 10 years suggest that an extra 16% should be budgeted on to you existing housing cost - i.e. lifting the 12% cited above to 14 %.

As the politicians say – 'very difficult to give you a figure', but that seems to be not too bad a guide, but is assembled from clients opinions rather than records.

Table 4. How does batch farrowing/weaning compare with other improvement strategies?

| | Cost | Improved growth | Reduced mortality | Drug and vet/med | Approx. payback time |
|---|---|---|---|---|---|
| All-in/All-out | Low | 1-7% | 4-6% | -25 to 45% | Variable** |
| **3/5 week batch farrowing/weaning** | **Low** | **12-15%** | **40-45%** | **-30 to 50%** | **20 months*** |
| PD*/ sow medication | Fair | 25-45% | 45-65% | -55 to 65% | 9-15 months |
| Full De-population | High | 30-40% | 65-85% | -70 to 90% | 14-26 months. |

*Partial Depoulation ** This depends on how out of date the farm was before All In/All Out *** After conversion period of 6 months.
**Sources:** Various from 2006, based on a survey by Kingston (2004) And Parity Segregation come in the above Table? At the time of writing there seem to be insufficient details to complete the columns for reliable reference – these will emerge with time.

## Some snags to batch farrowing.

Most breeders I have visited are happy with the system chosen, but a few were not The problems seem to have been:-

- Skilled staff are vital. Those new to the schedules must receive training from your veterinarian or the manufacturers of the progestagen. Your staff must also be enthusiastic about the concept –they should be, as they will benefit from more sociable working hours.
- Breeders not going into the calculations and requirements thoroughly enough before changing. Table 5 is just one of about 10 which must be studied in advance of changing – not part-way through the process! Fortunately there is plenty of well-set out literature from the manufacturers and the pig specialist veterinarians. A pre-visit from the former's specialist is wise and veterinary supervision from time to time during the transitional period is, I think, essential.

**Table 5. Group size in relation to herd size**

| System | No. of groups* | Size of sow herd | | | | | | | |
|---|---|---|---|---|---|---|---|---|---|
| | | *100* | *200* | *300* | *400* | *500* | *600* | *700* | *800* |
| Continuous | 21 | 5 | 10 | 14 | 19 | 24 | 29 | 33 | 38 |
| 3 week | 7 | 11 | 29 | 43 | 57 | 71 | 86 | 100 | 114 |
| 4 week | 5 | 20 | 40 | 60 | 80 | 100 | 120 | 140 | 160 |
| 5 week | 4 | 25 | 50 | 75 | 100 | 125 | 150 | 175 | 200 |

With acknowledgement to Janssen Animal Health (2010)
* Varying sow performance can disturb the batch 'rythym'

- Housing must not be too close. Each batch should have its own air space and each batch of pigs needs its own separate pens/accommodation. This will involve some extra housing costs, as described, which must be calculated before taking the decision to change. Power consumption at peak times (but normally not overall) will increase, for instance all the new weaners all need heat at the same times.
- There is some slippage of productivity (e.g. more empty days) during the conversion period (up to 6 months), but done as instructed this is not onerous and is an investment in better future economic performance of both animals and staff.
- A well-managed gilt pool is essential.
- Of course All-in/All-out and good records are essential, too.

To conclude, having visited many batch farrowers since the idea started, my main impression was **how contented the staff were** with the new routine.

This must count for a lot.

# REFERENCES

Armstrong, D. (2002) ' Is Batch Farrowing for You?' Pig World Aug. 2002.

Grey,S. (1999) 'Bunch Your Farrowings' (housing details) Pig Int. <u>29</u> No.6.25-27.

Janssen Animal Health (2010) 'Batch Farrowing Guides' (two) **Recommended**.

Jennings, D. (2002). 'A Simple Way to Synchronise Oestrus'. 'Pig World, ibid.

Kingston, N. (2004) 'Health Upgrades: Disease Reduction Strategies for Finishing Herds', Procs. R.A.C. Conf. Cirencester, England. Sept. 2004. Recommended.

Newsham Hybrid Pigs (2002). Batch Farrowing Manual.

# 21

# PARITY SEGREGATION

**Concept:** The intention is to minimize the chance of disease in the breeding herd by....

1. Keeping gilts and first-litter sows well away from the rest of the older sows in the herd in order to minimize potential disease-shedding from these young animals, whose immune status is not yet fully established.
2. Following a similar practice for the progeny of these young sows, keeping them in a separate nursery and not mixing them with the weaners from the rest of the herd until the nursery phase is completed.

This development should be of particular interest to pig breeders due to growing evidence of its influence on disease. While it has been discussed by academics for at least five years, only within the past 2 years has attention been drawn to the concept, almost exclusively in the USA by articles and at technical meetings. Elsewhere I find pig breeders are vague about what it entails and have assumed that it means raising sows separately, parity by parity.

" Crazy idea; how can that possibly be afforded in housing and labour?" is the common response I get from producers. That is not what is meant by Parity Segregation, which is a more pragmatic and sensible concept.

Parity Segregation involves organizing the breeding unit into two populations. Young immature sows, i.e.. the first two litters especially, and the rest of the herd.

## Why? And what are the benefits?

First, some background to the concept. Research into sow nutrition seems to have taken a bit of a back seat over the past 20 years, at least in comparison to baby pig and grower diets. This is especially true of the sow's trace mineral needs and protein/amino-acid intake.

But progressive farmers have already realized that nutritionally the gilt is a very different animal to the older – even the slightly older – sow. The Americans have an understandable production-line mentality where it is economic for them to do as much as possible on their large farms in the same way as possible. For example gilts and sows tend to be fed the same way – on the same diets. By not varying production methods like this so as to take advantage of the economy of scale, they have now run into the problem of a very short sow productive life. I suspect that the same is increasingly true all over the world although not everyone accords the importance to it that economics suggest.

It costs a great deal of hard-earned capital to replace a sow and get the replacement gilt to produce her first litter. Sure, this should be recovered 20 weeks or so later from the sale of this litter at slaughter.

But then to achieve only two more litters – instead of four or five more - squanders about 46% of that initial investment and foregoes subsequent income from batches of pigs which will never be born to that sow. It is like having to replace a new car which is worn out after a few thousand kilometers. A huge drain on replacement capital. Having to accept replacement rates of 45% and even 50% is uneconomic. It is a global problem for pig producers now.

## So why Parity Segregation?

The first litter female, Parity 1, (or P1, the gilt pregnancy parity being P0 in modern terms) is very vulnerable to protein loss down through that first lactation. Dean Boyd, an American researcher prominent in this field, tells us… "A 4 kg body **protein** loss (note: not total bodyweight loss) during first lactation is sufficient to reduce second litter size by 0.75 pigs, and weaning to oestrus interval increases in proportion to body protein loss. In contrast, limiting body protein loss to less than 2kg can result in a 1.0 increase in second litter size compared to the first".

And there could be other benefits, suspected but not fully confirmed yet, in respect to disease - PRRS, mycoplasma pneumonia, and piglet scours. Because the gilt and possibly the young P1 sow, and even the P2 sow, are such a source of infection due to their underdeveloped immune systems, medication costs are reduced in the more mature sows and their offspring by up to 50%, and a 20% reduction looks to be typical. Additionally there is evidence that Parity Segregation is much more likely to extend the sow productive life of the herd (SPL) - indeed many breeders who are already giving their gilts special attention are realizing this.

A future area of research associated with the concept could be that the progeny of young P1 and P2 – even possibly P3 sows if these latter are from a highly productive strain of female - could benefit from a different balance and/or amount of nutrients compared to the progeny of the more mature P4 to P7-P10 sows. Table 1 illustrates the big difference in micro-mineral intakes between the first two pregnancies and those of the more mature sows. If this is typical surely we can no longer afford to feed the same gestation diets to at least the P0 and P1 sows (and probably the P2 sows as well) compared to the rest of the herd which comprise those from P3 onwards? I have indicated this in the preceding chapter on feeding the modern gilt.

**Table 1. Progressive decline in micronutrient intake with advancing reproductive age**

| Micronutrient intake | Gilt | After... | Litter 1 | Litter 3 | Litter 5 | Litter 7 | Litter 9 |
|---|---|---|---|---|---|---|---|
| in pregnancy (g/kg bodyweight) | 39.2 | | 26.8 | 19.5 | 16.3 | 15.0 | 14.2 |

Extrapolated from Boyd and Hedges, 2003.

"The older (bigger) sow is increasingly at nutritional risk reproductively and immunologically."

*Boyd, 2007.*

## Segregated parity <u>rearing</u>?

It looks as if the 'high-risk' progeny of the young P1 and P2 sows will need specialized treatment both nutritionally and possibly with in-feed medication – raising the possibility of parity segregation carried on to the early rearing stage of their progeny as well (Table 2).

**Table 2. The problem of progeny from the young sow (Due to their lack of sufficient immune status)**

| | Gilt progeny | P2+ progeny |
|---|---|---|
| Wean weight (kg) | 5.30 | 5.74 |
| Nursery mortality (%) | 3.17 | 2.55 |
| Nursery weight gain (g/day) | 412 | 435 |
| Nursery medication cost(Can $) | 2.15 | 0.55 |
| Finishing mortality (%) | 4.31 | 2.35 |
| Finishing weight gain (g/day) | 735 | 763 |
| Finishing medication cost (Can $) | 1.82 | 1.01 |
| Enzootic pneumonia (%) | 31 | 11 |

**Source:** C Moore (2001).

## Is Parity Segregation for breeding females feasible?

It seems so. The better breeding units are already partially on the way there anyway. They have introduced a gilt pool section to the farm, and already feed a gilt developer diet .The next stage is to continue them on into a separate P1/P2 section of the farm in which a special young-sow lactation diet is used and after that, in the second pregnancy, the gilt developer diet can be used to avoid practical limitations, such as having a further diet in the young sow section of the farm. Having such a segregated area for both sows and weaners allows the sort of special care, nutrition and attention which progressive breeders are already realizing pays dividends  for the gilt. But Parity Segregation takes this on further, certainly to P2 and maybe to P3. The jury is out on P3 segregation at the time of writing but under close veterinary supervision from P0 to P2, it is possibly not necessary. One of the results of this trial work is interesting, in that most male pigs have a have higher mortality on into life than the females, especially when the health of the growing section of the unit becomes unstable. Why - I do not know.

## The 'High Risk' Concept

Dr. John Deen, a leading American pig specialist veterinarian is on record in saying " The presence of higher risk animals **does affect the rest of the population** and causes higher mortality than low risk animals".

We know that due to her immature immune defences, the gilt certainly, and the  P1/P2 sows usually, are high risk animals and that their progeny may, or will also be, disease-shedders. Thus to mix them with the older portion of the herd, both in the breeding and nursery units, is unwise in future planning decisions.

## So where are we now?

1.  Gilts have less pathogen protection. Figure 1 shows the likely immune status of the gilt compared to the third parity sow.
2.  Separation of gilt and sow flows on the farm have reduced disease.
3.  Gilt progeny have different medical needs after weaning.
4.  Vet-related costs in the older portion of the herd have fallen by at least 20%.
5.  Nutrition limits litter size in young sows (P1and P2) versus older sows but in different ways - such as protein and micronutrient levels. They will need different gestation and lactation diets. Parity segregation makes this possible.
6.  Sow productive life and thus weaning capacity is increased.

## The thinking behind parity segregation

If the gilts with litters at foot can be kept and managed separately from the rest of the herd then the chance of disease transmission to the existing herd is (much) reduced. There are suggestions that this could well be extended to the second litter sows as well, who may well not yet be up to their full immune defence shields, but after discussions with leading pig veterinarians on both sides of the Atlantic this could be influenced by how thoroughly the gilts have been managed up to and during their induction process, which I outlined in detail in the preceding chapter.

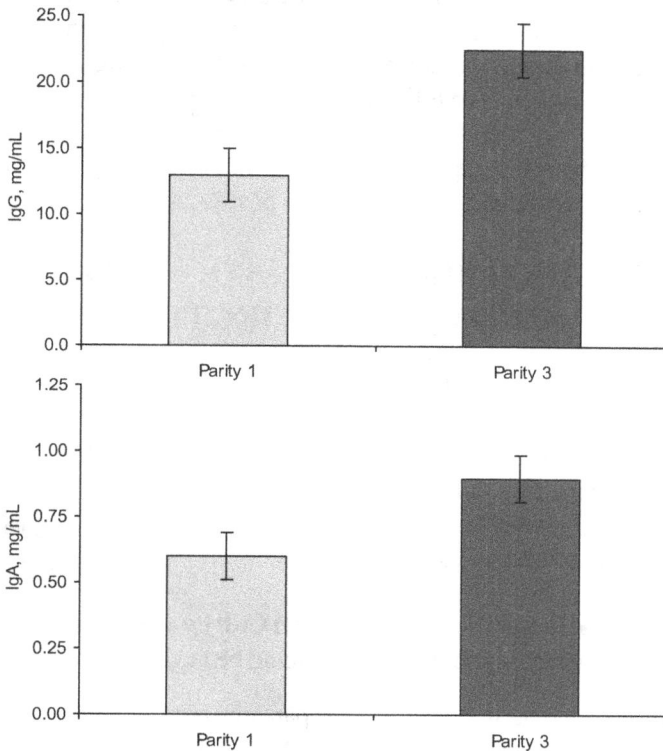

**Figure 1.** Circulating concentrations of IgG (top panel) and IgA (bottom panel) in gilts (Parity 1) and sows (Parity 3). Immunoglobulin concentrations were evaluated in serum obtained within 24 hours pre-farrowing. Source: Burkey *et al.* (2008)

Parity segregation also extends to the **progeny** of the first litters – and again possibly the second-litters too – which are housed in a separate nursery and preferably looked after by different stockpeople or by the same staff using different implements and overalls, boots etc, on a shift basis.

Boyd (2007) reports that on 368,000 pigs, the average mortality in a segregated (i.e.. 'disease stable') nursery compared to a 'disease unstable' conventional system was 1.72% to 3.03%, with 4 days less in the nursery before transfer weight and 5 days quicker to slaughter.

The thinking behind this logical extension of the concept is that the progeny of first and possibly second litter sows have also been sub-standard in acquired immunity from their respective dams and will also be mild or alternatively potent shedders of disease to the progeny of older sows with which they are placed in the same nursery environment. The mild shedders merely cause the weaners from the older sows to divert food into reinforcing their immune shield (see Immunity chapter, page 85) and thus grow more slowly. However those more severely challenged can contract clinical disease from these disease-shedding weaners from young sows to which their immune status is insufficient to fully protect them.

Segregating the weaners into separate nurseries and management protects both groups of vulnerable animals at this age and degree of post-weaning stress.

# IS PARITY SEGREGATION TOO EXPENSIVE?
# A COMMON OBJECTION

**The extra cost factor - existing older sows**

Like Computerised Wet Feeding (CWF) cost is a primary obstacle and I have attempted to put some figures to this from farms which have experimented with the concept – admittedly with less convincing results compared to CWF as there were far fewer experienced practitioners to interview.

* The cost of **the better diets needed raised CoP by 4%,** but as my observations on modern gilt feeding suggest – you should be feeding them like this anyway.

* The cost of keeping the older sows separately, out of contact with the young sow section, looks to be about **another 1.25%** on the total CoP of the herd.

**1.    The extra cost factor – high-risk females ( Parity 0 and 1)**

* Assuming the presence of a separate gilt pool, suggests an extra cost for the additional parity **gestation accommodation of 2.5%** and **1.4% for additional labour,** assuming current labour cost is 13% of total CoP.

* Special diets for the  gilts and young sows. Most progressive breeders should be feeding the gilt developer and gilt lactation diets, so there should be no extra cost. Discussions with one N American feed compounder indicated that these would be 16% more expensive  than conventional diets.

I calculate that as a proportion of a herd with a 5.5 litter average SPL, **such an dietary regime will raise CoP by 3.0%.**

**Thus at a rough stab at the figures available to me at the time of writing, it looks as if the concept could raise the breeding herd costs by around 12%.**

## 2. The extra cost factor - the progeny of young sows kept apart during the nursery stage

Interviews suggest that this depends whether the breeder has spare and separate nursery rooms dedicated to these animals and not shared or transferred to others, even after a thorough All-in/All-out cleaning process.

If so there should be no additional housing costs – just extra labour which will be lower than for breeding as there are none of the onerous breeding or farrowing tasks. Say **+ 0.5%** **on CoP**. (However if sufficient of a new dedicated nursery arrangement is needed, this extra accommodation cost built from new, one producer reported **another 6 % on his CoP**, spread over a 12 year amortization life). The American pioneers report that the hived-off weaners will need special vet/med attention until they are merged with the growers from the older sows post-nursery.

From American veterinarians figures, this looks to be around an **extra 2.0% on CoP.**

## And the returns?

While I am reasonably assured about the costs quoted above, the paybacks are much more difficult to assess.

This is because disease patterns on the units who have adopted parity segregation seem to rise and fall and the intensity of both clinical and sub-clinical infections vary too. Several N. American veterinarians are generally in favour of the concept, with a few cynics thinking that a typical breeder would be unable to maintain the undoubted discipline needed over time. Well maybe, but surely those breeders farsighted enough to invest in the concept are not likely to be considered 'typical' for a year or two?

We don't seem to have enough hard evidence at the time of writing.

All the same, with disease being the main influence on profit or loss (after the pig price) anything which can help mitigate its effect should be considered seriously.

## Advice

Perhaps the best advice for the time being is to understand the mechanisms of the concept and do all you can to minimise contact between the sows in the first two parities from those in subsequent ones and likewise for their offspring in the nurseries.

There are still unanswered questions. The costs of converting an existing breeding farm into two separate sections have not been fully explored and recorded - my efforts above are an attempt to reply to enquiries about the subject. I have found, not surprisingly, that costs will vary markedly according to the space available; whether a gilt pool is already in operation; and paybacks in savings from a longer sow productive life; lower nursery mortality; and the savings in medication costs.

However, anyone considering building a new unit, or extensively re-modelling an existing one, should think carefully about a separate section of the farm for the first two parities. For a really big organization – two or more separate farms. This could be the way things are going to move in the future.

## SOURCES USED...

Boyd, R Dean et al.(2007).'Segregated Parity Structure in sow farms to capture nutrition, management and health'. ADSA Discover Conference 13, Nashville, Indiana, September 2007.

Boyd, Dean. (2006) 'Split herd Feeding Helps The Senior Sow' Pig International March 2006 20-21.

Burkey, T E. et al. (2008) Does Dam Parity Affect Health? Ex- Pig Site Newsletter April 2008

Cunningham, G et al. 'Parity Segregated Production' Western Hog Jnl. Winter 2003pps 26-28.

National Hog Farmer (2004) Blueprint series No. 38. 'Parity-Based Management' Many pioneering articles.

# 22

# AVOIDING THE SEASONAL INFERTILITY HEADACHE

## SEASONAL INFERTILITY

A decline in reproductive efficiency predominantly across the late summer/ autumn period seen as:–

*   Delayed puberty in gilts
*   Problems with gilts cycling
*   Extended weaning to oestrus interval
*   Reduced farrowing rate
*   Shorter oestrus periods
*   Increase in number not in pig
*   Increase in mummies and stillbirths
*   Abortion storms

## INCIDENCE

70% of all abortions could well be due to this cause, while many farmers (seemingly mistakenly) believe abortions are largely infectious in origin. First noticed in the UK in 1970, it was particularly bad here in 1970's and seems to be getting worse, possibly due to more breeding being outdoors, better recording of farrowing rate and live births from season to season and a rise in peak summer temperatures.

*   Abortions occur.          *Herd Target: 1 in 100 served sows, rises to 13 or 14 or more over short periods. (Figure 1)*

*   Stillbirths increase.     *Herd Target: 3% true stillbirths, can rise to 8% or more.*

*   Mummies increase.      *Herd Target: small 0.5%, large 1.0%, rises to 3% or more.*

Watch for the following in autumn & winter especially ...

- Regular returns to service increase, particularly among gilts. *Herd Target: <10% rises to 20%+. (Figure 2)*
- Weaning – conception interval lengthens. *Herd Target: <6-9 days rises to 10-12 days.*

Thus numbers born alive fall, and herd targets of 10.5 – 11.0 can decrease to 10.0 or less over a six month period, which improves again in late winter and spring. Bad cases can cost 150-250 pigs per 100 services, a severe drain on cash-flow, reducing annual gross margin/sow by 18% and more.

Seasonality is not a genetically-controlled trait and cannot be influenced by selective breeding. We must therefore look at management measures to help lessen the effects of the complicated problems of both 'Summer' and 'Autumn' infertility.

## Summer or Autumn infertility?

Newcomers to pigs are sometimes confused by the two terms. This is not at all surprising, as the predisposing causes of a reproductive fallaway in autumn/early winter occur in spring/early summer, hence 'Summer Infertility', while the effects are noticed in autumn/early winter. Summer infertility is one component of 'Seasonal Infertility'. The other is Autumn infertility.

A lesser seasonal effect than that described as the main problem of 'Summer Infertility' also occurs in late winter when the primary causes have occurred in the autumn 3 months earlier, hence the term 'Autumn Infertility'.

So . . . In **Summer Infertility** the infertility occurs in the autumn/early winter but the cause(s) of it occur in the summer.

In **Autumn Infertility** fertility occurs later in the winter but the cause of it occurs in the autumn.

**A reflection:** Confusing isn't it! Whoever thought up the two definitions (in my opinion) was not thinking logically and certainly not from the viewpoint of the farmer! To associate occurrences with when the problems appear and are therefore noticed by the farmer is surely more logical and useful! No wonder the student is confused! But we are stuck with them now.

## Summer infertility

With more sows being bred outdoors the incidence of summer infertility is climbing due to heat stress in two decades of summers in the northern hemisphere which have recently been hotter, arguably due to global warming. Figure 1 shows a typical summer infertility pattern affecting abortions.

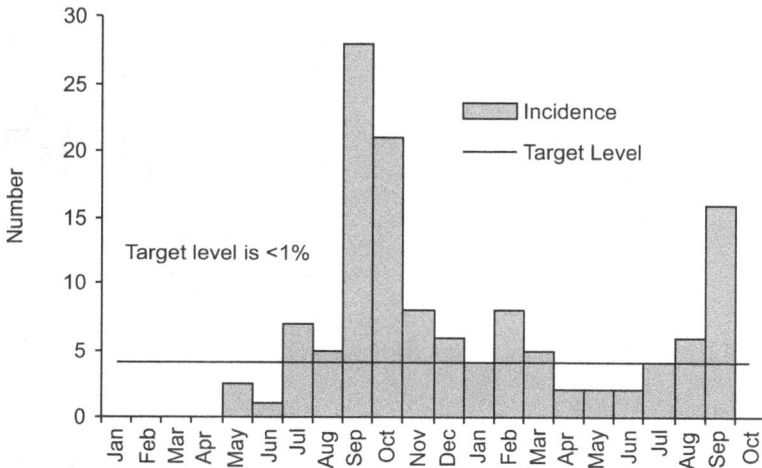

**Figure 1.** Abortions by time of year (500 sows). Northern Hemisphere.

Summer infertility is also common in indoor herds which suggests the primary cause is indeed temperature as many breeding barns can get very hot in the summer.

## Photoperiod

Photoperiod is the period of time during a 24 hours day that an 'organism' (i.e. a pig) is exposed to daylight or artificial light - also called daylength. A great deal of attention has been placed in the past on daylength being a primary cause. It could well play its part, especially during the autumn when shortening daylight hours triggers the 'feral factor' effect originating in the primeval sow so as to avoid her farrowing and raising a family in the inclement late winter months. The experts argue over daylength however, although my experience at the sharp end of pig farming suggests that it is involved, as advice on lighting when things seemed too dim around service time has often helped a great deal to improve conception rate. As Prof. Wiseman says "If day length is responsible (for summer infertility) why are no problems experienced before June (northern hemisphere)? And why are there year-on-year differences when day length is kept constant?"

Yes, it is indeed a puzzle! For example, Figure 2 shows how returns can vary from season to season, and again in Figure 3 similarly with farrowing rate.

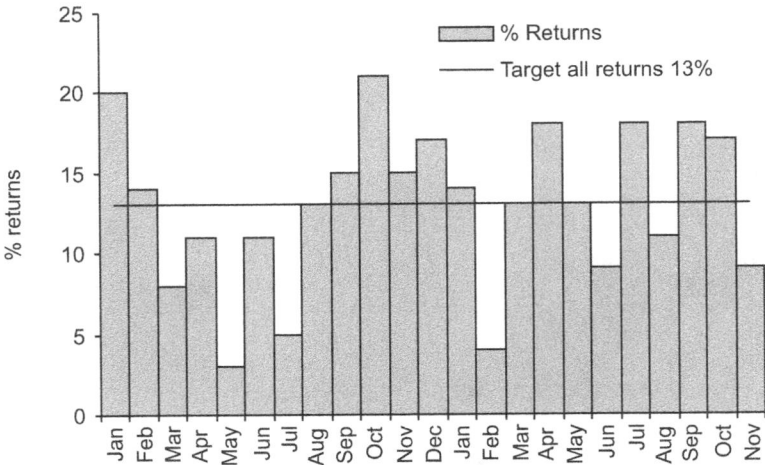

**Figure 2.** Seasonal infertility. Return rates allied to season. Northern Hemisphere.

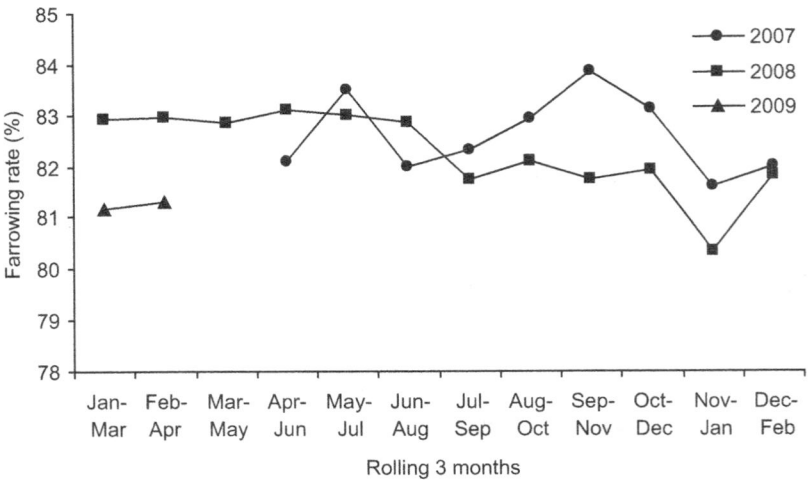

**Figure 3.** Seasonal effects on farrowing rate. Source: Nadis (2009).

Figures 2 and 3 and illustrate another puzzling factor - the variability of incidence from season to season. A hot summer does not always result in seasonal infertility in autumn nor have some cool, wet ones necessarily reduced it from the previous year.

# DEALING WITH SEASONAL INFERTILITY – A CHECKLIST OF LIKELY CAUSES

Seasonal Infertility cases are multi-factorial (many possible causes), complicated and not very well understood, although countries with distinct climatic seasonality – with bright springs and hot summers and rather dull, sometimes chilly winters (e.g. coastal Australia and Central USA) – are more experienced in dealing with the problem. Difficulties have arisen because some good quality research has tended to confound some of the current advice and experience.

## A BASIC CHECKLIST FOR SUMMER INFERTILITY

✓  High temperatures, especially affecting boars.

✓  Too much bright light in spring; decreasing light patterns in autumn; too little indoor light at all times.

✓  Nutritional stress in hot weather.

✓  A variety of other stress factors.

## A COMPLEX SUBJECT – AND A DILEMMA

While experience, particularly in Australia, Mid West USA and Spain – and now recently in the UK, is contributing to knowledge on Seasonal Infertility all the time, there is a good deal yet to discover. In fact some scientists are quite vehement in their criticisms of some aspects of current thinking about the subject. So when experts disagree ….!

Take the embryonic mortality aspect of seasonal infertility. Dr Phil Dziuk, an eminent researcher, wrote to me setting out an apt and amusing analogy thus . . .

"Embryonic loss is an elephant. It is the same elephant as described by each of a group of blind men. One blind man who grasped the leg in his arms said it was like a tree, another felt the trunk and said it was a large fire hose, a third touched the tusk and described it as a spear and the fourth thought it was a rope as he held the tail. They were each correct but they were each wrong. Embryonic loss in the eyes of the nutritionist is a result of improper feeding practice. The veterinarian declares that subclinical endometritis or an infectious organism is the cause, while an injection of the proper combination of hormones will cure it, according to the endocrinologist. An undesirable set of genes that can be selected against is responsible proclaims the geneticist. Maternal-embryo histoincompatibility explains it, says the immunologist. The cytogeneticist finds chromosomal aberrations and deduces that these errors are at fault. Each may be correct but possibly each is also wrong. The elephant of embryonic mortality may be even more complex when viewed individually by the many research workers who have studied it over the years, or it may be many factors acting through a relatively common mechanism to produce one result, loss of some embryos."

Good stuff! Hopefully, those of us at the sharp end working on farm problems (while as blind as anyone) have felt all over the elephant for a period of many years before coming to a conclusion and while still puzzled – and still blind – have had a lot of elephants on which to practise so as to form an opinion of their shape!

This chapter therefore outlines my own experience, and is based on what other practical people – farmers and veterinarians in particular – have found helps.

## ANTICIPATING SEASONAL INFERTILITY

Past records are a starting point, especially farrowing rate – the number of farrowings achieved from a given number of services.

As figures 1, 2 and 3 illustrate, the drop-offs in productivity can vary – from season to season and also from farm to farm, so it is important to know just how you are likely to suffer on your unit from your past records. Two are particularly useful.

(a)   A cusum graph (cumulative sum) on farrowing rate, plotting number of farrowings on one axis against effective services on the other. Your desired target is the 45° bisector. This ongoing weekly set of graphs will give you warning on when the farrowing rate starts to underachieve as it dips steadily below the 45° line.

(b)   Back this up with a farrowing rate graph against time, based on a 3 month retrospective rolling average, as in Figure 3. This shows you where your dip occurred historically and the degree of severity, as well as whether your previous countermeasures worked well or not.

Now you have the basic information on your unit to plan ahead what to do in advance so as to counterbalance what could happen in the months to come.

## Dealing with the two forms of seasonal infertility

<u>Autumn abortions</u> are the most common especially on outdoor units .In the northern hemisphere most abortions will occur in September,

Look at your past records to try to quantify the predicted losses from this cause and assuming it will happen again, plan back the 115 days for the extra services you will probably need to compensate, and so hopefully maintain your farrowing target,

Yes, things change. Some years, usually only occasionally as autumn abortions seem to be a fact of life these days, you will over-breed but this is not so serious as suffering those missed farrowings.

It is by no means easy to get it right but you must try each year. A good pig veterinarian can help as part of his service contract. Good pregnancy diagnosis is vital.

<u>Summer infertility.</u> Will normally start from northern hemisphere breedings in July (possibly earlier if the late spring days have been very bright and clear especially if your sows are outside in paddocks or yards) to as late as October if we have an warm and sunny autumn. This increase in pregnancy diagnosis failures and/or returns requires immediate compensatory action to fill the threatened fewer farrowings to come 115 days later in the winter around November to February. Aggravated too by the effects of the dormant `feral factor` still lurking in some genes (see page 493).

Two things:-

1. Order up the extra gilts you think you will need from your past records for the gilt pool well ahead of the normal causative breeding peak in July. This could mean as early as March to get them acclimatized and induced correctly by July (see the Gilt section) Work out an order schedule with your genetics supplier. They will help – it pays them to do so. This will mean you should be insured against a drop in farrowings in October/ November onwards. Remember - summer infertility (caused in <u>summer</u>) will hit you in <u>autumn</u>/ early winter if you do nothing.

2. Check, check, check all the many factors that can affect those summer breedings. Do everything you possibly can to have your sows and boars in tip-top condition. Dozens of things to keep in mind to fight off those summer returns and PD failures - which I describe in the next chapter.

# EXPANDING THE CHECKLIST

## Temperature

Boars : Temperatures over 27°C may well affect boar libido and, after excessive heat of 5 to 14 days, damage sperm quality for 4 to 6 weeks thereafter.

Sows : Temperatures over 22°C and especially towards 25°C and over affect appetite, particularly in lactation. This can throw the sow into negative energy balance, when she has to use her body tissues to an extent which may affect reproductive efficiency, even to the state of abortion.

Gilts : Gilts are better able to withstand heat than sows but are particularly susceptible to increased stress especially if water is short or if they are overcrowded in groups. Allow at least 3 m² (32ft²)per animal. The typical effect of temperature on litter size is shown in Figure 4.

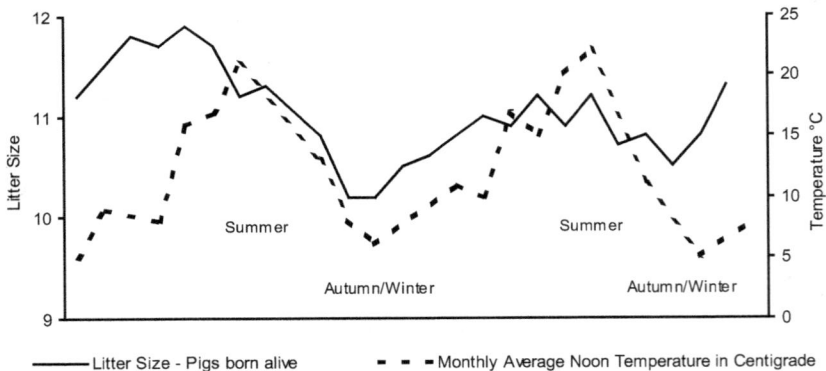

**Figure 4.** Seasonal infertility: Typical (actual) records showing effect on litter size of increased temperature.

## Staff

It is extremely important to check sows are in-pig during the summer months - one of the drawbacks to outdoor breeding is to be able to preg-check well and constantly - it takes time to restrain all but the most placid outdoor sows. Nevertheless outdoor stockpeople must be encouraged to do this. Renewed attention to AI technique in hot weather is also worthwhile indoors, but especially outside.

In order to encourage the dedication needed during hot weather, some breeders offer a bonus based on sow productivity in the autumn months based on the annual average farrowing rate, which is a good benchmark indicator of any effect on it of seasonal infertility. Any sow's body condition is important, especially on an outdoor unit and this depends on her eating sufficient food to sustain her exercise outdoors. On such farms a sow can walk several kilometres ever day.

Indoor units can get very hot, as I said, and adequate nutritional intake techniques must be explained to staff in that sows need to be fed differently (dietary composition and when it is best fed) in expected spells of hot weather, say an average daytime temperature of +8°C above normal. See figures 6 and 7 and the chapter on 'Hot Weather'.

Beware of staff summer holidays! The more senior, experienced workers often have prior claim on school summer holidays and care must be taken not to have all the skilled staff unavailable at this critical hot weather time. The same goes for Christmas and New Year vacations.

## Light

Having gone along with the latest doubts on daylength being a **primary** cause of both types of seasonal infertility, my own experience and a few farm trials with cooperating clients suggests to me that light **intensity** is a factor in reproduction in the female, and that this impinges on both types of seasonal infertility.

**Too much light:** Very bright light seen on those clear late spring/early summer days especially if the animals have access to direct sunlight in outside runs or outdoor paddocks. Runs should be shaded with 'Galebreaker' mesh covers and paddocks have a shaded lying area which sows are encouraged to use, both in the cool, clear days of spring and the hot, muggy days of summer.

Try to locate shaded areas near to tree belts or on windward slopes. A few sow nuts sprinkled inside shaded areas encourages occupation.

**Decreasing light patterns:** These are inevitable as mid-summer progresses through autumn, and result in a quite natural reduction in the hormones needed to maintain the pregnancies of sows and gilts mated during the peak summer day-length period. (This is known as the 'feral factor' achieved by sows in the wild who instinctively do not wish to farrow down and raise a litter in the depths of winter).

It is almost impossible to counteract this decreasing light pattern effect on outdoor breeding farms short of bringing the sows into a structure where day length can be to a certain extent controlled for 40-60 days after service. Cheap polythene and straw bale structures are used. But this can easily be done with sows housed indoors, where 'autumn infertility' can also be a problem.

Some researchers have questioned the amount of time and trouble currently spent on this area of the whole subject in relation to the other factors involved.

**Too little light:** This is could be a cause of 'Summer Infertility' in indoor herds, where the problems occur much earlier than 'Autumn Infertility'. Here poor stimulation into oestrus and conception due to a combination and progressive build-up effect of poor light, stress from cold and damp quarters and nutritional catabolism (essential within body nutrient destruction as distinct from nutrient construction) occurring towards the end of winter or after a 'late-spring'.

The later arrival of 'springtime' weather seems to be a recent (1990s onwards) phenomenon, resulting in 'longer' (even if milder) winters, due to heavier cloud cover in NW Europe. Maybe this 'winter shift' is the result of global warming.

To combat this, lighting should be at least 350 lux (about as bright as a 100 watt strip-lit kitchen, and certainly light enough to read the small print in a newspaper) shining into the sow's eyes (not behind the head). Opinions differ on the daylight/darkness (<25 lux) mix but the author has found 16-18 hours 'on' and 6-8 hours 'off' to be satisfactory as the following quotation suggests:

"John Gadd, the British-based consultant who writes regularly for Pork Journal, was responsible for New Zealand producer Neil Managh "seeing the light". Gadd, who was in NZ on a speaking engagement, visited Neil's piggery at Feilding. "He wasn't in the operation one minute before he told us about what we should be doing in our mating area, and it helped us no end," Neil said. "He told us to light up our mating area, using a timing clock and good lights, so it is as bright as daylight but better. We played around and just set it. Now they come on at 6 am whether it is winter or summer and they go off at 9pm. Since we did that we rarely get a return. We just don't get seasonal infertility. We can wean 14 sows in a week or thereabouts. We usually do that on a Thursday or a Friday and by Wednesday it is very rare that the whole lot are not completely mated and finished – and we put it down to those lights. It is around the farrowing parts and the mating area that the lighting has really helped us. The funny thing is that it cost so little, about $200 or $300 and we were up and running."

*Australian Pork Journal*

On the other hand, we are being told by good researchers in the seasonal infertility area from Australia and S. Africa, that reducing the light to darkness ratio to 10 hours light, 14 hours darkness "restores good oestrus in sows and gilts, as well as more advanced puberty attainment in gilts across the summer". (Janyk, personal communication 2002). Is this the effect of light intensity or photoperiod (light to darkness ratio)? Does it mainly apply to strains where the 'feral factor' is still strong? We need to watch the research and resulting advice on this, and may eventually have two sets of guidelines for summer/winter or seasonal infertility/no-seasonal infertility circumstances. Meanwhile, I still get plenty of comments similar to that in Table 1.

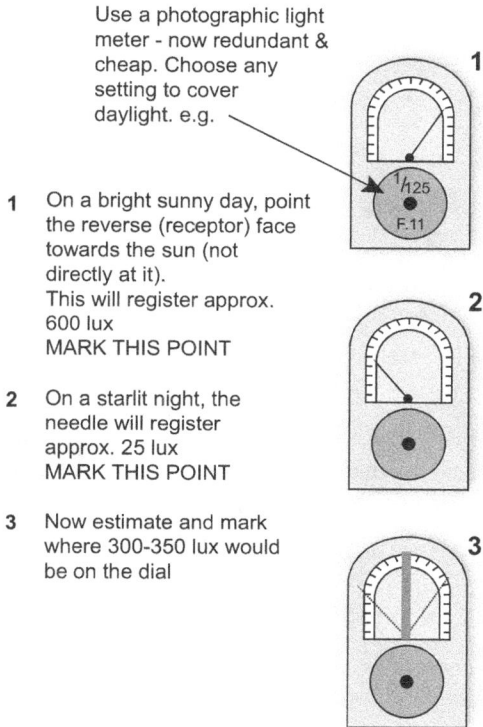

Use a photographic light meter - now redundant & cheap. Choose any setting to cover daylight. e.g.

1 On a bright sunny day, point the reverse (receptor) face towards the sun (not directly at it).
This will register approx. 600 lux
MARK THIS POINT

2 On a starlit night, the needle will register approx. 25 lux
MARK THIS POINT

3 Now estimate and mark where 300-350 lux would be on the dial

**Figure 5.** How to measure light intensity.

**Table 1. Extra light brought forward conception**

|  | *Control* | *Extra light* |
|---|---|---|
| Number of Sows | 164 | 163 |
| Average days to mating | 5.9 | 5.5 |
| Mated within 5 days (%) | 68.5 | 83 |
| Mated 6 to 10 days (%) | 26.8 | 10.9 |

Source: Author's library data from another farm

The ratio of positive comments from 111 farm visits on the subject since 1979 (90% of them followed up six months later or more) is 72 positive, 17 'don't really know' and 11 'no differences'.

## STRESS

Stress comes in many forms, but in the case of Summer Infertility, heat stress is paramount. Here are a list of things to consider to mitigate the effect of heat stress on breeding animals.

## CHECKLIST : COMBATING NUTRITIONAL STRESS IN HOT WEATHER

✓ Is your food fresh enough?

✓ Is it stored in a cool place?

✓ Have you adequate water available? Troughs are better than drinkers. Water flow must be at least 1.5l/min. if so.

✓ In the farrowing crate, a self-dispensing feeder is a good idea as the sow can eat when she feels like it during cooler periods evening, night, morning.

✓ Blow fresh outside air (even if it is hot) on to the food. The arrangement of a linking tube fed by a small 20 cm fan gently ticking over delivering air to a down-pipe over each feed trough in the farrowing pen works well (Table 2) as the air is fresh over the food which seems to tempt eating in hot countries.

✓ Consult a nutritionist to alter dietary ingredients (e.g. more fat, less cereals). See Hot Weather Nutrition checklists.

✓ Feed pregnant sows and also boars earlier in the day – well before the stockperson's breakfast break.

✓ Use barley straw, not wheat, as bedding.

✓ Feed wet by pipeline. Especially consider it in hot climates.

✓ Do not overfeed thinnish pregnant sows 14 days or so before farrowing as many producers are tempted to do.

✓ The author has noted benefits from feeding a yeast-based additive – Yea Sacc (Alltech Inc).

## AI OUTDOORS?

Some may question the practicalities of using AI on an outdoor herd or less intensive situations, but the problems can be overcome. Feed can be used to entice the sows onto a high-sided, low-loading trailer where they are inseminated. It helps entry if it is the trailer used to transport their feed every day, but I worry about disease transmission.

**Table 2. Before and after results from snout-cooling lactating sows (Phillippines 1993)**

|  | *Before* | *After* |
|---|---|---|
| N°. of farrowed sows recorded | 826 | 260 |
| Mean temperature*, °C | 25° | 26.1° |
| Airspeed m/sec* | 0.3 | 0.35 |
| RH% | 81 | 90 |
| Lactation feed intake, kg | 3.8 | 4.5 |
| Av. condition score at weaning | 2.1 | 2.8 |
| Litter size (b/a) | 9.1 | 9.9 |
| Sow weight loss in lactation, kg | n/d | 10.1 |

* Over the sow's back.  Source : Client's records

Note: While snout cooling hardly cools the sow at all, it supplies a gentle flow of fresh, uncontaminated air over the sow's nose, which seems to reduce latent stress. Snout-<u>cooling</u> is not an accurate name for this useful idea. Snout freshening?

Better is to construct a simple canvas or polystyrene tent where AI is carried out, especially seen in the centre of "wheel-type" paddock layouts. The AI materials are kept in an insulated box and the entire operation is carried out by one person. In the hottest weather when the sows were reluctant to move from the wallows or shade they were inseminated on the spot with no apparent problems (Sunderland, 1991) – but careful cleaning of the vulval area is essential! Insemination is often done in a central corral area, or by inseminating at feeding times, heat detection being done at the previous day's feeding.

# CHECKLIST:  MANAGEMENT TIPS FOR HOT WEATHER

✓  Keep sow groups as small as possible and don't mix them until after implantation (14-22 days).

✓  Handle movements to and from boars gently, and confine to essential moves only.

✓  Running a vasectomised boar with to-be-bred sows in hot weather helps stimulate oestrus dulled by heat.

✓  Check ventilation is adequate, especially fan capabilities. (See Hot Weather Chapter.)

✓  Drip cool farrowed sows, spray cool pregnant sows and boars (ibid.)

✓   Serve at either end of the day when it is cooler.

✓   Young boars rather than older ones seem to work better in extreme heat. If possible buy new boars in the spring.

✓   Increase the number of gilts to be served by 10-15% in a hot spell.

✓   Use supplementary AI especially when it is hot.

Several results known to the author average out at over 2.0 more total-borns per litter from boar to AI service over boar only services in hot weather (Reed, 1990). At present AI prices, only 0.8 more total-borns (0.6 more b/a's) will easily pay for the AI plus labour, which must be considered as an extra cost in hot weather.

✓   Feed when it is cooler early and later in the day.

✓   Serve females with two different boars in hot weather in case one is susceptible to heat stress.

✓   Mycotoxins thrive in hot weather, especially if 'hot-wet'. Do a mycotoxin audit (*See Mycotoxin Chapter*)

---

Having said this, my experience is that <u>outdoor</u> AI is most cost-effective when:–

•   The herd suffers from infertility problems. Higher-performing clients who tried it showed little benefit over their own skilled conventional mating management, using boars.

•   During periods of hot weather, when it is very valuable.

•   The outdoor stockmen attend the same hands-on training courses as their indoor colleagues.

Here are some further tips gathered from producers across the world who are getting on top of seasonal infertility on their outside units.

## WALLOWS AND SHADE

All pigs are prone to sunburn and this causes considerable stress leading to classic seasonal infertility. Mud wallows are an excellent sunburn protection and these must be kept topped up with water, but not diluted slurry effluent, when things are dry.

I have already mentioned shade as it affects too bright radiation in spring. Such days can catch the producer unawares, and I have driven past many paddocks near to roads and motorways where sows and gilts were out basking in the welcome warmth after a cold winter – and getting sunburnt in the process. It may take as little as 3 hours exposure for this to affect productivity!

In general shades are put up too late as, in Europe, these days can occur in early April; shade is not so much a protection against heat, but against radiation.

To minimise trouble, posts and T-bars can be fixed to huts so as to erect plastic mesh-netting quickly in such cases. The placing of shade between huts' entrances encourages use but clear early spring days often come with cold, penetrating wind so a straw bale windbreak may be needed on exposed sites to encourage occupancy of an otherwise too-cold area.

We have a similar problem with shaded areas outdoors in a heatwave – the sows must be encouraged to lie in the shaded area, so light bedding (permissible when it is dry) and some feed nuts in it may housetrain them at the start of a hot spell – they soon catch on. Again the between-hut mesh shading does the same job. Shading over the wallows is also used on some farms. Wind pressure can be a nuisance, however.

- In the USA wooden A-shaped outdoor huts are popular, with a window at the rear. This allows a through draft to cool the interior, which when closed in winter, makes it a snug enough refuge.

- Huts should be insulated.

## A NUTRITIONAL ASPECT OF SEASONAL INFERTILITY: BOTH INSIDE AND OUTDOORS

Altering the nutrient allowance in pregnancy has been shown to lessen the problem.

Conventional advice on pregnancy is shown in Figure 6 where changing to a lactation diet for high performing sows or a special gilt pre-farrowing (developer) diet is fed about 14 days before farrowing to maximise litter weight.

However case histories of Summer Infertility trials (Love *et al.*) seem to be helpful (Figure 7).

The Australian work suggests that relatively high post-service feed intakes (45 MJ DE/day compared to say 26-30 MJ/day) for group-housed dry sows minimised the adverse effects of summer infertility. In trials, groups of 22/23 sows were fed up to 25.8 MJ DE/day, compared to similar groups fed 43.5 MJ/DE day – both for the first 28 days after service. The higher intake sows had 80% positive preg-checks across the summer period as against only 57% in the others.

**NORMAL PRACTICE**
**Sows or gilts, but certain gilt genotypes, need a special gestation ration?**

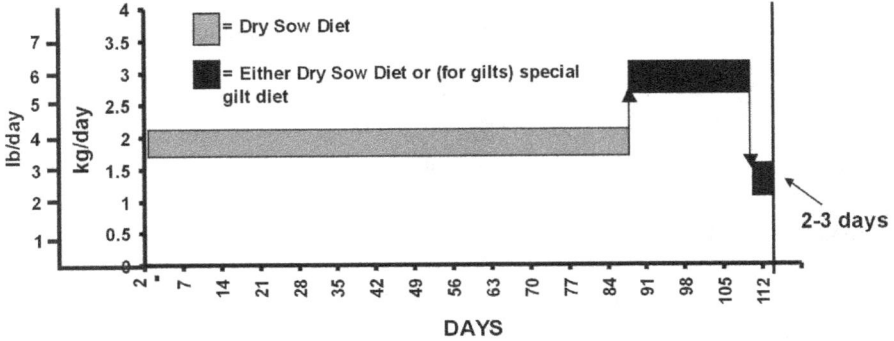

**Figure 6.**  How specialised feeding for seasonal infertility in pregnancy compares with normal practice

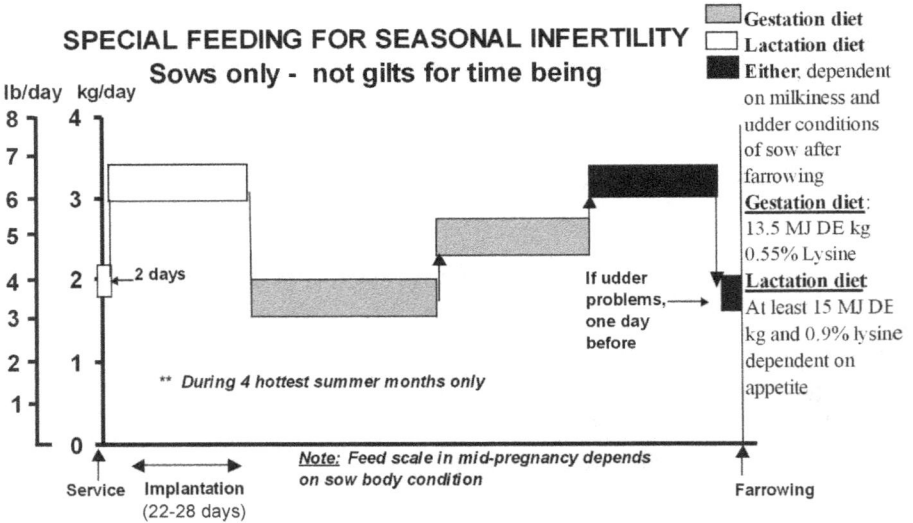

**Figure 7.**  Specialised feeding in pregnancy under seasonal infertility cases

After several years experience, the Australians have been advising 45 MJ/day i.e. 3.3 kg of a 13.6 MJ DE sow diet one month after service for all sows and gilts mated in Australia's 16 hottest weeks in Jan, Feb, Mar, Apr, which corresponds to our June, July, Aug, Sept. in the Northern Hemisphere.  Summer infertility used to cost them about £3 million (sterling) annually covering the national herd of 290,000 sows, or £10/sow).  I'm told now that this must have been halved – progress indeed!

# MODIFICATIONS?

Some American contacts in the big conglomerates tell me that farrowing rate from increased post-service summer feeding is improved significantly in the autumn/early winter. Past research (at normal temperatures) has concentrated on the effect of post-service feeding on litter size and ignored what happens to farrowing rate, as to achieve a confidence limit – worth translating into advice – on a 5% hike in FR needs figures from at least 400 sows in the treatment group over a period. The US integrators can easily observe data from much larger samples than this on their 5000+ sow units, where each unit could provide 1000 dry sows as replicates, and some operators have 20 or more such farms to select from. Thus current American advice on hot summer pregnancy feeding is also worthy of study. Consult your nutritionist.

So . . . as this aspect of summer sow nutrition is ongoing in all parts of the world, you should refer to a pig nutritionist for the latest advice in your area.

# EMBRYO MORTALITY?
# (ESPECIALLY IN GILTS)

These same sources advise a further modification, which is not to push the extra food in until perhaps 2 days after service which should take care, they believe, of any adverse effect on embryo survival in multiparous sows, although whether this takes care of the gilt's problem of embryo reduction if fed more than 35 MJ/DE day immediately post service isn't commented on.

This probably helps to account for the increases found in litter size across several parities. The moral for me is – don't let the sow 'nosedive' in lactation, and consider adding organic chromium to the sows diet all the year round, assuming local regulatory approval is in place, of course. (New nutrient additives take some years to obtain full approval as being safe and effective).

# MELATONIN – THE KEY TO THE DAY LENGTH
# PROBLEM?

Melatonin is a hormone released by the light receptor pineal gland only during the hours of darkness. Melatonin cuts down the production or release of gonadotrophin agents from the pituitary gland (*See Glossary*) which can interfere with the development of the follicles and ovaries. Extending the day length would therefore reduce the inhibitory effect of melatonin very early on in the reproductive process. Conversely an increase in darkness in autumn/early winter reactivates melatonin's interference with early reproductive processes, hence the sows 'natural' disinclination to breed successfully as daylight shortens.

## DO STRESSORS HAVE AN EFFECT?

Stress (anxiety, worry, fear, annoyance) is a poorly understood subject as it is so difficult to measure. Having been involved with pigs for so long convinces me that stress does interfere with many smooth metabolic processes and hormone pathways, and that in the case of seasonal infertility they are likely to involve, apart from disease…

Working in these hot conditions, as I find myself doing several times a year, suggests to me that what to do best for the gilt on farms with a seasonal infertility problem is a trade-off between summer infertility and embryo loss. A comparison between gilts' first farrowing performance in late autumn/early winter and how gilts do for the other 8 months could give a clue. Some countries have a further problem in that they don't know if it going to be a really hot summer/early autumn, so perhaps it is best to err on the safe side and not feed gilts too heavily for 3 weeks after service. The jury is still out on this one.

## STRESS CHECKLIST FOR SEASONAL INFERTILITY

Watch for . . .

✓    Heat stress, especially in the boar

✓    Stocking density, aggression, competitiveness and lack of adequate fleeing space.

✓    Putting small submissive sows (and gilts) next to dominant sows in the gestation stalls.

✓    Lack of access to shade, and where required, mud as a 'suncream'.

✓    Lack of access to water, both for wetting and drinking, including water/flow rate.

✓    Water temperature over 28°C.

✓    Indoor air heavily contaminated with gases.

✓    Rough floors, broken slats.

✓    Gestation crates too small.

✓ Diurnal temperature variation, especially causing night-time draughts.

✓ Noisy, rushed, non-empathic stockpeople.

✓ Noisy, unsettled gestation houses.

✓ Hunger – especially lack of gutfill.

✓ Unpalatable food; mould/mycotoxin presence.

✓ Constipation.

All these stressors – you can probably think of others – can tip the herd over into a seasonal infertility syndrome.

# TECHNICAL BACKGROUND TO SEASONAL INFERTILITY

While infertility problems in sows are rightly the province of the veterinarian, many management factors can be responsible on which the general adviser can help improve.

With global warming suspected to be on the increase, seasonal infertility is expected to become more of a problem as temperature stress and increased exposure to daylight are thought to be involved. In the UK, due to reduced capital costs and other benefits, outdoor breeding is increasing, hence the effect of season is likely to affect these problems too.

I find breeders adopt difficult remedial measures more rapidly if they understand a little of what is happening to their sows. First – as is often a good idea when dealing with animal problems – let's go back to nature, in this case the wild (feral) pig.

## THE FERAL FACTOR

The wild sow always was, and still remains, a seasonal animal. But the wild boar will still mate a willing sow whatever the season. Thousands of years of conditioning has programmed the sow's biology to decide, if made pregnant and the season ahead seems adverse, to call it a day and dispose of the litter.

Shortening day length seems to be the key to this process – this is the advance-warning built into the sow's consciousness wherever she lives, so with a reducing light pattern, chemical reactions are set in train involving the hormones maintaining pregnancy to reduce to levels which will not sustain the pregnancy process, even if satisfactorily mated.

Is the 'feral factor' still a problem today? The breeding companies tell me that it is of far less consequence now. However some of the internal chemistry still seems to be there.

Shorter day lengths start to predominate from midsummer into early autumn and wild sows mated during this period may well conceive, but because the forthcoming litter will then be born when food is short with warmth and shelter difficult in the depths of winter, the hormone progesterone (remember it as being aptly named – 'the pro-gestation-hormone'!) which is essential to maintain pregnancy, is dampened down by shorter day lengths.

Conversely as day length increases, the dampening effect is removed, and sows mated in mid to late winter are stimulated hormonally to farrow in the comfort of spring & early summer. The 'sensor' is the pineal gland which is a natural light receptor. Hidden away at the base of the skull, it is activated by light impulses from the eyes via the brain which then activates the adjacent pituary gland (*See Glossary*) to set in process a chain of hormone events affecting reproduction positively or negatively.

This is why so much attention has been paid to 'fooling' the pituitary, by providing extra light to extend day length artificially in autumn, into maintaining progesterone levels. Trouble is, this is easier said than done. And, as we have seen, other factors may interfere.

### Are these the reasons why seasonal infertility can be variable year-on-year?

These are nutritional stress, loss of body condition, social stress, extremes of temperature and disease. Trouble is, again, any one of these in combination with the others can tip the scales in favour of infertility. For example, a thinnish sow, slightly underfed, weaned into a pen of strangers, some of which are bullies, forced to lie out in a draught at night in a rather dirty resting area may not react – even to these insults – except in the late summer or early autumn, when the seasonal effects tip the balance.

So progesterone level may be threatened and nothing untoward results when such conditions apply, but the onset of shorter day length on top of too many of these adverse factors may precipitate abortions or reduced litter size. That is where management advice can help, in reducing the basic stressors.

## CAN WE INJECT PROGESTERONE?

This often doesn't work and would be expensive anyway – suggesting that the reaction may occur elsewhere in the hormonal pathway. Don't resort to hormones unless the veterinarian suggests it. A better solution – at least for the time being – is to do an overall audit of all the factors that are thought to affect the maintenance of progesterone by natural means and put them right if detected or suspected.

1. ***High feed levels in early pregnancy of the gilt may reduce embryo survival.*** This is because a high food level after service increases blood flow to the liver to deal with the arrival of these extra nutrients. This leads to a lower progesterone level. Meanwhile the newly formed embryos secrete a protein (RBP or Retinol Binding Protein) which helps them establish themselves in the uterus. If the progesterone level falls, the secretion of RPB is hindered and embryo mortality increased.

2. ***The same high food levels in early pregnancy in the second litter sow onwards*** do not seem to affect embryo survival or litter size to anything like the same extent, unless they are considerably overfed. Why?

Sows seem to have, naturally, a higher blood progesterone level after lactating. But a sow which has nose-dived in condition may not show this.

This seems to be happening with the hyperprolific gilt where her huge first litter (unless something is done to lift the load off her) renders her progesterone-deficient. Result, poor reproductivity next time.

The hormone insulin is probably vital in likely feed energy metabolism and reproduction. If the sow is energy deficient in lactation due for example to underfeeding or a larger litter, insulin is reduced and very soon progesterone production is compromised. Chromium (e.g. organic chromium from yeast) is probably best and is an important precursor of insulin production. It is only recently being added to breeding foods (at the low level of 200-300 parts per billion), which probably helps to account for the increase found in litter size across several parities.

---

# CHECKLIST: DEFENCES AGAINST SEASONAL INFERTILITY

---

Finally let's go back over some things you can do…

If outdoors…
- ✓ ***Provide shades***, erected early enough in Spring.
- ✓ ***Check for sunburn daily***; bring indoors.
- ✓ ***Provide insulated huts***, facing down the prevailing summer wind direction.
- ✓ ***Provide wallows,*** sited not too far away from the social areas.
- ✓ ***Ensure some pigmented genes*** are in your females.

Indoors and in general…

✓ *Use AI 24 hours after each natural service.*
A central service unit (or if outdoors an ecoshelter in the service paddock) is advisable to facilitate this, accentuate pheromone exchange and keep a check on services.

✓ *Serve in the cool* of the morning.

✓ *Keep boars present* for the first four to six weeks of pregnancy.

✓ *Up your service rate* by 10-15% if Seasonal Infertility is likely.

✓ *Do a mycotoxin audit* (See Mycotoxin section).

✓ *Check your indoor lighting adequacy* and diurnal light pattern.

✓ *Discuss with your breeding stock supplier* if he has strains where selection for decreased seasonality are available, e.g. percentage of pigmented genes, and stress-tolerant (docile) strains.

✓ *Discuss with your boar supplier* about doing a boar sub-fertility test during hot weather. Some boars are more hot weather resistant than others.

✓ *Record the maximum temperature* in the service area every day.

✓ *If the outside temperature rises past 23°C*, use AI top-up indoors or out in the summer/autumn.

✓ *Check temperature variation* within the service/breeding unit. In early spring and autumn this can cause abortions.

✓ *Provide lids* in winter or if the early autumn is unseasonably cold, for yarded groups to even out temperature variation, but make sure the run outside is well lit, and that the lids can be raised in summer or air blown through in on hot days (not at night).

✓ *Check feed intakes* in summer.

✓ *Watch you don't cut down on space allowances,* especially in summer.

✓ **Ensure you have organic** chromium, selenium and iron in the breeding feed.

# REFERENCES & FURTHER READING

Australasian Pig Service Assoc (1987-1999) Various annual volumes of 'Manipulating Pig Production' contain valuable research data.

Dzuik PJ (1987) Embryonic Loss in The Pig: An Enigma 'Manipulating Pig Prod' 1 28-39

Gardner JAA, Dunkin AC and Lloyd LC (eds) (1990) 'Pig Production in Australia' Butterworths (for Australian Pig Res. Council)

Kingston, N (2005) Plan for Summer Infertility. Farmers Weekly, Aug 5

Love et al 'Summer Infertility Effect of feeding regime in early gestation on pregnancy rates in sows' cited in 'Animal Talk' (eds Close & Cole) Sept 2001

Mackinnon J (1996) Autumn Infertility – Notes (Pers Comm)

PIC (2008-2009) Summer Infertility Series

Reed H (1991) Summer Infertility, Pig Farming June 1991 30-35

Sunderland, C (1991) Summer Infertility NAC Newsletter Oct 1991 pps 6-8

Wiseman, J and Walling, G (2009) Five Ways to Address Summer Infertility in Pigs

# 23

# DEALING WITH HOT WEATHER

Heat Stress: the body temperature of the pig must remain between certain limits to maximise production, safeguard welfare and resist disease.

When the pig's temperature rises beyond the upper limit it starts to pant. This is known as the Evaporative Critical Temperature (ECT). Panting increases the evaporative heat loss from the lungs as the pig starts to try to control its body temperature. Normally the pig breathes at 20 to 30 breaths per minute. ECT is estimated at around 50 to 60 breaths/minute, but this can rise to 200 breaths/minute as it crosses towards UCT - see below. The start of panting is a good indicator of heat stress and is a definite and urgent action level.

Upper Critical Temperature (UCT) is generally taken as being approximately 3 to 5°C above the warning signs of ECT.

Crossing beyond the UCT threshold immediately affects the pig's metabolism, seriously reduces appetite, lowers production and can jeopardise its defences against disease; it also compromises it's welfare.

As with LCT (Lower Critical Temperature) both ECT and UCT can vary across a 5 to 6°C ambient range dependent on, for example, the pig's age and weight, fat-cover, food and type of energy intake, protection from solar radiation, floor type, airspeed and skin wetness.

Table 1 gives an approximation of ECT and UCT for the different conditions shown.

*Advice:* **I advise the adoption of ECT as a warning, call-to-action threshold and UCT as crossing a definite danger boundary.**

**Table 1. The relationship between ECT and UCT under differing conditions**

| Stock | Age (wks) | Wt (kg) | Floor type | Air speed (m/sec) | Energy intake (MJ/d) | Skin* wetness (%) | Ambient temperature (°C) | |
|---|---|---|---|---|---|---|---|---|
| | | | | | | | ECT | UCT |
| *Piglets* | 1 | 2 | Mesh | Still | ad lib | 15 | 35 | 41 |
| | 4 | 5 | | | | | 33 | 39 |
| *Weaners* | 5 | 7 | Mesh | 0.1 | ad lib | 15 | 35 | 41 |
| | 6 | 10 | | | | | 33 | 39 |
| | 8 | 16 | | | | | 30 | 37 |
| | 5 | 7 | Concrete | 0.1 | ad lib | 15 | 36 | 42 |
| | 6 | 10 | | | | | 34 | 40 |
| | 8 | 16 | | | | | 31 | 38 |
| *Growers* | 9 | 20 | Concrete | 0.1 | ad lib | 15 | 30 | 38 |
| | 15 | 50 | | | | | 28 | 36 |
| | 21 | 90 | | | | | 27 | 36 |
| | 9 | 20 | Concrete | 0.5 | ad lib | 15 | 32 | 39 |
| | 15 | 50 | | | | | 30 | 37 |
| | 21 | 90 | | | | | 29 | 36 |
| | 9 | 20 | Concrete & spray | 0.5 | ad lib | 60 | 34 | 42 |
| | 15 | 50 | | | | | 32 | 40 |
| | 21 | 90 | | | | | 31 | 39 |
| *Dry sows* | | 150 | Concrete (one sow) | Still | 27 | 15 | 27 | 36 |
| | | | | 0.3 | 27 | 15 | 29 | 38 |
| | | | | + spray | 27 | 60 | 33 | 40 |
| | | | Concrete (group of 5 sows) | Still | 27 | 15 | 26 | 35 |
| | | | | 0.3 | 27 | 15 | 28 | 38 |
| | | | | + spray | 27 | 60 | 32 | 40 |
| *Lact. sows* | | 150 | Mesh | Still | ad lib | 15 | 22 | 32 |
| | | | | Still+drip | | 30 | 26 | 33 |
| | | | Concrete | Still | ad lib | 15 | 23 | 33 |
| | | | | Still+drip | | 30 | 25 | 34 |

Note:
- This table is not a set of recommendations. Use the values above as guidelines only and carefully observe pig behaviour.
- UCT data are estimates only. Beware of air temperatures approaching these UCTs. They signal danger; deaths could occur at temperatures at or above them.

*Skin wetness:    15% is normal wetness due to drinkers
30% is typical wetness of pigs during drip cooling
60% is an average wetness of pigs during spray cooling

Based on advice from the Australian Pig Research & Development Cooperation (who are arguably the world's leading experts on hot weather pig production) and from other sources.

# INITIAL CHECKLIST
# HOT WEATHER STRESS

✓ ECT (panting) is a warning sign of the commencement of profit loss.

✓ Check the ambient temperature over the pig with a clean digital thermometer.

✓ Consider the list of recommended actions shown in the following tables.

**Remember**

If the temperature climbs 2 to 3 °C above that measured when you notice panting, acute performance loss is likely. At UCT your pigs may enter the death zone. See Table 1. UCT is made much of by ag. environmentalists, however, *I consider ECT to be a more useful threshold than UCT* for two reasons:-

✓ ECT can often be identified by panting and respiration rate encouraging you (in fact instructing you) *to take remedial action at once*. Many times I have passed panting pigs at 60-80 breaths/min and the stockman did nothing. On instructions to wet the animal(s) or redirect airflow *now*, the stockman said "Do you mean *now*?" The answer is "Certainly, UCT is approaching or has been crossed and the danger point has arrived. Do not delay one minute."

✓ If the pigs have reached or exceeded UCT it could already be too late. Their metabolism is compromised and may have been damaged, taking some while to repair. Loss of productivity has been suffered, which in certain places in the breeding cycle for both boars and gilts/sows can be severe, i.e. litter size and farrowing rate in particular.

# HEAT LOSS

The pig can, to a certain extent, defend itself against the effects of heat stress by the following:-

*Radiation* (radiant energy): *The emission of energy waves from a source*, e.g. heat from a roof heated by the sun then striking the pig's body surface. Conversely the pig's body surface can radiate energy to the air around it which eventually warms the surfaces (roof, walls) around it, thus losing body heat.

Radiation typically accounts for 20% of the pig's heat loss in hot weather - but if the building surface temperatures are above that of the animal, there will be a net heat gain by the pig.

*Convection: The rise and removal of warm air from around a hot surface.* This is helped by moving the air over the pig's skin surface - as long as the air moved fast enough to dislodge the layer of still air which lies close to the skin. This is an effective means of cooling; but without wetting, only if the air velocity is at least 1m/sec and the temperature of the air 3°C below the pig's normal body temperature of 38.9°C. Thus, if the ambient temperature is between 26°C and 33°C, the pig can dissipate about 30% of its body heat to vigorous dry air movement of this standard.

*Evaporation: The conversion of liquid into vapour.* As it is the fastest moving molecules of heated liquid (in this case water) which escape from the pig's hot skin surface, the stored (kinetic) energy of the remaining water molecules is reduced, and therefore evaporation causes cooling. Air movement increases evaporative loss, so it is the combination of air speed and wetting which contributes most to skin cooling, not just **air movement alone**, or **wetting alone**.

Pigs can create their own evaporative heat loss by panting, wallowing and pen soiling, which are all early indicators of ECT.

For every 100 cc of water evaporated, 220 Btu (see Glossary) of heat is required. This means that in an ambient temperature of 27°C panting can, in theory, account for about 40% of a growing pig's heat loss. However, considerable and prolonged rapid respiration is needed to evaporate 100 cc of lung moisture, which is why skin wetting is preferable.

*Conduction: Heat transmission from places of higher to places of lower temperature.* The pig can alter its posture to provide greater contact with a cooler or wetter floor surface. But don't overrate it; conduction usually only accounts for 5-10% of the heat loss in hot weather as only 20% of the animal's skin can be in contact with a cooler floor surface.

# WAYS OF KEEPING PIGS COOLER

## Reducing radiant heat

Producers in **hot countries,** as well as those of us with more temperate climates now beginning to experience the effects of global warming, do not generally appreciate **the value of insulation** to anything like the extent of those who have to deal with cold weather.

Piggeries in the tropics often have little or no insulation, relying on air movement, solid floors and wetting to cool the pigs. An adequate insulation layer together with a white-painted outer surface to the roof alone can reduce the still air temperature inside by as much as 3°C. *Just painting the outside surfaced white can reduce internal solar radiation by 30%.*

Any benefits from other cooling methods follow on from this head start. A formula I brought back from Australia which is cheaper than white paint is a mixture of 4.5 litres of PVA emulsion, 22 litres of water, 20 kg hydrated lime and one handful of cement. Important: add the PVA to the water first and lime afterwards. This solution needs to be kept stirred.

Even in temperate zones it is surprising how much solar heat can penetrate a roof if the insulation is deficient. Recent measurements have shown solar gains of up to 30 Watts per square metre ($W/m^2$) coming through old roofs, and as much as 85 W/$m^2$ through single skin roofs. It is important to check the standard of your insulation and to bring it up to modern recommended standard U value of 0.4 $W/m^2/°C$. (See Glossary for full definitions of these terms).

## Shade cooling

Shades shield out the sun's rays (solar radiation) and provide a cool ground surface for the pigs to lie on. Shades can cut out the radiant heat from the sun by as much as 40%. Various commercial materials are available which can be set up in spring and taken down in autumn, but they need to be well-anchored and, as we can see in the section on Seasonal Infertility, *need to be erected soon enough* to catch the clear bright days of 10 hours sunlight in early spring - in this case not so much to produce shade cover but to distil the effects of bright light on reproduction and reduce sunburn.

Shades should, if possible, be sited on high ground to catch the breeze. On lower ground, locating them 50 metres downwind from a wood or lush vegetation helps cool what breeze there is. If shades are set high and sloped, e.g. from 2m rising to 2.5m with the higher end facing *away* from the sun's travel, this will help maximise radiant heat loss from the animals by exposure to 'cooler' northern (or southern) sky, according to what hemisphere they are in.

Always check the sun's angle when planning roof overhangs towards the sunward side of any tropical building where the side walls/blinds are removable to assist airflow (Figure 1). As well as protection from sunburn, the radiant heat emission from a concrete walkway fully exposed to the sun is considerable and either a permanent overhang or an extendable sunshade must be provided. Never use blinds to provide shade (except in a crisis sunburn situation) as this will disturb crossflow movement of air, and risk considerable elevations of ambient temperature despite the sun's exclusion.

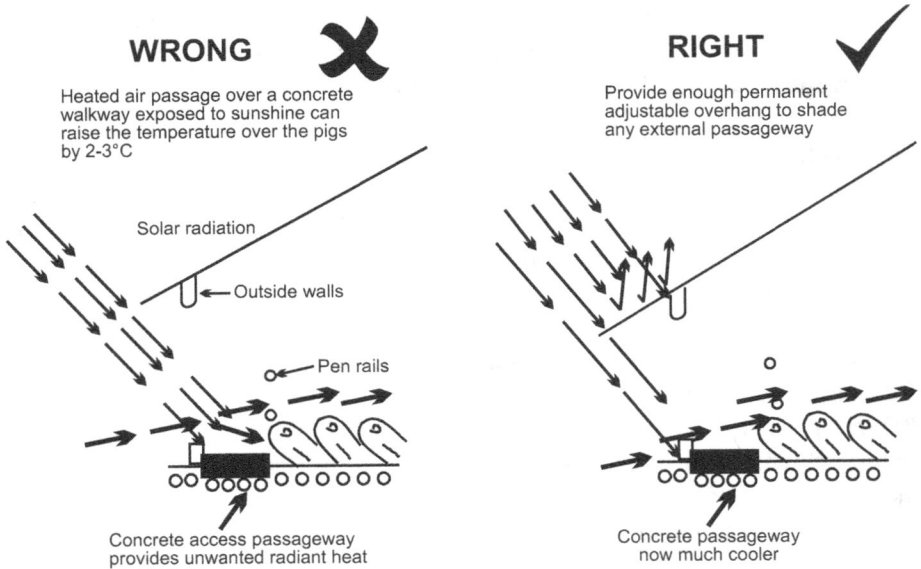

**WRONG** ✗

Heated air passage over a concrete walkway exposed to sunshine can raise the temperature over the pigs by 2-3°C

**RIGHT** ✓

Provide enough permanent adjustable overhang to shade any external passageway

Solar radiation

Outside walls

Pen rails

Concrete access passageway provides unwanted radiant heat

Concrete passageway now much cooler

**Figure 1.** Shading to prevent sunburn and lower house temperature.

## Ventilation in hot weather

Rapid air movement over the animal helps both convective and evaporative heat loss. Airspeed below 0.1m/sec is considered to be 'still air' and, as we have seen, an airspeed about 10 times more (1m/sec) and higher provides a cooling benefit for an unwetted skin surface. Thus *ventilation rate* and *air positioning* are the two main components of cooling by air movement.

Correct ventilation rate for hot conditions which is preferably measured in cubic metres moved per hour (m³/hr) is relatively easy to calculate (Table 2) and is often adequately incorporated into a hot weather ventilation system involving propeller fans. The converse is true where natural ventilation or Automatically Controlled Natural Ventilation (ACNV) is involved: these can rarely cope with outside temperatures over 28°C, where the heat produced by the pigs can raise the *Temperature Lift* - *the difference in temperature between the inside and outside (climatic) temperature* - within the building by 4-5°C, which often approaches or exceeds ECT and even UCT in some circumstances.

In practice, a temperature lift of 3-4°C is aimed for in hot weather conditions. However, the smaller the rise in lift, the greater is the airflow needed to achieve it; in a given situation a 3°C lift requires about a 32% increase in airflow than for 4°C.

**Table 2. How to calculate maximum ventilation rate**

| Maximum ventilaton needed | = | $\dfrac{\text{Heat output from the pigs in Watts}}{\text{Difference in temperature target between inside and outside} \times 0.35}$ |
|---|---|---|

*Reference data: Heat production in Watts*

| Weight (kg) | | At ECT (Watts) |
|---|---|---|
| 1 | | 4 |
| 6 | preweaning | 24 |
| 6 | postweaning | 13 |
| 10 | | 35 |
| 20 | | 51 |
| 40 | | 94 |
| 60 | | 121 |
| 80 | | 144 |
| 100 | | 163 |
| Dry sow 170 | kg | 142 |
| Farrowed sow 170 | kg | 272 |

**Worked example:** First stage grower to contain 500 growing pigs at maximum weight of 60kg. Ventilation to be designed to keep temperature lift down to 3°C.

Heat output of 500 × 60kg pigs is 500 × 121 Watts = 60,500W
Max Ventilation Rate needed is:- $\dfrac{65,500}{3 \times 0.35}$ = 62,381 m³/hr
(1040m³/min)

This sounds a lot, but remember 500 × 60kg pigs are involved, and several/many fans will be involved.

We now know how much air must be put into the building to maintain temperature lift no higher than the pig's ECT, assuming the pig's skin is not deliberately wetted.

# AIR POSITIONING

But what if the outside air is hot? There is then a limit to the degree of cooling which can be achieved by ventilating with outside air. The air must now be directed over the pigs themselves, so *air positioning criteria* are needed.

The principle (except in very hot conditions) is to have the airspeed over the pig's back at between 0.75 and 1m/sec dependent on the current ECT of the pigs - generally the higher level is advised but beyond 0.75m/sec, a lower airspeed with skin-wetting is more effective and uses less power.

In contrast to air positioning in cold conditions where we require a long air travel, visiting the passageway and dunging area first, good air mingling, then passing across the resting pig's back last (at 0.15 to 0.2 m/sec) - in the warm weather we need to direct the air directly over the resting pigs first and exiting over the dunging area (or in some cases the roof vent) last. Figure 2 in the Ventilation Section shows the two contrasting airflow patterns.

# HOW TO USE EVAPORATIVE COOLING

There are two types of evaporative cooling for pigs.

*Spray or mist cooling* for growers over 30 kg, finishers and dry sows

and

*Drip cooling* for farrowing sows, and also nursery pigs, say to 30 kg.

Spray or mist cooling is used for whole house cooling of pigs if the temperature is above 26°C for one hour, where the risk of chilling baby pigs and weaners is not present. Drip cooling, being more particulate, generally cuts in at 22°C for lactating sows and 30°C for weaners.

Spray cooling, with water droplets up to 20 times bigger than most mist droplets, need not necessarily use more water than misting (a common sales point of misting over spraying) if it is controlled properly as described in Table 3 and when so, I find it much more effective. However, many piggeries where spray cooling is used are badly operated. Here is a checklist on what to avoid.

**Table 3. Spray cooling growers, finishers, dry sows and boars. Note: Spray droplets, not fine mist.**

| | |
|---|---|
| Application rate | 330ml per hour per pig |
| Cycle time | 5 min on, 45 min off |
| Nozzle flow rate | *3 litres/hr × number of pigs/nozzle |
| Switch on temperature | 26-28°C (lower than this, just increase ventilation rate) |

| | |
|---|---|
| *Examples: | 1 spray nozzle in pen of 10 growers requires a nozzle flow rate of 30 l/hour while 2 spray nozzles in a pen of 10 growers requires a nozzle flow rate of 15 litres/hour |

Source: Kruger, Taylor and Crosling (1992)

# CHECKLIST ON EVAPORATIVE COOLING ERRORS

✓    Inadequate or over-use (waste) of water.

✓    Wetting and thus chilling (especially at night) sucklers and many newly-weaned pigs.

✓    **Spray**-wetting lactating sows. These should be **drip**-cooled, to the neck region only.

✓    Not applying forced air movement (under control) over wetted pigs in the hottest conditions.

✓    Not maintaining spray or drip cooling equipment.

✓    Not adjusting wetting times and evaporative intervals between them to suite climatic changes.

✓    Using spray cooling under 23-5°C for growers. In this temperature band increase the ventilation rate or raise/direct the airspeed more closely over the pigs.

# WHAT SPRAY COOLING DOES

Dependent on relative humidity and the amount of air movement over the body, a wetted pig takes up to an hour to dry. If its skin temperature is lowered by a short burst of spray of around 5 minutes followed by a break of 45 minutes to enable evaporation to take place during this time, even though the pig is living in a temperature of 25-27°C, it feels as if it is only at 20°C and its metabolism responds accordingly.

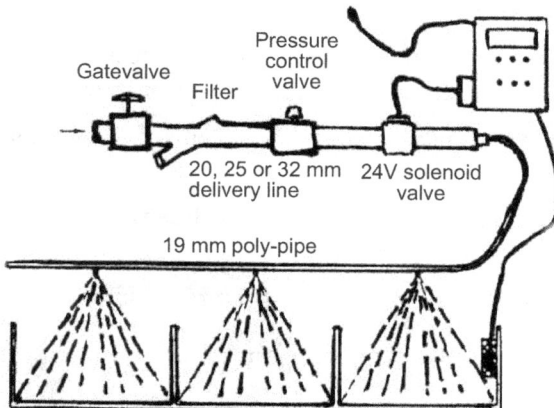

Source: PRDC (see references)

**Figure 2.** Spray cooling installation

## ADJUSTMENT MAY BE NEEDED

Individual producers may need to adjust the frequency and length of time the spray nozzles operate. Adjustments will be affected by several factors: type of building, ventilation, stocking density and local climatic conditions. In areas of high humidity, water will take longer to evaporate from the pig's skin, therefore the interval between spraying may need to be increased slightly.

## AN INSTALLATION CHECKLIST
## (SPRAY COOLING ONLY)

✓   Make sure the water flow can be turned on or off.

✓   Have one day's reserve water supply in header (reserve) tanks.

✓   Keep water source as cool as possible, as spraying with *conduction-heated* water stresses pigs.

✓   Water pressure is normally 140 kPa (20 *psi*).

✓   Use a 200 micron water filter capable of a flow rate of up to 1.4 litres/sec (see Glossary).

✓   Houses of up to 40 metres will need 20mm diameter delivery lines.

   40-60m will need 25mm delivery lines.

   60-100m will need 32mm delivery lines.

✓   For lateral lines use 19mm diameter tubes.

✓   Spray nozzles should deliver a uniform distribution of *large* droplets directed straight down.

✓   It is better not to use *mist* or *fog* nozzles as these increase humidity, have much less effect on shed temperature and are distorted by air movement.

✓   Do not use manually-controlled systems. Use solenoid valve automatic controllers.

✓   Use a control system (e.g. Farmex) which can integrate with curtain or shutter openings.

## AIR SPEED ACROSS THE PIG

It is important to stress that water cooling will not be effective unless there is some air movement over the pig. Water cooling must be done in conjunction with adequate ventilation. *A minimum air speed over the pig's back of 0.2 metres per second (5 seconds to cross one metre) is necessary.*

## Checkpoint

A spray cooling system working properly will result in pigs reverting to 20 to 30 breaths/minute within 25 to 30 minutes after the 'off' period. If respiration rate is climbing to twice as much at this time, shorten spray off-time and check air speed/ air positioning over that batch of pigs.

# DRIP COOLING

Drip cooling in farrowing crates where the sow can be wetted, leaving the sucklers largely untouched, is a cost-effective and efficient way to cool lactating sows in hot weather (Table 4).

**Table 4. The benefits of drip cooling on performance**

| *Sows 27-34°C* | *Drip* | *No drip* |
|---|---|---|
| Breaths per minute | 28.5 | 63.6 |
| Weaned per litter in sows lifetime* (kg) | 56.21 | 50.91 |
| Sow weight loss in lactation (kg) | 3.79 | 38.53 |
| Daily feed intake (kg) | 5.74 | 4.79 |
| Return to oestrus (days) | 5.0 | 5.0 |

Drip at 4 litres per hour per sow

*i.e. in this trial. Source: Murphy *et al.* (1988)

Table 4 is typical of the benefits in a hot country - piglets are weaned at heavier weight and sows maintain body condition better than non-cooled sows.

In more temperate climates where farrowing house summer temperatures cause the sow to be at ECT or near UCT levels for more than 12 hours/day the benefits have been 0.5 more piglets born alive, 300g heavier weaning weights and 1 kg/day more sow lactation food eaten.

**Table 5. Effect of drip cooling on sows incorporating stillborns and mortality**

| | | |
|---|---|---|
| Stillborns reduced 0.1/litter | $p < 0.05$ | ⎫ |
| 24 hours deaths reduced by 0.13 per litter | $p < 0.01$ | ⎬ Significance |
| Weaning weight increased by 400g/piglet | $p < 0.001$ | ⎪ |
| ADG increased by 14g/day to weaning | $p < 0.001$ | ⎭ |

Payback on equipment cost from these figures 1.69 years
Year = two hottest months of summer i.e. payback from 3.38 months use/year

Source: Cutler (1989)

Even allowing for the need for drip cooling in the four hottest summer months, the payback over cost from 290 kg more liveweight sold per sow/year at weaning and 2 days quicker growth to slaughter was . . . "At current prices this paid back the capital and installation costs on a 100 sow unit in on year" (Maxwell, 1989).

Table 6 details the design requirements for drip cooling sows and weaners and Figure 3 gives installation guidelines.

**Table 6. Design requirements for drip cooling**

|  | *Lactating sows* | *Weaners (only in extreme temperatures)* |
|---|---|---|
| Application rate | 330ml per hour per pig | 65ml per hour per weaner (5 weaners/dripper) |
| Cycle time | 1 min on, 10 min off | 1 min on, 10 min off |
| Dripper flow rate | 3-3.5 litres/hr | 3-3.5 litres/hr |
| Switch on temperature | 22-24°C | 32°C (larger weaners only) 35°C (small weaners) Avoid use in high air speeds |

Source: Kruger, Taylor and Crosling (1990)

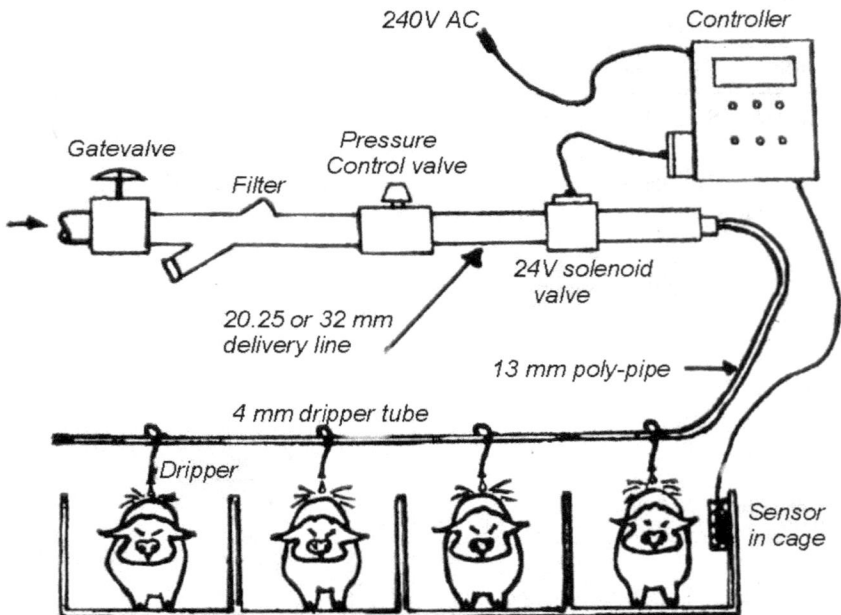

**Figure 3.** Drip cooling system (Source: PRDC, see references)

# A DRIP COOLING CHECKLIST

## General

✓ Provide ventilation. 0.2m/sec seems adequate (5 seconds to cross one metre).

✓ A filter is wise, capable of dealing with a flow rate of 0.8 l/sec at a 200 microns mesh.

✓ A pressure reducer is advisable to maintain even flow.

## Lactating sows

✓ Location is very important to minimise wetting the litter (and the sow's feed if fed dry nuts).

✓ Fix the 4mm dripper tube to the *top* of the 13mm delivery pipe so that the 4mm tube rises up and bends down. This ensures that when the water is turned off the dripper line empties quickly and there is no drip 'run-on'.

✓ Locate the drip over the shoulder within the circle dimensions described below - this assumes a perforated/slatted bed. If the bed is solid, to minimise the chance of mastitis, less-good but advisable is to locate the drip over the rump where the rear mesh/slats are.

✓ Do not install drippers where the water can flow into the creep area.

✓ Water must be able to be increased or decreased and turned off when the pen is empty or on cool nights.

Studies on how often the sow lies out flat in the farrowing crate show that the occupies this position for 94% of the time. In a normal 575-600mm wide farrowing crate with equal space outside the side rails, sows seem to have a preference and will normally lie on one side for 70% of the time and about 30% on the other. The head is displaced laterally by some 180mm between these two positions - which is not much. Thus the target area for wetting is a circle approximately 300mm radius, with the drip set at least 500mm back from the edge of the feed trough.

## Weaners

✓ Use one dripper for every 5 weaners, dripping over the slats.

✓ Avoid use in high air speeds, this will chill them - a maximum of 0.2 m/sec even in high temperatures.

Producers may care to use timed 'piggy showers' rather than drips, where a special shower area is available in each pen. This, while excellent, tends to be expensive, with paybacks calculated from installations of nearer 3 years.

# EVAPORATIVE PAD COOLERS

These are growing in popularity in hot countries especially where the climate is **hot, dry and dusty** in summer. The system works less well or hardly at all in areas of high humidity. The benefits in dry climates can be remarkable (+38 kg MTF, 12-105 kg) but **installation and design are critical**. Air is drawn through wet fibrous pads. Water is supplied to the pads by a water sump and overhead distribution system. Rigid plastic pipes or open rain gutters with evenly spaced holes allow water to drip uniformly over the pads. Pipe size, hole size, and spacing depend on water flow-rate and are sized for each system. For best evaporative efficiency, more water is supplied than is evaporated. To conserve unevaporated water, provide a filtered return to the sump. A sloped gutter below the pads collects and conveys unevaporated water back to the sump. Control the make-up water line with a shut-off float valve.

Protect the water distribution system from insects and debris. Filter recirculated water before it returns to the sump. Also, install a filter between the pump and distribution pipe or gutter. Control the system with a thermostat set to begin wetting pads at the desired cut-in temperature i.e. ECT. To reduce algae growth on the pad, stop the pump several minutes before the fans to dry pads after each use. Figure 4 illustrates one such design incorporating a cooling chamber.

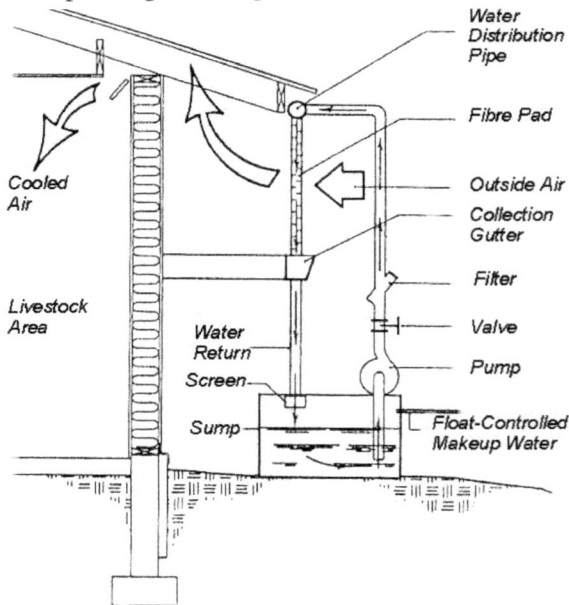

Ventilating exhaust fans pull hot air through wet fibre pads.
Air heat evaporates the water and lowers air temperatures.

**Figure 4.** Evaporative pad cooling system, based on PRDC advice

It is vital to have the installation designed by a ventilation engineer. There are formulae for water requirements, pad size and power loading needed, as well as the need to control the internal ventilation carefully between operant and non-operant times. There are home-made systems, often incorporating a cork pad taking up most of the end wall, with the house's ventilation system drawing air through. These often cause problems and soon get clogged up which lowers efficiency, raises power costs and under-ventilates the far pens in the room.

## EVAPORATIVE COOLING - A BASIC CHECKLIST

Evaporative cooling systems require regular maintenance for proper operation. Develop a maintenance schedule from the following guidelines:

✓ Replace woven fibres annually, so a swing-away wire mesh front is advisable.

✓ If the pad settles, add more pad material so air does not short-circuit through the open space(s).

✓ Hose pads off at least one every two months to wash away dust and sediment.

✓ Control algae build-up with a copper sulphate solution in the wash water. Light-tight enclosures around pads and water sump also help control algae.

✓ As water evaporates, salts and other impurities build up. Bleed off 5-10% of the water continuously to remove salts or flush the entire system every month. *Caution*, bleed-off water can be toxic - dispose of it properly.

✓ Remember, power costs rise by almost 25% during operation due to the pad's resistance to the drawn-in air. The engineer may advise extra fans to ensure adequate air is drawn in *and circulated*.

✓ For a 50 sow farrowing house or 200 pig nursery about 2 litres/minute flow rate will be needed around the system, which will have at least 5m² of pad area to be wetted, i.e. about 2.25 × 2.25 metres in size.

## SNOUT COOLING
### (*Sometimes called Zone Cooling*)

This is an excellent low cost system to improve appetite in hot weather and is almost always confined to lactating sows who can suffer very badly from lack of appetite in hot conditions, especially hot humid conditions.

There seem to be two principles involved:

1.  **Freshening the air over the feed trough**

    Here air from the outside - which may well be at 30°C or more - is positioned accurately and deposited to fall gently over the feed trough at about 0.25-0.30 m/sec. The idea is not to cool the sow at all but to keep the space near the sow's head, especially over the feed trough, free of gases. A 15-20cm plastic pipe conveys outside air, pushed in by a lightweight propeller fan, to 8 cm downpipes fixed to the front end of the farrowing crate. The downpipe contains an adjustable damper to alter exit air speed and stops over the feed trough just out of the sow's reach.

    Figures from farms who adopted this form of air freshening show, without question, lactation feed intake improves by 1 kg/day in temperate zone summers and up to 2 kg/day in the tropics.

    Remember, *snout air freshening does not cool the sow*, it encourages her to eat more and so improves body condition at weaning.

2.  **Zone cooling**

    In hot weather, most animals lose 60-70% of their heat through evaporation from the respiratory tract. Zone cooling the area around the head helps improve cooling. Zone cooling is generally used for stall- or crate-restrained animals and occasionally for animals in individual pens such as a boar pen. in farrowing buildings, zone cooling makes the sow more comfortable while allowing higher temperatures in the pig creep.

    Zone cooling, however, does not satisfy all of the hot weather ventilation needs, and a conventional system sized to provide adequate ventilation rate is still needed.

    In theory, zone cooling will not work so well if the outside air is very humid. This is because the high moisture content of the air hinders effective dissipation of the respired moisture from the lungs. However, the air freshening aspect is still there, and the higher air velocity to the head plays its part. Because of this higher velocity (Table 7) great care must be taken not to cause a draught in any side creep area, as the air can 'skid' around solid farrowing pen divisions and chill the piglets.

    A ventilation engineer will have internal distribution duct and downpipe duct dimensions to accommodate the air velocity recommendations in Table 7, as well as the fan power loadings needed.

Table 7. Recommended air flow rates for zone cooling of
breeding pigs

| Type of animal | Ventilation rate* (m³/minute) |
|---|---|
| Farrowed sow, 200 kg | 2 |
| Dry sow, 150 kg | 1 |
| Boar, 250 kg | 1.6 |

*Zone cooling systems should still be ventilated at the hot
weather ventilation rates given in Table 1. Source: Jones *et al.* (1990)

# THE USE OF CURTAINS (BLINDS) IN HOT WEATHER

Vertically rising reinforced canvas curtains or blinds are a common method of
ventilating hot weather buildings. ***Their design and operation leave much to be
desired, however.***

Curtains on 90% of the farms with daytime temperatures usually between 26 and
34°C are designed for such a daytime temperature band, where a reasonable flow of
air is required across the pigs, along with spray cooling.

However, thought is rarely given to late evening and night time temperatures which
can often fall sufficiently to cause chilling. This is especially dangerous with young
pigs in the 20-40 kg band. There can be a problem with food conversion - pigs growing
between 20-40 kg will need to eat 18g food more/day for each 1°C below their LCT
to maintain growth. So if appetite is limited, growth is 12g/day slower, which results
in 2 days longer to slaughter.

***Even so the real danger from slightly low temperatures is chilling.*** Chilling causes
stress and stress lowers the animal's immune defences to many diseases, especially
respiratory ones. Chilling is an additional effect to the surrounding air temperature,
as the floors are often wet from the day's spraying to keep the pigs cool. Wet floors
increase thermal conductivity so the pig loses additional heat, especially at night, not
only to the cooler air over it, but also to the cold surface under it. Some critical organs
(live and kidneys) are in direct contact with this too-cold surface.

## Manual curtain alteration is never frequent enough

As the warm evening turns to a cooler, chilly or even cold night, the curtain is the
main device to retain the correct inside air temperature - just above the pig's LCT.

These, being manually operated, are usually lowered too late; as a result the pigs are chilled and stressed as the outside temperature falls.

I have measured such falls in the tropics between 7pm and midnight. In each case I have asked the owner what he thought the range of fall over this period was, the reply being "4-5°C", yet the instruments showed it was between 8 and 12°C. Only one curtain adjustment meant that all pigs were well outside the correct range across 4 hours and from midnight onwards were frequently below LCT by 2°C (older pigs) and as much as 5°C below LCT for weaners.

If you get more virus and respiratory disease than you would like, then crude and inefficient curtain operation is probably a major reason.

# SOLVING THE PROBLEM

## 1.   Automate the curtain

No stockman, however, diligent, can keep pace with the diurnal temperature changes twice a day (morning and evening), and may be further changes during the day or night if the weather changes. An automatic controller makes over 30 adjustments in a normal day, and sometimes more.

A series of temperature sensors (normally 3 for a building 10 metres wide × 30-40 metres long) linked to a controller which instructs a curtain motor (or motors, see below) will keep the temperature bands close to the pigs' LCT (this can be preset into the controller accordingly to the age of the pig, type of floor surface, degree of roof insulation and feed density etc) so that the pig is not chilled even if the outside temperature falls either quickly or slowly (Figure 5).

## THE DISADVANTAGE OF MANUALLY OPERATED CURTAINS

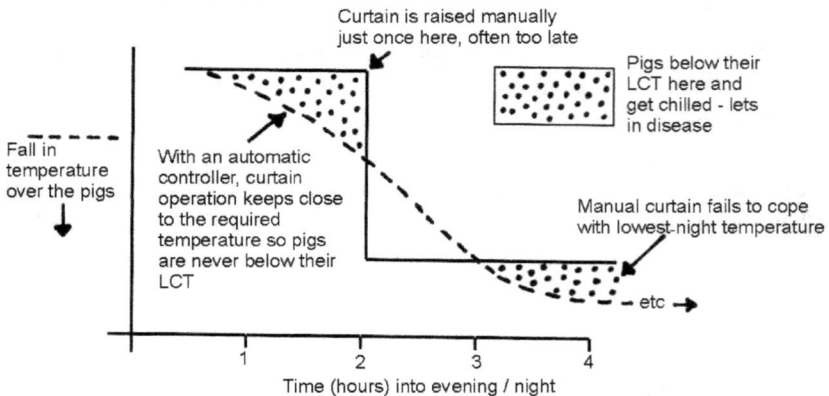

**Figure 5.** The disadvantages of manually operated curtains

## 2. Make the curtains fit snugly

The top should be rigid and fit into a slot. The sides should 'fold' into a box, or if on rollers should rise and fall via a narrow slot. Both are to reduce draughts, and this especially important for weaners to about 20 kg.

In non-tropical areas solid louvres are better than curtains, but are some three times more expensive to construct.

## 3. Quartile the building

'Quartiling' is to divide the building into four squares or broad rectangles served by *two* curtains down the side of the building, not just one (Figure 6).

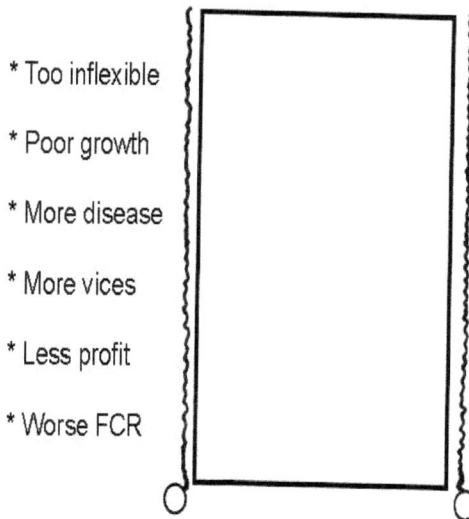

* Too inflexible

* Poor growth

* More disease

* More vices

* Less profit

* Worse FCR

**Figure 6.** Wrong: Two long blinds running all down one side of the building

Just having one blind down one side is too inflexible as it:

- Cannot cope with the sun's daytime movement from horizon to horizon.
- Cannot sufficiently cope with changes in wind direction.
- Curtains are often too long, so the upper gap tends, especially after time, to be unequal down its length.

This is not so important during the day, but can be critical at dusk and during the night. Maximum operational length at the time of writing is about 30-35 metres. If the building is longer, use several blinds (Figure 7).

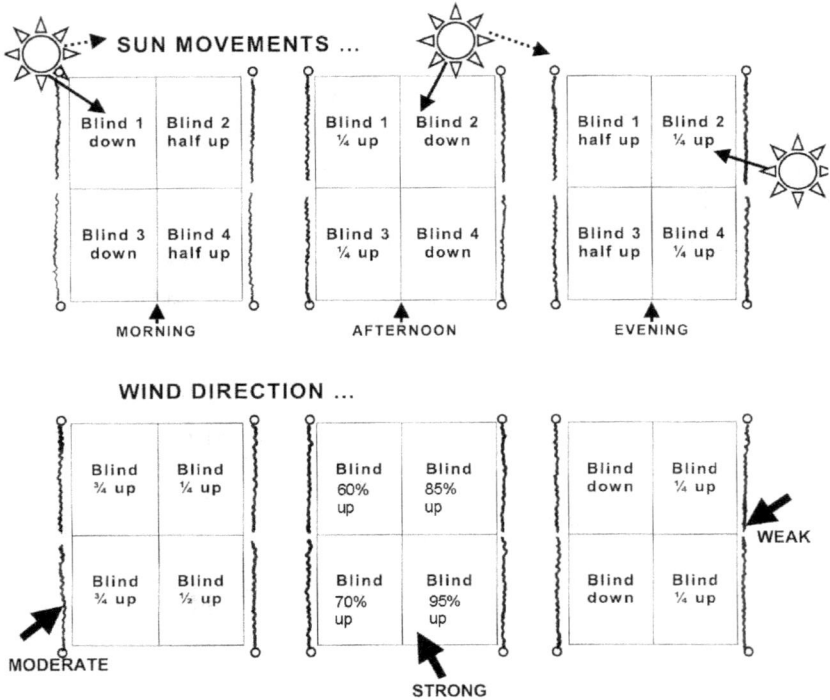

**Figure 7.** Correct: 4 shorter blinds quartiling the building.

# THE BOTTOM LINE - QUARTILING PAYS

Table 8 shows that in growing pigs, due to faster growth and better food conversion, the food saved was 26.4 kg/pig mainly due to faster growth. This translates into a massive 162 tonnes of feed/year for this 300 sow unit.

**Table 8. Effect of quartiling curtains**

|  | *Before* | *After* |
|---|---|---|
| FCR 20-90 kg | 3.2 | 3.0 |
| Days to 90 kg | 161 | 153 |
| % dirtying pens | 36 (at 29.3°C) | 12 (at 29.2°C) |
| Food saved on 300 sow farm/year (t) | - | 162 |
| After deduction for one-off cost of auto installation (t equiv) |  | 130 |
| PAYBACK (months) |  | 10 |

Source: Client's records

The only difference in the above table was that two buildings (nursery and finishing) were converted from manually operated curtains to quartiled automatically-controlled curtains. The cost of conversion was 80% of the cost of the food saved, thus payback was in 10 months. Not surprisingly the farm is now fully converted to automatic curtaining. Further exercises have shown the payback to be 1.2-2.1 years on food savings alone, not counting the cost of reduced disease. One farm reported a marked improvement in the mortality and morbidity level of nursery pigs after 1½ years. This alone paid for the alterations in 6½ months.

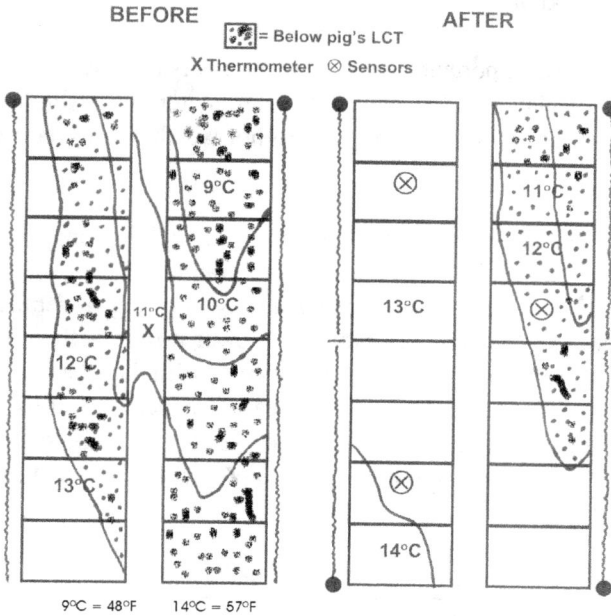

**Figure 8.** Temperature bands before and after quartiling (nighttime). Note the correct positioning of sensors

## THE BOAR IN HOT WEATHER

The testes are usually 2.5°C cooler than normal body temperature, and this temperature must be maintained for optimum boar fertility. The most significant effect of "overheating" is a reduction in the numbers of motile sperm ejaculated at mating. Motility enables sperm to reach and fertilise eggs after mating, with 95% of ejaculated sperm normally motile. This percentage declines at air temperatures above 30°C, to the extent that on one trial, motility of sperm collected from boars heated to 40°C fell to *below 5%*.

So, at air temperatures above 33°C, in addition to depressed motility, other influences on fertility such as total sperm numbers, percentage of live sperm and percentage of abnormal sperm ejaculated can be expected to deteriorate; *this effect appears between 2-5 weeks after heat exposure*, with the shorter time period likely to apply to those boars mating more frequently.

## Boar libido

Australian research has demonstrated boar sex drive doesn't appear to become depressed until air temperatures approach 38°C, with no residual effect on libido following the period of heat stress.

Boar libido plays an important role in enabling boars to actively seek out females in-oestrus, and display an adequate level of stimulatory courtship behaviour prior to completing an active, quality mating. So hot weather may reduce preliminary courtship.

The activity of "flank nosing" by boars within the first 30 seconds before mounting females appears to be an important behavioural trait to ensure good fertility. Boars displaying high levels of nosing activity during courting generally have higher conception rates than those boars exhibiting lower nosing activity (Table 9). It is thought that increased nosing activity by boars may stimulate the release of oxytocin in the female, which in turn may lead to increased chances of fertilisation occurring through increases in uterine contractions and sperm transport along the oviduct. It is believed 'tasting' the saliva of boars is as important to the female as smelling his pheromones.

**Table 9. Serving - length of act and contentment**

| Score | Conception rate (%) | 'Courting' (mins) | Mounting attempts | Intromission (mins) |
|-------|---------------------|-------------------|-------------------|---------------------|
| 1. Poor<br>2. Moderate | } 80.5 (75-86%) | 1.8 | 1.9 | 2.1 |
| 3. Good<br>4. Excellent | } 83.4 (75-91.8%) | 0.42 | 1.1 | 3.1* |

*Litter size was increased by 0.48 piglets for each minute's increase in duration of intromission

Correct receptivity: Correct surroundings: Quiet unhurried handling

Source: Rikard-Bell (1994)

# WORKLOADS

The effects of summer heat on boars may be lowered by reducing the frequency of boars mating during periods of hot weather. The effects of heat on sperm for production and the determination of the optimum number of sperm for fertilisation to occur show that following short-term heat exposure, boars can maintain fertility provided they are used less frequently.

The recommendations below are drawn from Australian work and are:-

**Table 10. Mating frequency guidelines in hot weather**

| Temperature to which boars are exposed for a single 12 hour period | Maximum mating frequency 1-6 weeks following exposure |
|---|---|
| Below 30°C | Twice/day |
| 30° | Once/day |
| 33° | Four times/week |
| 36° | Twice/week |
| 40° | Infertile for most of the period |

It is possible that some boars will be capable of maintaining their fertility at higher mating frequencies than those recommended. All other things being equal, I suspect these boars are likely to be those with larger testicles, but this needs scientific confirmation.

Regardless of workload, it is recommended whenever possible that matings be conducting early in the day (when temperatures are likely to be lower) and ***prior to feeding***, in order to maintain boar libido (Table 11).

**Table 11. Comparison of early morning *vs* all day service routine in hot, humid weather**

| | Sows served between 5am-7am (temperature 24-26°C) | Sows served all day (temperature 27-34°C) |
|---|---|---|
| Farrowing rate (%) | 88 | 72 |
| % repeats | 13 | 23 |
| Born alive/litter | 9.87 | 8.91 |

# REFERENCES AND FURTHER READING

Readers may be concerned that some of the tables and references in this chapter which cite those of the 1990s and before, may look to be outdated. Do not be!

This is because thermodynamics is a science based on physics whose laws are immutable, thus the calculations involved have not changed with time. Much of the basic information we use today, based on original research into pig thermodynamics is still a reliable source, and just because it carries an early date is due to the fact that the researcher was first to establish/describe it!

Indeed. I am grateful to several of the housing and environmentalist pioneers, now retired, who over the years have generously guided me. The first was John Randall (who wrote to me saying, politely, that from my writing I seemed to know little about thermodynamics and ventilation and to 'come up to the NIAE* for a day or two and we will put you straight'.

I did, and they did! Other pioneers whose groundbreaking work is in this book were George Carpenter, Jeff Owen, Leonard Mouseley, and Chris Boon – using a mountaineering term, true 'first-ascenters.'

*National Institute for Agricultural Engineering, Bedford, England.

Assistance with this section is gratefully acknowledged from the Pig Research & Development Corporation, Australia, whom I consider the world's leading experts on hot weather production. Their publication in the Australian Pig Housing Series "Summer Cooling" (1992) is still a major source of valuable practical information (cf Figures 2 and 3) and I am grateful to the PRDA for permission to quote from selected advice therein. Despite its date of publication, the advice given has stood the test of nearly 20 years subsequent experience. The Australians do some fine research - very practical and to the point.

BPEX UK (2009) *Action for Productivity* sheets. Heat Stress - Indoor Herds; Heat Stress - Outdoor Herds.

Jones, D.D. *et al* (eds) (1990) *Heating, Cooling and Tempering Air.* Monograph published by Iowa State University, USA pp 12-15 and 24-28.

Kruger, Taylor and Crosling (eds) (1992). Summer Cooling. In: *Australian Pig Housing Series*. ISBN: 07305 98926.

Murphy, Nicols and Robbins (1989) Drip Cooling of Lactating Sows. *Pigs*, May 1989, pp 8-9. Elsevier Publications.

# 24

# DIRTY PENS AND THEIR PREVENTION

Pigs soiling their resting/sleeping area with urine and/or faeces.

## *TARGET*

A dirty pen floor is one of the most annoying and time-consuming problems facing pig producers. I see bad cases on one in twenty of all farm visits – with a third of them left unattended when a simple solution is all that is needed to rectify matters. The target must be zero.

Pen fouling can occur at any time – from sucklers voiding in their 'creep' areas where they are supposed to lie warm and dry in between meal-times, to finishing pigs making a quagmire out of their sleeping and socialising areas. Dry sows in stalls can mess up their underbelly area too.

Strange as it may seem, I have always enjoyed being asked to deal with a dirty pen floor problem as it is relatively easy to cure. If you are observant, the reasons are always there to be seen.

## THE BABY PIGLET

### Creep-fouling

Nearly always due to giving the newborn pigs too much space in the shut-off creep area. Faced with a lying space which is too generous, they naturally treat one end of it as a lavatory.

Sure cure: allow no more than 1½ times the supine[1] body area of the total number of piglets in the litter; or 2 times the area once a creep hopper is placed in the creep. Vary the area by putting in a wooden dividing board, a plastic can or wooden box to shut off or take up the unwanted space until the pigs get bigger (Figure 1).

---

[1] Supine = The space occupied by the head, body and extended legs of the piglet when lying on its side as distinct from the sternum position where the pig lies full-length on its chest.

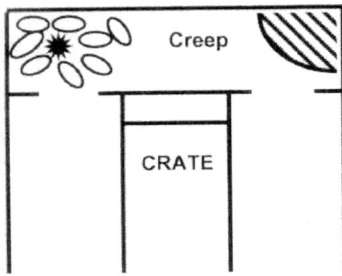

Giving sucklers too much room in a forward creep encourages them to use the excess space as a lavatory.

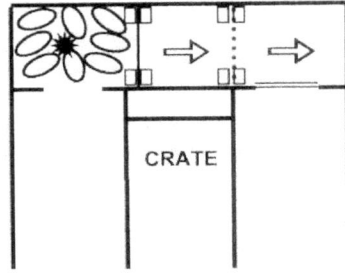

**Remedy:** Close off one pophole until the end of the suckling period. Fix vertical battens at 1/3 and 2/3 of the creep box, and place a crusher-board at either location as the piglets grow.

By training them to accept that the only area to rest in is that allowed for sleeping supine, they will not foul the box – unless the space under the lamp or over the heated pad is uncomfortably hot.

**Figure 1.** Dealing with a fouled forward creep box in a farrowing pen

## Fouling near to the sow's head

With a solid floor this area – on either side of the sow's head – can be filled in with a concrete rounded 'hill', rather like a woman's breast, which will deter elimination. Alternatively put a grille or drainage area in these 'dead' places in the farrowing pen. If the sow spills a lot of water in this position, consider how this can be lessened – provide a good deep 4-litre drinking trough, for example. Another solution – but at the farrowing house design stage – is to eliminate this 'dead' space by allowing the creep area in the adjacent farrowing pen to occupy it. Remember, all pigs prefer to eliminate in corners, so don't have obvious corners at the sow's head – only down at the other wet end of the pen.

## Adequate temperature gradient

Even a newborn pig can be instinctively drawn to eliminate in a colder area. Trouble is it needs an adequate temperature gradient between the resting/ exercising area and the elimination areas to encourage it to do so – my own experience suggests a drop of over 3°C, at least. This is often not provided in the open-plan farrowing house, as insufficient use is made of flaps or deflectors to 'knock down' some of the cooler incoming air towards the sow's rear end. Not a draught, but just a gentle downward drift of cooler incoming air. This is easily achieved with flaps set at the correct angle up on the ceiling, together with the use of simple smoke test tubes to see where the incoming air is going and how fast it falls. But hardly anyone uses this essential aid!

# OLDER PIGS

Anthropomorphism (the attribution of human characteristics to non-human objects) is common among pig producers, probably because the pig is, in many ways, rather similar to us. *We* regard wrong-mucking as a 'bad habit', but the pig probably considers it to be a natural reaction to alleviate an uncomfortable situation it finds itself in, often through no fault of its own. Or it could be just pure laziness! Let me explain.

There seems to me to be five main reasons why a pig eliminates in the wrong place.

1.   **Because it is too hot, and wishes to start to cool itself by lying in urine and faeces to conduct away more heat from its body.**

   Examples:

a)   The house and ventilation design which cannot get rid of an excessive temperature rise towards and beyond the pig's ECT (Evaporative Critical Temperature).

b)   Overstocking, which causes a localised temperature rise.

c)   Poor ventilation, especially in centre pens, pens exposed to mid-day sunshine, and troughs or solid partitions which deflect cooling airflow.

2.   **Because the pig has too much space in which to void.** We may provide a suitable elimination area, but if the free space between its preferred (dry) resting area and where we wish it to void is too generous, it may decide to relocate its voiding area closer to 'home'. Anthropomorphically we call it 'laziness', to the pig it is merely common-sense!

3.   **Because it is difficult to get to the voiding area.** The above problem is made much worse by making the access to the dunging area difficult (narrow pass-ways, excessive change of floor height (steps), slippery surfaces, insufficient voiding area as the pig gets longer with age, other pigs lying in the way). Providing unpleasant conditions where *we* want the pig to dung can persuade the pig to disagree with us. For example poor slat dimensions and cracked or pitted solid floors, darkness in the slatted area and excessive temperature differences between resting and voiding areas.

4.   **Overstocking :**  Overcrowded pigs can also be forced to eliminate wrongly because they are prevented from establishing their usual social pattern of behaviour. What they see is no clear voiding area, or other pigs in the way

preventing the lower orders in the group from getting to it. Usually overcrowding results in over-hot or airless, gas-filled conditions and the pig's natural answer is to wet (cool) itself which is a purely defensive reaction, and urine is the most convenient source of liquid available to it. If the 'common sense' answer is to wet it's bed - so be it!

5.     **Habit :**  The classic case is where a young pig has been kept in a hot nursery with a totally wire or slatted floor and is then moved to a *solid or part-solid* pen. Not surprisingly it has been encouraged to void where it likes in the all-wire pen, especially if there is poor air placement.  It may continue to do so in the new quarters.  This can be overcome by ensuring that :–

a)     There is a distinct air pattern provided in the all-wire room (often a flat-deck) where cooler air falls on to the preferred dunging area (as far away from the resting/feeding areas as convenient and with drinkers sited within this cooler zone).  The temperature difference between lying and voiding area should be at least 3°C dependent on the LCT (Lower Critical Temperature) of the pigs. An airbag trunk connected to outside air placed over the dunging area often solves this problem.

b)     Training nursery pigs on an all-wire floor not to void on a solid area when they are eventually moved by putting a solid comfort-board in front of the feeder in the all-wire pen so that they get used to solidity being a 'clean' surface.

c)     The space allowed in the part-solid follow-on pens should not be too generous. It often is, as small pigs are put in pens big enough for them at finish weight. The answer is to use crusher boards or variable-geometry pen fronts to close the space at first so as to allow the pigs to grow *with* the space available rather than grow *into* it.  Housing designers – please think about this!

d)     The slatted/void to solid/resting ratio should be adequate, at 1 to 3 or 1 to 4 maximum.  Again, it often isn't, especially in the long narrow pens seen often with pipeline wet-feed units or in broad finishing piggeries where too narrow wet/dry feeder pens make best economic use of total area.

# CHECKLIST
## THINGS TO LOOK FOR WHEN NOTICING DIRTY PENS

✓     *Where* are they in relation to end/outside walls, centre of the building, ventilation inlets/ extractor fans. Where are the *dry* ones? What's the difference – this will help you decide what the cause is.

✓     Is there anything making the pigs in dirty pens too hot?  Or their air supply too gaseous? (Use a gas detector phial.)

✓ Have you lost control of air movement – summer/winter or day/night ? In other words, can you measure it? (Use a smoke 'candle'.)

✓ Check stocking density. If you've got dirty pens up to maximum stocking density, try destocking by 15% (1 in 7).

✓ Have you got the tools to assist you? A proper clean calibration-checked thermometer set just above the pigs back? Smoke tubes? Draeger ammonia phials and suction bulb (10-50 ppm $NH_3$ range is adequate using ammonia as a marker gas). Deflector boards hung on wire or string? Airbag trunking? And lastly – elbow grease plus a pail of water and yard brush!

## CHECKLIST WHEN WRONG-MUCKING OCCURS IN PART-SLATTED GROWER/FINISHER PENS

Pigs prefer to eliminate in corners – we need to tell them which corners!

✓ Check stocking density for summer/winter conditions, sometimes day/night conditions. (See stocking density section).

✓ Dunging area 3°C colder than lying area? In the tropics a difference of 2°C is usually adequate.

✓ Pens should not be longer than x2 the width, with 1½ times better. Use a crusher board to adjust balance, *i.e.* shape.

✓ Feeders in among, or beyond the lying area are problems – check that they are quite close to drinkers in dunging areas as long as there is at least 1.65m gap to the opposite division across the pen – otherwise move the feeders to the front wall/rails.

✓ Always site pen furniture so that the pigs follow this primary natural pattern – awakening, feeding, drinking, eliminating, social congress, resting – in a clockwise or anti-clockwise direction. Too many crossed paths may result in vices, aggravated by high stocking density and inadequate ventilation control.

✓ The dunging area should be along the short side of the pen, of sufficient depth to accommodate a finished pig's length (1 metre"). Do not close the dunging:sleeping ratio to less than 1:3.

✓ High temperatures - Is it hot or airless at the following temperatures ?

*Up to 23°C\**

Encourage better ventilation – provide natural cross-flow ventilation *over* kennels, for example, to remove the airlessness inside. Put an offset hip to any kennel roof to encourage rotary airflow inside.

*23°C to 26°C\*)*

To put in air-bag ventilation (so as to position greater air movement/turbulence over the pig's resting area). This can be shut off at night if required (see page 529 on how to do this).

*26°C+\**

To merit spraying if the circumstances allow.

\*    *Dependent on the E.C.T. of the pigs in question. You   should consult a ventilationist as it can vary by as much as about 4°C from these approximate problem thresholds.*

# DEEP STRAWED PENS

Groups of pigs have to eliminate somewhere in a yard, so provide a concrete scrape-through area. This is commonly seen, but what is almost never done is *to provide downflow ventilation over it as well*. Generally a simple airbag hung on wire loops with a single row of holes on the underside is adequate. Growing pigs especially will often also eliminate in the strawed corners, and some sows will too. The airbag provides that 3°-5°C vital difference, even in summer, when in the hottest weather the bag can be replaced with one possessing a row of dual outlets (called sipes – see page 529) set at 45 degrees on either side of the bag to push air across both the scrape-through passage *and* strawed resting yards. In this way the animals are less stressed and are encouraged to dung/urinate in the customary areas.

Clients have often been impressed by this simple, low-cost precaution, which in cold weather can be shut off at night. Many pig producers have ignored the airbag concept – a serious error. Details on how to make your own are available - it is not difficult.

The airbag concept is so useful because . . .

•    It is flexible.

•    It is cheap.

•    It can position air either by a row of holes (sipes) down the tube(s), or by rotating them. Simple, as the airbag tube is clamped to a cylindrical air straightener which can be loosened, the tube with its exit holes repositioned, and then reclamped again. The tube(s) are suspended by metal rings, so can be easily rotated by hand to position the air just where you want it.

An airbag duct. Photo: Harcourt

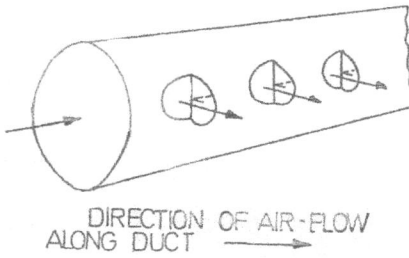

DIRECTION OF AIR-FLOW
ALONG DUCT

**Effect of 'D' flaps on air discharge.**

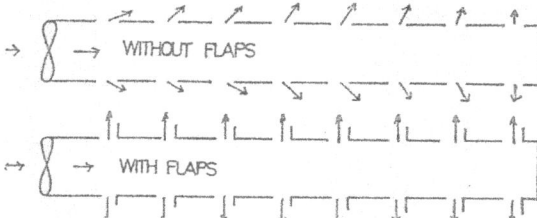

WITHOUT FLAPS

WITH FLAPS

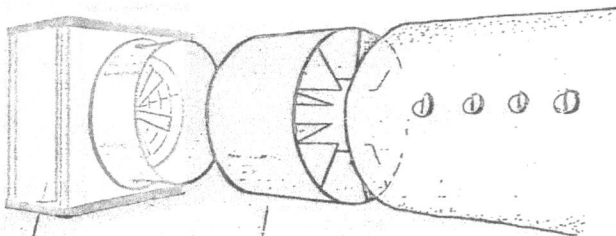

Filter box (optional, used in hot, dusty climates) and fan

Air straightener to prevent the polythene duct from twisting due to the fan's torque

# 25

# HOW TO AVOID TAIL BITING AND OTHER VICES

Chewing of the tail end which, if unchecked, can lead to severe erosion of the whole tail, sepsis, stress and even death.

## *TARGETS*

Incidence should be nil. It has been suggested that 14% of all condemnations in Europe are due to tail biting, so the problem is a major one. I see evidence of tail biting (past or present) on 1 in 8 of the farms I visit even today.

## THE EXTENT OF THE PROBLEM

NADIS UK, which operates a disease survey programme, reported in 2007 that in our quite sophisticated pig industry . . .

* The overall incidence from 400,000 pigs surveyed by 14 pig veterinarians was 1.2% (a later survey by BPEX/RSPCA in 2009 reported a 3.96% incidence).

* That 3-5% of pigs may be affected week in, week out. Of these 1% were badly enough damaged to merit euthanasia.

* A further 1% were condemned at slaughter, usually reported as 'pyaemia' (see Glossary) on a processor's condemnation sheet.

* At this level on a 300 sow herd this cost 4.4% of the annual gross margin/sow (140 pigs lost) but this excluded the costs of treatment, care, isolation and lost growth.

* Pigs on slatted floors showed the greatest incidence (see Figure 1).

* There is a widely held belief that tail biting is not a problem on straw. NADIS reports have shown that it is likely to be from 20% to 60% of the incidence on slats but with considerable regional variation, and 30% of all cases recorded (Figure 1).

A more recent British survey (NPA, 2010) involving 104,400 finishers on slats, 204,350 on straw and 10,200 finishers outside, attracting responses from 13.8% of the UK herd in 2009, showed that 56% did not have a tail biting problem while 44% did.

I notice from my travels that the incidence must be much higher in Eastern bloc countries and in Central America.

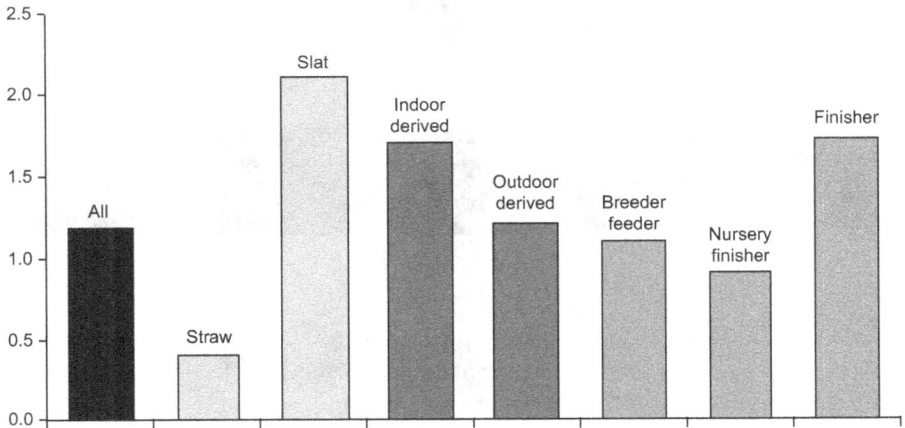

**Figure 1.** Tail biting - % of UK weaners and growers affected.
(Source: NADIS (UK) 2007).

# CHECKLIST

The checklist gives an indication of what are the most common causes. (This is just one man's experience of witnessing or being asked to help alleviate several hundred cases)

Over 45 years I have seen a trend away from nutritional causes to those of environmental effects and possibly that of genetics.

We are feeding pigs better now, but maybe overcrowding them too much. Our pigs are growing very fast, and could be less docile than they were decades ago.

These are the main areas you should examine when tail biting occurs.

Analyses of 283 outbreaks 1961 – 2008* (25% over the past 10 years)

| *Areas of Attention* | | *Importance Rating* |
|---|---|---|
| ✓   Overstocking | | 60% |
| ✓   Ventilation | - inadequate | 50% |
| | - wrongly positioned | 50% |
| | - gases ($CO_2$, $NH_3$) | 15% |
| | - low speed cold draughts at night | 40% |

✓ Badly placed pen furniture causing aggression      10%[1]

✓ Uneven mixing      18%[1]

✓ Pigs moved off bedding (straw) to solid floors/slats      20%

✓ Poor trough design   15%[1]

✓ Sick pigs not removed promptly      60%

✓ Genetics (tendency to lose docility?)      15%[1]

✓ Nutritional      - salt      20%[2]

                - diet which pigs dislike      15%[3]

✓ Water inadequacy (all variants of including fouled drinkers)      20%[1]

✓ 'Boredom'. Over the past 10 years, providing occupational material has reduced incidence, in my experience, by 50%

At least 10 nutritional factors are said to be involved (like low protein, etc.) but I classify them still as largely unproven/undemonstrable. We feed pigs well these days – look elsewhere!

[1] These cases/incidences I suspect are more important than the rankings indicate, at least since 1985. Very little research on them, I find.

[2] Salt is strange. While the salt levels in tail-bitten pigs' diets are found to be adequate (0.4%), raising the level to 0.8% (Muirhead, 1989) *with plenty of water* does help cure outbreaks. But why? Could it be the presence of other unpalatable ingredients masked by an increase in salt?

[3] Includes sudden change, and/or mycotoxin presence.

Note: These are subjective findings : what one man's experience has found to have stopped the outbreaks when they were attended to, in order of approximate ranking when followed up later. However, how many of these cases would have cleared up anyway?

## Early signs

✓ Agitation/disturbance/restlessness in pens.
✓ One (usually small but vigorous) pig molesting others.
✓ All pigs tucking their tails down close to their rumps.

═══════════════════════════════════════════════

After all, is tail biting so strange? All parents have witnessed the tendency for children to chew things as their teeth and gums develop. Is tail biting in young pigs the same, but misplaced, tendency?

# PRIMARY CAUSE – SOMETHING TO DO

Pigs have a natural rooting instinct. They are inquisitive, curious and aware animals, so when not eating, drinking and sleeping are looking for 'something to do'.

Yet we keep them in 'overcrowded' and 'unnatural' conditions. Modern production means we have to do this – compared to the wild, anyway. We have to ensure that we do not overstep the *overcrowding* or *unnaturalness* boundaries.

Both cause restlessness or low-level NES (see Stress Chapter) stress (anxiety). Happy pigs rest for up to 82% of their time, dozing away the hours, so 18% of their time is spent up-and-about being inquisitive and looking for something to do. And if they can't find something to do, then they'll look for trouble!

So we must give any confined grower something to do – and if rooting comes naturally, something to root – a large ball, old 6" diameter ball-cocks from lavatory cisterns, a heavy log, some greens, even a sod of earth (very short-lived!) Not chains, these swing about, slap other pigs in the face and tend to raise the restlessness level! In the wild, pigs do not graze branches like cattle – they *root*. So give them 'rootability'. A very good diversion is a 2 metre (2 yard) long piece of toughened alkathene tubing, as pigs like to chew across an object as well as pushing it.

Next, keep the pigs resting – comfort and a feeling of well-being helps enormously. ***Contented pigs do not tailbite!*** I've seen it time and time again. If your pigs aren't dozing away 20 hours out of the 24 you should hold a restlessness audit! As for yourself, what you are doing, by mistake, to raise the restlessness level?

# ENVIRONMENTAL ENRICHMENT

The idea behind environmental enrichment is well summarized in the UK Welfare Code guidance, which states:-

" *Environmental enrichment provides pigs with the opportunity to root, investigate, chew and play.*
*Straw is an excellent material for environmental enrichment as it can satisfy many of the pigs behavioural and physical needs. It provides a fibrous material which the pig can eat. The pig is able to root in and play with long straw and when used as bedding, straw provides the pig with physical and thermal comfort. Objects such as footballs and chains can satisfy some of the pigs behavioural needs, but can quickly lose the novelty factor…….. and… are not recommended unless changed on a weekly basis.*"

## Manipulative materials

Here are some thoughts on the subject. Straw can be expensive and not available to all. **Alternatives in the form of 'toys'** have been…

Ropes and rags, plastic drums, old Wellington boots*, a variety of large balls, logs, fresh tree branches*, plastic pipe, suspended chains (but only with wood or portions of rubber conveyor belt attached), old tyres*, ball salt licks*, disused feed bags - all used with a variety of success, suggesting that they may be farm-specific rather than universally successful. For further discussion, see the Stress Chapter, p. 135.

Those marked with an asterisk* are considered unwise for various safety-to-the-pigs reasons, for example wire in the tyres, and salt poisoning. And in creating a nuisance – pieces of chewed boots and rubber ending up in the slurry pits and causing emptying problems.

**Alternatives to straw as bedding** or rootable materials have been…

Peat and spent mushroom compost (see below)
The seven other materials as in Table 1
Shavings* (some woods can contain toxic resins).

## Peat and spent mushroom compost

While peat on offer is an excellent deterrent, environmental constraints preclude its use. Spent mushroom compost is equally good and is a by-product needing disposal anyway, so a word on this material for any of you near to a mushroom farm.

Pioneer work in Northern Ireland, where initially shallow trays of peat or mushroom compost were placed in pens where tail biting was a problem, proved effective. However this tended to take up a lot of pen space even though the results were good in relation to other 'rootable' materials (Table 1).

Since then racks 1.8 x 0.6 m have been suspended 60cm above the floor rising to 75cm towards slaughter weight for 18 pigs/pen 35-95 kg. The racks are lined with chickenwire at a 30mm$^2$ grid. Spent mushroom compost is added each day and the pigs pull it through on to the floor, where it is rooted through, played with and most of it eaten. The cost is under £0.25/ finishing pig. While costs of tail biting are difficult to pin down,with tail biting probably costing in the region of £1.15/pig (Robertson 2008) this potential 4:1 return looks promising.

In one trial where tail biting was as high as 11%, this dropped to 1% once compost racks were installed.

**Table 1.** Time spent rooting in various materials (%)

| Peat | Spent mushroom compost | Sand | Tree bark | Forest bark | Barley straw | Sawdust |
|------|------------------------|------|-----------|-------------|--------------|---------|
| 70 | 63 | 46 | 45 | 43 | 34 | 11 |

Source: Beattie et al. Agr Res.Inst. N. I. (2008)

Adding carbon dioxide under pressure to the pigs water supply also has had remarkable success but is very expensive - the ingested gas seems to quieten the pig

## Straw, and other materials

There is no doubt in my mind that pigs on straw are far less likely to tail bite. Unfortunately straw is not always available and even if it is, after a difficult harvest can be expensive. In such cases producers must redouble their efforts to seek other palliatives, temporary if necessary like discarded paper bags (but not dried bracken which is poisonous, or coarse sawdust which can aggravate the lungs) Sand in my experience, tends to wreck food conversion! Tree bark from a tree surgeon is good but avoid any inclusion of conifer, laurel and cupressus. Forest bark is widely used by the garden centres and so the price is now prohibitive.

Meanwhile take another look at the likely origins of the trouble previously suggested in this chapter  - with overstocking the most likely cause.

# HOW TO APPROACH A TAIL BITING OUTBREAK

While the effect of tail biting is restlessness and stress, the causes could be many and interrelated.  Because of this combination of factors it is difficult for the stockperson to pinpoint the one reason which has pushed the pigs over the brink.

Most producers are haphazard in their approach.  The problem is all too visible and needs urgent attention, so a wide variety of remedies are applied at once. Instead:

Short term:

(1)    Remove all bitten pigs immediately.

(2)    Spray mark suspected biters to help identify the culprits.

(3)  Remove those tail bitten to a convalescent pen.

(4)  Do not delay - act at once or it will get worse.

(5)  Add chewable objects. Pigs like to 'nose' and if possible they try to demolish such items, such as paper bags, really thick rope, length of thick plastic piping etc.

Longer term:

(6)  Try to prioritise the ***most likely*** causes from the advice given in the following pages and attend to them first.

(7)  Only try one or two solutions at a time, and allow time – about 2 days, for any effects to show.

(8)  Record dates, pen numbers, weather changes, numbers of pigs affected and any noticeable changes in behaviour.

The purpose of this planned approach is to locate the cause and solve the problem permanently. It works!

## THE FOLLOWING CAUSES HAVE ALL BEEN IMPLICATED IN TAIL BITING

### Overstocking

Correct stocking density for pigs (this is covered in detail in the Stocking Density section) is unfortunately not as simple as the recommended published allowances suggest. This is because the pig uses some parts of the pen as a living area and other parts which he will tend to avoid, such as wet or mucky floors, cold, draughty spots and areas where he feels constricted and may find it difficult to move away from if attacked. Long narrow pens are an example – fewer pigs (it is suggested –15%) must be housed in pens which are more than 2½ times as long as they are broad. A square or a 1½:1 rectangle is preferable. In any case, do not exceed the rule of thumb of 115 kg liveweight per $m^2$; this is particularly important when penning entires together, which can be more prone to tail biting when overcrowded.

Table 2 indicates the effect of overstocking.

### Pen furniture

It has been suggested that pigs, on awakening or from rest, tend to eat first, then drink, then urinate/ defecate, then 'socialise', then return to rest. It is perhaps best to locate these various areas so that the pigs 'rotate' around the pen to left or right rather than cross across each other and so invite antagonism (Fig. 2). How important this is in reducing tail biting is not known, and it is probably less important than ensuring the correct stocking density level.

**Table 2. Likely costs incurred by overstocking a nursery and a finishing house by 15 per cent**

| | Pigs 6-35 kg | | Pigs 36-100 kg | |
| --- | --- | --- | --- | --- |
| | *Correct density* | *+15 per cent* | *Correct density* | *+15 per cent* |
| Daily gain, g | 518 | 480 | 844 | 848 |
| Days in pen | 56 | 60 | 77 | 77 |
| Overhead costs @ 24p/day (£) | 13.44 | 14.40 | 18.46 | 18.46* |
| FCR | 2.02 | 2.12 | 2.42 | 2.63 |
| Total food eaten in period (kg) | 58.6 | 61.5 | 157.3 | 171.0 |
| Total food cost (£) | 11.13 | 11.69 | 27.53 | 29.93 |
| ***Extra costs/pig (£)*** | ***1.85 plus*** | ***2.40*** | ***Total 4.20*** | |

Savings in 15% less housing cost per pig (at £8.20/pig)**Savings £1.23**
**Costs £4.20**

The average REO for deliberately destocking to guideline levels on all three farms was 3.5:1

Even if stocking density and pen furniture are correct – the positioning of pen furniture can calm or inflame aggression

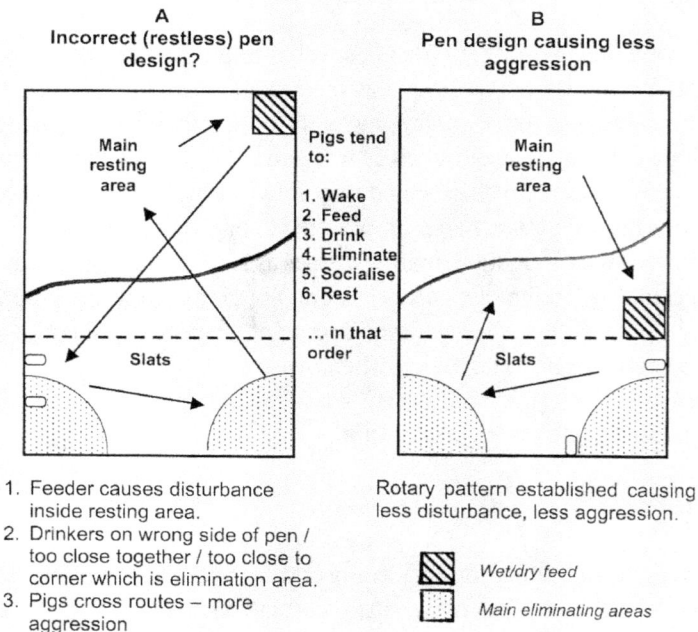

A
Incorrect (restless) pen design?

Main resting area

Pigs tend to:

1. Wake
2. Feed
3. Drink
4. Eliminate
5. Socialise
6. Rest

... in that order

Slats

B
Pen design causing less aggression

Main resting area

Slats

1. Feeder causes disturbance inside resting area.
2. Drinkers on wrong side of pen / too close together / too close to corner which is elimination area.
3. Pigs cross routes – more aggression

Rotary pattern established causing less disturbance, less aggression.

Wet/dry feed

Main eliminating areas

**Figure 2.** Arrows indicate progression within the pen

## Temperature and environmental factors

In my experience tail biting does not necessarily follow from the pigs being obviously too hot or obviously too cold. Much more likely is the overall number of other stressors associated with too hot/too cold conditions. Pigs transported from other farms need correct temperatures, especially fluctuations and excessive air speed can start vices in transported newcomers. Again a stress audit is advisable –
stocking density, temperature fluctuations, sufficient air changes, presence of new companions, dietary change, unpalatable/stale food, presence of a cold draught (especially at night) can themselves precipitate tail biting when if absent, sheer temperature errors alone may not.

Most of my tail biting visits have been during the warmer working hours, but I've noticed that after what must have been a cold, frosty night, the pigs must have been too cold, and that might have been the cause.

## So . . . check temperature variation across 24 hours

One of the commonest causes is diurnal (day to night) temperature *fluctuation*. The pig has a 'comfort zone' inside which there is little thermal stress if any. The comfort zone is the temperature above the LCT (Lower Critical Temperature) and below the ECT (Evaporative Critical Temperature). This band varies in width from 5C° in the weaner to about 11C° in the finisher so is quite narrow, and I have found regular excursions outside the relevant band for the age/weight of the pigs housed over a 24 hours period often triggers tail biting. This may be the reason why runt, thin or 'nervy' weaners/early growers often instigate the process, as they are just not warm enough.

Always check the pigs are within their comfort zone *day and night* so as to avoid too much temperature fluctuation.

## Draughts

I believe night-time draughts are a primary cause. Robertson (1999) recommends that to contain tail biting an air speed of 0.15 to 0.3 m/sec is needed in buildings with temperatures below 20°C, and at above 28°C the velocity should be higher, from between 0.74 to 1.3 m/sec.

In winter and especially at night, walls tend to be cold, so cold, heavy air will flow downwards on to a resting pig lying close up against them. In volume of (cold) air and in downward air speed these are not great, but over a period of time are sufficient to cause sufficient discomfort (stress) to precipitate tail biting in the affected pigs which become restless. Nailing a 3 cm wooden batten, triangular in cross-section, at 1 to 1.5 m internally on such (outside) walls deflects the cold air onto the rising stream of warm air over the resting group and the draught is dissipated. Several puzzling cases of tail biting have been cured once this simple and cheap device was installed.

## Gases

High ammonia ($NH_3$) and carbon dioxide ($CO_2$) levels cause irritable behaviour, including tailbiting. This is seen in the cold winter months when ventilation adequacy is reduced in order to keep the pigs warmer/lower power bills. Pigs can be affected at $NH_3$ levels of only 10 ppm which is very easy to exceed. Tests have shown that winter month levels of 25 to 30 ppm $NH_3$ are quite commonplace in some grower houses/nurseries and a reduction to 12 ppm has stopped tail biting.

So apart from attention to ventilation adequacy, keeping pens cleaner and drier (especially bedding), ensuring slurry channels are less than two-thirds full, *i.e.* 15 cm from the slat under-surface, having slurry pit baffle boards to prevent cold up-draughts and the inclusion of a yucca ammonia-inhibitor like DeOdorase (Alltech) in the feed or in the slurry itself will all help. Dust is also a vehicle for ammonia particles (see p. 135).

## Ventilation rate too low

The incidence of tail biting in finished pigs reported by processors is highest in the early summer quarter. If this damage started in 3 to 4 month-old pigs, the worse season on-farm could be Jan-Mar. This period coincides (in the northern hemisphere) with cold weather conditions which encourage low ventilation rates.

## Tailbiting in kennels

Most kennels provide a dry and warm sleeping area yet are often the source of tailbiting. Why? Most are designed for cheapness and so have flat roofs, which contain a small ventilator and/or can be raised at the front. (Figure 3). In cold weather however kennel lids are shut down and only the ventilator used. This is insufficient as the air does not circulate, but hangs around at lid-height accumulating gases. The air depends mainly on the pigs' movement to agitate it, which at night can be insufficient or even non-existent over the sleeping animals.

Far better is to spend (about 18%) more on a staggered hipped roof. (Figure 4). This gets the air rotating naturally and with the ventilation open or part-open, the gases eventually disperse. Table 3 demonstrates the value over the years of such a design *à propos* tailbiting. The same sort of figures are probably equally true of streptococcal meningitis, another problem brought out by hot, airless conditions.

**Correct**: Strip ventilator in lidded section of hipped roof reduces draughts. Rotary air pattern passing across strip ventilator gently sanitizes air

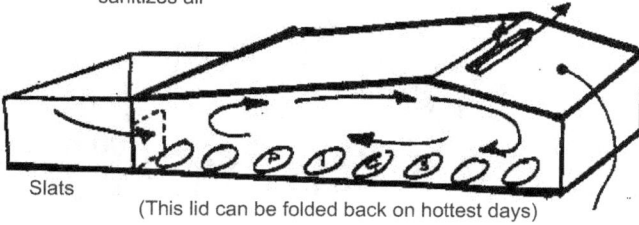

Slats

(This lid can be folded back on hottest days)

**Incorrect**: Flat cover with only one raisable lid to ventilate causes through draughts at night or in cold weather and/or build-up of stale air and gases, as well as micro-organisms, under flat roof

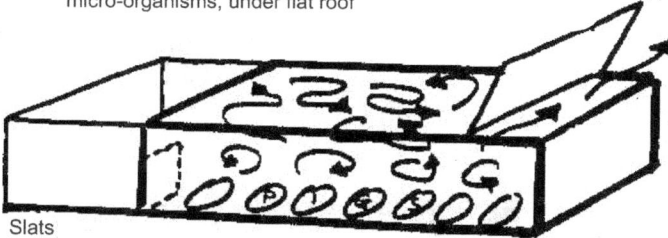

Slats

**Figure 3.** Natural ventilation of kennels.

250mm high hip to roof ensures continuous thermal buoyancy as pigs rest

Whole roof slides to one third its length in warm weather

Simple letter-box slide, 150mm deep, acts as cold weather control & allows gases and foul air to be filtered off

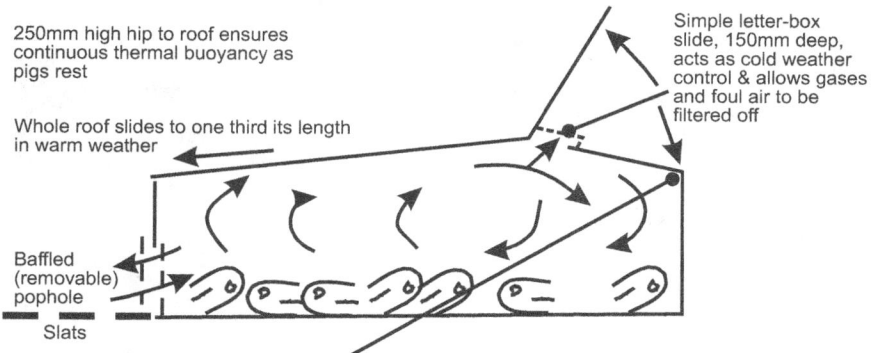

Baffled (removable) pophole

Slats

Inspection lid (one third of roof area) is 'chocked' as intermediate ventilation (See Figure 5)

**Figure 4.** Avoiding stale air in kennels.

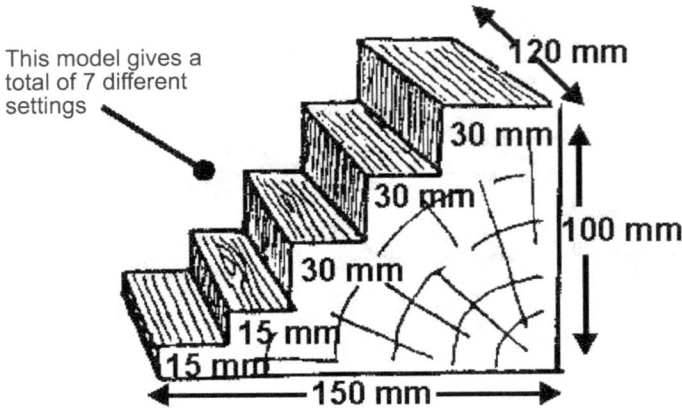

**Figure 5.** Example of a 'chock'.

**Table 3.** Incidence of tail biting in flat and hipped roofed kennels

|  | Flat roof used | Hipped roof used |
|---|---|---|
| Number of pens affected | 27 | 5 |
| Where roof line was changed to hipped | 15 | 2 |

Clients' records

Tailbiting is a most distressing occurrence for both pigs and stockpeople. The bitten must be removed to a hospital pen and treated to heal the wounds. An attempt should be made to identify the biter(s) which is/are removed to another place where plenty of manipulable materials are available. The problem occurs on when or whether to re-introduce the segregated biter to his previous companions - after 10 days or so? Sometimes it worked, sometimes not. If not, we sent the offender off to the local butcher and had done with it.

**Conclusions**

(1) Tail biting from whatever cause could be up to 5 times more likely in flat-roofed kennels.
(2) Where the roofline was changed to hipped, tail biting was reduced sevenfold.

# Lighting

Does bright light or very low light encourage tailbiting? Reduced light intensity has helped to reduce vices in poultry, and the same may be true of pigs (not proven). Pigs do seem to need about a 4 to 6 hour dark period every 24 hours (<25 lux - 15 lux preferable), while continuous light over 60 lux may cause irritability? A largely unexplored area. EU welfare regulations will stipulate 40 lux for 8 hours/day in the near future.

# Feeding

Nutritional errors have long been thought to be a precursor of tailbiting but this has probably been overrated. Some 40 years ago diets were less well balanced than they are now and manufacture is more consistent.

The following nutritional areas may be implicated:–

*Salt*

Check that the level is at least 0.5% (0.2% sodium).
If tailbiting is present, raising the salt level to 0.75% or even 0.8% is permissible for a time (5-10 days) ***providing adequate water is available***.

*Other minerals*

• Total phosphorus should be greater than 0.5%.

• Ca:P ratio should be less than 1.25:1.

• Magnesium could be increased to 0.08.% (American data). I'm told they use magnesium oxide at 1 to 2 kg/t.

• A proportion of the trace element provision should be of bioplex origin (organic mineral proteinates).

• Some people use salt and special anti-tailbiting blocks (Frank Wright Ltd.)

## Other findings are

- Low soya bean meal (<5%) may aggravate it.

- Spent mushroom compost (surface casein layer removed) cured outbreaks at 0.5 kg pig/day placed behind a wire mesh grid (Hillsborough, 2000). Some people have used fibrous peat.

- Energy deficiency/imbalance could be involved in young fast-growing pigs; provide extra energy (Kyriazakis, 1996).

- Increase tryptophan level (Aherne 1997).

- Fibre should be at 3% (Scottish Quality Pork, 2009).

- Queuing at single space feeders at peak feeding times, due to stress caused by enforced and premature withdrawal, increases tail biting. Refer to the feeder space allowances in the Food Conversion chapter.

## Single space feeders

Queues at the SSFs are a common contributory cause, stimulating frustration and anxiety. Robertson (1999) suggests up to 70% of feeding visits can be cut short by forced withdrawal from the feeder. The writer has observed such action carefully and noted mild tail-chewing within 3 minutes among those pigs which were chased away.

Fitting dividers or 'stalls' to feeding positions "causes a significant decrease in this effect and had a subsequent effect on reducing the incidence of tail biting," says Robertson.

In the case of the SSF, providing an extra feed station is the only alternative, apart from destocking below the accepted levels, which is often impractical, and can be costly.

## Tail-docking

In Europe approximately 80% of intensively-housed pigs are tail-docked, where the last 33% or more of the tail is removed soon after birth, up to 12 hours old, but not in piglets more than six hours old so as to allow the piglet to have at least 4 good, uninhibited, colostral suckles. Alternatively, but less-favoured, in older pigs approximately 1-2 cm of the tail-tip only is removed. Cambac, a UK pig research consortium, after surveying tail-docking on over 40,000 pigs in a postal survey, concluded that from this evidence that tail biting is three times more likely where tails have not been docked – 9.4% of undocked pigs suffered incidents of tail-biting while 3.3% of docked pigs were affected. We do not know how future EU welfare regulations will view tail-docking.

Veterinarians believe pain following tail-docking of very young pigs is transitory and research evidence (Noonan et al, 1994) suggests that done correctly it is no more distressing than simple handling.

As to hygiene - clean hands and instruments. I used a mixture of 50% French chalk and 50% sulphamezathine powder in which to dip the stumps so as to dry them up.

Adverse effects may result from poor technique however, therefore stockpersons performing this task must be properly trained, preferably by a veterinarian.

Legislation on tail docking is expected in several countries. Cauterising shears are available, but need skilful use.

## TAIL-DOCKING CHECKLIST

If the process is allowed in future . . .

✓ Two people are better than one.

✓ 6 to 16 hours old to 3 days. 12-48 hours is best in the author's experience.

✓ Get trained in the technique and materials to use.

✓ Leave approximately 16 mm of tail length (Muirhead, 1997), about 2/3rds of an inch.

✓ The instrument must be sharp. Alternatively use a gas cauterizer.

Do *not* use a burdizzo ring as for lambs' tails.

✓ Preferably dip the wound in iodine cow teat dip antiseptic. In less developed countries a mixture of chemist-grade French Chalk and sulphamezathine powder 50/50 can be used, but do not let the powder mix get congealed, so renew frequently.

✓ Bleeding, if any, should stop in 30 seconds.

✓ Check for bleeding after 5 minutes. If it continues, apply a strong tourniquet – get your vet to show you how. Remove after 10-15 minutes.

Tail-docking can be governed by animal welfare legislation in certain countries. Check what you can and cannot do by law in your locality. Some areas of the world require veterinary presence or permission for certain simple operations.

# FLANK GNAWING

In my experience this is much less common than tail biting (my records suggest about nine times less). It is quite often seen in hot humid climates, and in N.W. Europe in summer rather than winter.

Young pigs which have been inclined to the process called 'PINT' (Persistent Inguinal Nose Thrusting) while suckling the sow or when weaned, on other pen-mates, may start gnawing flanks later on in life – up to 18 weeks old in my experience. PINT (also called snout-rubbing) seems to originate when sucklers have missed out on a feed and try to massage the udder after milk let-down has ceased. They may then develop the action as a habit and go on to nose-thrust the flanks of their pen-mates in the nursery, which is rarely objected to, oddly enough. Licking the flank, usually the inner back leg area of a reclining weaner can also occur. Wherever the action, the area eventually becomes sore in the attacked piglets and any resultant exudate encourages other (non-PINT-inclined) piglets to nibble at the spot, eventually causing a wound.

The writer has seemingly cured several bouts of PINT reported to him as a farm problem, in subsequent litters on the same farm, by fostering-off smaller pigs in the litter who have started to PINT (who may be the instigators) on to placid sows with a smaller litter and with plenty of milk to spare.

# EAR CHEWING IN SMALL PIGS

Ear chewing or nibbling may have the same origin as tail biting, but the writer has noticed it more often in suckling pigs where the ear could be regarded as a substitute nipple for small pigs forced off the suckle, and in weaners in hot conditions, where the back of a companion's ear can exude moisture which either smells or tastes good. In these cases the back of the ear, not the tip, is made sore.

It is essential, as with tail biters and ear-chewed pigs, to hospitalise them immediately before the wound becomes serious/infected.

Good, alert stockmanship is essential, *e.g.* check functional teat availability.

# BAR-GNAWING IN STALLED SOWS

This is pure boredom, some producers and all welfarists get very worried about it, which seems to be rather unnecessary. The sow is only trying to entertain itself, akin to watching television in our case? Thus, like head swaying, it is a relief of stress caused by the very act of stalling sows, which is now increasingly being banned in several countries on welfare grounds. Both, to my mind, are a good thing which doesn't make me popular overseas!

If you are still allowed to stall your sows – don't worry about these odd behavioural relaxations/ releases; instead turn your attention to keeping them warm enough at night with ample water to drink at all times – two common faults seen when I've been called in to advise on bar-gnawing!

## VULVA BITING

A most frustrating vice, and little understood.

Erosion of the sow's vulval area by biting from other group-housed sows. Often occurs in late pregnancy and in time can spread to the majority of sows in the group. In the worst (unattended) cases the whole vulva can be destroyed and lives lost.

## CHECKLIST

✓ Check vulvas and for bloody noses daily, especially towards the end of pregnancy when the vulva swells and becomes noticeable.

✓ Remove a bitten sow to a separate pen.

✓ Initially there is usually one offender – if noticed, remove her.

✓ Have your sows enough floor space? 2.7m²/sow is advised.

✓ The problem is more common now ESF systems are used. This could be due to sows getting impatient when waiting to feed, so a well-designed plan where resting sows can see the feeder(s) and where the imminent waiting area is not constricted by a wall or angle will help reduce its likelihood.

## URINE DRINKING
### UK and Australian experience

This does not seem to be associated with tailbiting/flank gnawing as cases appear quite independently of each other. It may be associated in pigs under 6 kg weight with pizzle and ear sucking, but it is much more common in older pigs (12-20 kg) where it was not often associated with the latter (<10%).

**Likely causes**

- Water deprivation. *i.e.* below 2.2 litres/day/pig.
- Difficulty in obtaining water, *i.e.* flow rates for 5 kg pigs less than 250 ml/min and for 12-35 kg pigs of 500 ml/min.
- Hot weather, especially if drinkers are poorly sited or the pens narrow.
- Overcrowding.
- Water quality *i.e.* slime/balantidium coli.
- Poor drainage, especially if any of the above points are borderline or below.
- High relative humidity in covered pens, especially if the ventilation rate is well below recommended levels. In these cases it can occur in relatively colder conditions. However, cases (in Australia) have occurred in hot but very low RH conditions and cured at once when the pigs were misted.

All cases studies responded to the alteration of one or more of the above criteria.

The following cases may also be implicated from a study of the rather subjective literature on the subject:

- Too low/too high salt level in the diet (one authority has suggested urine drinking is a precursor to 'salt poisoning').
- Too coarse grinding, barley, wheat etc and general gut disturbance due to harsh feed texture. (In my experience this is quite likely to result in aggression/ tailbiting, ulcers and colitis).

# SAVAGING

There is no pig textbook, save one, from the 30 or so on my shelves which even mentions this alarming and annoying occurrence – let alone what to do about it.

That surprises me, as that great pig-specialist veterinarian, the late Mike Muirhead once told me that he thought the incidence could be as high as between 1% and 3% in replacement gilts, so it must be relatively common. I have no figure to give you for older sows. It is predominantly a problem with gilts and occasionally young sows.

I first came across savaging in the 1980s. For seven years we had a breed at Dean's Grove farm whose sows were noted for their milkiness and mothering qualities, even if their progeny were a bit slow-growing compared to those of several competitors. In the end we decided to try some lines from a breed whose progeny were said to be faster and leaner growers.

## Black looks

We soon found these sows were rather highly-strung in comparison to the placid old things we were used to and we got some black looks from our manager Gordon, who was a very good stockman, and who found the new gilts a handful to manage and train.

From years of hardly a case of savaging, we experienced quite a lot, which got us thinking that it could be a breed difference. But other breeders who had the same breed – but not from the same source, necessarily – said they had little of this problem from their gilts. This suggested it was more of a line within the breed rather than the breed itself. Anyway we were stuck with these flighty gilts and had to do the best we could with them.

This initial experience plus those from the 30 or so call-outs I've been involved in since then leads me to suggest the following likely causes....

## A SAVAGING CHECKLIST

✓ **Nervous breed or line within the breed**. Watch suspect lines very carefully for the first 3 days after farrowing. If the suggestions below don`t work then you may have to change the breed emphasis.

✓ **Harsh environment and nutritional inadequacy.** I was told years ago that these could be aggravating factors in nervous sows. But surely these days with gilts being such an expensive animals they are looked after well in these respects – no, I think the reasons are closer to home.

✓ **Lack of empathy.** Even before the 1980s when I was helping to select thousands of replacement gilts a year for the huge Taymix farm and for local farmers nearby, I often did my `clap test` close to the ears to note *any* degree of alarm, or conversely unconcern. Talking to our own stockpeople as well as those farmers putting forward their gilts for selection it seemed most likely that the quietest, least nervous gilts were those whose attendants had frequently moved among them/ talked to them/ played music as routine.

✓ **Large groups in straw yards.** There is more of this these days – 100 to 150 animals together. No real pecking order exists and gilts can become apprehensive and develop what I call a 'nervous hostility syndrome' to their companions which could well manifest itself upon their new-borns when the time comes. Not everyone is large enough to have such huge gilt pools, but the same condition could develop in smaller replacement gilt pens (the peck-order is established but confinement stress replaces it) especially if the gilts do get stressed and the stockperson is too busy to walk among them and get them used to him/her.

Such disturbed animals develop a 'wild-eyed' look in late pregnancy and especially after farrowing, and while pregnant may need segregating into another yard with really ample straw, dimly lit (75 lux) and frequent movement among them by their stockperson with the occasional pig nut scattered in front of individuals. This encourages empathy and reduces the feeling of strangeness/anxiety in the new entrants.

✓    **Some other things you can do**. Try putting a nice quiet sow next to a potentially nervous gilt in the farrowing pen.

Get her used to confinement. A nervous gilt may turn into a savager just because she objects to the sudden confinement of the farrowing crate. The solution could be to put such a 'wild-eye' suspect into a gestation crate for a while (14 days) next to a placid sow in the same condition before either of them farrow. Once the bureaucrats ban the gestation crate this opportunity will disappear!

Once a gilt savages, try rubbing the mouth area with a rag and wiping all her litter with it. – but this is better done under mild sedation – see below.

✓    **What the vet can do**. He can use an anti-hysteria drug on a long-term basis.

Or he can show you how to use a tranquilliser like 'Stresnil' - carefully using the correct dose for weight, and to time it correctly so that the gilt, when she is sufficiently sedated, her litter can be moved out of harm's way and smeared with the gilt's mouth area before she fully comes-to.

Vis-à-vis other drugs, some people have told me prostaglandins on the 113th day could help reduce savaging – but is this so? Surely these merely help to get all the gilt farrowings into working hours so that supervision is more likely and a 'wild-eye' more likely to be noticed and a Stresnil shot for her given? Discuss with your vet, please.

# REFERENCES

Aherne, F. (1997) News and Views. *Western Hog Journal*. Spring 1997. p 24.

Hillsborough, E., Beattie, V. (2000) Mushroom Compost May Cut Tailbiting. Report on Agr. Res. Inst of Northern Ireland trial. *Farmer's Weekly*, 31 March 2000, p 47.

Kyriazakis, I. (1996) Tailbiting Can be Caused by Extra Energy. *Pigs*, **12**, p 40.

Muirhead, M. (1989) Personal communication, but see also in Muirhead and Alexander, *Managing Pig Health*, 5M Enterprises, 466-467.

NADIS Report (2009) Tailbiting: Vice with a £10,000 or more price tag.

Noonon, G. et al. (1994) Behavioural Observations of Pigs Undergoing Taildocking. *Appl. An. Behaviour Science*, **39**, 203-213.

# FURTHER READING

I give below some suggestions on the more useful journals, books and internet sources which anyone wishing to improve his knowledge of the rapidly-moving field of pig technology might consider for his bookshelf. They are not in any order of importance, but are the ones – among over 50 possible sources – where interesting, groundbreaking and up-to-date information on pigs will be found.

## MAGAZINES

Sadly, in the past 7 years several excellent ones have ceased publication. The doyen of them all, which is still flourishing, is *PORC MAGAZINE* (French). This is a quite remarkable publication, but only for those of us who can read French! Between 100-150 pages each month, with 100 or more colour photos and diagrams. No pig journal in the world can touch it; the French and French-Canadians are so lucky. Other pig journals – please copy!

*PIG PROGRESS* (Dutch, in English) is an excellent monthly, with good, solid, informative guest-written articles. They have a weekly blog column from several authorities, on pigprogess.net which are well worth reading.

*NATIONAL HOG FARMER* (USA). The best of the American monthlies. The magazine is weighted towards US economics and internal strategies, but it is the only magazine to publish, one or two times a year, a series of 'Blueprint' editions where one subject – e.g. disease, nutrition, management – is covered in depth. Worth subscribing just for these authoritative statements of where we are at these days.

*WESTERN HOG JOURNAL* (Canadian). This bi-monthly has developed into a first-class general pig journal over the past three years, supported by the excellent practical academics at the agricultural and university centres in Alberta and Saskatchewan in particular, and edited by a well-known swine consultant and farm manager.

*PIG INTERNATIONAL.* Good articles on world-wide pig production. You need to see what the rest of the world is doing: this journal does it well. Unfortunately the hard copy is now only available to American subscribers, but a digital edition is provided via the internet. I find this less convenient and hard on the eyes.

*PIG WORLD*.  Now that the 50 year old (UK) 'Pig Farming' is no more as a separate entity, it's successor is 'Pig World'. This is a much-improved monthly source of important information on the British pig scene. It has now been adopted as the official organ of the UK's National Pig Association.

## TECHNICAL JOURNALS

There is a wide range of these to choose from. All are expensive and the majority publish research papers about other livestock species too.  Most are biennial or quarterly.

*PIG VETERINARY SOCIETY JOURNAL* (British) is really for pig veterinarians but it does contain good material on management/disease-associated items. Twice-yearly.

It is difficult to recommend one technical journal for the pig producer, but both the *JOURNAL OF ANIMAL SCIENCE* and *ANIMAL PRODUCTION* provide ground-breaking research papers, the former primarily nutrition and the latter on many aspects of farm animal technology. Pigs however, comprise only some of the papers.

## BOUND PROCEEDINGS

There are two important international pig conferences (outside the large annual American circuit) whose annual proceedings are worthy of study. The proceedings of the Banff (Canada) Pork Seminars *"ADVANCES IN PORK PRODUCTION"* and the Australian Pig Science Association *"MANIPULATING PIG PRODUCTION"* are important updates on current pig technology.

## INTERNET SOURCES

There are a wide number of company-issued newsletters specific to their products and interests, too many to list here (I find difficulty in finding the time to scan many of them anyway).  However, one not to be missed is the weekly 'PigSite' published by 5M Publishing (Thepigsite.com). An impressive, and free, compendium of global news, prices and technical articles culled from the net, as well as an electronic reference section of authoritative subjects. Not to be missed. There are also a variety of (quite expensive) DVD pig courses, some better than others, as some are surprisingly rather out-of-date. If the publishers are going to charge (understandably enough) for these training sessions, surely they should update them at least every two years?

# MY TOP TEN BOOKS ON PIGS

Arguably the best book on pigs ever written is Mike Muirhead and Tom Alexander's 600-page volume "*Managing Pig Health and the Treatment of Disease*". While specialising in veterinary aspects of pig production, it engages itself deeply into practical management, as it should. While expensive, no pig farm's office should be without a copy.

Another first class book, this time on nutrition, is Close and Cole's '*The Nutrition of Sows and Boars*' (Nottingham University Press). Easy to read and understand, with clear directions on diets and feed levels.

David Taylor's regularly updated book '*Pig Diseases*', now in its 7th edition, is a valuable reference source, and not expensive. Published by Dr Taylor. Excellent that it is kept up-to-date every 3 years or so with the changing disease scene these days.

For the slaughter pig, English, Fowler, Baxter and Smith's '*The Growing and Finishing Pig*' (Farming Press) is still a valuable source, if a little dated now (1988) but we need an update, please, someone!

Prof Colin Whittemore's majestic work '*The Science & Practice of Pig Production*' (Longman Scientific & Technical) is authoritative, but personally I prefer his shorter '*Elements of Pig Science*' (Longman's) concisely written by a real word-craftsman, rare among leading academics.

Paul Hughes and Mike Varley's '*Reproduction in the Pig*' (Butterworths) gives an excellent grounding in this critical and easily 'got-wrong' area.

On housing, Gerry Brent's '*Housing the Pig*' (Farming Press) while somewhat aged now (1986) is still a remarkably good reference source on getting dimensions and construction basics right. I understand a new book on pig production from this consultant is imminent.

There is no good reference work I know of on the important subject of ventilating piggeries, but a remarkably good monograph is the now out-of-print UK Farm Electric Centre booklet '*Controlled Environments for Livestock.*' If you ever find a copy, grab it! It is absolutely first class and desperately needs a reprint, but the FEC has been disbanded on cost grounds along with many of our vital establishments in Britain. So short-sighted of our politicians.

Co-products (by-products) can play a part in reducing feed costs so '*Co-Product Feeds*' by Robin Crawshaw (Nottingham University Press) is the most comprehensive I've seen.

I have not come across a really good book on baby pig management and weaner nutrition. Badly needed and the information is there, but has tended to be commercially guarded. I have tried to lift the lid a little on the chapters on 'Creep Feeding' and 'The Post-Weaning Check to Growth'.

Good textbooks on outdoor pig-keeping, stockmanship/man management as well as slurry disposal also exist. I buy these off Amazon.

For 30 years I've been looking for a book on interpreting statistics which didn't lose me after the first 10 pages. At last I've found one – Derek Rowntree's '*Statistics Without Tears*' (Penguin). This is a gem, and even the academics don't sniff at its simple language. Another one by Lucy Tucker 'Simplistic Statistics', is nearly as good but shorter and thus cheaper. With these two the newcomer to statistics need not be frightened - as I was!

# MISCELLANEOUS

Students and practitioners of pig production should keep their eyes open for good commercial monographs and booklets. The UK British Pig Executive and the advisory branches of Government organisations (Universities in the USA) like DEFRA/BPEX in the UK from time to time produce important monographs. For example the 6 ADAS monographs on group housing of sows referred to in this book and the very recent Veterinary Bulletins on PMWS/PDNS. The superb Annual Reports of the Danish Ministry of Agriculture (National Committee for Pig Production) show what this vibrant nation is doing in its pig research, and are helpfully available in English. Several commercial firms [Farmweld (USA); Eli Lilly/Elanco (USA/Europe); DuPont International;  Intervet (UK); JSR Genetics (UK); Alltech (USA); PIC (USA) etc] have excellent information and well-balanced and not overtly commercial booklets on their spheres of interest.

In the field of nutritional biotechnology, i.e. the use of naturally-sourced feed additive products, Alltech stands supreme in its literature. '*Nutritional Biotechnology in the Feed and Food Industries*' is an annual volume incorporating the annual Pig Science Symposium which shows all of us in pigs the cutting edge of this area of growing interest in, and importance to, global pig production. If you ask them very nicely they will send you a CD of the latest symposium, which usually contains much new information on pigs. Contact your local Alltech office.

Porc magazine
(in French)

Editions Boisbaudry
CS 77711.35577
Cesson-Sevigné cedex
France

E-mail: lviel@editionsduboisbaudry.fr

Pig Progress

Reed Business International
PO Box 4
7000 BA
Doetinchem
The Netherlands

E-mail: vincent.ter.beek@reedbusiness.nl

National Hog Farmer

Primedia Business Magazines Inc
9600 Metcalfe Avenue
Overland Park
KS 6612-2215
USA

E-mail: primediabusiness.com
Editor's E-mail:
dpmiller@primediabusiness.com
Website: www.nationalhogfarmer.com

Western Hog Journal

Alberta Pork
4828-89 Street
Edmonton
Alberta
Canada TG3 SK1

E-mail: whj@albertapork.com

Pig World

Pig World Benniworth
PO Box 1000
Market Rasen
LN8 6LE
UK

E-mail: office@pigworld.co.uk

Animal Production

British Society for Animal Science
PO Box 3
Penicuik
Midlothian
EH26 0RZ   UK

E-mail: bsas@ed.sac.ac.uk

# FROM A PERSONAL POINT OF VIEW - SOME SUBJECTS I HAVE HAD TO LEAVE OUT!

It is impossible to include many important subjects on profitable pig production in one volume or it would become unwieldy and expensive. The reader may be interested in consulting the subjects below which have appeared in detail in the author's two previous publications and which, for reasons of space, now needed fully to discuss the latest developments in profitable pig production, are not described so comprehensively, and some not at all, in this latest volume. Both previous books were also published by Nottingham University Press in 2003 and 2005.  Look at www.nup.com for prices and availability.

### Pig Production Problems - A Guide to Their Solutions
*"A remarkable and very useful book"*

A detailed treatise on all aspects of piggery ventilation • The cost of empty days • How to ensure sound legs • Culling strategy - when to cull and when to not • Risk management in pig farming • Negotiating a successful deal • Franchising - one of several ways out of trouble • Dealing with a cash crisis • Tips on selling your pigs well • Getting to the top of the loan queue - what a bank needs to know from you • Making a partnership work • Fermented liquid feeding and other future production prospects • Pig and carcase identification - and many others

### Pig Production: What the Textbooks Don't Tell You
*"So many answers here I didn't know about"*

How to buy pig feed • How to buy santising materials • Buying skills • Key questions to ask salespersons • How to negotiate a good deal • Specificity and prioritisation • The problem of wasted food and energy • Medication - feed or water? • Condensation - prevention and cure • How to do a farm trial which means something • Using a statistician to help you decide on product claims • Water basics • Stomach tubing • Taping and double-taping splays • Heat detection skills • Being there at farrowing • Using a vet properly - and many others.

# A GLOSSARY OF 450 TECHNICAL TERMS IN PIG PRODUCTION

## WHAT YOU NEED TO KNOW BUT WERE AFRAID TO ASK

Attention to problems on the pig farm often involve the producer reading, or being quoted, technical terms in reports from veterinarians and other advisers. These are often couched in scientific jargon (to facilitate accuracy) which may or may not help the farmer and his stockpeople take decisions. In addition, many research papers are similarly peppered with technical terms understood by the scientist but not always by a lay reader.

This glossary is not exhaustive by any means, and lists the sort of terms and definitions I may have had to look up over the past 45 years. Thus it may also be of help to the reader – to whom I apologise if I have insulted your intelligence with some of them! But we are all different, and I wish someone had done this for myself years ago! I have consulted a wide variety of sources, both printed and verbal, to whom I am grateful, but the interpretations are largely my own. If you disagree, can put it better or can suggest omissions – I'd be glad to know.

| | |
|---|---|
| **Absorption** | to take in or assimilate substances e.g. *into* tissues. Commonly confused with adsorption (qv) |
| **Acral** | (*vet*) affecting the extremities |
| **Acronym** | Word or term formed from the intitial letters of other words (PRRS, FCR) |
| **Acute** | 1. Sharp |
| | 2. Of recent onset – rather than depicting severity (although it is commonly used in this latter form) |
| ***Ad libitum*** | (*ad lib*) without restraint |
| **Adipose** | fatty |
| **Adjuvant** | a material that aids another, i.e. in a vaccine to increase antigen potency |
| **Adsorption** | attracting and holding other substances *onto* surfaces |
| **Adventitious** | (*vet*) acquired, not in the correct place (accidental) |

| | |
|---|---|
| **Aetiology** | (*vet*) the science dealing with the causes of disease |
| **Aerobe** | (*adj* aerobic) a micro-organism which needs oxygen to function fully |
| **Agalactia** | (*vet*) partial or complete lack of milk |
| **Agonistic** | tensing of muscle(s) (not antagonistic = aggressive) |
| **Aitchbone** | the hip bone |
| **Algesia** | (*vet*) sensitivity to pain, hence analgesic = pain-killer |
| **Ambient** (temperature) | temperature close to the pig's body |
| **Amino-acid** | (*nutr*) protein building block |
| **Ampere** | (*constr*) the unit measuring the strength of an electric current |
| **Anabolism** | (*adj* anabolic) the formative stage of metabolism (qv) |
| **Anaerobe** | (*adj* anaerobic) a micro-organism which can grow in the absence of oxygen |
| **Analgesia** | absence of sensibility to pain (analgesic = painkiller) |
| **Analogous** | resembles, similar to |
| **Anaphylaxis** | (*adj* anaphylactic) severe or unusual allergic shock reaction |
| **ANFs** | Anti-Nutritional Factors. Materials present in certain feed raw materials which interfere with digestive or metabolic pathways |
| **Androgen** | male hormone |
| **Androstenone** | (*see* Skatole) |
| **Anoestrus** | no oestrus, lack of oestrus |
| **Anoxia** | (*vet*) (*adj* anoxic) interference with (lack of) oxygen supply |
| **ANS** | Automatic Nervous System, involved in 'Flight or Fight' stress |
| **Anterior** | towards the front |
| **Anthropomorphism** | attribution of human characteristics to animals |
| **Antibody** | specialised proteins produced by lymphocytes (white blood cells) in response to presence of an antigen (qv) |
| **Antigen** | (*adj* antigenic) any substance foreign to an organism (i.e. a pig) reacting with an antibody so as to produce an immune response within the organism/pig |
| **Antioxidant** | (*nutr*) material which inhibits the oxidation (qv) of compounds e.g. prevents rancidity of fats |
| **Arthritis** | (*vet*) inflammation of a joint |
| **Aspirate** | (*vet*) Suck out. (However, aspiration can mean inhalation) |
| **Astringent** | (*vet*) causing contraction |
| **Ataxia** | (*vet*) muscle incoordination |
| **Atresia** | (*vet*) (*adj* atresic) closure of a structure |
| **Atrophy** | (*vet*) wasting; shrinking |
| **Attenuation** | (*vet*) reduction; thinning (diluted) |
| **Attrition** | wearing away |

| | |
|---|---|
| **Atypical** | not typical, irregular |
| **Audit** | a systematic review |
| **Autogenous** | (*vacc*) self-generating; originating within the body |
| **Autolysis** | self destruction of a cell |
| **Autonomic** | not subject to voluntary control (e.g. Autonomic Nervous System) |
| **Axis** | a line about which a figure, curve or body is symmetrical (or about which it rotates) |

The New Terminology – 'A'

**AIV : Annual Investment Value** The number of times per year the savings or improved income from the extra investment is turned over in relation to the investment cost per tonne, sq metre, per pig, etc. Highest value best.

**AMF : Absolute Mortality Figure** How many piglets were lost relative to those born alive. A much better figure than % mortality. Expressed as 'AMF 0.9 of 12 BA' (BA = born alive). **Target AMFs** vary from 0.6 for 8 BA to 1.2 for 14 BA

| | |
|---|---|
| **Bacterin** | a vaccine made up from killed bacteria |
| **Bacteriocide** | substance which destroys bacteria |
| **Bacteriostat** | substance which inhibits but does not destroy bacteria |
| **Bang out** | Hitting a kennel cover to ensure any sleepy or sick pigs exit for inspection |
| **Batch farrowing** | farrowing sows in deliberately formed groups to facilitate workload and supervision, also to make disease in the offspring easier to contain |
| **Bentonite** | a pure clay capable of absorbing moisture |
| **Berkshire** | a pig breed noted for its meat carcase quality |
| **Beta carotene** | (*nutr*) Vitamin A |
| **Bile** | fluid produced by the liver which breaks up large fat globules for digestion by enzymes. Stored in the gall bladder |
| **Billion** | one thousand million and is commonly shown as $10^9$. (Not one million million) |
| **Bioactive** | secures a response from living tissue |
| **Bioassay** | testing the power of a drug on a living organism |
| **Biopsy** | (*vet*) removal (for microscopic examination and testing) tissue from the living body |
| **Biosecurity** | **all** the measures taken to preserve health and defend against disease, not just sanitation measures |
| **Biotechnology** | the application of scientific biological principles for industrial purposes (e.g. genetic engineering, pharmaceuticals, etc but also from wholly natural sources, e.g. yeast by-products) |

| | |
|---|---|
| **Biotin** | (Vitamin H) Vitamin B complex, involved in hoof strength |
| **Birthweights** | target on at least 15 to 17 kg of living neonates (qv) per litter. ***Target 1.5 kg piglet. Action level <1.2 kg*** |
| **Blastocyst** | early stage of embryo formation (from 'blast', a bud) |
| **Blind teat** | milk may be present in the mammary gland but the teat canal is blocked. In gilts can be a genetic defect |
| **Bloat** | (*gastric*) distension with gas, common with feeding (hot) whey |
| **Blood poisoning** | common term for blood infected with bacteria or their toxins |
| **BOD** | Biochemical Oxygen Demand, used to measure the potency of effluent and is the amount of oxygen needed over a specific time to decompose organic matter at 20°C |
| **(Body) condition score** | a subjective method of estimating the fat cover of (usually) a sow, across a range of 1 (emaciated) to 5 (obese) |
| **Born alives** | those piglets which drew at least one breath, confirmed by the bucket test – did the lungs float or sink quickly? ***Target: born deads < 5%*** |
| **Breech presentation** | foetus buttock-first at parturition |
| **Brewer's grains** | feedstuff residue after starch fermentation |
| **Brewer's yeast** | brewing by-product after harvesting and drying saccharomyces cerevisiae yeast |
| **Brooder** | substantial cover over heat source used early-on in wean-to-finish housing |
| **Brown fat** | fatty tissue which gives off heat. Pigs with higher (genetic) levels of brown fat 'burn off' food – a useful effect in us greedy humans but which raises FCR in pigs |
| **BTU** | British Thermal Unit |
| **Buffer/buffering agent** | material in solution which increases the amount of acid or alkali needed to produce a unit change in pH (qv) |
| **Bulk density** | the density of a granular substance (e.g. animal feed) calculated as the unit volume of the substance including the spaces between the particles/grain (see density) |
| **Bump feeding** | Feeding extra 2 to 3 weeks before farrowing |
| **Bursa** | (*vet*) small fluid-filled cavity the body produces to protect against friction |
| **Bursitis** | inflammation of a bursa |
| **Caecum** | a small pouch between the small intestine and colon containing cellulose-splitting bacteria. Poorly developed in pigs and humans compared to ruminants |
| **Calculi** | (*vet*) accretal stores (as in kidney stones) |

| | |
|---|---|
| **Calcaemia** | *(vet)* excessive blood calcium |
| **Calorie** | *(nutr)* the amount of heat needed to raise 1 gramme of water by 1°C (1 calorie = 4.187 joules) |
| **Calpain** | an enzyme which breaks down muscle structure thus improving tenderness. Calpostatin is an inhibitor and increases with stress. |
| **Capacitor** | instrument for storing charges of electricity |
| **Capsid/capsomer** | shell which protects a virus nucleus. Made up of capsomes |
| **Carbohydrate** | the simplest carbohydrates are the sugars (saccharides). More complex are the polysaccharides (e.g. starch and cellulose). Sugars (e.g. glucose) are intermediates in the conversion of food to energy; polysaccharides serve as energy stores in plants and seeds, potatoes, etc. Cellulose, lignin etc provide supporting cell walls and woody tissue in plants thus are not very digestible. |
| **Carcinogen** | *(vet)* cancer-causing substance |
| **Cardiovascular** | *(vet)* to do with the heart blood vessels |
| **Caries** | *(vet)* decay |
| **Casualty** | any pig slaughtered in an emergency due to disease, injury or distress. Casualties must be distinguished from culls |
| **Catabolism** | *(adj* catabolic) procedure within the body where complex structures are broken down into simpler compounds with the release of energy |
| **Catalyst** | *(n* catalysis) a substance which assists/speeds up a reaction but which is not used up during the process. |
| **Cathartic** | causing bowel evacuation |
| **Caudal** | *(vet)* towards the tail |
| **Cell-mediated** | affected by the cells in the body rather than chemicals |
| **Cellulose** | *see carbohydrate* |
| **Celsius** | 0°C freezing, 100°C boiling. Centigrade = 100 steps. Celsius to Fahrenheit (qv) °F = (°C x 9/5)+32 i.e. °C x 9 ÷ 5 + 32 = °F. |
| **Centimetre** | (cm) 100th of a metre, 0.3937 ins |
| **Cervix** | *(vet) (adj* cervical) neck, or narrow part of an organ. In the female between the uterus and the vagina. A safety valve to protect the uterus from foreign bodies. |
| **Chelate** | *(nutr)* claw (pronounced kee-late). Inert substance which holds a trace element (mineral) until the right digestive conditions release it for digestion. |
| **Chemotherapy** | *(vet)* medication |
| **Chitterlings** | *(nutr)* deep-fried delicacy; made from sections of the pigs large intestine |
| **Chromosome** | contains coiled DNA. In animal cells, determines sex and |

| | |
|---|---|
| | transmits genetic information |
| **Chronic** | *(vet)* in existence for a long time and causing less of a reaction than acute (qv) |
| **Cilia** | tiny hair-like substances which move the cell or move mucus over it |
| **Circadian (-rhythm)** | body activities which occur at regular intervals irrespective of light or dark influences ('biological clock') |
| **Clinical** | obvious disease (sub-clinical; less obvious or undercurrent) |
| **Coander (Coanda) effect** | 'skidding' (water or) air along a flat surface thus lessening resistance and helping the direction of flow. |
| **Coefficient** | (of variation CV) *(stats)* the change between the variation of certain factors usually expressed as a percentage. The lower the percentage, the more close together the data points are to each other and vice versa. Calculated by dividing the Standard Deviation (q.v.) by the mean (q.v.), then multiplying by 100. |
| **Cohort** | *(stats)* a group of animals with similar characteristics used in a research trial |
| **Colitis** | *(vet)* inflammation of the colon |
| **Colon** | the large intestine between the caecum and rectum |
| **Colonisation** | *(bact)* the ability of bacteria to adhere to a living surface and then multiply |
| **Commensal** | (usually *bact*) living on or inside another organism but causing no harm to it |
| **Condensation** | (vaporous) the change of a vapour into a liquid (see Dewpoint) |
| **Condition Scoring** | *(see* 'Body') |
| **Conduction** | the movement of energy (sound, heat or electricity) by the agitation of molecules inter alia |
| **Congenital disease** | present at birth |
| **Congenital** | evident from birth |
| **Consultant** | an ordinary guy a long way from home, or who has left salaried employment on grounds of economics or age, and whose quality of life improves markedly thereafter. |
| **Convection** | the movement of heat through a liquid or gas. Heat expands portions of the material, they become less dense and rise; their place is taken by colder portions, thus setting up convection currents. |
| **Convex** | curving outwards (concave is curving inwards) |
| **Correlation** | *(stats)* the degree of association of variables. Linkage, i.e. age of the pig can be correlated or linked to increased fat cover |
| **Cortex** | an outer layer |

| | |
|---|---|
| **COSHH** | Control of Substances Hazardous to Health. UK regulations (1989) providing one set of stipulations for all occupational health risks |
| **Cost/benefit analysis** | taking account of social costs/benefits as well as purely financial ones |
| **Costive** | constipated |
| **Critical temperature(s)** | see Temperature |
| **Crate** | Correctly only used for confinement over and after *farrowing*, while 'stall' (q.v.) is correctly only used for confinement in *pregnancy* e.g. 'farrowing crate' and 'gestation stall'. (The same applies to the latest developments of 'free access', applied to both measures, where 'crate' and 'stall' still apply - hence 'free access crate' and 'free access stall'. |
| **Crude fibre** | *(nutr)* the non-digestible cellulose, hemi-cellulose and lignin portions of a feed (see also Fibre) |
| **Cycle** | not the same as parity (qv). Cycle denotes a time from event to event e.g. birth to birth or breeding to breeding |
| **Cystitis** | *(vet)* inflammation of the bladder |
| **Deadweight** | carcase weight dressed to a specific standard |
| **Deamination** | *(nutr)* processing of surplus protein to waste material |
| **Deliquescent** | absorbing moisture from the air, e.g. copper sulphate is deliquescent, tends to soften/liquify |
| **Denature** | *(nutr)* to produce a structural change in a protein which causes it to reduce its biological properties |
| **Density** | the ratio of the mass (weight) of a substance to its volume |
| **Dermal** | *(vet)* to do with the skin. Dermatitis = skin disease |
| **Dessication** | drying |
| **Deviation** | *(stats)* a measure of dispersion, indicating variability from the *Standard deviation (SD)* average. Data which have a normal distribution of 66% of the data points are within 1 standard deviation and 95% fall within 2 standard deviations. So two standard deviations take in 95% of the pigs and three standard deviations 99% of the pigs. *See also* 'Significance' |
| **Dewpoint** | the temperature at which water vapour present in the air saturates the air and begins to condense to form water deposits, i.e. dew begins to form |
| **Diagnosis** | the identification of disease. Clinical diagnosis is the identification from clinical signs during life backed by by laboratory tests (see also prognosis) |
| **Dietetic** | *(nutr)* to do with the diet |

| | |
|---|---|
| **Differential diagnosis** | *(vet)* using the differences in diseases derived from symptoms backed up by epidemiological (qv) tests to select a diagnosis most suited to the evidence |
| **Digestible (DE)** | Energy the gross energy eaten less that voided in the faeces.  1 MJ (megajoule) DE = 239 Kcals.  (See also Metabolisable Energy and Net Energy) |
| **Dilation** | *(vet)* stretching |
| **Discrete** | separate. (Note: discreet = tactful) |
| **Disseminated** | scattered |
| **Dissociation** | *(nutr)* separate, as in a nutrient passing through the gut wall |
| **Distal** | *(vet)* remote; far end of |
| **Distribution** | *(stats)* 'normal distribution' is a graphical curve appearing like a bell, symmetrical on both sides of the vertical axis, increasing to peak incidence on the left side and decreasing to zero incidence on the right side |
| **Diuretic** | increasing urine amounts, also a product to do the same |
| **Diurnal** | day to night/night to day |
| **Dose-response curve** | *(stats)* shows how a drug responds in its effects to increased dosage |
| **Dräger tube** | hand held gas detector |
| **Dressing percentage** | (Killing Out % : Yield, USA) deadweight (qv) Expressed as a percentage of liveweight shortly before slaughter. |
| **Ductile** | drawing out without breaking |
| **Duodenum** | first organ leading out of the stomach which primarily digests fats.  The bile and pancreas empty into it |
| **Dynamics** | (e.g. thermodynamics) the reaction of bodies to force (in this case heat) |
| **Dysplasia** | *(vet)* abnormal development |
| **Dyspnoea** | *(vet)* difficult breathing |
| **Dystocia** | *(vet)* abnormally laboured farrowing (foetal d. = due to the foetus; maternal d. = due to the sow) |
| **Dystrophy** | disorder due to faulty nutrition |
| **Eclampsia** | *(vet)* post-natal convulsions |
| **Econometrics** | the measurement of cost-effectiveness |
| **Ectoparasite** | a parasite living on the host's body, e.g. fleas (endo = inside the body, e.g. worms) |
| **Electrolyte** | a substance, normally a mineral salt, which allows the intestine to insorb water at the same time it may be exsorbing it (i.e. during diarrhoea) thus deterring dehydration. |
| **Elisa Test** | enzyme-based test for the degree of immunity to detect and |

| | measure either an antigen or the antibody to it (qv). Useful in specific circumstances |
|---|---|
| **Embryo** | from the time the organism develops a long axis to the time the major limbs etc have started to develop, when it becomes a foetus |
| **Emetic** | *(vet)* used to induce vomiting |
| **Empirical** | simple, basic |
| **Empty days** | (Open Days; Non-productive days) the number of days per year or per litter the sow is not carrying or feeding piglets. *Targets* per year 28-30; per litter 12-13 |
| **Emulsion** | two unmixable liquids where one is dispersed with the other both in small droplet form. These can settle out and many emulsions need shaking before use |
| **Encephalomyelitis** | *(vet)* inflammation of both brain and spinal cord |
| **Endemic** | present at all times |
| **Endocrine** | hormonal |
| **Endogenous** | *(vet)* produced by the body or organism itself |
| **Endometrium** | the lining of the womb |
| **Enteritis** | *(vet)* inflammation of the intestine |
| **Entire** | both testicles descended. (As distinct from cryptorchid where neither have descended and monorchid, only one) |
| **Entopic** | occurring in the correct or expected location |
| **Enzyme** | *(nutr)* a protein which acts as a catalyst (qv). A chemical 'go-between' facilitating metabolism. The pig may contain 13,000 different enzymes. Addition of some to the feed can improve performance/reduce pollution |
| **Epidemic – epizootic** | *(vet)* disease attacking many subjects at the same time |
| **Epidemiological** | the study of diseases and their causes |
| **Epidermis** | outermost skin layer(s) |
| **Epithelial** | *(vet)* to do with cell formation in the body (see epithelium) |
| **Epithelium** | the cell covering of external and internal body surfaces |
| **Epizootic** | [1] widely diffused and rapidly spreading [2] concerning an epidemic |
| **Erythrocyte** | red blood cell (corpuscle) |
| **Ethology** | study of animal behaviour |
| **Excipient** | adding a filler or carrier |
| **Exogenous** | outside the body |
| **Exponential** | (growth) *(stats)* ever-increasing |
| **Expression** | 1. to squeeze out 2. *(Gen)* The manifestation of a heritable trait (q.v.) in an individual carrying the gene or genes which characterise it |

| | |
|---|---|
| **Extrapolation** | (extrapolated from) *(stats)* inferred or deduced from the data presented |
| **Extrinsic** | outside (opp: intrinsic) |
| **Exudate** | *(vet)* fluid emanating from a wound or irritant |
| **Eyponym** | name including a person's name e.g. Aujeszky's disease |
| | |
| **F1** | First filial generation or first cross, terms used in genetics |
| **F2** | Second filial generation, ibid |
| **Fahrenheit scale** | Freezing point is at 32°F and boiling point of water at 212°F. Fahrenheit to Celsius °C = (°F − 32) x 5/9 |
| **Falciform** | *(bact)* sickle shaped |
| **Farrowing index** | number of farrowings a sow achieves per year. *Target* 2.4. Can be used on a herd basis i.e. *Total number of farrowings in a given year divided by the average sow inventory for that year* |
| **Farrowing interval** | Time between farrowings *Target* 152 days in normal circumstances |
| **Farrowing rate** | number of farrowings to a given number of services. *Target*: 87-92% (indoors) |
| **Farrowing fever** | *(vet)* MMA syndrome (qv) |
| **Fermentation** | enzyme conversion of carbohydrates etc to simpler substances (like lactic acid). Done artificially, i.e. for the animal, it helps digestive efficiency |
| **Fibre** | crude fibre (qv) is today considered a largely meaningless term. Neutral Detergent Fibre is a better term (qv) but still has limitations as it does not quantify Non-Starch Polysaccharides (NSP) (qv)which themselves can compromise pig performance. Levels, and treatment of, NSP can however improve the digestibility of fibre. Fibre quality and amounts offered can be deficient in modern sow diets (constipation; lack of gutfill; stress) |
| **Fibroma** | *(vet)* fibrous tumour or swelling |
| **Filamentous** | a long threadlike structure (as in a filamentous blastocyst whose 'arms' attach themselves to the womb wall at implantation (qv) |
| **Fimbriate** | *(bact)* fringed border |
| **First litter sow** | female pig between the date of the first effective service and the date of the next effective service (following the successful completion of the first pregnancy) |
| **Fixed costs** | *(econ)* labour, contractors' costs, buildings and rent, machinery and equipment, finance charges, stock leasing, feeds, insurance, sundries |

| | |
|---|---|
| **FLF** | *(nutr)* Fermented Liquid Feeding. A process where either the complete feed or critical starch ingredients (such as wheat) are fermented by the addition of a starter culture (and possibly enzymes) plus heat to help predigest the feed. |
| **Flocculation** | settling out of particles in a liquid. Usually soft particles as distinct from harder ones (sedimentation). Can happen in wet-feed circuits between feeds. |
| **Fomite** | inanimate material which carries disease – bedding, dust, faeces, etc |
| **Fructose** | a sugar found in fruits and honey (lactose = milk sugar; mannose = yeast sugar) |
| **FSD** | *(nutr)* farm-specific diets, i.e. designed for a specific farm or range of buildings |
| **Futures** | *(econ)* specified quantity of products guaranteed for delivering at a specific future date at a contracted price |
| **Galacto** | pertaining to milk |
| **Gastritis** | *(vet)* inflammation of the stomach lining |
| **Gastroenteritis** | *(vet)* inflammation of the stomach lining ***and intestine*** |
| **Gene** | unit of heredity comprising a simple segment of DNA molecule that makes up a chromosome. Two copies of each gene, one from each parent, are present in every cell. |
| **Generic** | *(name)* the name of a drug not protected by trademark, usually describing the drug's chemical makeup |
| **Genome** | correctly; the complete inventory of hereditary traits contained in a half-set of chromosomes, or in layman's terms - all the genes in an organism |
| **Genomics** | The mapping, sequencing and analysis of all genes in a given species (*See also* 'nutrigenomics) |
| **Genotype** | the entire genetic makeup of an animal (see also phenotype) |
| **Gestation** | 110-116 days in the sow, from fertilisation to birth |
| **Gilt** | correctly, a young female pig which has not yet had a litter of pigs, rather than one before first conception. She then becomes, ***after parturition***, a first-litter sow |
| **Glässers disease** | *(vet)* contagious disease of young pigs caused by the haemophilus organism |
| **Glucosinolates** | found in brassica crops, interfere with iodine uptake |
| **Glutaraldehyde** | a popular disinfectant, superseded by Peracetic Acid and Peroxygens (qv) against viruses |
| **Gnotobiotic** | an animal which has been born germ-free (see also SPF) |
| **Glycomics** | the metabolic activity of complex sugar molecules |
| **Grain** | (high moisture) contains 22% to 40% moisture but must be ensiled anaerobically |

| | |
|---|---|
| **Gravid** | pregnant |
| **Gutfill** | [1]correctly, the amount of the pig's total body weight which is comprised of food, digesta and faeces [2]Also describes satiation |
| **Gross margin** | *(econ)* net output minus total feed costs and other variable costs (qv) |
| **Habituation** | an aspect of learning in which repeated applications of a stimulus results in decreased responsiveness - such as a clock ticking in a room |
| **HACCP** | (pron. 'Hassap') Hazard Analysis Critical Control Points - a structured approach to identifying (food) production safety problems and controlling those problems |
| **Haemoglobin** | *(nutr)* component of blood which transports oxygen to and from muscles. (Neutralised by too much carbon monoxide being breathed in) |
| **Halothane** | an anaesthetic (Halothane Test on pigs identifies individuals susceptible to certain stressors) |
| **Header tank** | *(env)* water reservoir at the top of a gravity water line |
| **Hepatitis** | *(vet)* inflammation of the liver |
| **Heterogeneous** | not uniform: dissimilar *Note: pronounced heterogeneous* |
| **Heterogenous** | from another source; not originating in the body e.g. 'foreign body', *Note:pronounced heterogeenous* |
| **Heterosis** | *(gen)* a first generation offspring showing greater vigour than either parent |
| **Histology** | *(vet)* science involving the structure of tissues |
| **Holostic** | *(stats)* complete/comprehensive |
| **Homeopathy** | *(med)* curing disease using substances similar to those causing it, given in small/minute doses over time |
| **Homeostasis** | stabilising mechanism within the body akin subjected to changing conditions (e.g. disease / stress / hunger) so as to mitigate their effect. Can also mean 'normality' |
| **Homeotherm** | warm-blooded |
| **Hormones** | a wide variety of chemical messengers with specific functions. Work via the bloodstream |
| **Humidity** | the amount of moisture in the air (see Dewpoint) |
| **Humoral** | processes carried out by the body's fluids |
| **Hybrid vigour** | *(gen)* heterosis (qv) better performance and viability in the first generation from matings of parents of different breeds. The advantage is quickly lost if hybrids are then interbred |
| **Hydatid** | *(vet)* Cyst-like |
| **Hydrochloric acid** | secreted by cells lining the stomach, vital (especially in the weaned pig) to 'sanitise' the ingesta prior to it being passed on to the duodenum (qv) |

| | |
|---|---|
| **Hyper-** and **hypo-** | 'hyper-' indicates extremeness, excessivity, e.g. Hyperactive; while 'hypo' indicates deficiency, beneath or under, e.g. Hypoglycaemia (low blood sugar) and hypodermic (under the skin) |
| **Hypertrophy** | *(vet)* increase in size of an organ due to excessive cell production |
| **Ibid** | Latin - 'As before', used to refer to a previous reference or statement |
| **–itis** | *(vet)* words ending in –itis refer to the inflammation of a particular organ (Enteritis: inflammation of the intestines) |
| **Ig** | prefix denoting any one of the 5 immunoglobulins (qv), IgA, IgD, IgE, IgG, IgM |
| **Ileitis** | inflammation of the ileum (qv) |
| **Ileum** | latter part of the small intestine |
| **Immunity** | the condition of being secure from a particular disease. A list of all the immunological terms and their definitions will be found in the 'Immunity' Chapter, Table 9. |
| **Immunoglobulins** | a specialised protein usually produced following exposure to an antigen (qv) |
| **Immunomodulation** | adjustment of the immune response to a certain level, as in immunopotentiation, or immunosuppression |
| **Implantation** | the attachment of the fertilized egg to the womb wall (endometrium) usually between day 10 to 30 after service |
| *In vitro* | seen in a test-tube or artificial environment (e.g. laboratory work/research) |
| *In vivo* | in the living body (research/trial work done on-farm) |
| **Incremental** | (costs) *(econ)* added or increasing costs |
| **Infarct** | *(vet)* area interrupting the blood supply |
| **Inflammation** | *(vet)* the normal healing reaction of the body to an injury i.e. protective walling off of an injured area from the cause |
| **Ingestion** | eating/swallowing material (ingesta) |
| **Inguinal** | *(vet)* the groin area |
| **Integrity** | *(vet)* e.g. tissue integrity. Condition/soundness of tissue |
| **Intrinsic** | *(vet)* inside |
| **Intumescence** | *(vet)* swelling |
| **Iodophor** | skin disinfectant |
| **Ishigami system** | specifically designed very cheap double skinned polypropylene growout housing based on 300-400 mm deep, re-used composted sawdust. Called 'pipehouses' in Japan from the arched supports. Unexploited in Europe. |

The New Terminology – 'I'

**ILR : Income to Life Ratio** a variant of PLR (qv). Preferred by some producers as there are many forms of *profit* while *income* is a singular, finite figure and thus easier to use.
**ISMT:** Immuno Suppressive Mycotoxin/Mycotoxicosis

| | |
|---|---|
| **Joule** | unit of energy. The energy needed to move one newton over 1 metre. 1 joule = 0.2388 calories. (A newton is a measure of applied force) |
| **Juxta-** | (prefix) near or next to |
| **K value** | (*Constr*) Thermal Conductivity of a Material (used in insulation calculations). Measured in Watts per metre per °C (W/M°C). Range of typical insulation materials is 0.02 to 0.2 with straw 0.07 (and copper wire 200). In insulation terms lowest is best; in electrical conductivity highest is best |
| **Kcal** | (*nutr*) Kilocalorie. 1000 calories (qv) or one Calorie |
| **Keratin** | (*nutr*) primary constituent of horn, hoof, hair, nails and secondary constituent of tooth enamel and skin outer layer |
| **Label Rouge** | French farm food quality designation – 'excellent' |
| **Labrum** | (*vet*) edge |
| **Lactogenic** | stimulating milk production |
| **Lacunae** | (*vet*) small cavities / hollows |
| **Laminar / laminated** | (*constr*) layered |
| **Lard** | commercially rendered pig fat |
| **Latent** | concealed, not obvious |
| **Laxative** | mild = aperient, strong = purgative/cathartic |
| **LD50** | poison indicator. The dose which will kill 50% of those tested |
| **Lesion** | (*vet*) wound, ulcer, sore, tumour, bite, scratch, etc. A deviation from the normal in a body |
| **<** | symbol for less than (remember it is < for less; more than is >) |
| **Lights** | butcher's name for lungs |
| **Lipase** | (*nutr*) fat-splitting enzyme |
| **Lipids** | fats, greases, oily and waxy compounds |
| **Lipoproteins** | how fats are moved in the blood e.g. HDL = high density lipoproteins and conversely LDLs |
| **Litter scatter** | (*econ*) the percentage of litters with more or less than a specified number of liveborn piglets. A litter scatter |

average of <8 is usually taken as a performance warning. ***Target*** 10% <8

**Livid**      correctly ***discoloured***, not necessarily red. Also black/blue

**Lochia**      *(vet)* discharge after parturition

**Lumen**      correctly, the inside of any tube, e.g. the lumen of the gut/intestinal lining. Also a measure of the flow (flux) of light

**Lux**      Measure of illumination. One lumen (q.v) distributed evenly over 1m². E.g. darkness = 10-15 lux, bright sunshine outdoors = 500 lux+. Breeding units need at least 2/3 300-350 lux (author's opinion) intermittent with 1/3 diurnal darkness per 24 hrs for best results

**Lymphocyte**      white blood cell, T cell

**Macerate**      soften by wetting

**Macrolide**      a specific type of antibiotic e.g. tylosin

**Mandatory**      required by law, essential

**Marbling**      intramuscular fat, improving eating quality when adequate (research long-overdue)

**Marker gene**      *(repr)* many genes are extremely difficult to find on the chromosome but can be associated with others found in greater numbers – marker genes

**Masking**      *(mycotoxins)* when a mycotoxin is bound to another substance in the feed, making it impossible to detect by conventional analytical methods

**Mastication**      chewing

**ME** *(nutr)*      Metabolizable Energy. The gross energy less that used in the faeces digestible energy (DE) and the urine and gases. Gross energy to DE losses in the pig are about 16% and DE to ME about 5%. See also NE (Net energy)

**Mean**      *(stats)* average, midway between two extremes. It provides no indication of the variability (q.v.) of, say, weights within a group

**Medial**      mid position, midline

**Median**      determined by aligning all pig weights in order of magnitude, then selecting the middle observation. If the distribution is 'normal' then the mean (average) will be very similar, if not identical

**Medulla**      the inner portion of any organ. Core

**Melatonin**      hormone released by the pineal gland (q.v.) controls the development of follicles and ovaries

**Meningitis**      *(vet)* inflammation of the lining of the brain, the meninges

**Mesentery**      *(vet)* membrane(s) attaching various organs to the body structure

| | |
|---|---|
| **Metabolism** | *(nutr)* all the processes which lead to the build-up of the body (anabolism) and the breakdown of body molecules to provide energy (catabolism) |
| **Metritis** | *(vet)* inflammation of the uterus |
| **Microgram** | one millionth of a gram or one thousandth of a milligram. Written as µg |
| **Microingredient** | *(nutr)* a nutrient only needed in ppm, milligrams, micrograms/ tonne |
| **Micron** | one thousandth of a millimetre |
| **Micronutrient** | element *(nutr)* a trace element, e.g. Se, Fe, Cu, Zn, etc |
| **Milligram** | one thousandth of a gram. Written as mg |
| **Millimetre** | one thousandth of a metre, written as mm |
| **MMA** | *(vet)* Mastitis-Metritis-Agalactia syndrome, known also as Farrowing Fever |
| **Modulation** | how a cell adapts to its environment |
| **Morbidity** | diseased |
| **Morphology** | the form and structure of an organism, cell etc |
| **MOS** | Mannan oligosaccharide. A naturally sourced (yeast) cell used as a feed growth enhancer and immunity modulator |
| **Motility** | movement |
| **Mould** | common name for a fungal organism/fungus |
| **Mummified** | degenerate (discoloured and shrivelled) piglets which died before farrowing |
| **Mutagenic** | *(gen)* inducing mutation |
| **Mutation** | *(gen)* the structural alteration of the DNA strand giving rise to a different genotype (qv) |
| **Mycosis** | *(bact)* any disease caused by a fungus |
| **Mycotoxin** | fungal poison |

The New Terminology – 'M'

**MTF (Saleable): Meat produced per Tonne (Ton) of Feed fed.** A more useful measure for the working farmer than FCR as it relates primary income (from lean meat after dressing out) against the primary cost of feed. It is also easier for the working farmer to secure the necessary data, and simple to convert to an equivalent cost per tonne (ton) of feed basis (see PPTE)

**Mortality – AMF (Actual Mortality Figure).** Better than % mortality as it more accurately expresses losses in relation to litter size, where % mortality can mislead.

**MSC: More for the Same Cost.** There are two practical ways of making profit – either produce More for the Same Cost (MSC) or produce the Same for Less

Cost (SLC). Theoretically SLC is better because if all producers produced more at the same cost the pig price may fall due to oversupply, while producing the same for less cost secures profit without overloading demand, thus stabilising pig price. The ideal of producing More for Less Cost (MLC) is usually unattainable in pig production!

| | |
|---|---|
| **Nano** | one thousandth millionth |
| **Nares** | openings of the nose |
| **Nascent** | (just born) more commonly just emerging from a chemical reaction |
| **Necrosis** | *(vet)* cell death *(adj* necrotic) |
| **Nematode** | roundworm |
| **Neonatal** | just born (usually up to one week). Neonate (n) |
| **Neoplasm** | a tumour |
| **Nephritis** | *(vet)* inflammation of a kidney |
| **NES** | Neuro-Endocrine System, involved in stress and immunity |
| **Net energy** | *(nutr)* Net Energy, NE, considers the amount of energy used in digestion and deducts this from ME (q.v.) to leave the amount available for growth and maintenance. NE thus provides a more accurate estimation of 'true' energy in a feed ingredient. |
| **Net** or **nett** | *(econ & nutr)* total amount (remaining) e.g. net energy. No further deductions made |
| **Net margin** | *(econ)* gross margin (qv) less fixed costs (qv) |
| **Net output** | *(econ)* sales plus credits, less purchases plus valuation charge (closing valuation minus opening valuation) |
| **Neuritis** | *(vet)* inflammation of a nerve |
| **Neutral detergent fibre NDF** | *(nutr)* the amount of cellulose, hemi-cellulose and lignins in the diet, all indigestible components, but useful in sow diets, much less so in others. |
| **Nervous system** | Central Nervous System involves the brain and spinal cord. Autonomic Nervous System is not subject to voluntary control |
| **Neutraceutical(s)** | *(nutr)* health-enhancing product or strategy |
| **Non-starch polysaccharides NSP** | *(nutr)* a constituent of NDF fibre which may have anti-nutrient properties, especially for the young pig |
| **Nostrum** | *(vet)* a quack remedy |
| **NVQ** | National Vocational Qualification. UK award for stockmanship skills etc |
| **Nucleotide** | *(nutr)* building block of DNA (as distinct from amino-acid = building block of a protein) |
| **Nutrigenomics** | *(gen)* the ability of nutrients to alter/improve gene expression e.g. selenium's effect on cancer reduction |

| | |
|---|---|
| **Objective** | (as distinct from subjective) based on sound evidence. Realistic |
| **Obstetrics** | *(med)* science of pregnancy and birth |
| **Occluded** | *(vet)* closed (or sometimes, severely obstructed) |
| **Oedema** | *(vet)* build up of fluids in a body (e.g. Bowel Oedema) |
| **Oesophagus** | the throat to the stomach passage (i.e. gullet) |
| **Oestrogen** | female hormone |
| **Ohm** | *(constr)* the unit of electrical resistance |
| **Oligosaccharides** | *(nutr)* complex carbohydrates which act as pre̲biotics (q.v.) and stimulate the growth of pro̲biotics (q.v.) |
| **Olfactory** | by smell |
| **Oocyst** | the highly resistant stage of a coccidia's life cycle |
| **Opportunity cost** | *(econ)* earmarking money put by for one project to be be spent on another should it appear opportunely |
| **Orchitis** | *(vet)* inflammation of a testis |
| **Organic** | produced only with assistance from materials harvested from living organisms and/or vegetable or animal fertilisers rather than those coming from synthetic chemicals |
| **Orthopaedics** | *(med)* the practice of muscular / skeletal surgery |
| **Osteitis** | *(vet)* bone inflammation |
| **Overheads** | *(econ)* fixed costs (qv) |
| **Oxytocin** | hormone which acts as a stimulant towards pregnancy and releases milk (together with the suckling stimulus) in lactation |
| **Oxidation** | *(nutr)* replacement of negative charges (electrons) on a molecule by positive charges (protons). The opposite of reduction. (See also anti-oxidant) |
| **P1 / P3 / P2** | optical probe backfat measurements in mm at two fixed points over the loin. 4.5 cm ($P_1$) and 8 cm ($P_3$) from the midline of the back of the last rib. The two added together describe the degree of fatness. Generally however, $P_2$ is the most commonly used, at 6½ cm |
| **Palpate** | examine by touch |
| **Pancreas** | organ which produces enzymes to break down proteins, carbohydrates and fats (pancreatic 'juice') |
| **Papilloma** | wart |
| **Para** | *(word element)* alongside, beside (but also, confusingly, apart from, against) |
| **Paradox** | quite different to what is expected |
| **Parakeratosis** | *(vet)* thickening and cracking of the skin, in pigs due to Zn deficiency |

| | |
|---|---|
| **Parameter** | a measurement which can be expressed numerically |
| **Parenteral** | *(vet)* administered not through the alimentary canal i.e. by injection |
| **Parietal** | *(vet)* referring to the walls of an organ |
| **Parity** | [1] similarity; [2] in the sow, the number of times a sow has farrowed e.g. a gilt is in parity 0 and a sow which has farrowed 4 times is in her 4th parity |
| **Parturient** | giving birth or related to birth |
| **Passive** | external stimulus (as distinct from 'active' where the animal responds spontaneously / originates the response) |
| **Pathogenic** | *(bact)* disease producing (pathogenicity = level of disease) |
| **Pathogenesis** | *(vet)* how a disease develops |
| **Pathology** | the study of disease |
| **PD** | 1. Partial Depopulation |
| | 2. Pregnancy Detection |
| **Pectoral** | *(vet)* the chest region |
| **Pellucid** | translucent |
| **Peptide** | *(nutr)* a precursor of protein containing several amino acids |
| **Peplomer** | Protein structure strengthening the outer wall of a virus. (*See also* Capsomer) |
| **Peracetic acid / peroxygen** | new and powerful virucide disinfectants (e.g. Virkon S) capable of very quickly penetrating the various protective layers of many viruses to destroy the nucleus |
| **Peracute** | *(vet)* very acute but shortlived |
| **Peri-** | *prefix*; around or close to, e.g. perinatal (as distinct from neonatal = just after) |
| **Peripheral** | near to the edge (of). (Periphery = outer edge or outside the central object) |
| **Peristalsis** | *(nutr)* the involuntary wavelike motion on down the gut |
| **Permeable** | permitting passage of a substance |
| **Pervasive** | widely-occurring |
| **Petecheal** | *(vet)* tiny blood-blistering |
| **pH** | measure of alkalinity / acidity. $<7$ = acid. $>7$ = alkaline. 7 = neutral (Range 1-14). |
| **Phagocyte** | cell which eats micro-organisms (pathogens) and other foreign particles |
| **Phenotype** | the outward appearance of an animal in expressing its inheritance (as distinct from genotype = its whole genetic make-up) |
| **Photoperiod** | the length of exposure to daylight, or artificial light |

| | |
|---|---|
| **Phytogenics** | the study of molecular relationships between nutrition and the response of positive genes (making good genes work harder) |
| **Pili** | hair-like structures found on the surface of bacteria helping them adhere to internal surfaces e.g. a gutwall. Also called fimbria |
| **Piling** | huddling together |
| **Pineal gland** | in mammals, a light receptor |
| **PINT** | Persistent Inguinal (q.v.) Nose-Thrusting |
| **Pint** | (*imperial*) = 586 ml (American 473 ml). Both approx |
| **Pipeline feeding** | (*nutr*) food mixed to a gruel and piped to a trough or feeder |
| **Pituitary (gland)** | situated at the base of the brain, which secretes several important hormones, and acts as a reservoir for others if deficient |
| **Plasma** | blood fluid containing the corpuscles (*see also* serum) |
| **Plasma protein** | (*nutr*) protein-rich fraction of blood also containing immunoglobulins, especially IgG |
| **Poly (prefix)** | many, numerous |
| **Polypeptide** | (*nutr*) a protein substance containing two (dipeptide) three (tripeptide) or more (polypeptide) linked amino-acids |
| **Polysaccharide** | see carbohydrate |
| **Postpartum** | after farrowing. Also 'postparturient' |
| **Potentiation** | the effects of two combinations being greater than the sum of either two alone |
| **Power** | (to the power of) mathematical symbol to simplify the display of very large (or very small) numbers, e.g. $1000 = 10^3$ or $1 \times 10 \times 10 \times 10$ (kilogram, kilowatt). Small numbers have a minus prefix e.g. $0.001 = 10^{-3}$ or $1 \div 10 \div 10 \div 10$ (millilitre, milliamp). Thus one billion (one thousand million, 1,000,000,000) is simplified as $10^9$ (ten to the power of nine) |
| **Prebiotics** | act on gut conditions or precondition nutrients or capture hostile organisms e.g. oligosaccharides. Different from probiotics (qv) |
| **Precursor** | forerunner, usually leading to another more active result |
| **Premature farrowing** | before day 110 of pregnancy but where some foetuses have survived, nevertheless, for 24 hours |
| **Primiparous** | a sow which has had at least one pregnancy resulting in viable offspring |
| **Probability** | (*stats*) P = a measure of likely reoccurrence. The number of times an event did occur divided by the number of times it might have occurred. (The number of times it might have occurred is defined as the adding together of all the positive and negative outcomes) |

| | |
|---|---|
| **Probe** | 1. a fat measuring instrument; |
| | 2. the measurement itself (see P$_2$) |
| **Probiotics** | beneficial organisms which colonise the gut surface rather than pathogens e.g. lactobacilli. Competitive exclusion |
| **Prognosis** | *(vet)* a forecast of the likely effects of a disease and its cure prospects. Diagnosis is the identification of a disease |
| **Proliferation** | increase, multiplication |
| **Prophylaxis** | *(vet)* fending off disease, prevention *(adj* prophylactic) |
| **Prosthesis** | *(vet)* an artificial body part replacement |
| **Protocol** | an action plan, set of guidelines |
| **Provitamin** | a substance from which an animal can form a vitamin |
| **Pseudorabies** | American name for Aujeszky's Disease |
| **PSS** | Porcine Stress Syndrome. Sudden death especially after transportation, fighting etc, associated with PSE (Pale Soft Exudative pork) |
| **Pulmonary** | pertaining to the lungs (or pulmonary artery) |
| **Pyrexia** | *(vet)* elevated body temperature, in the case of the pig above 40°C (103.5°F) |

The New Terminology – 'P'

**PPTE : Price Per Tonne (Ton) Equivalent.** Rightly or wrongly pig producers still make econometric judgements on price per tonne (ton) of feed. PPTE is a simple calculation which can convert economic advantages into a per tonne of feed equivalent figure i.e. how much the advantage would reduce the per tonne cost of feed across the feeding period.

**PLR : Profit to Life Ratio** further refines REO (qv) incorporating the *time* it takes to obtain the 'Return' part of the REO figure. PLR quantifies payback. Used principally in longer-term transactions, *i.e.* use of equipment and building refurbishment. See also ILR (Income to Life Ratio)

| | |
|---|---|
| **Quadrant** | one quarter of the circumference of a circle |
| **Quadrate** | square, four-sided *(adj* quadratic) |
| **Qualitative** | non-numerical description, e.g. colourful, small, etc |
| **Quantitative** | numerical description e.g. fourth, two kilometre, 1000, etc |
| **Quartile** | *(constr)* one fourth of a dimension plan or structure. Mainly used in ventilation design |

| | |
|---|---|
| **Random** | *(stats)* unplanned *(random variable, see* variable) |
| **Random (variable(s))** | *(stats)* a group of different, unplanned values |
| **Receptor** | *(vet)* molecule on or in a cell that recognises and binds with other specific molecules producing an effect within the cell (e.g. an oligosaccharide (q.v.) capturing a mycotoxin) |

| | |
|---|---|
| **Replacement cost** | *(econ)* value of breeding stock purchased together with their valuation charge (closing valuation – opening valuation) |
| **R-factor** | *(constr)* thermal resistance of a material (as distinct from thermal conductivity of a material – k value, qv). Measured in m² per °C per Watt (M² °C/W). Range 0.12-0.55 with highest values best |
| **Radiant** | emitting heat from a surface |
| **Ratio** | the relationship between two quantities |
| **Reagent** | *(chem)* material used to produce a chemical response so as to detect and measure other materials |
| **Recessive** | *(gen)* a gene which only functions when it is provided by both parents |
| **Reduction** | *(nutr)* see 'oxidation', its opposite reaction |
| **Reed-bed** | an effective and underused method of small-scale effluent disposal by organic plant absorbtion |
| **Regression** | *(stats)* the relationship between two or more random variables (q.v.) (usually drawn as a straight or curved line through the data points) |
| **Replication** | *(stats)* the repetition of an experiment to improve statistical accuracy |
| **Resorption** | reabsorption |
| **Retroactive** | (response) requiring stimulation to act. The opposite of proactive, which is to initiate an action |
| **Return to service** | see 'Service' |
| **Rhinitis** | inflammation of nasal lining |
| **Rideal-Walker Nº** | effectiveness of a disinfectant compared to phenol |

The New Terminology – 'R'

**REO : Return on Extra Outlay Ratio** a useful measurement of value for money (i.e. added value) enabling the producer to prioritise use of his capital and the vendor to justify the expense of good quality in a product

| | |
|---|---|
| **Saline** | salty |
| **Sandcrack** | a crack in the claw of a pig running in the direction from the coronet to the toe |
| **Sanitizer** | *(bact)* correctly, a combined detergent ***and*** disinfectant product (less good than using them separately) |
| **Saprophyte** | *(bact)* an organism which lives on dead tissue |
| **Sarcoma** | *(vet)* a malignant tumour capable of very fast growth |
| **Satiety** | *(nutr)* complete hunger satisfaction |
| **Sebaceous gland** | secretes sebum, an oily substance, around the hair follicles (over-secretion dries as dandruff) |

| | |
|---|---|
| **SED** | *(stats)* Standard Error of Difference. Measures the deviation between two means (q.v.) used to establish significance (q.v.) |
| **Sedentary** | inactive, lazy |
| **Selection intensity** | precise genetic measurement used by geneticists, not farmers |
| **Selection rate** | the number of (gilts) finally selected for breeding compared to the number that were originally viewed. |
| **Semi-sternum** | sitting/lying upright on the chest as distinct from lying supine (qv) |
| **Sensitivity** | *(bact)* susceptibility of an organism to a compound e.g. an antibiotic |
| **Sensitivity analysis** | *(stats)* the comparison of the performance of one or more actions to a common mean |
| **Septicaemia** | *(vet)* blood poisoning |
| **Septum** | the partition in the pig's snout |
| **Seroconversion** | *(vet)* the emergence of specific antibodies (q.v.) |
| **Serum** | usually involves the clear portion of blood plasma that does not contain blood cells.  Blood serum from pigs containing antibodies to a disease is called antiserum, and is used to provide temporary immunity to that disease when injected |
| **Service** | Normal or regular return to oestrus evident 18-24 days after previous service, measured from the first day of mating. Irregular return to as above but oestrus occurs after 24 days |
| **Sesqui-** | *prefix* meaning one and a half |
| **Shedding** | *(bact)* releasing (pathogenic) bacteria |
| **Sib** | *(gen)* correctly a blood relative but can be an abbreviation of sibling |
| **Sibling** | *(gen)* brother or sister |
| **Sigmoid** | *(stats)* S-shaped |
| **Significance** | *(stats)* determination of probability. 5% (1 in 20 chance of the treatment effect being a coincidence); 1% (1 in 100 chance); 0.1% (1 in 1000 chance). Commonly written as $P<0.05$, $P<0.01$; $P<0.001$, or as *, ** and *** respectively. P refers to probability |
| **Skatole** | a chemical constituting part of boar taint odour, along with androstenone |
| **SMEDI** | (Stillbirths, Mummifications, Embryonic Deaths, Infertility) *(vet)* due to Enteroviruses causing high piglet mortality |
| **Somatic** | *(vet)* whole body tissue rather than the cells which make it up i.e. muscles, skin etc |
| **SPF** | Specific Pathogen Free pigs reared disease-free (gnotobiotic) for pig trial research purposes. Also used to indicate high quality pork in Japan |

| | |
|---|---|
| **Specific gravity** | the weight of a liquid related to the specific gravity of water, which is 1.0 |
| **Squamous** | *(vet)* scaly or platelike |
| SSF | Solid State Fermentation. The growth of microorganisms, on or within water-soluble substrates which copy enzyme 'cocktails' at a lower cost, so as to improve digestion and help protect against certain Anti-Nutrient Factors (ANFs) |
| **Stable** | Scandinavian term for a piggery |
| **Stall** | term applied to the confinement device during pregnancy, as distinct from 'crate' (q.v.) which refers to farrowing confinement only, either total confinement or free-access |
| **Standard deviation (SD)** | see deviation |
| **Stasis** | Cessation or slowing |
| **Stenosis** | narrowing |
| **Sterotypies / sterotypic** | abnormal behaviour(s) characterised by rapidly-repeated actions to no fixed purpose/directed at inappropriate objects |
| **Sternum** | the breast bone |
| **Stillborn** | correctly, piglets which did not draw breath once expelled, as with born-deads. Can be confirmed by the 'bucket-test'. (see born-alives) |
| **Strain** | *(vet)* the outward manifestation of stressors (fright, aggression, vices etc) |
| **Straw-flow** | bedded flooring design, steeply sloped, where gravity and the pig's feet gradually move the fouled straw to a collecting channel outside the pen |
| **Streaming** | raising growing pigs which have recovered from a disease separately from their peers untouched by the disease |
| **Stress** | 1. conditions and reactions (stressors) affecting the wellbeing, mental and physiological, of the pig<br>2. compression, tension (structures) |
| **Striated** | streaked |
| **Subclinical** | (see clinical) |
| **Subcutaneous** | under the skin |
| **Subjective** | an unconfirmed, personal, opinion. The opposite of objective |
| **Supernatant** | the liquid lying above a layer of deposited insoluble material |
| **Supine** | lying flat on its side |
| **Surfactant** | substance which reduces surface tension, thus releasing bound particles e.g. soap, detergent |
| **Symptomatic** | indicating a symptom (of) |
| **Synchrony** | *(adj* synchronous; synchronism) occurring at the same time |
| **Syndrome** | *(vet)* a pattern or total of clinical signs constituting a whole picture |

| | |
|---|---|
| **Synergy** | (*adj* synergistic; non-synergism) combined action so that the total effect is greater than that of the two separately |
| **Syntax** | (*econ*) the rules of a language or computer program |
| **Systemic** | affecting the whole body system. Comprehensive |
| **Single Space Feeding (SSF)** | see wet/dry Feeding |

The New Terminology – 'S'

**SLC : Producing the Same for Less Cost** see MSC
**SPL: Sow Productive Life.** The number of litters produced in a sow's lifetime; current target 5 to 6, herd average.

| | |
|---|---|
| **T Cell** | (*bact*) lymphocyte, white blood cell |
| **T$_2$ toxin** | a mycotoxin |
| **Tare** | (*weight*) the weight of a vehicle and fuel less its load |
| **TDN** | **Total Digestible Nutrients** (*nutr*) Now discontinued measure of energy |
| **Temperature;** | ***Lower Critical Temperature (LCT)*** is the ambient temperature below which the pig needs to divert food energy into keeping warm. |
| | ***Evaporative Critical Temperature (ECT)*** is the ambient temperature at which panting occurs and urgent cooling action is needed. Respiration rate is usually more than 60 breaths per minute. |
| | ***Upper Critical Temperature (UCT)*** is the ambient temperature which, when exceeded, the animal's life could be in danger |
| **Terminal crossbreeding** | (*gen*) continuing breeding from a first cross without crossing further |
| **Therapeutic** | (*vet*) treating disease, curing, alleviating (*n* therapy) |
| **Therm** | heat required to raise 1000 kg of water 1°C. 1 therm = 1000 Kcal = 106 megajoules (MJ) |
| **Titre** | (*bact*) the amount of one substance required to react with another, used in determining antibody levels present |
| **Tomography** | used in radiology to visualise fat/lean deposition by scanning a cross section through the pig's carcase. Might be developed to encompass weighing? |
| **Topical** | (*vet*) a localised area |
| **Topping** | Removing to market of the heaviest animals in a pen of finishing pigs which can lead to performance benefits in those remaining |

| | |
|---|---|
| **Torsion** | twisting . As in gastric torsion (bloat) |
| **Trait** | 1. *(gen)* any genetically determined condition |
| | 2. distinctive behavioural pattern |
| **Transducer** | *(constr)* a device which converts pressure, temperature etc to an electric pulse |
| **Transverse** | *(constr)* side to side, across |
| **TTS (or 2TS)** | Two-Tonne Sow. Colloquial (UK) term for a target of each sow producing 2000 kg of pigmeat per year (currently the EU average in 2010) |
| **Type** | *(bact)* to identify an organism or blood group, etc |
| **Ultrasound** | used in pregnancy diagnosis by equipment capable of emitting radiant energy at over 20,000 cycles per second |
| **U-Value** | *(constr)* measure of thermal transmittance of a material, used in insulation calculations. The amount of heat which passes through 1 $m^2$ of a structure where the outside/inside temperature differs by 1°C. Expressed as Watts (q.v.) per sq metre per °C (W/m² °C). Typical range 0.5-5.5 with lowest values best. |
| **Variable** | *(stats)* different measurement(s). Random variable(s). A group or quantity which exhibit various values, each of varying probability |
| **Variable costs** | *(econ)* costs which are likely to vary frequently |
| **Vascular** | (system) to do with the blood vessels / blood supply |
| **Vector** | *(constr)* an effluent sprayer covering a defined area |
| **Vegetative** | its most common meaning is resting i.e. vegetative state |
| **Vein** | blood vessel leading from various organs back to the heart in contrast to an artery which carries blood from the heart to various organs and the extremities |
| **Velocity** | speed (air movement) vital in correct air placement in a piggery |
| **Ventral** | *(vet)* abdominal area; *(constr)* towards |
| **Vermifuge** | expels worms. A vermicide is a substance that kills them |
| **Vesicular** | *(vet)* blistered; pustules |
| **Viable** | correct (rational, acceptable) |
| **Villus** | *(pl* villi) a microscopic, very sensitive thread-like growth covering the intestinal surface which absorbs nutrients from the digesta, thus increasing the absorption area many thousandfold. |
| **Viraemia** | *(vet)* blood infected by a virus |
| **Virulence** | the degree of pathogenicity of an organism |

| | |
|---|---|
| **Viscera (visceral)** | large internal organs, such as the abdomen, liver etc |
| **Viscous** | (*adj* viscosity) sticky, thick liquid |
| **Volatile** | evaporates easily and quickly, also rapidly changing |
| **Volt** | (*constr*) the unit of electric movement or force (one ampere of current versus one ohm of resistance) |
| **Vomitoxin** | a mycotoxin causing vomiting, especially in young pigs |
| **Watery pork** | PSE pork (see PSS) |
| **Watt** | (*constr*) a measure of electric force i.e. the work done at 1 joule per second. Equal to 1 ampere under the pressure of 1 volt |
| **Wet-dry feeding** | (*nutr*) technique where a small amount of meal is nudged by the pig into a receptacle which is then moistened once the pig activates a drinker nozzle over it. Also called (inaccurately) Single Space Feeding |
| **Wet feeding** | (*nutr*) pipeline feeding (qv). Also known as Liquid Feeding (see also FLF) |
| **Wiltshire cure** | keeping bacon in a dry cool environment after a 3 to 4 day soak in a brine solution |
| **Withholding period** | mandatory period of withdrawal of a drug from animal treatment before that animal can be used for food |
| **Working capital** | (*econ*) the capital required for the daily operation of a business |
| **Weaning to service interval** | the time between the date of weaning and date of first mating. Date of weaning is day 0. |

The New Terminology – 'W'

**Weaning capacity:** The ability of a sow to produce a target weight of weaners in her lifetime, currently 500 kg is suggested

**WWSY : (Weaner Weight per Sow per Year)** The commonly-used term of pigs (weaners) per sow per year (PPS) is a less indicative figure than WWSY because it gives no indication of the weight of weaner achieved. Much better to express this as weaner **weight**, not weaner **numbers**, produced over a period when used in the critical economic assessment of a breeding farm – the sow inventory

| | |
|---|---|
| **Yield** | dressed carcase weight |
| **York Ham** | ham first pickled, then stored over a long period in dry salt |
| **Yorkshire** | (*gen*) term for the Large White breed (in Europe) used elsewhere in the world |
| **Zearalenone** | an oestrogenic mycotoxin (qv) particularly dangerous to pigs, especially gilts and breeding sows |

**Zinc oxide**  (*bact*) a useful anti-diarrhoeal used at high levels in the food of weaned pigs for a short period, but causes Zn build-up in soil

**Zoonose**  an animal disease transmissible to man

**Zygote**  (*gen*) the fertilised ovum just before first cleavage

# INDEX

### The 70 Checklists are shown in green

**NOTE: Many worked examples from trial work are in the New Terminology chapter, page 258.**

www.ingramcontent.com/pod-product-compliance
Lightning Source LLC
Chambersburg PA
CBHW061228220326
41599CB00028B/5371